HPE Aruba Networking Certified Professional - Campus Access Architect

OFFICIAL CERTIFICATION STUDY GUIDE
(EXAM HPE7-A11)

First Edition

Igor Chinchak, Tyler McMinn

HPE Press
660 4th Street, #802
San Francisco, CA 94107

HPE Network Campus Access Architect

Official Certification Study Guide (Exam HPE7-A11)

Igor Chinchak, Tyler McMinn

Published by:

Hewlett Packard Enterprise Press

660 4th Street, #802

San Francisco, CA 94107

ISBN: 978-1-960058-50-8

Printed in Mexico

WARNING AND DISCLAIMER

This book provides information about the topics covered in the HPE Network Campus Access Architect certification exam (HPE7-A11). Every effort has been made to make this book as com-plete and as accurate as possible, but no warranty or fitness is implied.

The information is provided on an "as is" basis. The author, and Hewlett Packard Enterprise Press, shall have neither liability nor responsibility to any person or entity with respect to any loss or damages arising from the information contained in this book or from the use of the discs or programs that may accompany it.

The opinions expressed in this book belong to the author and are not necessarily those of Hewlett Packard Enterprise Press.

FEEDBACK INFORMATION

At HPE Press, our goal is to create in-depth reference books of the best quality and value. Each book is crafted with care and precision, undergoing rigorous development that involves the expert-ise of members from the professional technical community.

Readers' feedback is a continuation of the process. If you have any comments regarding how we could improve the quality of this book, or otherwise alter it to better suit your needs, you can con-tact us through email at hpepress@epac.com. Please make sure to include the book title and ISBN in your message.

We appreciate your feedback.

Publisher: Hewlett Packard Enterprise Press
HPE Press Program Manager: Michael Bishop

About the Authors

Igor Chinchak is an expert HPE Aruba Networking Certified Instructor, a lead Computer Network Architect, and DevOps Software Engineer. He delivers the full suite of HPE Aruba Networking certification courses. His extensive credentials include multiple expert certifications in HPE Aruba Networking and Juniper Networks technologies.

Tyler McMinn is a highly qualified HPE Aruba Networking Certified Instructor who brings extensive expertise in network infrastructure and automation technologies. His credentials cover multiple certifications, including HPE Network Security Expert, HPE Network Campus Access Architect, HPE Network Data Center Professional, HPE Network Switching Professional, HPE Networking ClearPass, and Aruba Certified Design Expert (ACDX).

Introduction

The HPE Network Campus Access Architect study guide will teach you the advanced skills necessary to implement and operate enterprise-level HPE Networking switching solutions. You will build on the skills you learned at the Professional level to configure and manage modern, open standards-based networking solutions using HPE Aruba Networking CX Switch Operating System (AOS-CX) routing and switching technologies. In this guide, you will learn about AOS-CX switch technologies:

In this guide, you will go on a journey to design and implement a campus access solution based on customer needs.

You start off in **chapter 1** just like an actual design projects starts, with gathering information – key stakeholders, physical site locations, network evaluations, and project scopes. This information helps to inform the architecture, covered in **chapter 2** with things like pertinent HPE Aruba Networking products, management platforms and key architectural principles.

You dive into Radio Frequency (RF) planning in **chapter 3** with deployment models, site surveys, and validating coverage.You move into high-density design in **chapter 4** – challenges, solutions, coverage, and tactics for optimizing high-density deployments.

Then you move on to wired network design in **chapter 5**, which covers 2 and 3-tier architectures, closet layouts, cabling, and more. Then you improve performance and manageability with network segmentation in **chapter 6**, along with ideas related to IP addressing design, routing protocols, and VRF lite.

Now you've got a good network design, let's lock it down. It is all about campus security in **chapter 7**, especially as relates to leveraging ClearPass, Onboarding, traffic security, profiling, and HPE Aruba Networking Secure Access Service Edge (SASE).

Now you improve resiliency and flexibility with overlay networks in **chapter 8**. This includes HPE Aruba Networking gateways and NetConductor.

With a good, solid, and secure network design, the last thing to do is optimize it. In **chapter 9** you explore things like QoS best practices, strategies for Wi-Fi, and HPE Aruba Networking User Experience Insight (UXI).

Certification and Learning

Hewlett Packard Enterprise Certification and Learning provides end-to-end continuous learning programs and professional certifications that can help you open doors and accelerate your career.

We provide

- **Professional sales and technical training and certifications** to give you the critical skills needed to design, manage and implement the most sought-after IT disciplines; and
- **Continuous learning activities and job-role based learning plans** to help you keep pace with the demands of the dynamic, fast paced IT industry
- **Advanced training** to help you navigate and seize opportunities within the top IT transformation areas that enable business advantage today.

As a Certification and Learning certified member, your skills, knowledge, and real-world experience are recognized and valued in the marketplace. To continue your professional and career growth, you have access to our large HPE community of world-class IT professionals, trend-makers and decision-makers. Share ideas, best practices, business insights, and challenges as you gain professional connections globally.

To learn more about HPE Certification and Learning certifications along with continuous learning programs, please visit

http://certification-learning.hpe.com

Audience

This book is designed for those who have experience in advanced level implementation and maintenance of wired solutions. They possess experience assessing and interpreting existing networks and network design documentation. They regularly troubleshoot, resolve issues, and perform ongoing support of large network environments. They have familiarity with wired network security best practices and perform duties independently with at least five years of experience working as a team leader of the team that supports and maintains complex campus, edge branches, and data center networks.

Assumed Knowledge

This is a expert level guide for senior network engineers. It is assumed that you have a thorough understanding of HPE Networking switching implementations in small-to-medium businesses and enterprise environments.

We recommend students to have completed the Campus Access Fundamentals, Implementing Campus Access course or have equivalent on-the-job experience before reviewing this guide.

Minimum Qualifications

Note that there are no prerequisites for taking the HPE7-A11 exam. However, we recommend having already achieved the HPE Networking Certified Professional - Switching exam (HPE7-A08.)

Relevant Certifications

After you pass this exam, your achievement may be applicable toward more than one certification. To determine which certifications can be credited with this achievement, log in to The Learning Center and view the certifications listed on the exam's More Details tab. You might be on your way to achieving additional certifications.

Preparing for Exam HPE7-A11

This self-study guide does not guarantee that you will have all the knowledge you need to pass the exam. It is expected that you will also draw on real-world experience and would benefit from completing the hands-on lab activities provided in the instructor-led training.

Recommended HPE Training

Recommended training to prepare for each exam is accessible from the exam's page in The Learning Center. See the exam attachment, "Supporting courses," to view and register for the courses.

Obtain Hands-on Experience

You are not required to take the recommended, supported courses, and completion of training does not guarantee that you will pass the exam. Hewlett Packard Enterprise strongly recommends a combination of training, thorough review of courseware and additional study references, and sufficient on-the-job experience prior to taking an exam.

Exam Registration

To register for an exam, go to https://certification-learning.hpe.com/tr/learn_more_about_exams.html

CONTENTS

Chapter 1: Information Gathering 1
Objectives 1
Key Stakeholders 1
Key stakeholders of a deployment project 1
Interview stakeholders 3
Physical Environment 4
Physical sites 4
Building 100: floorplan example 5
Building 100: network closet 7
Current Network Usage 8
Users 8
Roaming user type examples 9
User device needs 10
Device needs 11
Documentation & administrative information 12
Wireless RF/network questionnaire 14
Wired network questionnaire 15
Scope of the Project 17
Anticipated growth 17
Assess application requirements 17
Assess roaming requirements 19
Basic wired connectivity requirements 21
Campus external network connectivity needs 22
Remote user connection needs 23
Assess security requirements 23
Security needs 24
Assess the need for redundancy 26
Management needs: consider admin access 27
Migrating to AOS-10 possible? 28
Learning check 29
Question #1 29
Question #2 29
Question #3 29

Chapter 2: HPE Aruba Networking security-first, AI-powered networking 31
Objectives 31
HPE Aruba Networking security- first, AI- powered networking Campus Architecture 31
HPE Aruba Networking architecture 31
HPE Aruba Networking security first, AI-powered networking overview 32
HPE Aruba Networking campus design 34
HPE Aruba Networking security-first, AI-powered networking services 35
Network connectivity 35
Network security policy 36
Network services 36
HPE Aruba Networking security- first, AI- powered networking campus reference architecture overview 38
Campus IT staff challenges 39
HPE Aruba Networking security- first, AI- powered networking .edge services 39
These are the design goals 40
HPE GreenLake for Networking service pack examples 41
HPE Aruba Networking Portfolio 43
Product line 43
HPE Aruba Networking CX switching advantages 45
AOS-CX switch VXLAN tunnels & virtualization 46
HPE Aruba Networking Central Management Platform 49
HPE Aruba Networking Central 49
Cloud-managed 50
Flexible and cost-effective cloud networking 51
HPE Aruba Networking Central for AOS-CX management 52
HPE Aruba Networking Central Client Insights: introduction 52
HPE Aruba Networking Central Client Insights: capabilities 54
AI Insights 54
HPE Aruba Networking Central On-Premises 55
HPE Aruba Networking Central On- Premises versus Cloud : Feature Comparison 55
HPE Aruba Networking Central On-Premises Architectures 56
HPE Aruba Networking Central On-Premises Software Licenses 57
HPE Aruba Networking Central Subscriptions 58
HPE Aruba Networking Central subscriptions for AOS-10 gateways 58
HPE Aruba Networking Central Foundation and Advanced licenses for APs and SWs 60

AOS-10 Architecture 61
Evolution of HPE Aruba Networking APs and gateways with AOS-10 61
AOS-10: overview 62
AOS-10: architecture 62
AOS-10: scalability 63
Device to HPE Aruba Networking Central communication 64
HPE Aruba Networking Central groups and sites 65
Learning check 66
Question #1 66
Question #2 66
Question #3 67
Chapter 3: RF Planning 69
Objectives 69
Deployment Models 69
Review common WLAN types 69
Wireless coverage considerations 71
Review coverage vs. capacity 71
6 GHz/5 GHz versus 2.4 GHz coverage 73
Target RF signal levels 74
Types of wireless deployments 75
Typical wireless bandwidth capacity in an office environment 76
Sources of attenuation / concentration 76
RF design 77
AP mounting considerations 78
Open office environments 79
Deploy APs in rooms – not in hallways 80
AP placement summary 81
Avoid mixing different AP models 82
Transmit power settings 83
Transmit and basic data rates 84
Estimating AP count for 5 GHz and 6GHz* 84
AP count: additional factors 85
Summary of AP/client design 86
Selecting APs and Antennas 87
Wireless LAN design overview: WLAN elements 88
HPE Aruba Networking wireless LAN design overview 88
HPE Aruba Networking wireless LAN design overview: APs 89
HPE Aruba Networking Indoor Access Points 90

HPE Aruba Networking Outdoor/ruggedized Access Points 91
Indoor 7xx series Wi-Fi 7 91
HPE Aruba Networking Wi-Fi 7 AP portfolio comparison 93
Indoor 600 series Wi-Fi 6E 94
HPE Aruba Networking Wi-Fi 6 AP portfolio comparison 96
Hospitality and Remote series APs 97
Outdoor 700, 600 and 500 series APs 99
APs for hazardous locations 101
Cable category 102
AP mounting kits 103
APs mounting kits comparison 104
Ethernet cable length for 1 GbE connections 105
AP Deployment Technologies 106
Where to start? 106
Survey methods 107
Physical building materials 107
Selecting a survey type 108
Onsite passive survey example 110
Onsite active survey example 111
Onsite survey with temporary placed APs 112
Spectrum clearing methodology 114
WLAN RF planning process 114
Site survey heat map generator 116
Floor plan: add calculated APs / removing APs 118
Floor plan: recalculating APs 119
Creating a Bill of Materials with Iris 120
HPE Aruba Networking AirMatch 122
AirMonitor mode for security 124
Review AM deployment recommendations for rogue AP detection 125
Roaming recommendations 126
ClientMatch 127
ClientMatch methodology 128
HPE Aruba Networking AP professional services 130
Learning check 130
Question #1 130
Question #2 130
Question #3 131
Supplemental Section: Review RF Fundamentals 131

IEEE 802.11 standards 131
2.4 GHz band and channels 134
5 GHz allowed channels in FCC countries 134
Usable 6 GHz channels in Americas & Europe/CEPT 135
6 GHz: massive increase in capacity 136
802.11n / 802.11ac summary 137
802.11ax summary 138
802.11be summary 138
Wi-Fi 6 enhancements: High Efficiency 140
Wi-Fi 6 enhancements: OFDMA 141
Wi-Fi 6 enhancements: MU-MIMO 142
Wi-Fi 6 enhancements: Transmit Beamforming and TWT 143
Considerations when upgrading to Wi-Fi 6E 144
Enabling all three radio bands 145
Device classes in 6GHz 145
Multi-User MIMO (MU-MIMO) 148
MU-MIMO spatial streams 149
802.11ax data rates by client capability 150
Minimum RSSI requirements by data rate 151
802.11ax selected rates (Mbps, short GI) 152
Modulation versus cell size 152
Wireless collision domain 153
HPE Aruba Networking Air Slice 154
BSS coloring 156
Outdoor APs: link budget 156
Chapter 4: Wireless High-Density Design 159
Objectives 159
High-Density Challenges 159
High-density (HD) WLANs today 159
What is a very high-density WLAN? 160
How do 802.11ax technologies affect high-density deployments? 162
How do 802.11be technologies affect high-density deployments? 164
Aesthetics 165
HD & VHD RF Coverage 167
Higher-density, open-office layouts 167
Power settings 168
RF coverage strategies 168
Overhead coverage 169

No RF spatial reuse with omnidirectional antennas 171
Examples – overhead coverage 172
Side coverage 172
Selecting APs for VHD 174
Review HPE Aruba Networking VHD design methodology 174
Step 1- key design criteria for typical VHD WLAN 174
Step 2 – estimate associated device capacity 176
Step 3 – ADC address design examples 176
Step 4 – estimate AP count 178
Step 5 – dimension gateways 179
Step 6 – edge design 180
Step 7 – core design 181
Step 8 – server design 182
Step 9 – WAN edge design 183
Deployment Examples 183
Example scenario: typical multi-auditorium 183
Step 1: understanding load in auditoriums 184
Step 2/3 - estimate associated device capacity 185
Step 4 - estimate the AP count using a AP-730 186
Step 5 – calculate system throughput 186
Learning check 188
Question #1 188
Question #2 188
Question #3 188
Question #4 189
Chapter 5: Wired Network Design 191
Objectives 191
2-Tier versus 3-Tier Design 191
2-tier versus 3-tier topology 191
Benefits of a 2-tier topology 192
When 3-tier may be required 193
Options for implementing routing 194
VSX pair placement 195
VSX pair placement in campus fabric 196
VSX design considerations 197
2-tier and 3-tier wired 199
2-tier campus LAN 200
3-tier redundancy core 201

When to use VSX in the core 203
Core layer 204
HPE Aruba Networking CX product portfolio 206
HPE Aruba Networking CX 10000 Series and 10400 switch series 207
HPE Aruba Networking CX 9300 Switch Series 207
HPE Aruba Networking CX 8400 Switch Series 209
HPE Aruba Networking CX 8360 Switch Series 210
HPE Aruba Networking CX 8325 Switch Series 211
HPE Aruba Networking CX 8325H switch series 213
HPE Aruba Networking CX 8320 Switch Series 214
HPE Aruba Networking CX 8100 Switch Series 215
HPE Aruba Networking CX 6400 Switch Series 216
HPE Aruba Networking CX 5420 switch 217
HPE Aruba Networking CX 6300 Switch Series 219
HPE Aruba Networking CX 6300L switch series 220
HPE Aruba Networking CX 6200 Switch Series 221
HPE Aruba Networking CX 6200 Switch Series modular and fixed power models 223
HPE Aruba Networking CX 6100 Switch Series and CX 6000 series 224
HPE Aruba Networking CX 4100i Switch Series 225
Closets and Cables 226
Network distribution 226
IDF / MDF closets 227
Rack unit (RU) 227
Patch panels and cable organizers 229
Consider the media: copper 229
Smart Rate 230
Quick terms 232
Switch to switch - direct attach cable (DAC) 233
HPE Networking breakout cables for high-speed ports 234
Switch to switch - direct attach cable (DAC) 235
Wired connection using fiber optics 236
Cables and transceivers: SFP 237
Cables and transceivers: SFP+ 238
Cables and transceivers: SFP28 and SFP56 240
Cables and transceivers: QSFP 241
Access layer 241
Identify AP-7xx series access requirements 241

Identify AP-6xx series access requirements 242
Identify AP-5xx series access requirements 243
Uplink requirements and oversubscription 244
Factors that affect appropriate oversubscription 245
Example designs for fixed-port switch ports 246
Example designs for 6300/6400 switches at the access layer 247
Uplink plan for VSF fabrics 249
Using current traffic information to assess plan's validity 249
Upstream VSF bandwidth allocation 251
Downstream VSF bandwidth allocation 252
VSF - In-Service Software Upgrade (ISSU) 252
Edge port PoE 253
Considerations for the PoE budget 254
PoE and PoE+ review 255
Core and Aggregation Layers 256
Design the core for a 2-tier topology 257
Consider the wired traffic flow in a 2-tier topology 258
Plan appropriate oversubscription 259
Consider hardware requirements at the core 261
Recommended 2-tier collapsed core switches 262
Design a 3-tier network 263
Example medium-large 3-tier design 264
Example large 3-tier design 265
Plan VSX links at the aggregation layer and core 266
Plan aggregation & core bandwidth 267
Learning check 268
Question #1 268
Question #2 268
Question #3 269
Question #4 269
Supplemental: Switch management & control plane 269
HPE Aruba Networking Central for AOS-CX management 269
Network automation with AOS-CX 270
Dynamic Segmentation with centralized overlay: problem solved 271
Dynamic Segmentation: operation overview 272
Network Segmentation 273
VLANs, performance, and security 274
Spanning-Tree 275

Spanning-Tree root bridge selection 276
Spanning-Tree features 277
Access Control Lists (ACLs) - ingress 279
Ensuring resiliency in a traditional design 280
Static routes vs dynamic routing 283
OSPF routing 284
OSPF routing benefits 286
Single OSPF area: limited scalability 286
Multi-area OSPF: large scalability 288
Special area types 289
Ensuring routing resiliency 289
IP multicast 291
IP multicast PIM-SM 293
Supplemental: Stacking technologies 294
Switch virtualization technologies 294
Virtual Switching Framework (VSF) 295
Switch virtualization solutions 296
AOS-CX switch virtualization: VSX vs VSF 297
VSX benefits 298
VSX components overview 298
VSX - data plane virtualization 299
VSX redundancy 301
VSX benefits: layer 3 303
Data plane virtualization 303
Chapter 6: Network Segmentation 305
Objectives 305
Segmentation with VLANs and VRFs 305
Benefits of VLAN Segmentation for wired networks 305
Out of band management network 306
Isolate traffic for access control 307
Group traffic for specialized devices 310
Minimize wired broadcast domain 310
Example: assign logical VLAN IDs 312
Assign logical subnet addresses 314
Fit into an existing solution 316
Detect and address issues with large broadcast domains 317
Example 2-tier wired VLAN design 318
Example 3-tier wired VLAN design 319

IP Addressing Design for Wireless Networks 320
Advantages of single VLAN 320
Recommendation for single VLAN 322
Key considerations 323
HPE Aruba Networking wireless broadcast / multicast solutions 325
Multicast transmission optimization 326
Dynamic multicast optimization 327
Single VLAN optimizes ipv6 traffic 328
Identity-based HPE Aruba Networking firewall 329
IP Routing Protocols and VRF Lite 330
OSPF overview 330
OSPF routing sample topology 331
BGP overview 333
Multi-homed ISP connections 334
VRF lite overview 335
Learning check 336
Question #1 337
Question #2 337
Question #3 337
Supplemental: Multi-area OSPF 337
Multi-area OSPF: large scalability 337
Need for route and LSA aggregation between areas 338
Problem solved: ABR route aggregation 339
Special area types 341
Chapter 7: Campus Security 343
Objectives 343
HPE Aruba Networking Security 343
HPE Aruba Networking design tenets 344
HPE Aruba Networking AIOps: Overview 344
HPE Aruba Networking AIOps: Key benefits 346
HPE Aruba Networking Zero Trust Security components 347
HPE Aruba Networking solutions that support ZTS 349
HPE Aruba Networking campus security policy concepts 350
Why network authentication 350
HPE Aruba Networking ZTS in action 351
ClearPass Policy Manager & Cloud Auth 352
ClearPass Policy Manager overview 352
ClearPass Policy Manager appliance options 353

ClearPass Policy Manager 6.12 licensing 354
ClearPass Policy Manager sizing & scaling 357
ClearPass Policy Manager sizing & scaling: specifications 360
ClearPass Policy Manager sizing & scaling: performance 360
ClearPass Policy Manager sizing & scaling: virtual appliances 362
ClearPass Policy Manager hardware and virtual appliances 363
CPPM Publisher node sizing 364
CPPM Subscriber node guidelines 365
ClearPass Policy Manager Insight: report configuration 366
ClearPass Policy Manager Insight: placement considerations 366
CPPM server placement: centralized or distributed 367
CPPM features 368
ClearPass Policy Manager NAC functions 370
Introduction to CPPM services 371
Role-Based Access Control 372
The evolution of device visibility 372
Sources of usable context 373
Introducing the HPE Aruba Networking 360 Security Exchange 374
The benefits of integration 375
ClearPass Policy Manager guest access 377
Review of cloud authentication & policy for HPE Aruba Networking ClearPass 377
Cloud Authentication key benefits and features 379
Onboarding and PKI 379
Why use ClearPass Onboard Software? 380
Onboarding vs Guest Access 380
Options for onboard certificate authorities 381
Securing Client Traffic 383
Need for secure wireless communications 383
Review WLAN security options 383
HPE Aruba Networking MPSK ClearPass 385
Visitor wireless features 386
User roles overview 389
Role-based AOS GW/AP firewall 390
Designing campus role-based policy 391
Review application visibility 394
Dynamic Segmentation with centralized overlay 395
Customizing authenticated user access 396

Access point tunnel forwarding mode 397
Access point bridge forwarding mode 398
Access point mixed forwarding mode 399
Profiling and NAC 400
ClearPass Policy Manager endpoint profiling 400
Profiling with HPE Aruba Networking Central Client Insights 402
HPE Aruba Networking AI-powered Client Insights 403
HPE Aruba Networking Client Insights 404
Microsoft Endpoint Manager 405
HPE Aruba Networking SASE 407
Unified SASE with HPE Aruba Networking 407
How HPE Aruba Networking SSE is modernizing remote access 407
Secure Service Edge (SSE) 409
HPE Aruba Networking SSE management console 410
HPE Aruba Networking SSE and Identity Providers (IdPs) 411
HPE Aruba Networking SSE service structure 413
EdgeConnect SD-Branch 413
EdgeConnect SD-Branch secure connection 414
Learning check 415
Question #1 415
Question #2 416
Question #3 416
Supplemental: AAA Overview 416
AAA services 416
Review Authentication, Authorization, Accounting (AAA) 417
Controlling network access 419
Layer 2 vs layer 3 authentication 419
Port-based authentication use cases 420
Common authentication methods 421
MAC-auth overview 423
Captive portal authentication 424
802.1X authentication 426
Examples of EAP methods 427
EAP-PEAP considerations 428
EAP-TLS considerations 429
Validating domain users are on domain computers with EAP-TLS 430
Validating users are on domain computers with TEAP 431
WPA/WPA2-personal issues 432

Wireless security: WPA3 ... 434
WPA3-Personal SAE ... 435
WPA3-Personal benefits ... 436
WPA/WPA2/WPA3-enterprise ... 437
WPA3-enterprise: overview ... 438
How the RADIUS server sends dynamic settings ... 438
WPA/WPA2-personal ... 440
OnConnect for wired non-RADIUS enforcement ... 440
What is a digital or X.509 certificate? ... 441
Certificate usage ... 442
Example certificate ... 443
CAs and the public key infrastructure ... 444
Trusted issuers ... 445
Chain of trust ... 447
Validating that a certificate is trusted ... 448
Types of ClearPass server certificates ... 449
HPE Aruba Networking network infrastructure certificate options ... 451
CA-signed certificates for RadSec ... 452
Certificates within a windows domain ... 454
Certificates in a Microsoft Azure AD domain ... 454
Options for deploying certificates on user devices ... 455
Supplemental: Network Device Hardening ... 456
Network infrastructure security overview ... 456
Best practices checklist ... 459
Review management access on AOS-CX switches ... 459
Review management authentication options ... 460
Best practices for authentication methods ... 461
Manager roles ... 463
Understanding how credentials are stored in the config ... 464
Use secure protocols ... 465
Set up SSH securely ... 466
Managing certificates with Enrollment over Secure Transport (EST) ... 468
Disable SNMPv1/v2 ... 469
AOS-CX: set up SNMPv3 ... 470
Review RADIUS vs RadSec ... 470
Review RADIUS Dynamic Authorization ... 472
RADIUS Interim Accounting ... 473
Standard ports for network device and CPPM communications ... 474

Summary of Spanning Tree protections 474
DHCP attacks 475
ARP protection 476
Results of ARP protection 478
DHCP snooping and ARP protection 479
Chapter 8: Overlay Networks 481
Objectives 481
HPE Aruba Networking Gateways 481
When to use gateways 481
HPE Aruba Networking gateways 482
HPE Aruba Networking 9000 series gateway (9004) 483
HPE Aruba Networking 9000 series gateway (9012) 485
HPE Aruba Networking 9100 series gateway (9106) 485
HPE Aruba Networking 9100 series gateway (9114) 486
HPE Aruba Networking 9200 series gateway (9240) 487
9200 licensing Base, Silver, and Gold 488
HPE Aruba Networking virtual gateways 490
Gateway review 491
Secondary cluster redundancy 492
Centralized Overlay Networks with HPE Aruba Networking Gateways 493
Traditional network design 493
Network design with centralized overlays 494
Multi-cluster deployment - MultiZone 495
Centralized overlays – switch UBTs 496
PAPI communication 496
UBT reserved VLAN mode: default 497
UBT extend VLAN mode 498
Distributed Overlay networks with HPE Aruba Networking Central NetConductor 499
Network design with distributed overlays 499
Static VXLAN vs EVPN 501
VXLAN packet structure 503
Dynamic Segmentation with GBP 504
HPE Aruba Networking Central NetConductor overview 506
HPE Aruba Networking Central NetConductor orchestration 507
HPE Aruba Networking Central NetConductor provisioning 509
EVPN-VXLAN fabric considerations 511
Distributed fabric personas 512

Supported devices for distributed fabric personas 513
Learning check 514
Question #1 514
Question #2 514
Question #3 515
Supplemental: VXLANs 515
VXLAN review 515
VXLAN terminology 516
VXLAN VTEP 518
VXLAN tunnel 519
VNI "Virtual Switch" 520
Chapter 9: Network Optimization 521
Objectives 521
QoS Best Practices 521
Application classification 521
Quality of Service overview 523
Voice requirements for low loss, latency, and jitter 524
Summary of application requirements by class 526
Meet application requirements 527
Traffic prioritization 528
Capacity planning 529
Over-provisioning 530
Wireless QoS 531
Issues that prevent efficient use of wireless capacity 531
How WMM works 533
Features to optimize wireless capacity 533
UXI 534
The cost of not monitoring end user digital experience 534
HPE Aruba Networking User Experience Insight (UXI) 535
HPE Aruba Networking User Experience Insight (UXI) Agent 536
UXI components 538
UXI features 538
UXI sensor portfolio 541
UXI use cases 541
Learning check 542
Question #1 542
Question #2 542
Question #3 542

Supplemental: Wired QoS 543
Using L2 and L3 classification 543
DiffServ Code Point (DSCP) 544
DSCP relationship with per-hop behaviors 545
Example: six apps divided across a four class LAN queueing model 547
Traffic prioritization and queuing 547
Traffic classification and marking: WMM 548
Wi-Fi Multimedia (WMM) details 549
Wi-Fi Multimedia operation 550
BSS coloring 551
WLAN multicast and broadcast — DMO 553
Broadcast to unicast conversion 555
What is Intelligent Call Handling (ICH) 556
How ICH works 557
Conclusion 559
Next steps 559
Get certified 559
Advance your knowledge 559
HPE7-A11 practice exam 561
Objectives 561
Exam overview 561
Ideal candidate for this exam 561
Exam contents 561
Advice to help you take this exam 562
Exam objectives and exam information 562
HPE7-A11 exam objectives 562
Exam information 563
Practice exam 564
Practice exam questions and answers 564
Questions 564
Answers 586
Learning check answers 609
Overview 609
Chapter 1: Information Gathering 609
Chapter 2: HPE Aruba Networking security-first, AI-powered networking 609
Chapter 3: RF Planning 610
Chapter 4: Wireless High-Density Design 610
Chapter 5: Wired Network Design 610

Chapter 6: Network Segmentation 611
Chapter 7: Campus Security 611
Chapter 8: Overlay Networks 611
Chapter 9: Network Optimization 612

1 Information Gathering

Welcome back! This is chapter 1, all about gathering required information for your design efforts.

OBJECTIVES

After completing this chapter, you should be able to:

- ✓ Determine the key stakeholders.
- ✓ Assess physical site locations.
- ✓ Evaluate the network usage.
- ✓ Determine the scope of the project.

Key Stakeholders

Let's start by looking at the key stakeholders.

Key stakeholders of a deployment project

The first step in any project is defining the Scope of Work of that project. Knowing what it means to complete a project can only be determined by first understanding the customer's goals. The organization's key stakeholders must be identified to clarify their expectations. This simple step of establishing who determines goals and defining the scope of work will set the project up for success from day one.

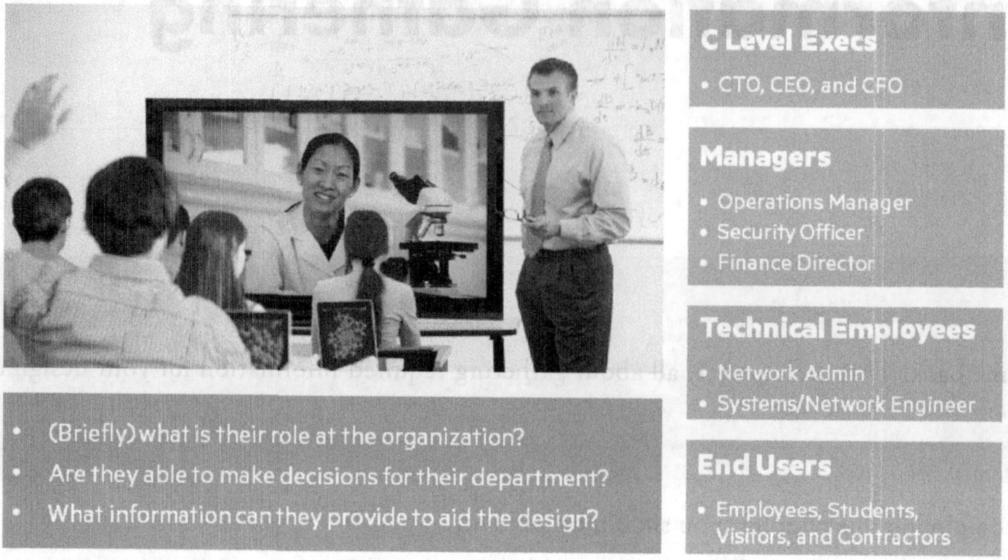

Gathering information for a network design project will require pulling data from different sources in the customer organization. Consideration must be given to from the top "C" or Chief level executives and network managers, down to the technical systems engineers and admins supporting the network, and of course, to the end-users themselves. Each group has pain points that need to be identified and addressed. Furthermore, each group holds vital information about how priority should be assigned to applications and devices. These findings will dictate how the HPE Networking solution should be defined to service those needs and address existing concerns.

Consider the following questions when evaluating a new customer contact for the project:

- (Briefly) what is their role at the organization?
- Are they able to make decisions for their department?
- What information can they provide to aid the design?

A CTO or Chief Technology Officer will have concerns regarding costs such as capital expenditure (CapEx) versus return on investment (ROI). Yet you would likely avoid going into technical discussions, such as whether to deploy layer 3 routing to the access layer.

Managers can provide you with critical business requirements and projected growth of end-users and employees. They also may have the flexibility to make purchasing decisions within their department and may have some technical resources such as access to floor plans.

Technical Employees such as network admins are an excellent resource for technical requirements and information about the existing network design and pain points to be resolved. Often, these employees are unaware of budgetary concerns and will require the CTO or their manager to sign off on any purchases or changes.

End users should never be overlooked. The investment of a network design is to serve end-users such as employees, visitors, students, and more. While managers and executives may dictate the purchase of a network, the end users should see an improvement in pain points that make that investment worthwhile. Consider even a simple end-user survey as to the usability of the network before and after the network is deployed.

Note

Many large and complex network design projects will require a long sales cycle, and it might include a POC (proof of concept) and a pilot of the new network systems to show the value they can bring to an organization. You can often define the project's scope in an RFP (Request for Proposal) or RFQ (Request for Quote); you can often expect a highly competitive sales process. The price and benefits of the new network design and solving the business needs are paramount when discussing deployment projects.

Interview stakeholders

Once a network design project is underway, you will want to begin the information-gathering process by talking to the organization's stakeholders and clarifying their expectations. It is essential to define the project's scope by discussing the current network deployment and problems with those systems.

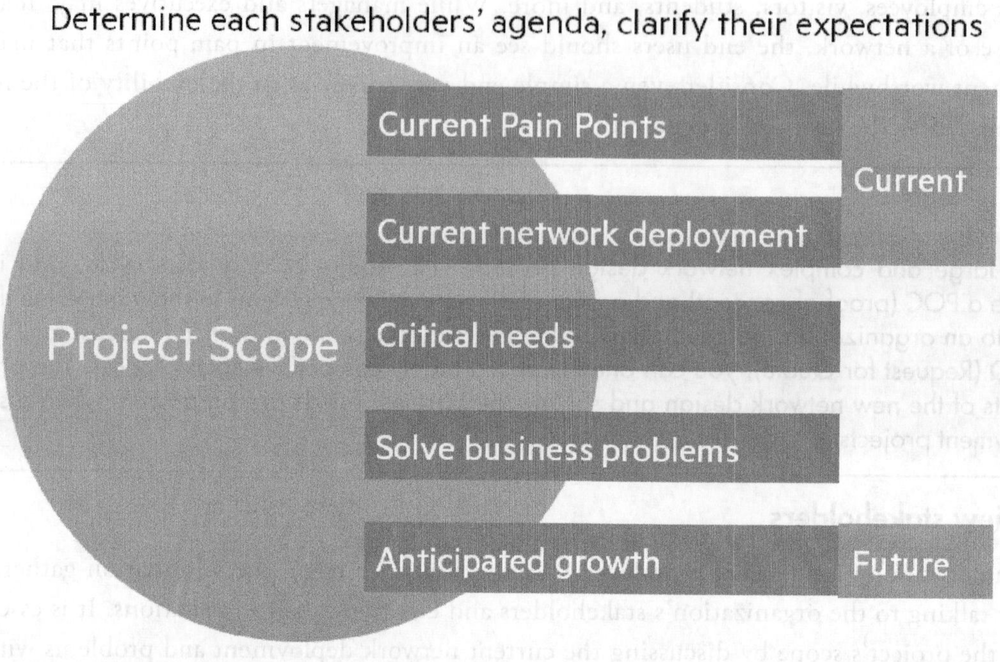

During the interview process, you should discuss the existing network's critical needs and pain points. The organization's anticipated growth includes additional users, devices, or expanded locations.

Consider the following stakeholder interview questions:

- What is the scope of the project?
- What is the primary purpose of the new network deployment?
- What business problems are you trying to solve?
- What are your most critical needs?
- What are your pain points or most important issues?
- What is your anticipated growth of the environment?

Physical Environment

Next, let's examine the physical environment of the location.

Physical sites

You need to document the number of physical sites or locations that are within the scope of the project.

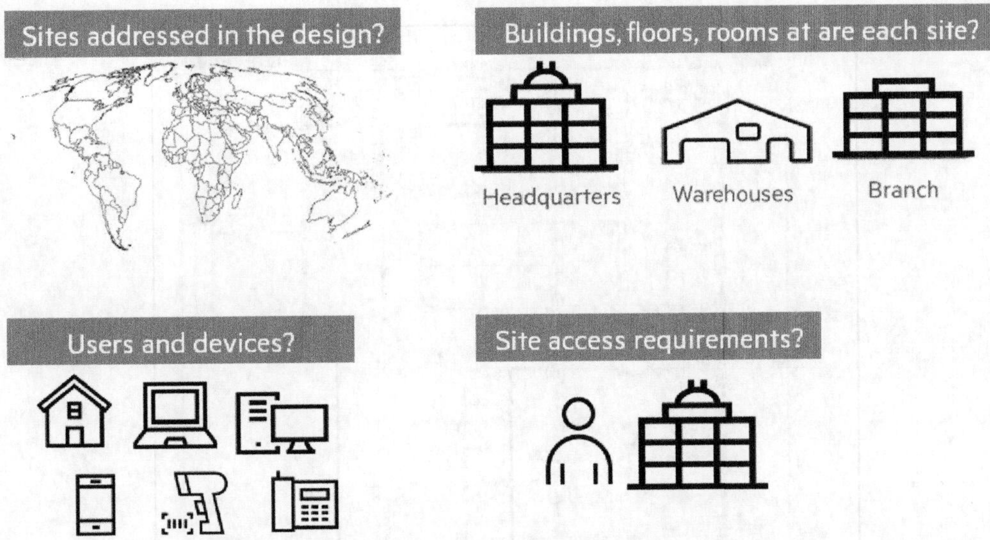

It is a good idea to identify the purpose of the network in each location and the types of users the network will serve, such as the main campus or HQ location, branch office, or remote workers.

- What sites are being addressed in the network design?
- Countries involved?
- What buildings, floors, and rooms are each site?
- Building title and significant function?
- Civic address of any buildings?
- Users and devices?
- Estimate of total users by site?
- Remote workers?
- How are users segmented by type, time of day, and roles?
- Average device counts by users?
- Business critical device types such as work laptops and mobile devices?
- Site access requirements ?

Building 100: floorplan example

Be sure to note the critical features of floorplans when they are provided. First, be clear what the exact location where the floorplan is from. Which region and campus is this taken from? What is the building's civic address? Where are the corner X and Y coordinates for latitude and longitude if provided? Which floor number is being shown, and for which building number or name?

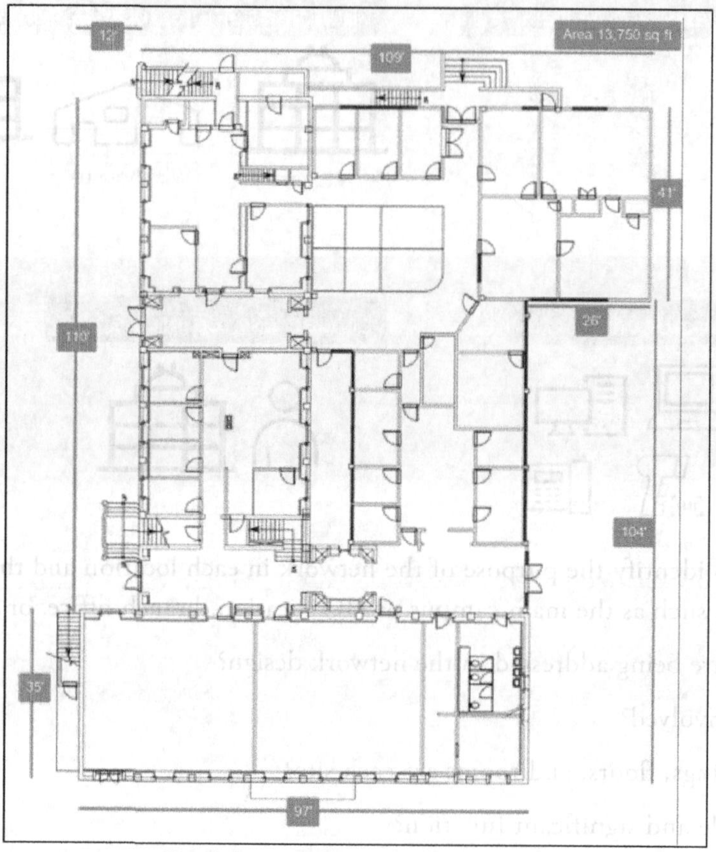

Building 100 1st Floor

Next is the critical step of scaling the image to match the actual floor space that is being described. Mistakes in scaling will distort any future location awareness or heat mapping done with tools such as Floor Plan in HPE Aruba Networking CentralBe sure to note the critical features of floorplans when they are provided. First, be clear what the exact location where the floorplan is from. Which region and campus is this taken from? What is the building's civic address? Where are the corner X and Y coordinates for latitude and longitude if provided? Which floor number is being shown, and for which building number or name?

Next is the critical step of scaling the image to match the actual floor space that is being described. Mistakes in scaling will distort any future location awareness or heat mapping done with tools such as Floor Plan in HPE Aruba Networking Central. Confirm the measurements are in feet or meters depending on the customer region.

Building materials are another factor that should be considered when providing a floor map. Taking the time to note the internal and external wall-building materials can go a long way in future attenuation and RF heatmap calculations. Perhaps more importantly, the AP count is impacted by over or underestimating the required number of APs for optimal floor coverage. For example, with interior walls of newer plaster, the radio signal may propagate further than expected, causing potential co-channel inference between neighboring APs. The opposite issue of holes in coverage may arise if the walls are made of brick and concrete, typically found in older buildings.

Review any floor plans provided:

- Do they include the civic address of the building?
- Latitude and longitude?
- Floor number?
- Measurements in square feet or square meters?
- Ceiling height, atriums, indoor and outdoor?
- Building materials: walls, doors, windows, or any notable furniture that may impede wireless propagation.
- Critical areas of high density or areas without coverage?
- Any other points of concern, such as proximity to radar interference or co-location issues?
- Confirm the measurements are in feet or meters depending on the customer region.

Physical address is called "Street address" in HPE Aruba Networking Central Floor Plan. The site represents an individual building, as per HPE Aruba Networking Central Floor Plan requirements.

Building 100: network closet

You should understand the current RF environment before creating an RF plan and determining the number of APs in each location or floor. Ideally, this would include a recent site survey.

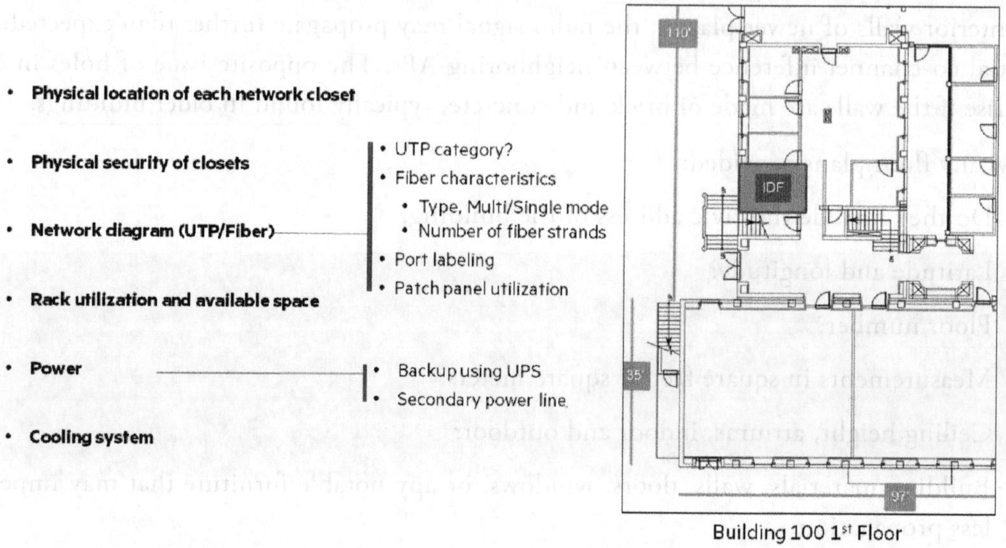

You will need a building floor plan file. This file can be in DWG (CAD files), JPG, PNG, GIF, or other electronic formats to import into Floor Plan or other planning tools. You also need to know the scale of the floor plan and the building materials, so you can begin to plan wireless coverage. In addition, determine the business use of the floorspace (e.g., retail, hospital, or education).

To complete the wired network design, you should obtain a floor plan or indoor/outdoor schematic of the building, including the locations of wiring closets and power and cooling capabilities. The customer should also provide a wired network diagram and wireless access point drops.

Current Network Usage

In this section, you will discover current network usage for the campus.

Users

You need to document the number and type of users for each location in the project's scope. For example, the application needs of doctors needing wireless coverage at the bedside will differ from the wireless needs of nurses using voice devices to communicate.

Number of users per location

Types of users

Mobility requirements/usage

Stationary requirements/usage

User density

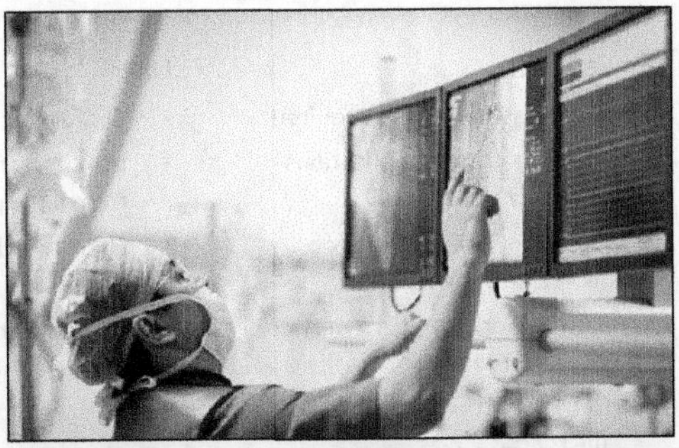

The characteristics of mobile devices connecting to a WLAN affect the configuration of the network infrastructure. Mobile devices are, by nature, more portable than non-mobile devices. When a WLAN must support portable devices, it requires robust roaming support as the devices move through the WLAN. Roaming support allows for a seamless transition from one access point to another. The user does not have to reconnect, reassociate, or re-authenticate.

Roaming user type examples

Document the number of users for each location in the project scope. Consider general user types by roaming, application requirements, number of devices, working hours, or any other differentiating factors that may impact their presence on the network. This may also align by user department, making the task easier.

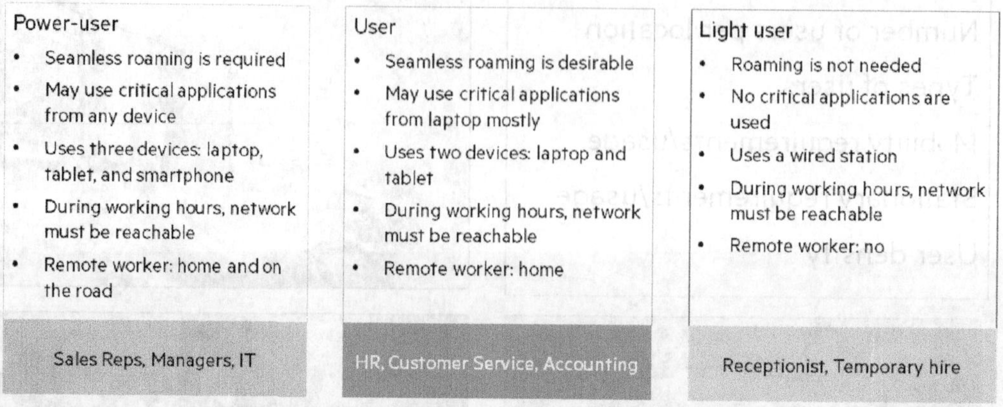

The characteristics of mobile devices connecting to a WLAN affect the configuration of the network infrastructure. Mobile devices are, by nature, more portable than non-mobile devices. When a WLAN supports critical applications on many portable devices, it requires robust roaming support as the devices move through the WLAN. Roaming support allows for a seamless transition from one access point to another. The user can avoid having to reconnect, reassociate, or reauthenticate as they move through the building.

User device needs

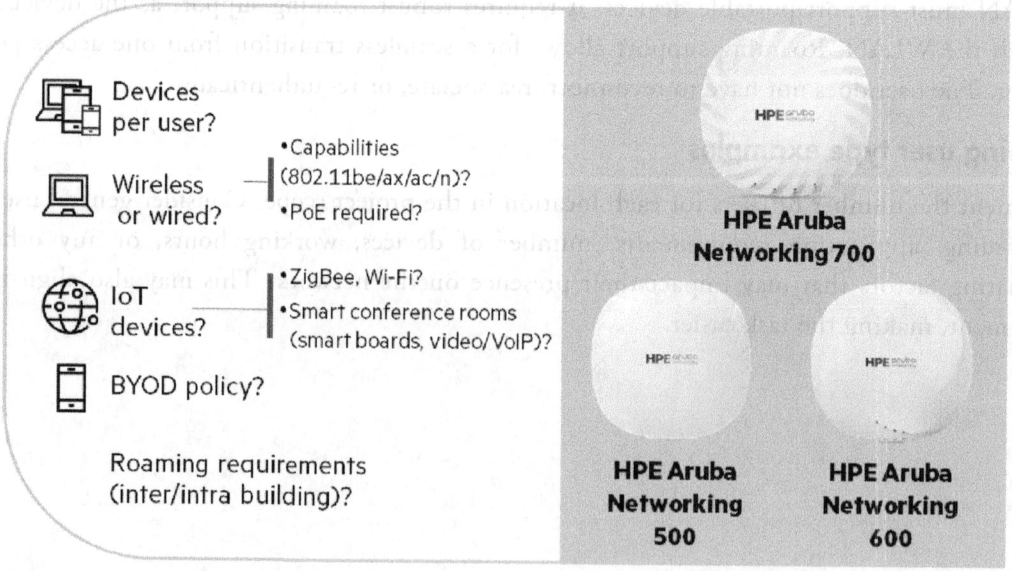

Determining the user device needs is critical for accurately designing a wireless network. Begin by asking, "How many devices will each user have?" HPE Networking recommends that you plan for three to four devices per user in a medium campus: a laptop, a tablet, a smartphone, and perhaps another personal device. Looking at the entry-level 500 series HPE Aruba Networking AP, moderate capacity could be obtained by multiplying the estimated number of users by the number of expected devices. Utilizing the HPE Aruba Networking 650 series AP would allow for not only 2-3x the throughput but also double the capacity, along with other advantages such as layering user traffic across all three radio bands to further improve the user experience. Support Wi-Fi 7 deployments with HPE Aruba Networking 700 series APs.

Zigbee is a low-rate wireless personal area network based on 2.4 GHz and lower (country-dependent) band frequencies between 20 and 250 kbps at a range of about 10-20m. For additional references that can be used to develop a proper definition of HPE Networking's utilization of Zigbee:

https://community.arubanetworks.com/discussion/zigbee-addressing-support-question
https://iot-utilities.arubademo.de/docs/app/zigbee.html

Device needs

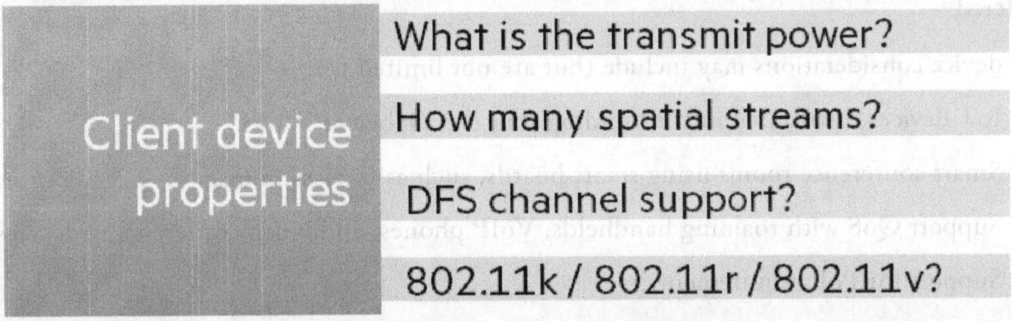

Other important considerations about the device capability should include support for the 6 GHz bands, max transmit power, number of supported spatial streams and OFDMA features, and support for extended DFS channels. Determining the maximum number of devices desired for each AP is essential. Typically, HPE Networking recommends of 75-100 devices per radio for 600 series APs and 700 series. This number may be higher or lower depending on traffic type (voice or data), expected load, and connection type (802.11be/ax/ac/n). Note that the published maximums range from 256 to 512 maximum per radio, but this should be avoided except in very high-density deployments covered in module 4 of this guide.

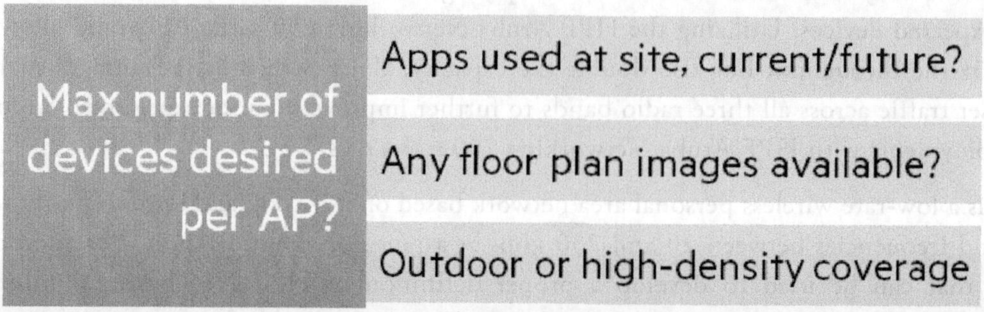

You must ask the customer for their security requirements, not only the technical kind but also regulatory. We should also gather all their information regarding regulatory requirements for their environment. For example, Wi-Fi 6E and Wi-Fi 7 support 6 GHz but only for indoor deployments pending further regulatory change. This is to avoid interference with the spectrum reserved for satellite communication but may change in the future if GPS location validation of the AP is considered.

Other device considerations may include (but are not limited to):

- IoT devices distributed in the building that use Zigbee or Wi-Fi.
- Smart conference rooms using smart boards, such as in classrooms.
- Support QoS with roaming handhelds, VoIP phones, and video conferencing devices.
- Support for PoE requirements?
- What is the maximum transmit power of the least-capable standard device in the network?
- How many transmit, receive, and spatial streams do the most common devices support?
- Device support for the U-NII-2 extended band and the 6 GHz as well?
- If DFS channels are being considered, do the devices most commonly used in the network support DFS channels?

Documentation & administrative information

Determining the user device needs is critical for accurately designing a wireless network. To better refine customer needs around an AP model and the number of APs, ask, "How many devices will each user have?" HPE Networking recommends that you plan for at least three devices per user: a laptop, a tablet, and a smartphone.

Tech docs and floorplans	How old is the info? Is it still relevant going forward?
Config version control system	Change tracking, who apply changes? Who approves them, is there a process for it?
IT dept device access	Campus: yes Data center: limited
Tech experience: switching, wireless, security	HPE Aruba Networking certified administrators? Reliance on a third-party vendor? HPE Aruba Networking experience? Which HPE Networking products and solutions?

Other important considerations about the device capability should include max transmit power, the number of supported spatial streams, and support for DFS channels.

Next, it is essential to determine the maximum number of devices desired for each AP.

Before starting an RF plan, you should also gather any floor plans or image files from the customer and determine the scale of the floor plans.

You must ask the customer for their security requirements, not only the technical kind but also regulatory. This is true for both wired and wireless. We should also gather all their information regarding regulatory requirements for their environment.

Customer User Device Needs Job Aid:

- How many devices will each user have? HPE Networking recommends that you plan for at least three devices per user: a laptop, a tablet, and a smartphone. The number of devices per user also has ramifications in the design of VLANs and subnets. You should also consider if all devices will run simultaneously, as this will impact AP density.
- What is the maximum number of devices desired for each AP? What applications will be in use at the site, both presently and in the future? Bandwidth requirements help determine coverage vs. capacity requirements.
- Are any floor plan images available? Floor Plan supports directly importing JPEG, GIF, PNG, PDF, and CAD (.dwg and .dwf) files for floor plan formats.

- What is the maximum transmit power of the least-capable common device in the network?
- How many transmit, receive, and unique streams do the most common devices support?
- If DFS channels are being considered, do the devices most commonly used in the network support DFS channels?

Wireless RF/network questionnaire

A standard method of gathering information is a questionnaire to ask about the network environment and customer needs. Before creating a design, you should determine which 802.11 standards the customer devices support and client device upgrades planned for the near future.

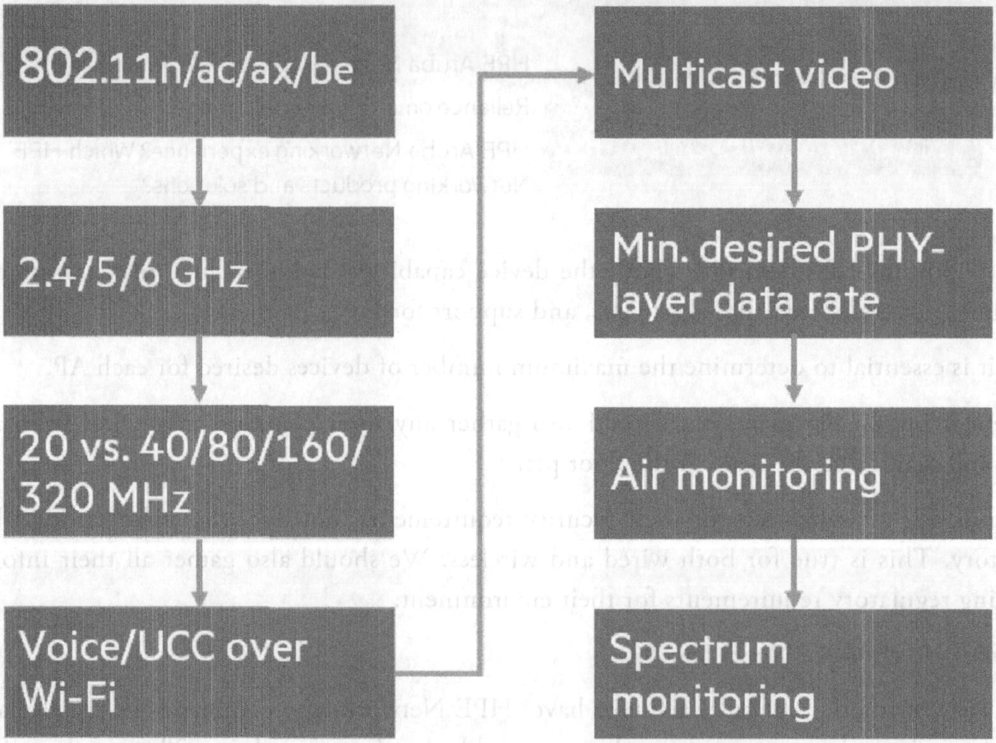

Almost all new enterprise deployments will be based upon 5 GHz and 6 GHz radio frequency channels unless you only require 2.4 GHz coverage. You may also have to perform an on-premises walkthrough and spectrum analysis to determine the feasibility of 40 and 80 MHz (or wider) channel use.

The primary use of the network, such as voice, video, or data, is critical when designing an RF network. Knowing a targeted minimum data rate is also helpful before starting the design process. Lastly, it is essential to ask the customer if air monitoring/WIPS with containment or spectrum monitoring services are necessary for the network.

Remember that many customers might not know the answers to these questions. Many environments feature clients introduced by the users, and some customers lack insight into all the applications in use. As a general best practice, you should implement a design to support 802.11be/ax and voice and video applications to future-proof.

Wireless RF/Network Questionnaire

- What 802.11 PHY types are required during the WLAN lifecycle (802.11ac, 802.11ax or 802.11be)?
- Which RF bands will be used (2.4 GHz, 5 GHz, and 6 GHz)? Plan to use both bands due to increases in client density.
- What channel width (40 MHz vs. 80 MHz) will be used in each band? Typically, 20 MHz channels are used in 2.4 GHz, 40 MHz channels are in the 5 GHz band, and 40 MHz to 80 MHz are in the 6 GHz band.
- Will voice-over Wi-Fi be used? This answer will affect your planning for roaming and AP signal strength calculations.
- Will multicast video over Wi-Fi be used? Dynamic multicast optimization (DMO) should be implemented to provide QoS.
- What minimum desired PHY-layer data traffic rate must be available throughout the coverage area? Do some areas have different minimum data rate needs?
- What are the desired air monitoring rates? Are dedicated air monitors required for security or compliance purposes?
- Are there any known interference sources? Is there a need in spectrum monitoring?

Additional questions could be added, for example: Do clients support UNII-4 channels?

Wired network questionnaire

You should also investigate the customers' wired network needs by asking: how many wired ports the customer requires for users and devices? What types of applications run over the network?

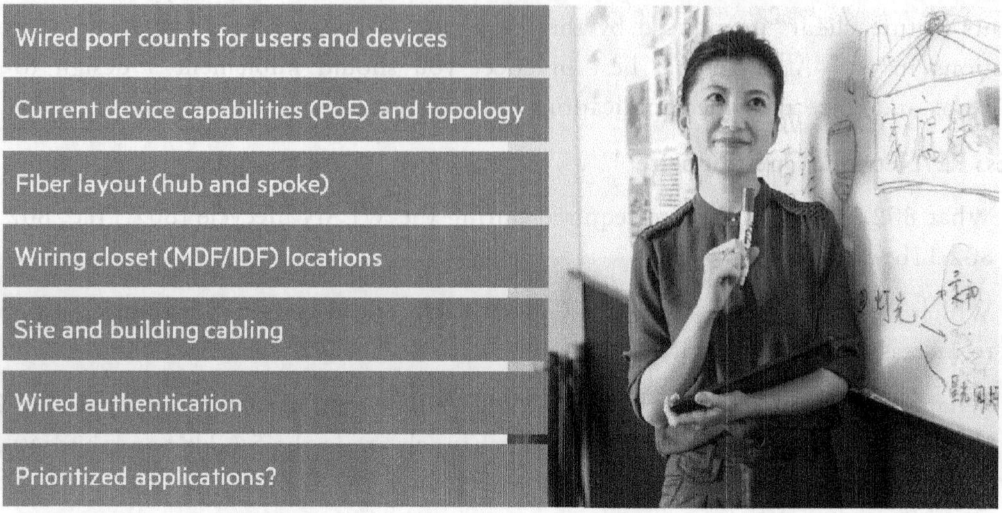

It is also essential to ask the customer about device capabilities (including PoE), logical topology, physical wiring closet locations, and site cabling. For example, what room or closet does the networking cable leave the building to reach the internet? This is typically called the Main Distribution Frame (MDF). Or you may hear the term Intermediate Distribution Frame (IDF) for a closet where cabling comes in and leaves a single floor. Other people may replace these terms with "main computer room" or "data room."

Other questions to ask the customer concerning their wired network include how many wired ports the customer needs to support. This includes ports for wired users and other devices such as printers and Internet of Things (IoT) devices. You should also ask what applications users run on the wired network. Ask whether the customer can share information about the current topology and device capabilities.

You will also need to assess the physical requirements. In addition to closet locations, you must gather information about the site's cabling. This will help you select the correct switches and transceivers, and it could affect whether you propose switches that support Smart Rate ports.

You should also remember that the customer's information about the topology might not be up to date. If the customer is open to you passively collecting some information about the network, you could use additional tools to map the network.

Wired Network Questionnaire:

- How many wired ports for users and devices?
- What types of applications are used on the network?
- What are the current device capabilities and topology?

- Where are closets located?
- How is the site cabled?
- What are wired authentication requirements?
- What QoS prioritization required?

Scope of the Project

It is vital to understand the scope of work for the project.

Anticipated growth

During the customer interview, you need to factor in the anticipated growth of the organization as well.

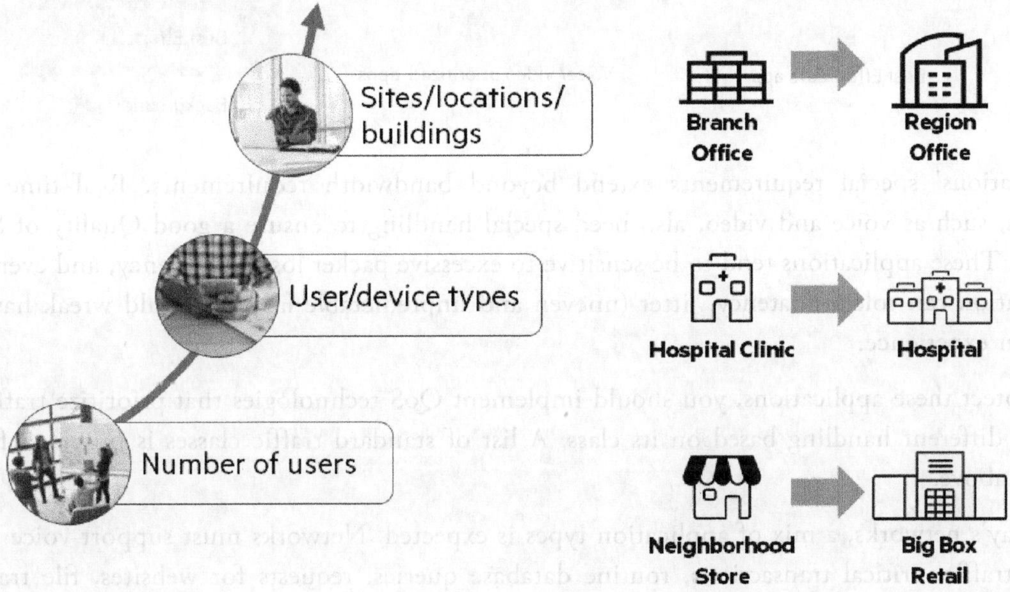

If the organization is planning a new location, a significant increase in the number of users, or a change to a new device type, then you should factor these into the design plan from the beginning. Also, it would be best if you discussed any anticipated changes to the network.

Assess application requirements

Different applications have different requirements, and understanding applications' bandwidth requirements can help you plan for capacity. You should also consider whether applications send bursts of traffic or a constant flow.

- What are the throughput requirements across multiple users over the long term?
- Does traffic flow more heavily from the core to the edge than the reverse? Also, does traffic flow east-to-west, north-to-south, or vice versa?

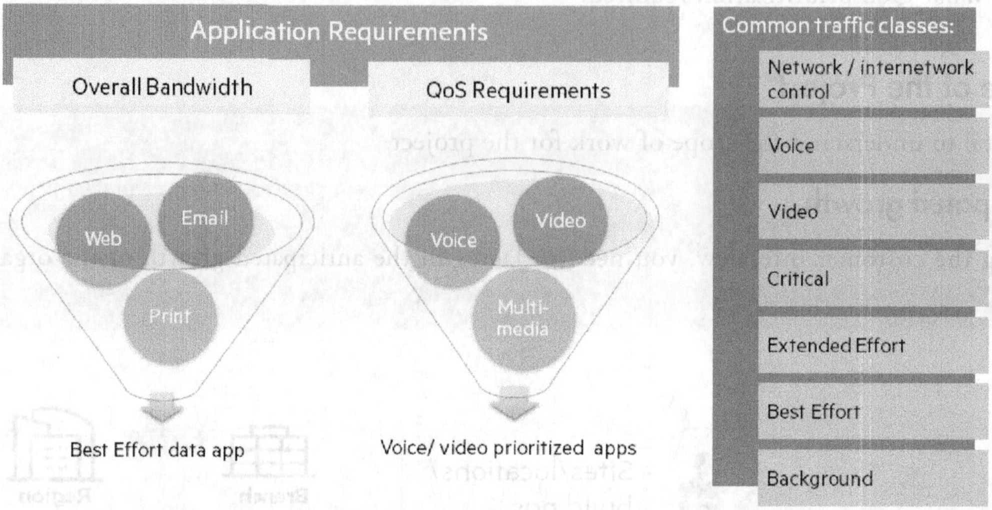

Applications' special requirements extend beyond bandwidth requirements. Real-time applications, such as voice and video, also need special handling to ensure a good Quality of Service (QoS). These applications tend to be sensitive to excessive packet loss and latency, and even if the application can tolerate latency, jitter (uneven and unpredictable latency) could wreak havoc on the user experience.

To protect these applications, you should implement QoS technologies that prioritize traffic and give it different handling based on its class. A list of standard traffic classes is provided for reference above.

In today's networks, a mix of application types is expected. Networks must support voice traffic, video traffic, critical transactions, routine database queries, requests for websites, file transfers, and more. Each of these applications must compete for network bandwidth. Ensuring availability and good performance for mission-critical and business-critical applications can present significant challenges, particularly in an environment that extends across wireless and wired media.

The first key to ensuring that users have a good experience running all applications is providing adequate bandwidth from end to end.

Understanding applications' bandwidth requirements can help you plan for capacity. You should also consider whether applications send bursts of traffic or a constant flow. What are the throughput requirements across multiple users over the long term? Does traffic flow more heavily from the core to the edge than the reverse?

Only a thorough examination of the throughput requirements can ensure that your plan will provide the customers with the required experience. In the Ethernet network, you must scope the expected traffic loads on each physical link and plan appropriately. Planning capacity for the wireless network can be challenging, with multiple users sharing the wireless media (although multi-user MIMO [MU-MIMO] introduces another wrinkle). Later modules in this guide will help you plan wireless cells for typical and high-capacity use cases.

Applications' special requirements can extend beyond bandwidth requirements. Real-time applications, such as voice and video, also need special handling to ensure a good quality of service (QoS). These applications tend to be sensitive to excessive packet loss and latency, and even if the application can tolerate latency, jitter can wreak havoc on the user experience. It can occur even in networks with adequate provisioning. QoS technologies help to prioritize the correct traffic at these times.

Work with the customer to define traffic classes. Traffic classes suggested by IEEE 802.1Q-2005 include network and internetwork control traffic (the packets exchanged between network infrastructure devices for layer-2 and -3 protocols), voice, video, mission-critical applications, essential applications that need extended effort, best effort (regular traffic), and background traffic (less important traffic or traffic that can tolerate delays). Later in this guide, you will learn more about creating a plan to provide the correct service to each class using the appropriate HPE Networking network technologies.

Assess roaming requirements

In a wireless network, users want to roam seamlessly, so you must determine the extent of roaming domains for your customers.

For example, should users be able to roam seamlessly throughout a building without a session interruption?

Should coverage and roaming extend between buildings? You should also discuss what roaming means to the customer.

A user working a laptop and then taking that laptop elsewhere introduces less rigorous roaming requirements than users talking over a wireless voice app on the go, and some manufacturing and warehouse sites have specialized needs.

The customers' roaming requirements will affect your RF plan and solution architecture.

You need to determine the extent of roaming domains for your customers. For example, should users be able to roam seamlessly throughout a building without a session interruption? Should coverage and roaming extend between buildings? If the latter, you will need to plan the outdoor coverage, and if roaming domains extend far, you will need to consider how to cluster controllers to meet the needs.

You should also clarify with customers what they mean by roaming.

When users access the network through a wireless connection, they expect to be able to roam. First, you need to clarify what users mean by "roaming." A user working a laptop and then taking that laptop elsewhere introduces less rigorous roaming requirements than users talking over a wireless voice app on the go. Voice over WLAN requires roaming to occur under 50ms to remain undetectable to the user. This is a different consideration from the ITU 150ms end-to-end delay of voice calls for satisfactory MOS (Mean Opinion Score).

You might also need to meet special requirements for manufacturing or warehouse environments.

Basic wired connectivity requirements

You should now have a good sense of the users and devices the customer's unified wired and wireless network must support, both now and in the future. Next, you will examine wired connectivity requirements.

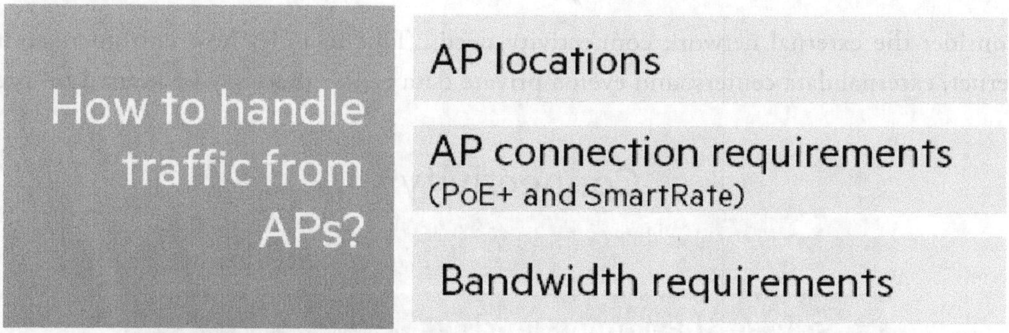

The wired network will almost always function as the backhaul for a wireless network, connecting wireless users and APs to their controllers. You will need to consider the locations for APs and the requirements (e.g., PoE+) for their wired connections. You must also determine the type of cable because PoE+ should have at least CAT5e cable, but CAT6a cables and above are preferable in modern networks.

Information you collect about user density and application requirements will help you assess how much bandwidth you need to plan on the APs' switch uplinks and across the network. Customers might require wired connectivity for traditional workstations or devices like printers. Again, port count and application requirements can help you to plan the uplink bandwidth.

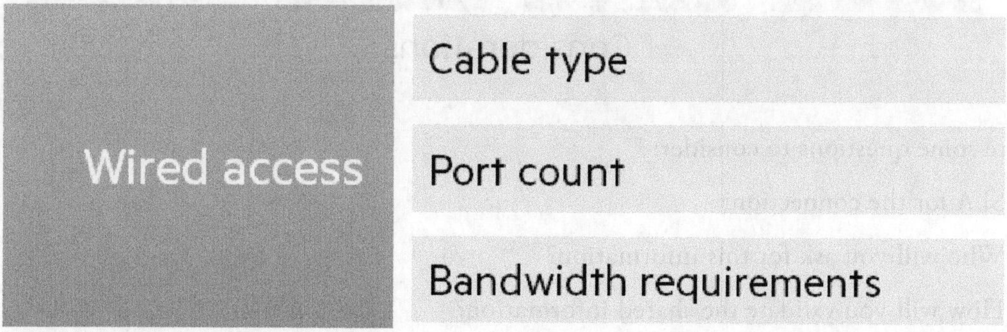

Customers typically require wired connectivity for some devices, including traditional workstations and printers. As customers use wireless connectivity more for their employees, many are repurposing Ethernet ports for Internet of Things (IoT) devices like smart lights, HVAC systems, and more. You should get a port count for all devices, and again, application requirements can help you plan uplink bandwidth.

High availability requirements for the wireless network: use access point models with two PoE ports or distribute access points connections between multiple switches or VSF stack members to ensure reliable and uninterrupted wireless network access, even in cause of PoE switch failure.

Campus external network connectivity needs

Also, consider the external network connectivity needs. This includes how customers connect to the internet, external data centers, and even a private data center that may be housed on premises.

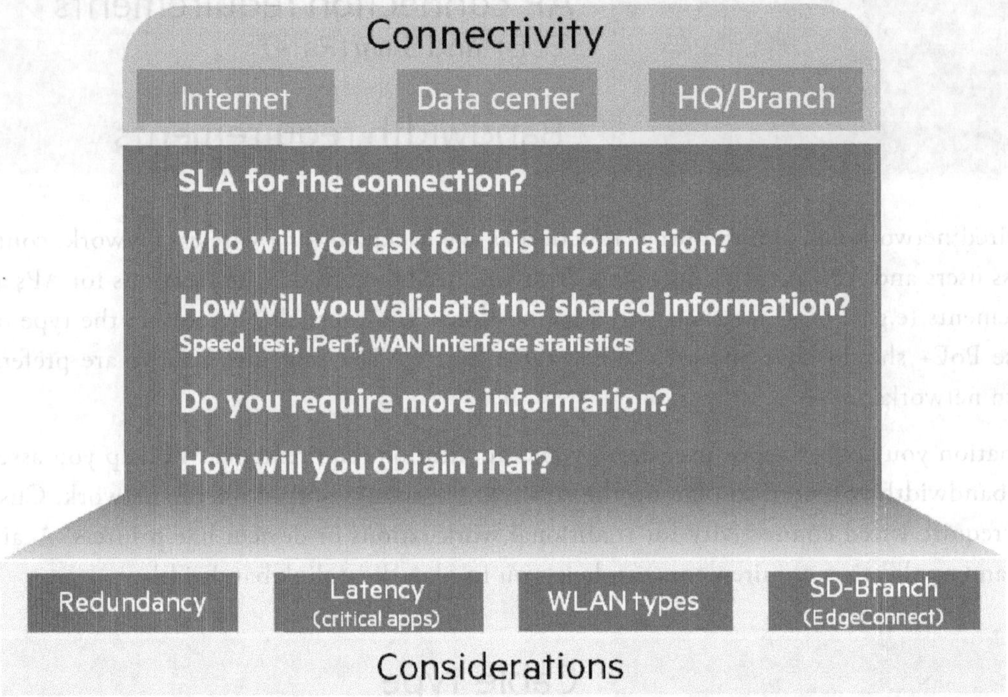

Here are some questions to consider:

- SLA for the connection?
- Who will you ask for this information?
- How will you validate the shared information?
- Speed test, iPerf, WAN Interface statistics
- Do you require more information?
- How will you obtain that?

Remote user connection needs

- Level of support for end-user remote access to campus network?
- Is the network prepared for massive teleworker traffic?
- What is expected from the connection? (from branch office; from anywhere)

You should also investigate the customers' wired network needs by asking how many wired ports the customer requires for users and devices and what types of applications run over the network. Other important questions for the customer include device capabilities (including PoE), logical topology, physical wiring closet locations, and site cabling.

Typical questions to ask the customer about the wired network are how many wired ports need support, including ports for wired users and other devices such as printers and Internet of Things (IoT) devices. You should also ask what applications users run on the wired network and whether the customer can share information about the current topology and device capabilities. You will also need to assess the physical requirements. In addition to closet locations, you must gather information about the site's cabling. This will help you select the correct switches and transceivers and might also affect whether you propose switches that support Smart Rate ports.

Assess security requirements

A company's vulnerability to these risks depends on various factors, including the types of network access offered, the types of users at the site, and the types of devices in use. Wireless access is inherently more open than wired access, but you can plan solutions that appropriately control wireless and wired access.

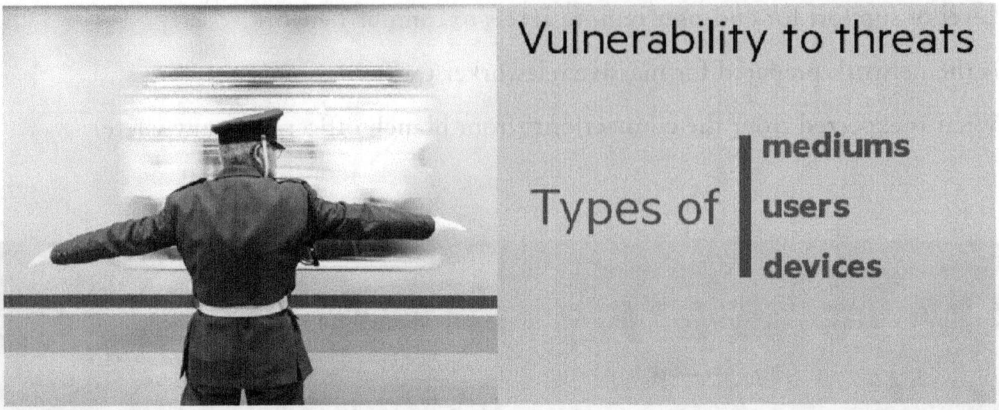

Because many companies implement 802.1X on wireless (but not wired) connections, wireless connectivity could be more secure than Ethernet on a campus without physical monitoring. You should discuss with customers whether they would like to integrate their wired connections with the security options available for wireless connectivity.

HPE Networking has designed its solutions especially to offer role-based control at a deep level. You should work with the customers to understand which categories of users they need to control. You should also discuss whether they need a solution for handling Bring Your Own Device (BYOD), which you can offer with ClearPass Onboard.

Some customers want to classify users as employees, contractors, and guests, or in an education setting, students and faculty. Other customers will want to take full advantage of the role-based firewall and classify their users more granularly.

As you discuss the types of devices that the solution needs to control, assess whether the customer has vulnerable IoT devices. This can affect whether you recommend a solution such as ClearPass Device Insight through HPE Aruba Networking Central to assist with full spectrum visibility.

Security needs

You should discuss security needs with C-level executives so you can plan to provide the proper access security solution for the customer. Unauthorized network access exposes companies to data breaches, malware infections, and interrupted services due to Denial of Service (DoS) attacks.

Corp is aware of unsecured network cost?	Regulatory penalties, loss of intellectual property, downtime, cost of remediation and restitution, loss of reputation and business
Corp has defined security goals?	Ask about the current security policies, how to manage incidents, protocols to deal with that, etc.
Current configuration?	What security technologies and best practices were implemented? Current RADIUS server? Basic PKI infrastructure?

Customers should understand what they stand to lose. Data breaches can lead to regulatory penalties or loss of intellectual property, and unplanned downtime causes lost revenue. These issues can lead to costs to remediate the problem and possibly to provide restitution if customers' data is compromised. Less tangible and critical effects include a loss of reputation and future business.

Customers should understand that an enterprise wireless network should not use a single password but the user-based authentication provided by 802.1X. This approach provides better protection against hackers and enables the Aruba solution to control authorized users differently with a role-based and stateful firewall. Recent attacks should show customers that internal threats can pose real risks to their operations and bottom line.

In more detail, risks include regulatory penalties. The Health Insurance Portability and Accountability Act (HIPAA), the Payment Card Industry Data Security Standard (PCI DSS 2.0), and other federal and private regulations require particular security measures. These might extend well beyond personal firewalls and local data encryption into extensive network security requirements. While regulations may not precisely state network security requirements, companies will want to add extra layers of protection for applicable servers and data to protect themselves from security breaches that may result in fines or other penalties. These fines can add up quickly since each compromised piece of data can incur a fine. One security breach may compromise thousands of records and incur fines that can risk an organization's financial stability.

Legal protections aside, most companies are interested in ensuring their proprietary data and company plans do not fall into the wrong hands. While legal measures for restitution exist, most companies would rather avoid the legal hassle by ensuring that their intellectual property is free from snoopers.

Security breaches, denial of service (DoS) attacks, and viruses can cause disruptions or failures in the network that result in downtime, and downtime represents revenue loss to most companies.

Other costs include hiring IT security professionals to assess a breach, estimate its scope, and plan a response, as well as a way to avoid problems in the future. When there is a breach in regulated data, the company must also spend money notifying users whose records have been compromised and receiving legal advice about the implications. The company might need to offer restitution in the form of identity protection to those whose records were compromised.

Security breaches can cause companies to lose status or reputation with their customers and partners. A loss in reputation can impact a company's bottom line and result in revenue loss that can continue for years after the security breach.

Network requirements? AAA, alerts, Wireless IDS, secure remote connections, device visibility, segmentation, IPsec tunnels created by the edge firewall.

Security concerns?
Challenging granular policies
Lack of security policies

A company's vulnerability to these risks depends on various factors, including the types of network access offered, the types of users at the site, and the types of devices in use. Wireless access is inherently more open than wired access, but you can plan solutions that appropriately control wireless and wired access. Because many companies implement 802.1X on wireless (but not wired) connections, wireless connectivity could be more secure than Ethernet on campus without physical monitoring. You should discuss with customers whether they would like to integrate their wired connections with the security options available for wireless connectivity.

Aruba has designed its solutions especially to offer role-based control at a deep level. You should work with the customers to understand which categories of users they need to control. You should also discuss whether they need a solution for handling Bring Your Own Device (BYOD), which you can offer with ClearPass Onboard.

Assess the need for redundancy

Because customers rely on their network services, they need consistent availability. Still, various issues can interrupt services, from a cut cable to a configuration error, a power failure, or more. As you will learn in a later module, you can plan Aruba solutions to provide varying levels of availability against this issue. Higher degrees of availability requires more equipment (e.g., an additional controller for redundancy) and technologies for detecting and responding to failures.

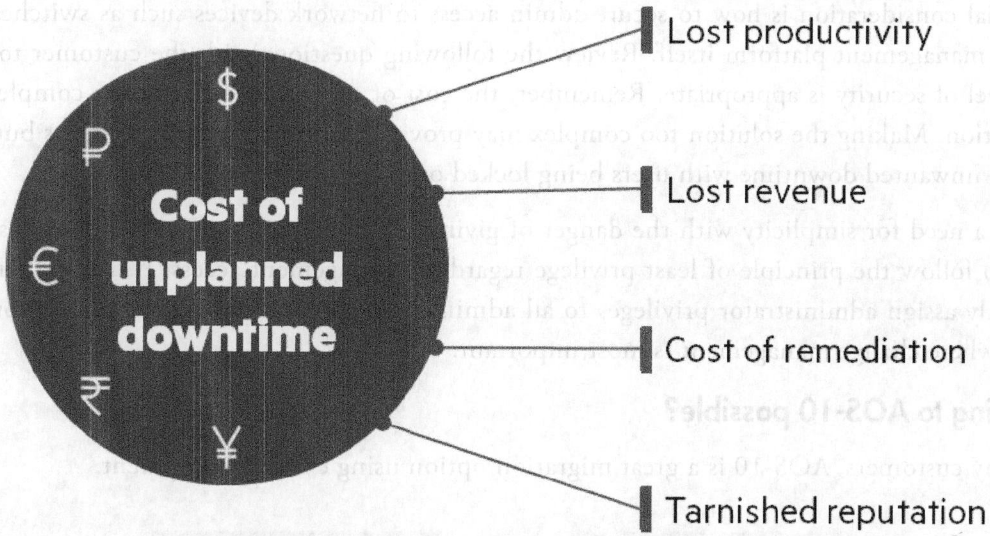

Because these requirements translate to additional costs and complexity, you need to assess the availability level this particular customer requires based on unplanned downtime costs. The needed level typically depends on how vital a service is to the customer's mission and the risks they run from losing that service. These risks include lost productivity, lost revenue, the cost of remediating the issue, and hidden costs, such as the blow to the customer's reputation.

For example, if the customer relies on wireless voice applications, the wireless network will need to provide higher availability than if the customer uses wireless as a convenience for guests. A hospital that uses the network for supporting critical services during surgery will have much higher availability requirements than a typical office.

Management needs: consider admin access

- How does IT staff currently manage the network?
- Do you have accounting for admin access?
- What authentication sources are you using?
- What is expected from the management platform?
- Who will you ask for this information?
- How will you validate the shared information?
- Do you require more info? How will you get it?

Of crucial consideration is how to secure admin access to network devices such as switches, APs, and the management platform itself. Review the following questions with the customer to gauge what level of security is appropriate. Remember, the cost of security is often in the complexity of the solution. Making the solution too complex may provide additional security benefits but could result in unwanted downtime with users being locked out.

Balance a need for simplicity with the danger of giving too much access. The standard best practice is to follow the principle of least privilege regarding management access. It may be tempting to quickly assign administrator privileges to all admins, but this would enforce policy anonymity exactly where change management is most important.

Migrating to AOS-10 possible?

For many customers, AOS-10 is a great migration option using existing equipment.

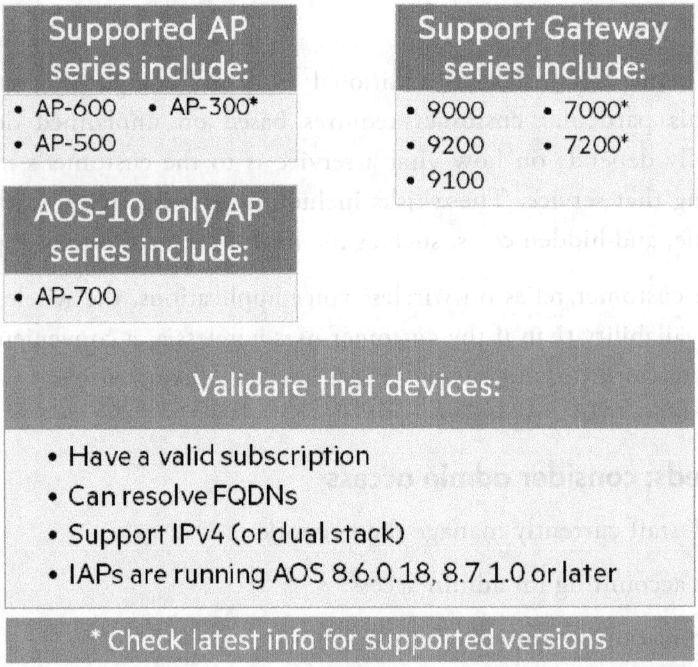

For a successful brownfield deployment with AOS-10, first, assess whether the current HPE Aruba Networking APs and Gateways are supported by referring to the summary list shown or checking the following links:

https://arubanetworking.hpe.com/techdocs/central/latest/content/aos10x/aos10x-overview/supported-devices-aos10.htm

https://arubanetworking.hpe.com/techdocs/central/latest/content/aos10x/cfg/aps/ap-migration.htm

Learning check

Check your understanding of gathering information by answering the following questions.

Question #1

When discussing the project scope of work with key stakeholders, which would be considered a question to determine "pain points"?

- ○ What wireless standards are you most interested in?
- ○ What is your anticipated business growth for the next three years?
- ○ What network issues are impacting the business the most?
- ○ How soon should the project be completed?

Question #2

What features of the floorplan are important to be collected and used for the wireless network design (select four)?

- ☐ The floorplan must be in an AutoCAD (.DWG) format
- ☐ Civic address and coordinates
- ☐ Included floor number
- ☐ Building materials provided for walls, doors, and windows
- ☐ Noted ceiling heights and atrium locations
- ☐ Details about emergency exits and escape routes

Question #3

Which of the following is true regarding device capabilities?

- ○ You should determine if devices support 2.4 GHz or 5 GHz. No consideration is yet required for 6 GHz as it has yet to be ratified.
- ○ 500 series APs support 6 GHz band.
- ○ Wi-Fi 6E allows to use 320 MHz channels in the 6 GHz band.
- ○ Migrating to AOS-10 is possible with 600, 500, and some 300 series APs.

2 HPE Aruba Networking security-first, AI-powered networking

Understanding HPE Aruba Networking security-first, AI-powered networking Architecture enables a best-in-class design for a campus network. Customers can benefit from an evolved solution over traditional methods of network management and deployment.

Note

HPE Aruba Networking security-first, AI-powered networking was formerly known as HPE Aruba Networking ESP.

OBJECTIVES

After completing this chapter, you should be able to:

- ✓ Examine the HPE Aruba Networking security-first, AI-powered networking campus architecture.
- ✓ Disseminate the HPE Aruba Networking portfolio.
- ✓ Plan HPE Aruba Networking Central as a management platform.
- ✓ Choose the HPE Aruba Networking Central subscriptions.
- ✓ Determine how AOS-10 and AOS-CX architecture fits the campus design.

HPE Aruba Networking security-first, AI-powered networking Campus Architecture

Let's start with an overview of HPE Aruba Networking security-first, AI-powered networking campus architecture.

HPE Aruba Networking architecture

HPE Aruba Networking security-first, AI-powered networking development was driven by customer needs and the failings of legacy systems. This AI-powered, cloud-native platform automates, unifies, and protects the Intelligent Edge. You get several benefits:

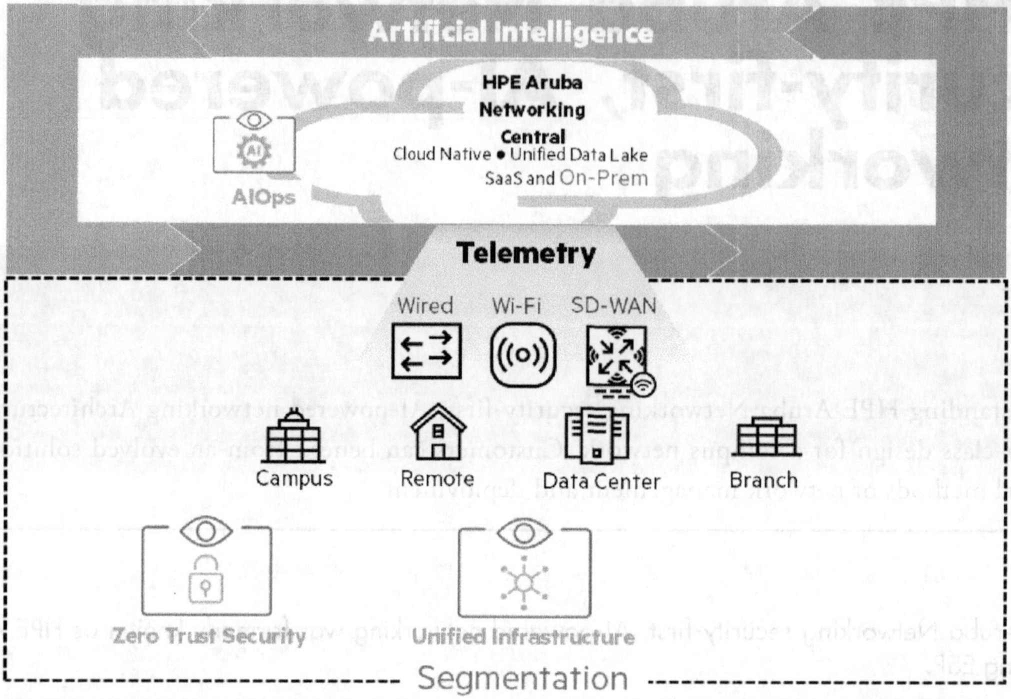

Simple, flexible, and highly reliable network designed to allow efficient access to required services, applications, and data for only authorized users.

Supports programmability and ease of upgradeability with best-in-class enterprise networking hardware.

HPE Networking's evolutionary end-to-end architecture is managed using cloud-based HPE Aruba Networking Central, providing Machine Learning adaptability and Artificial Intelligence Operations (AIOps) to improve user experience, minimize downtime, and achieve Zero Trust Security.

Although HPE Aruba Networking Central is a cloud-native solution, there is also a "HPE Aruba Networking Central On Premises" (COP) solution.

HPE Aruba Networking security first, AI-powered networking overview

HPE Aruba Networking security-first, AI-powered networking is based on four core attributes:

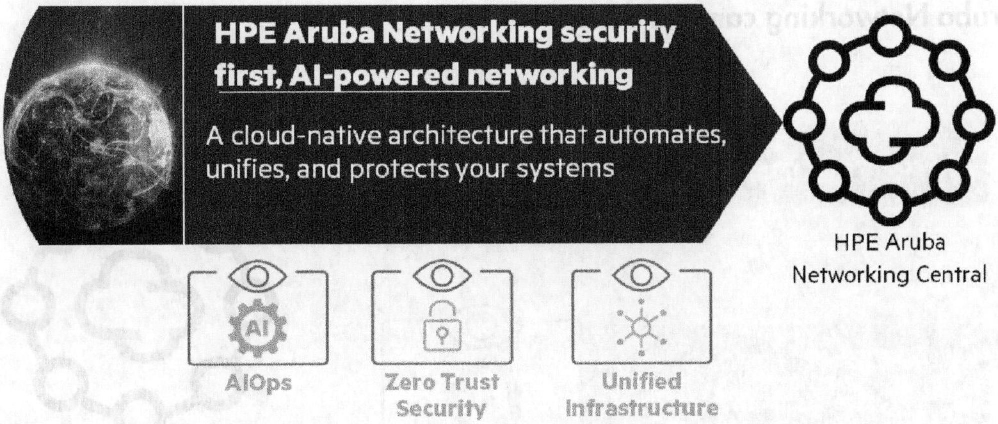

- HPE Aruba Networking Central is a Software-as-a-Service(SaaS) management solution, ready to monitor and manage any company of any size. HPE Aruba Networking Central's flexibility gives you greater choice and control over how you consume and pay for technology to accomplish your goals.
- **Artificial intelligence operations** - gives the network a 6th sense, resolving issues before they impact the business or at least resolve them faster
- **Unified Infrastructure** - means you can manage different types of devices in different locations using a single interface. It eliminates silos and streamlines operations.
- **Zero Trust Networking** – Every device plugs into the network gets authenticated and authorized. You are not inherently trusted, and zero trust security includes role-based security to minimize the need for manual configuration.

HPE Aruba Networking campus design

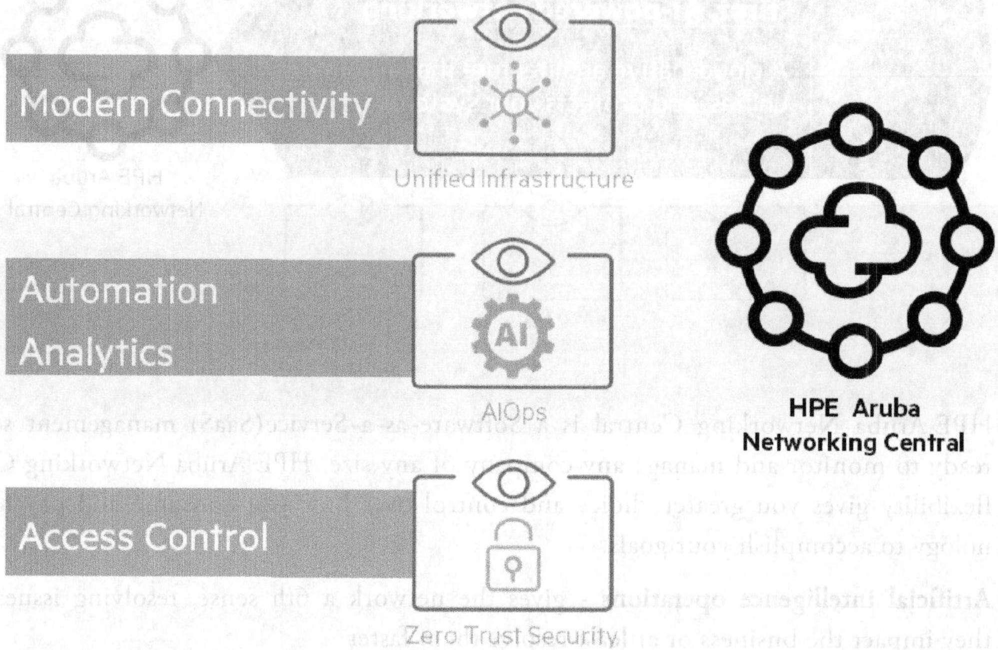

Modern Connectivity

HPE Aruba Networking security-first, AI-powered networking starts with a modern switching infrastructure. HPE Aruba Networking OS-CX switches offer scalability with a wide choice of port densities, speeds, and availability using HPE Aruba Networking switch models.

Automation

On-premises solutions such as ClearPass and HPE Aruba Networking Fabric Composer (AFC) as well as cloud-based with HPE Aruba Networking Central providing hosted cloud solutions such as HPE Aruba Networking ClearPass Device Insight and HPE Aruba Networking Central NetConductor (ACN), HPE Aruba Networking Meridianand HPE Networking User Experience Insight (UXI).

Analytics

On-the-box HPE Aruba Networking Network Analytics Engine (NAE) along with machine learning and AI operations (ML/AIOps) cloud analytics, alerts, and reporting to quickly find the root cause of an issue and resolve it with minimal downtime.

Access Control

ACL, deep packet inspection (DPI), and session-aware firewalls extended ubiquitously across wired and wirelessly connected clients, all while leveraging centralized policy enforcement with ClearPass.

HPE Aruba Networking security-first, AI-powered networking services

HPE Aruba Networking security-first, AI-powered networking architecture can be better understood when broken into separate layers. The base infrastructure for a modern edge services platform starts with robust and centrally managed AOS-CX switches, APs, and Gateways. A unified infrastructure provides consistent programmable management and access across wired and wireless connectivity.

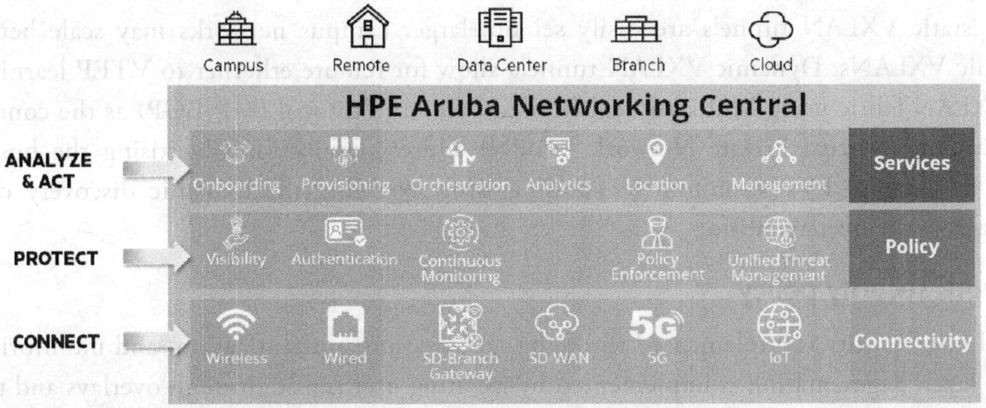

The policy layer directs traffic based on user and device roles to provide automated traffic enforcement. Orchestrated tunneling between the access layer to central services or from access points to gateway services enables a zero-trust network. The services layer provides an interface for administrators to manage, analyze, and orchestrate the network. This is aided by modern AIOps, allowing for natural language search, dynamic network baseline tools, and self-healing insights.

Network connectivity

Take a deeper look at the Unified Infrastructure layer and the use of overlays. Overlay tunneling to HPE Aruba Networking Gateways uses Generic Routing Encapsulation (GRE) tunnels. This is a carry-over technology from previous AP to HPE Aruba Networking Controller tunneling.

Another kind of tunneling is distributed overlays that run between switches. These switch-to-switch tunnels use virtual extension LANs or VXLANs to enable layer 2 reachability over an existing layer 3 underlay. Distributed overlays leverage symmetric Integrated Routing and Bridging (IRB). Almost unlimited VNIs or VXLAN Network Identifiers can segment user traffic across the overlay in place of VLANs. VXLAN Tunnel End Point (VTEP) is the on-ramp and off-ramp switches at either end of the tunnel. These are typically employed at the access layer. If used, they perform the actual encapsulation and policy enforcement of Group Base Policies (GBP).

While static VXLAN tunnels are easily set up, larger campus networks may scale better using dynamic VXLANs. Dynamic VXLAN tunnels allow for remote ethernet to VTEP learning across the VXLAN fabric using Multiprotocol Border Gateway Protocol (MP-BGP) as the control plane and Ethernet Virtual Private Network (EVPN) address-family for advertising the host IP and MAC addressing. EVPNs minimize flooding while enabling the dynamic discovery of remote hosts across the overlay fabric.

Network security policy

Zero Trust Security is implemented through segmentation, authentication, and monitoring in the policy layer. Segmentation is implemented by isolating user traffic through overlays and traffic filtering. Dynamic segmentation using user-based tunneling (UBTs) provides centralized overlays to a gateway or gateway cluster. VXLANs, either static or dynamic (EVPNs), offer distributed overlays.

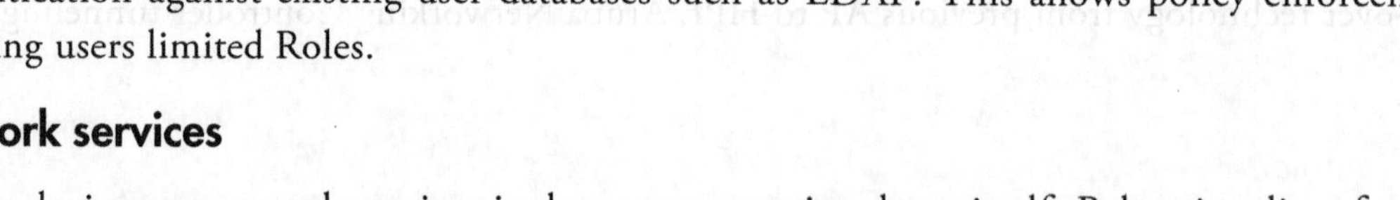

The HPE Aruba Networking ClearPass Policy Manager Platform provides RADIUS/TACACS+ authentication against existing user databases such as LDAP. This allows policy enforcement by assigning users limited Roles.

Network services

The final piece to network services is the campus services layer itself. Below is a list of common services HPE Aruba Networking offers for the campus:

HPE Aruba Networking Central as a Cloud Platform: Cloud management of HPE Aruba Networking gateways, switches, and APs.

HPE Aruba Networking Central Management: Deploy, manage, and optimize the WLAN, LAN, VPN, and even SD-WAN for the campus.

- HPE Aruba Networking Central AIOps: AI combines network and user-centric analytics to identify anomalies and report them to IT staff.
- ClearPass Policy Manager: Role-based secure network access control.
- HPE Aruba Networking Client Insights: Utilizes infrastructure telemetry from clients, APs, switches, and gateways without requiring physical collectors or agents.
- HPE Aruba Networking User Experience Insight (UXI): Cloud-based 24/7 service assurance solution for users, networks, and applications. UXI sensors enable continuous monitoring and testing across wired, wireless, and WAN.
- Live Upgrade: Using wireless telemetry data through AirMatch, a coordinated upgrade can be rolled across APs that minimizes impacting clients. This reduces downtime and reliance on maintenance windows.
- AI Insights: HPE Aruba Networking Central service that identifies, categorizes, and provides resolution to issues that affect client onboarding and connectivity.
- AI Assist: Leverages event-driven automation to collect troubleshooting information on identified issues before they impact user and business applications.
- AirGroup: Works with multicast DNS (mDNS) and simple service discovery protocol (SSDP) across VLANs to enable these devices to be discoverable from any client location or VLAN.
- Air Slice: Enables HPE Aruba Networking Wi-Fi 6 APs to prioritize client traffic at the physical radio level. Air Slice is transparent to the client, does not have standard requirements, and tightly integrates into the DPI firewall capabilities of HPE Aruba Networking APs.
- AirMatch: Uses gather telemetry data from APs for RF events and measurements, channel range, transmit power range, and overall operational conditions.

- ClientMatch: Optimizes the client association between channels and AP radios to avoid sticky clients.

HPE Aruba Networking security-first, AI-powered networking campus reference architecture overview

HPE GreenLake for HPE Aruba Networking provides:

- Access to HPE GreenLake services using a subscription-based model.
- IT Service Management (ITSM) dashboard for single-pane management.
- Intelligent operations options with proactive 24x7 remote management and monitoring by HPE Networking experts.
- Delivers agile Network as a Service (NaaS)
- Customer Experience Management (CEM)

- Delivers digital insights and solution expertise to continuously analyze and assess your networking environment, optimize performance and security, lower risk, and accelerate return on investment (ROI).

Campus IT staff challenges

- Large scale network environment
- Need to support legacy infrastructure
- Human factor affects root cause analysis

Today's IT staff face challenges in campus networks. They typically need to support a large, extensive network service with expanding requirements for new technologies. They may be using legacy or siloed infrastructure needing manual support and maintenance, all the while trying to do traditional root cause analytics by hand, which can be difficult, time-consuming, and error-prone.

HPE Aruba Networking security-first, AI-powered networking edge services

The HPE Aruba Networking security-first, AI-powered networking campus uses modern Edge Services to enable a secure and consistent network regardless of organization size. All the following technologies are backed by artificial intelligence for analytics and best practice guidance.

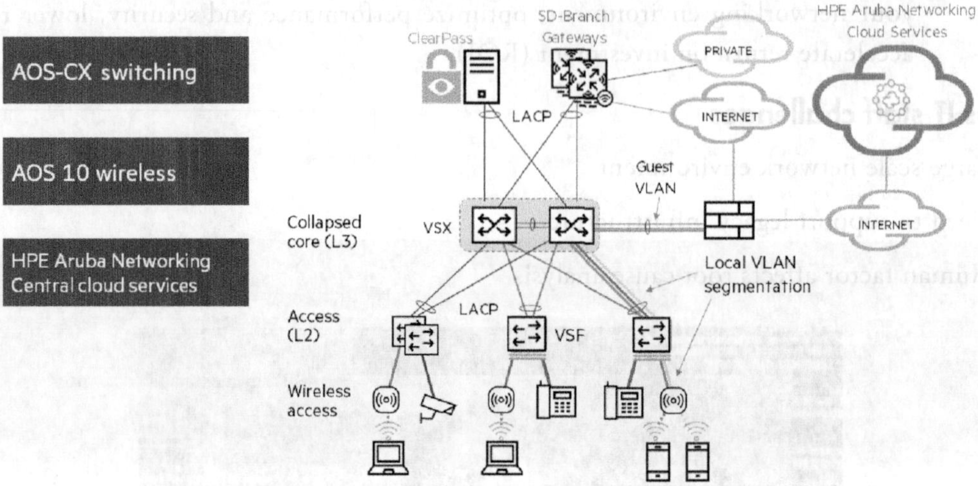

Leverage HPE Aruba Networking AOS-CX for a joint switching operation across campus, data center, or branch while managed on-premises or through HPE Aruba Networking Central in the cloud.

Wireless support through AOS-10 provides a cloud-native, enterprise, wireless operation system for APs and Gateways.

HPE Aruba Networking Central expands customer visibility across a reliable and accessible cloud platform. HHPE Aruba Networking Central management, policy, and analytics for wired and wireless HPE Networking devices.

These are the design goals

The overall goal of good campus design is the following:

- Produce a simple, scalable design that can be replicated across sites.
- Settling upon a set of specific products and performing all required operations and maintenance functions.
- Achieve customer requirements; we will revisit this in a later module.

HPE GreenLake for Networking service pack examples

One option to streamline purchasing of HPE Aruba Networking products is with HPE GreenLake Service Packs. These are packaged solutions to ensure a customer can deploy hardware with assurance that the hardware will be supported, licensed, and warranted.

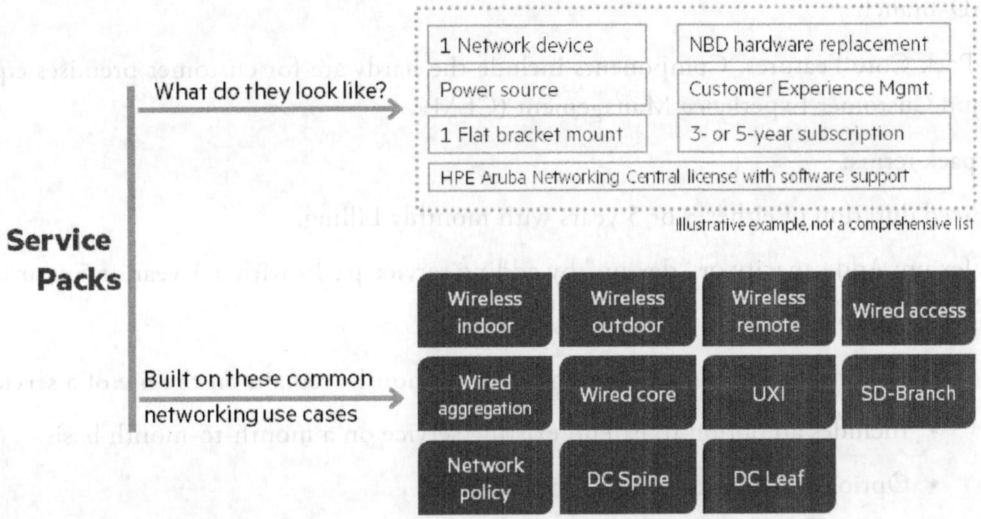

A Service Pack may include:

- 1 Network Device
- 1 Flat bracket mount
- 1 HPE Aruba Networking Central license
- 3 or 5-year subscription
- Customer Experience Management (CEM)
- Hardware and software support
- Power source

Example Wireless Indoor Service Pack: Pre-configured with an HPE Aruba Networking AP-510 series AP. Note that this can be tailored to use an AP-535 if requested.

Service Pack options include:

- Wireless indoor
- Wireless outdoor
- Wireless remote
- Wired access
- Wired aggregation
- Wired Core
- UXI
- SD-Branch

Service Pack Core Features: Components include the hardware for customer premises equipment (CPE) and Customer Experience Management (CEM)

Service pack terms:

- Fixed duration of either 3 or 5 years with monthly billing.
- Flex up: Add capacity or "flex up" by adding service packs with a 3-year or 5-year term.
- End of term:
  - Provided by HPE Networking when 12 months remain on the life of a service pack.
  - Includes an option to extend existing service on a month-to-month basis.
  - Option to renew for another 3 or 5 years.
  - Option to return equipment.
- Flex down:
  - Return 10% of the total contract value (TCV) during the contract term.
  - Customers who choose the flex down option at the time of order can return with no penalty after the first 12 months and before the last two months of the Service Pack term.
  - 60-day notification required to flex down.
- Pass:
  - Enables HPE Aruba Networking equipment's safe and sustainable return at the end of the Service Pack Term.
  - Includes a prepaid return service, including features such as packing and shipping, documentation, and claim resolution.

HPE Aruba Networking Portfolio

Let's look now at the HPE Aruba Networking portfolio of products.

Product line

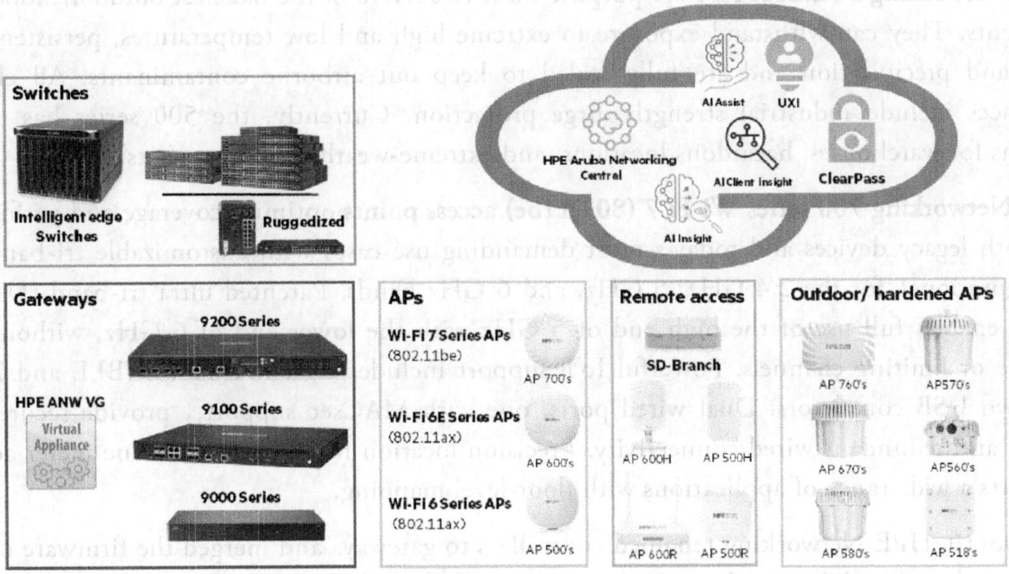

Here is an overview of the HPE Aruba Networking products and their specifications:

The 500 series APs support Wi-Fi 6 (802.11ax) and provide differentiated capabilities, including AI-powered Wi-Fi performance assurance, always-on connectivity, and WPA3-certified security. Green AP mode offers AI-powered energy savings of up to 70%. Additionally, OFDMA and MU-MIMO efficiently and simultaneously service multiple clients to provide a good experience in high-density environments. These technologies will be covered in the upcoming module 3. HPE Aruba Networking APs also support IoT and leverage your Wi-Fi infrastructure to support Zigbee and Bluetooth devices and applications.

The AP-600 series expands upon the existing features found in the AP-5xx series. These indoor APs support Wi-Fi 6E based on the 802.11ax standard and offer the advantage of the 6 GHz band to provide far greater speeds, wider channels for multi-gigabit traffic, and less interference. These include 802.11mc Wi-Fi Round Trip Time (RTT), allowing precise indoor positioning and interoperability with Google, Apple, and Bing Maps. This provides an alternative to the current method of triangulating location based on RSSI.

HPE Aruba Networking has several options for remote access that fall under the categories **SD-Branch and Microbranch**. SD-Branch includes deploying a gateway at the remote location. This scales for locations with over 5,000 users or more than 500 APs. For Microbranch locations such as a small branch or home office, good candidates for APs are the 500H, 600H, 500R and 600R series. The "H" stands for hospitality, and the "R" for remote. These vary in supported standards, number of ports, and whether the use case requires PoE PD and PSE features.

HPE Networking's outdoor APs are purpose-built to survive in the harshest outdoor/indoor environments. They can withstand exposure to extreme high and low temperatures, persistent moisture, and precipitation and are fully sealed to keep out airborne contaminants. All electrical interfaces include industrial strength surge protection. Currently, the 500 series has outdoor options for warehouses, hazardous locations, and extreme-weather environments.

HPE Networking 700 series Wi-Fi 7 (802.11be) access points optimize coverage and performance for both legacy devices and today's most demanding use cases with customizable tri-band radio configurations1 for the 2.4 GHz, 5 GHz, and 6 GHz bands. Patented ultra tri-band (UTB) filtering enables full use of the high end of 5 GHz with the lower end of 6 GHz, without interference or limiting channels. Powerful IoT support includes dual IoT radios (BLE and Zigbee) and two USB connectors. Dual wired ports, one with MACsec support,1 provide secure, high-speed, and redundant wired connectivity. Precision location feature with sub-one meter accuracy supports a wide range of applications with floor-level mapping.

In AOS-10, HPE Networking renamed controllers to gateways and merged the firmware code for SD-Branch and mobility implementations into one firmware platform.

Intelligent edge switches combine a modern, fully programmable OS(AOS-CX) with carrier-grade hardware, leading performance, and incorporate the industry-first Network Analytics Engine to easily monitor and troubleshoot network, system, application, and security-related issues.

HPE Aruba Networking Central is a cloud-based management system integrated with the HPE GreenLake edge-to-cloud platform. HPE GreenLake delivers everything as a service, allowing the features to tie in with HPE Aruba Networking Central. These include AI Insight, AI Assist, AI Client Insights, UXI, and Activate.

HPE Aruba Networking User Experience Insight (UXI) allows for proactive monitoring, troubleshooting, and addressing application and network performance issues. UXI gives near real-time data insights into your network and application performance by deploying UXI sensors or end-point agents for Zebra.

HPE Aruba Networking Activate is a web service that enables zero-touch provisioning of HPE Aruba Networking devices.

ClearPass provides network control, access security, and advanced features such as Captive Portal, guest login, self-registration, and onboarding of employee-owned devices.

HPE Aruba Networking SD-WAN provides flexible WAN management options for retail, hospitality, and healthcare organizations, moving branch traffic off private WAN links in favor of direct Internet access.

HPE Aruba Networking Central AIOps is a narrow AI focused on supporting and troubleshooting HPE Aruba Networking networks. The HPE GreenLake hosted services provide the following.

AI Insight enhances network monitoring by providing a service level baseline of WLAN, switching, and WAN performance.

AI Assist: When an issue arises, this provides an automated collection of log files and troubleshooting data while generating HPE Aruba Networking TAC engagement.

AI Client Insights- Profiles clients on your network by leveraging native infrastructure telemetry from existing APs, switches, gateways, and clients without requiring physical collectors or agents.

UXI- HPE Aruba Networking User Experience Insight provides an AIOps-powered incident detection feature. This leverages AI to analyze network issues for a more efficient MTTR workflow.

HPE Aruba Networking CX switching advantages

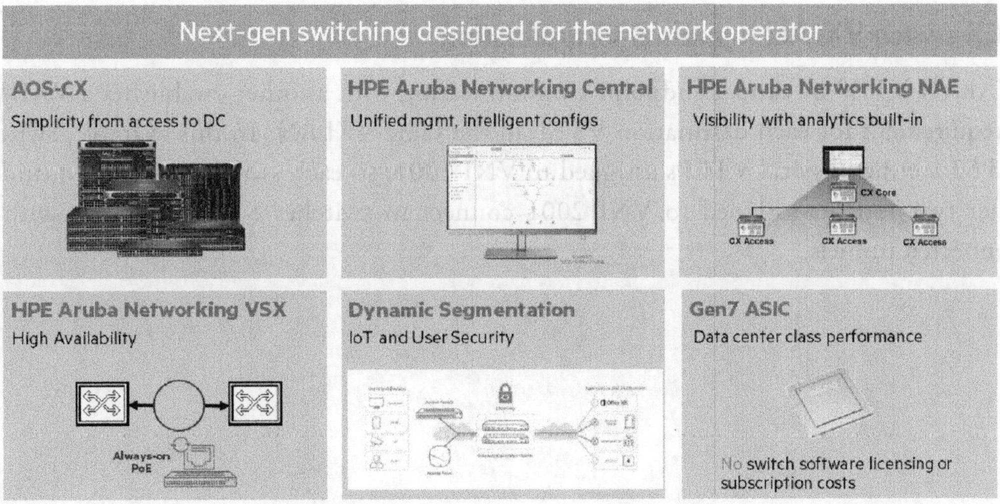

The HPE Aruba Networking AOS-CX switching advantages:

A single operating system across your entire environment spanning from edge to data center.

HPE Aruba Networking Central is a management tool. It is a unified infrastructure, a single-pane-of-glass to manage and monitor your HPE Aruba Networking wired and wireless devices. HPE Aruba Networking Central supports a multi-editor mode for advanced configurations, which allows you to orchestrate common configuration changes across one or multiple CX switches.

HPE Aruba Networking Network Analytics Engine (NAE) is a tool for monitoring and troubleshooting. NAE provides network operators with distributed, real-time insights into network health for a faster, more proactive way of monitoring, detecting, and troubleshooting IT and user-impacting issues.

HPE Aruba Networking Virtual Switching Extension (VSX): technology bundles two physical switches into a single logical switch. This ensures high availability and support for live upgrades, eliminating maintenance windows and reducing downtime.

Dynamic segmentation unifies network access and security regardless of whether the user uses a wired or WiFi connection.

The top-shelf Gen7 ASIC is a microprocessor specialized in processing network traffic. It offers programmable, data center class performance with no bottlenecks and no switch software licensing or subscriptions.

AOS-CX switch VXLAN tunnels & virtualization

One VXLAN VTEP or tunnel endpoint communicating with another establishes a VXLAN tunnel, a requirement for each destination VTEP in the same VXLAN. In this example, switches S1, S2, and S3 are the physical VTEPs assigned to VNI 1001, so each switch maintains a tunnel with the other two. Clients assigned to VNI 2001 connect to switches S1, S3, and S4, with similar interconnected tunnels.

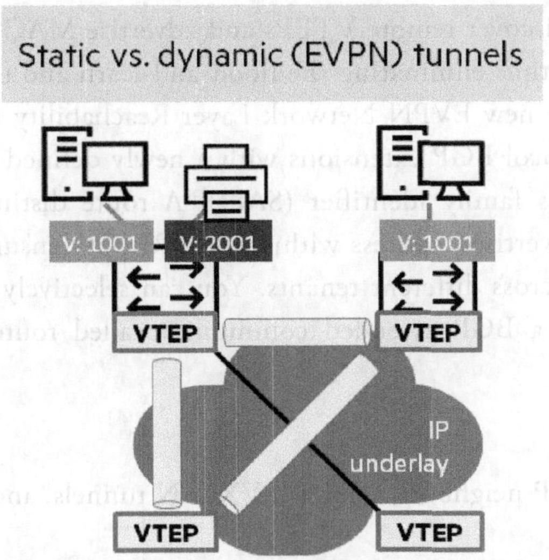

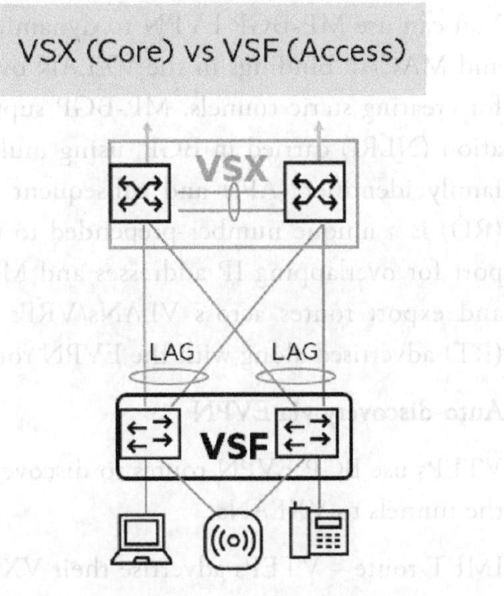

The tunnel is merely a mechanism of transport through the IP network. Multiple VNIs can use one tunnel because each VTEP encapsulated packet contains the associated VNI. The network then uses this VNI to send packets to the appropriate tunnel endpoints. This is why traffic from multiple VXLANs can use the same tunnel. The underlying IP transport network uses IP unicasts to deliver broadcast, multicast, and unknown unicast frames for each VXLAN.

There are two methods you can use to deploy a VXLAN infrastructure:

- Static VXLANs
- Ethernet VPN (EVPN) over a VXLAN

Let's introduce the two methods.

Static VXLANs

Static VXLANs use flood-and-learn (or ingress replication) to learn the MAC addresses of the remote hosts, which involves manual configurations of remote VTEPs in the flood list. Tunnels must be manually created, which is challenging in large networks. It is also difficult to maintain. Static VXLANs, therefore, are typically used in small deployments.

Ethernet VPN (EVPN) over VXLAN

Ethernet VPN (EVPN) is a standards-based BGP control plane that advertises MAC addresses, MAC and IP bindings, IP prefixes, etc. The initial EVPN standard (RFC 7432) defined the BGP EVPN control plane and specified an MPLS data plane. The control plane, containing an MPLS data plane, extends to consider additional data plane encapsulation models, including VXLAN, NVGRE, and MPLS over GRE. These are covered in detail in IETF RFC 8365. This section focuses on EVPN and its operation with a VXLAN data plane for building overlay networks.

You can use MP-BGP EVPN to dynamically discover remote VTEPs and advertise MAC address and MAC/IP bindings in the VXLAN overlay, thus eliminating the flood-and-learn and the need for creating static tunnels. MP-BGP supports a new EVPN Network Layer Reachability Information (NLRI) carried in BGP, using multiprotocol BGP extensions with a newly defined address family identifier (AFI) and subsequent address family identifier (SAFI). A route distinguisher (RD) is a unique number prepended to the advertised address within the VRF that ensures support for overlapping IP addresses and MACs across different tenants. You can selectively import and export routes across VLANs/VRFs using a BGP-extended community called route target (RT) advertised along with the EVPN routes.

Auto-discovery via EVPN

VTEPs use BGP EVPN routes to discover VTEP neighbors, establish VXLAN tunnels, and assign the tunnels to VXLANs.

IMET route – VTEPs advertise their VXLAN IDs through IMET routes. If two VTEPs have the same VXLAN ID, they automatically establish a VXLAN tunnel and assign it to the VXLAN. This method can create the replication list for forwarding broadcast, unknown, unicast, and multicast (BUM) traffic.

MAC/IP advertisement route and IP prefix advertisement route – In the EVPN deployment, VTEPs advertise MAC/IP advertisement routes or IP prefix advertisement routes, which carry export targets. When a VTEP receives a route, it compares the export targets of the route with the local import targets.

You will return to the topic of dynamic overlays in module 9. However, the full coverage and configuration of EVPNs with VXLANs are beyond the scope of this guide.

Virtual Switching eXtension (VSX) and Virtual Switching Framework (VSF) are similar to each other, and even to earlier virtualization techniques like Virtual Router Redundancy Protocol (VRRP). With these virtualization techniques multiple devices are perceived as a single device. Switch virtualization allows you to eliminate Spanning Tree Protocol (STP) and supports active/active traffic forwarding so all uplinks are being utilized at the same time. This improves resiliency while reduces configuration and management overhead. But there are distinct differences between VSX and VSF, as summarized in the figure. You use VSX at the campus network aggregation and core layers. You use VSF primarily at the access layer. The differences in features and function are intentional since priorities differ between the access and core layers.

Configuring VSF stacking is a plug-and-play affair: a switch, at default settings, automatically unites with another switch. This ease of deployment is especially important at the access layer: you may have dozens, or hundreds of access switches and this auto-join process simplifies the deployment of new switches.

Unlike VSF, you must manually configure VSX on both switches. VSX maintains separate control plane and management planes, and much of the data plane. This mitigates the shared fate nature of common planes, and thus provides for improved uptimes: you get high availability and redundancy. This is especially important when aggregating all those access layer and/or aggregation layer switches. VSX is easy to manage because you can synchronize the management planes, and optionally, other features like the MAC and ARP addresses, table information, and configuration information. With few exceptions, it uses an opt-in process: you select what you want synchronized between the two switches, giving you complete control over the solution.

Consider MTU requirements for VXLAN.

HPE Aruba Networking Central Management Platform

This section is about using HPE Aruba Networking Central as a management platform.

HPE Aruba Networking Central

HPE Aruba Networking Central is a cloud-based platform for managing HPE Networking-based wireless, WAN, and wired networks. HPE Aruba Networking Central supports group configuration to provision and manage similar devices with less work. Continuous monitoring and AI-based analytics provide real-time visibility and insight. Machine Learning leverages a growing pool of network data and experience. It is very secure, with HTTPS connections, certificate-based and cloud authentication, and policies.

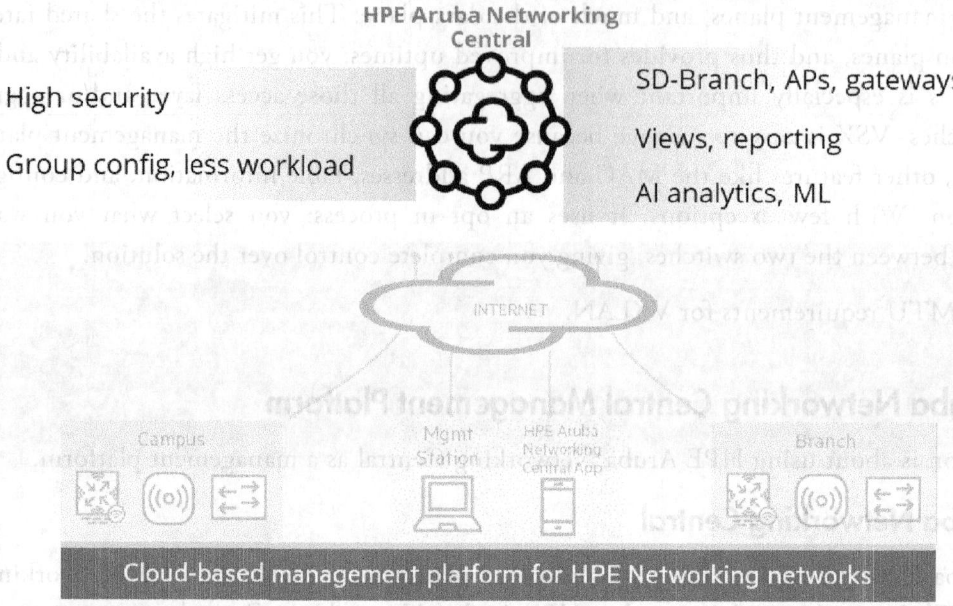

You get comprehensive network views, device status and health, and application usage. You can monitor, identify, and address issues with a data- driven dashboard, alerts, reports, and troubleshooting workflows.

Note

HPE Aruba Networking Central runs part of the control plane for some network devices and solutions, using the HPE Aruba Networking SD-Branch solution and HPE Aruba Networking APs and gateways running AOS-10.

Cloud-managed

HPE Aruba Networking Central is simple, with no hardware or software requirements on the customer's end. It's in the cloud, with all the resources needed for performance and visibility. HPE Aruba Networking Central is reliable, with guaranteed uptime and local management in case your connectivity to HPE Aruba Networking Central experiences issues or failures.

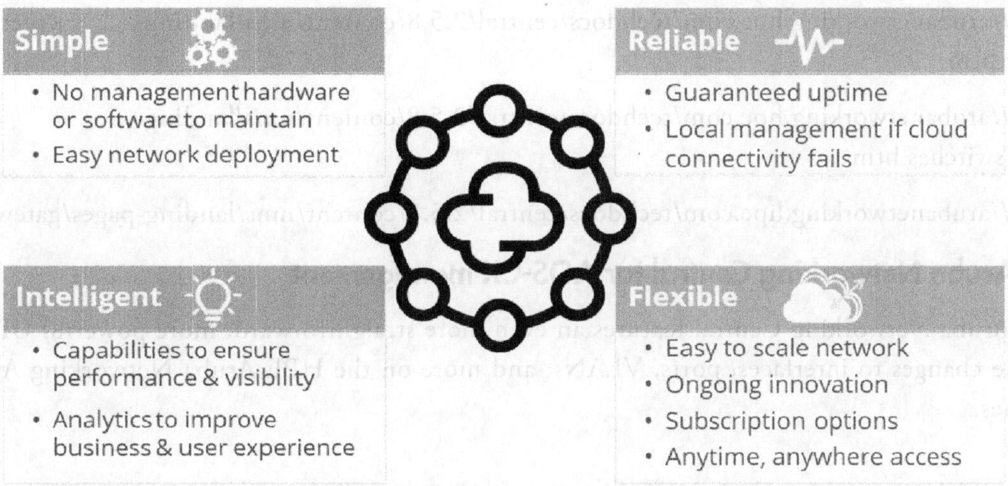

Plus, you get intelligent features to improve performance and visibility, with analytics to improve the business and user experience. And flexibility? It is easy to scale up as needed to accommodate growing systems and ongoing innovations, with simple subscription options and access anytime, anywhere.

Flexible and cost-effective cloud networking

HPE Aruba Networking Central supports management, monitoring, and configuration for the HPE Aruba Networking IAPs, AOS and AOS-CX switches, SD-Branch gateways, and AOS-10 APs and gateways. And since it's a cloud-based management system, it only requires a simple 1, 3, 5, 7, or 10-year subscription license, which includes tech support, with more additional services available.

https://arubanetworking.hpe.com/techdocs/central/2.5.8/content/nms/landing- pages/access-points.htm

https://arubanetworking.hpe.com/techdocs/central/2.5.8/content/nms/landing-pages/switches.htm

https://arubanetworking.hpe.com/techdocs/central/2.5.8/content/nms/landing-pages/gateway.htm

HPE Aruba Networking Central for AOS-CX management

HPE Aruba Networking Central features an even more straightforward, more powerful UI to ease routine changes to interfaces, ports, VLANs, and more on the HPE Aruba Networking AOS-CX switches.

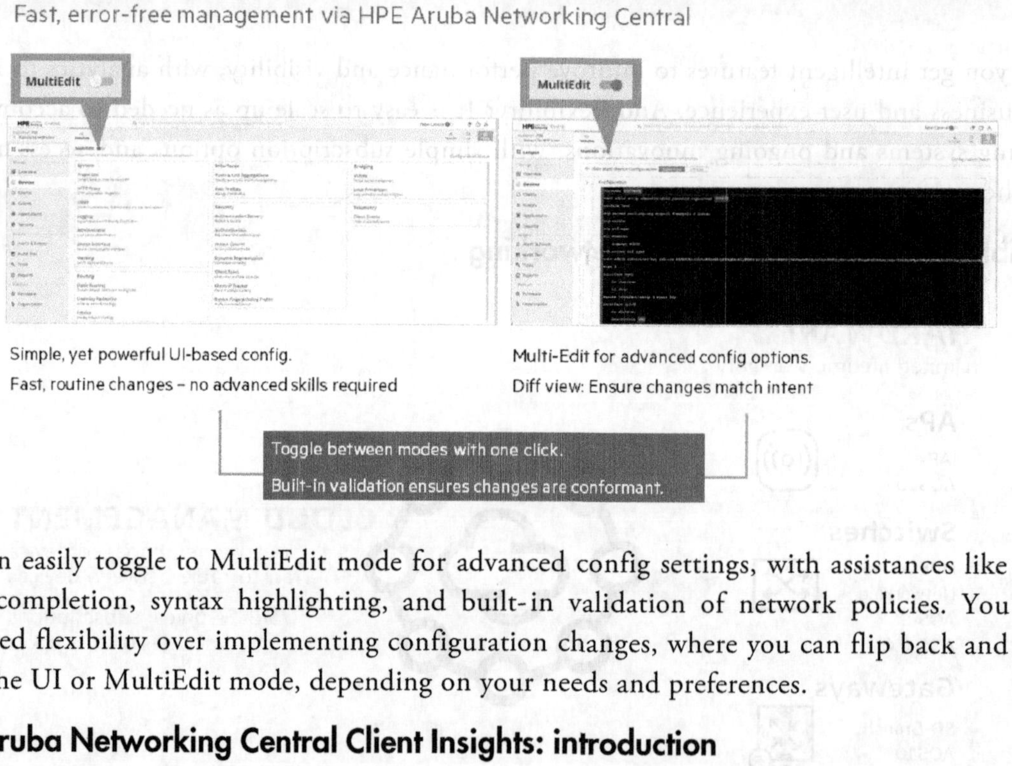

You can easily toggle to MultiEdit mode for advanced config settings, with assistances like command completion, syntax highlighting, and built-in validation of network policies. You have improved flexibility over implementing configuration changes, where you can flip back and forth using the UI or MultiEdit mode, depending on your needs and preferences.

HPE Aruba Networking Central Client Insights: introduction

HPE Aruba Networking Central Client Insights integrates with HPE Aruba Networking Central and ClearPass Policy Manager to enable comprehensive policy control and real-time enforcement.

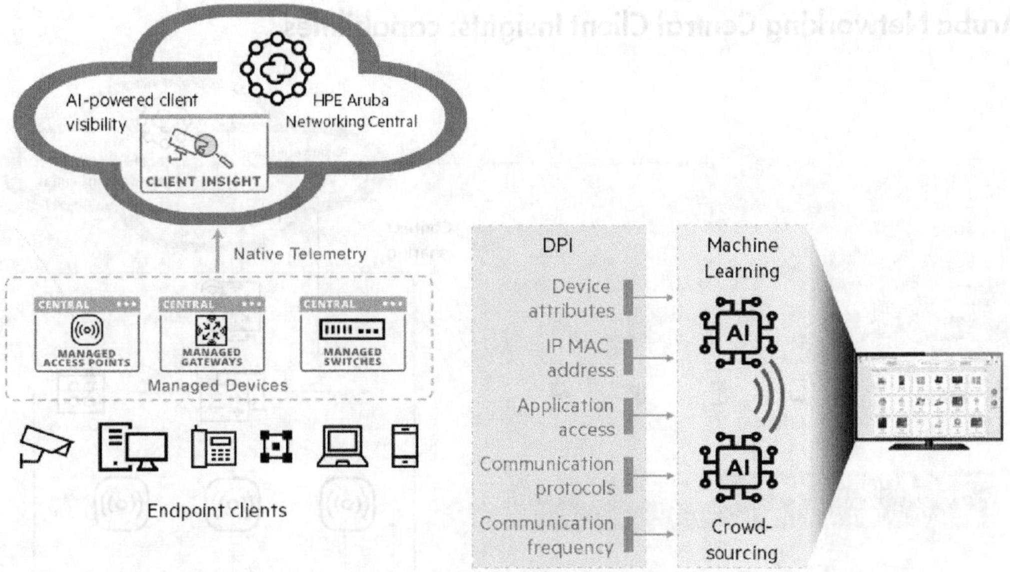

Client Insights is hosted through HPE Aruba Networking Central but gathers client information from packet flows through switches and deep packet inspection DPI from AP and gateway traffic flows. This meta-data is processed through machine learning and crowd-sourcing inClient Insights.

HPE Aruba Networking ClearPass Device Insight (CPDI) is a cloud application that enables network and security administrators to discover, monitor, and automatically classify new and existing devices that connect to a network. CPDI provides a full spectrum of visibility across the network by intelligently discovering and profiling all connected devices. CPDI allows to scan for CVE day-1 threats (National Vulnerability Database) integration and its integral part of Client Posture.

HPE Aruba Networking Central Client Insights: capabilities

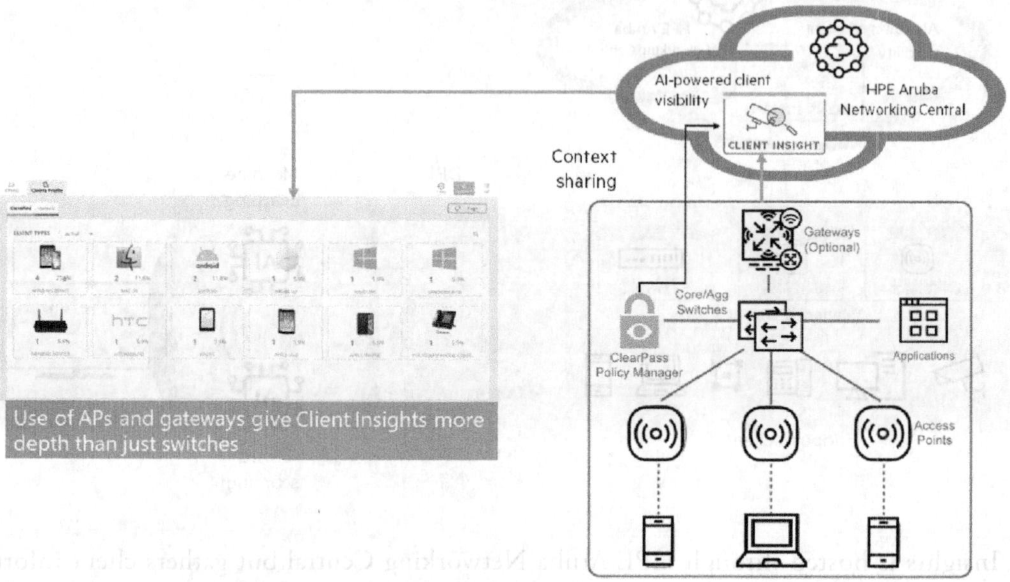

HPE Aruba Networking Central Client Insights:

- Detects device attributes such as device type, vendor, hardware version, and behaviors.
- Integrates with HPE Aruba Networking Central and ClearPass Policy manager to enable comprehensive policy control and real-time enforcement.
- Use only switches (minimal, but more insights to be included).
- Use of APs and Gateways (UBT) give more in-depth.
- Check the HPE Aruba Networking Central Cloud Client Insights faq for more information.

AI Insights

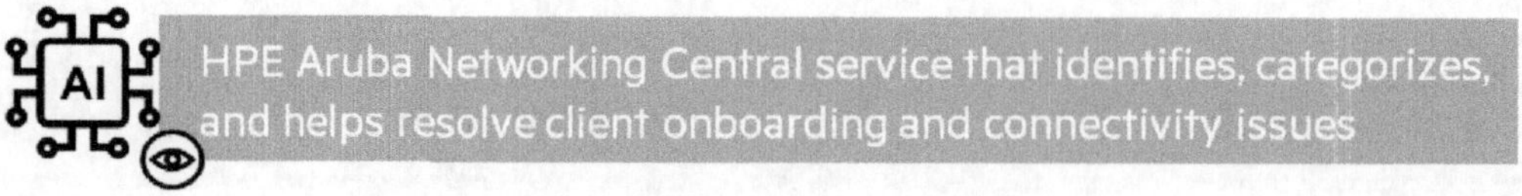

HPE Aruba Networking Central service that identifies, categorizes, and provides resolution to issues that affect client onboarding and connectivity.

Allows AI-driven network optimization using ML-based models that leverage HPE Networking's extensive Wi-Fi expertise to analyze network connectivity and performance.

- Delivers clear descriptions of detected issues.
- Data visualizations, context, and scope of impact for issues.
- Recommended fixes
- Data sources include wireless infrastructure, DHCP, Auth servers, and on-site data collectors.
- Data collected is sent securely via a secure tunnel to the AI insights cloud instance.
- Web-based dashboard provides insights and root cause analysis.
- Insights gives recommendations to fix current and foreseeable issues impacting network performance.
- Uses predictive analytics and ML to identify usage patterns matching "normal" customer behavior based on that customer type.

For example, after a brief learning period, AI Insights can predict when network demand for wireless will start and stop. This allows HPE Aruba Networking 5xx and 6xx APs to power up when demand starts and power down when demand begins to drop off.

AI Insights monitor connectivity, RF management, client roaming, airtime utilization, and wired and SD-WAN overall performance. This reduces troubleshooting ticket times or removes tickets from needing to be created in the first place. It also helps to ensure that service-level agreements (SLAs) are better met regarding network performance, uptime, and availability.

HPE Aruba Networking Central On-Premises

This next section approaches the topic of HPE Aruba Networking Central On-Premises as a management platform.

HPE Aruba Networking Central On-Premises versus Cloud : Feature Comparison

While HPE Aruba Networking Central On-Premises is designed to deliver the same experience as the cloud version, some of the Cloud features of HPE Aruba Networking Central are not supported by HPE Aruba Networking Central On-Premises. These features are heavily relied on the cloud's computing power and are therefore offered by HPE Aruba Networking Central's cloud-based network management solution.

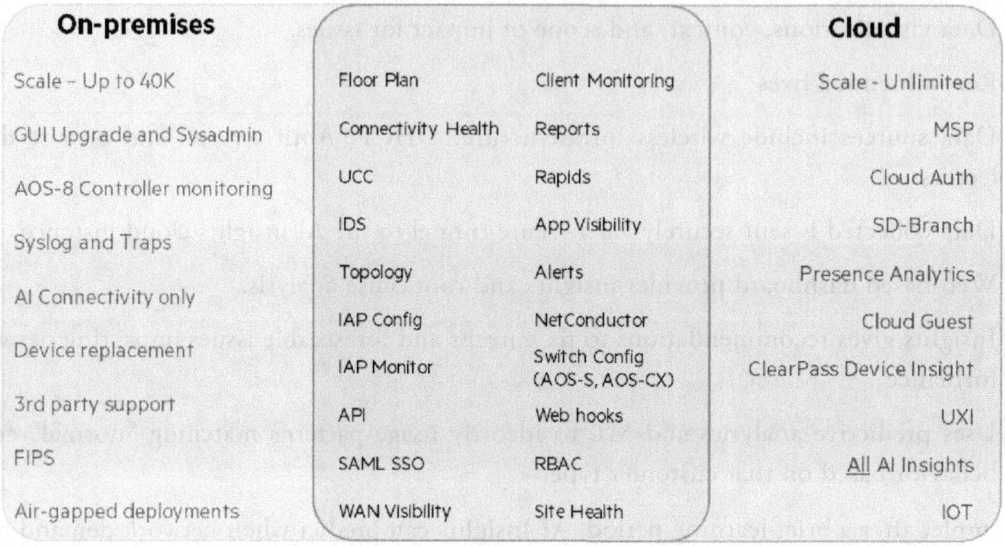

There are just a few differences between Cloud and On-Premises solutions:

- Scale: While cloud HPE Aruba Networking Central does not limit the number of managed devices, the On-Premises solution is scalable up to 40k managed devices.
- AI insights: Cloud computing power is required for full AI features. Check the HPE Aruba Networking Central On-Premises Guide to check the AI features supported by the On-Premises solution.
- SD-Branch and Cloud guest require internet connectivity. COP was primarily targeted at campus and data center environments, not WAN/cloud-connected use cases.

Note

"Presence Analytics" is a component of HPE Aruba Networking Central Location Engine (CLE).

https://arubanetworking.hpe.com/techdocs/aos/loc-serv/cle-dg/

https://arubanetworking.hpe.com/techdocs/aos/loc-serv/cle-dg/architecture/

HPE Aruba Networking Central On-Premises Architectures

Powered by server appliances with versatile deployment options

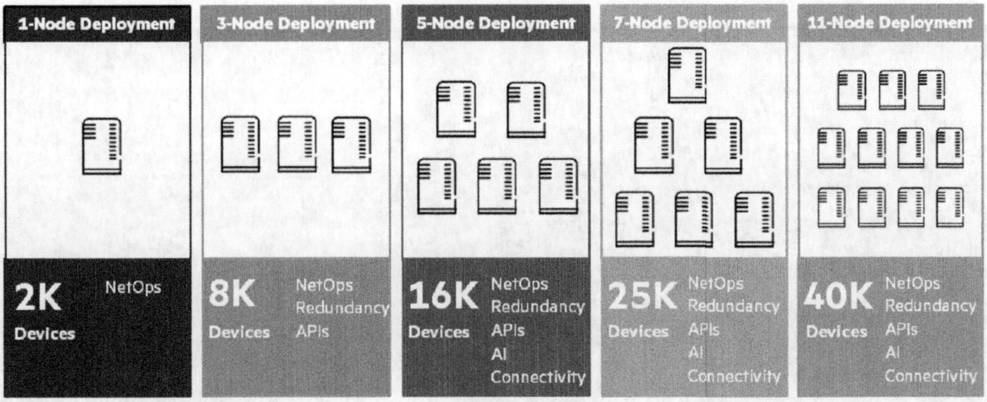

To address enterprise scale and resiliency needs, deployment options include one, three, five, seven and eleven-node server clusters. A single cluster of dedicated server appliances provides built-in redundancy – if an appliance goes offline, the cluster continues to monitor network devices without any service interruption. For smaller networks that consist of 2,000 devices or fewer, a one-node server deployment is also available that enables all primary network management capabilities. Five to eleven-node clusters are required for more advanced management features, such as AI Insights for wireless connectivity.

HPE Aruba Networking Central On-Premises Software Licenses

HPE Aruba Networking Central On-Premises software subscriptions are required to enable all the primary management capabilities for HPE Aruba Networking wireless access points, mobility controllers, and switches. Each network device managed through HPE Aruba Networking Central On-Premises requires a separate subscription. These subscriptions are available in 1-, 3-, 5-, 7-, and 10-year terms to support flexible consumption and budgeting. Each device subscription includes embedded software support that provides access to software updates, 24x7 troubleshooting and deployment assistance, and more. With software support embedded, customers gain immediate access to new features upon installing the latest firmware update.

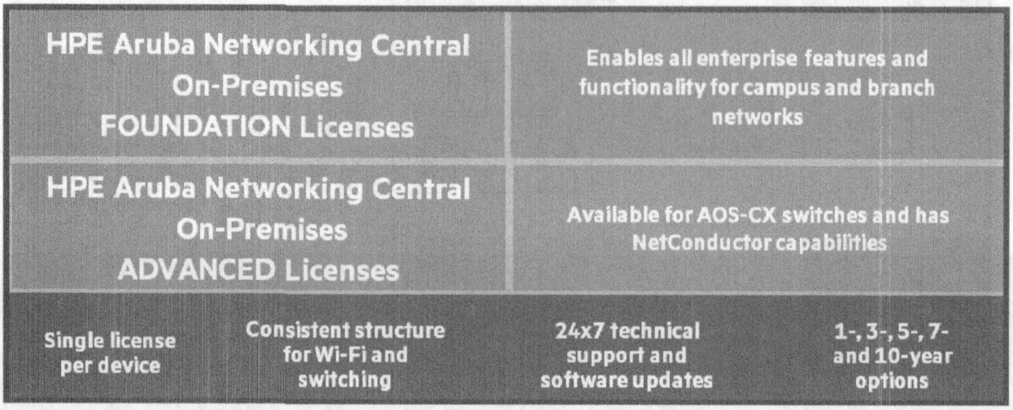

The purchase of a separate, perpetual software license is required to configure and monitor access points from HPE Aruba Networking Mobility Controllers on a per-device basis. Optional add-ons include Policy Enforcement Firewall (PEF) and RFProtect (RFP) capabilities. Subscriptions for HPE Aruba Networking Central On-Premises only cover the software support portion of AP/PEF/RFP/Mobility Conductor/Controller licenses.

For HPE Aruba Networking switches, note that a separate device subscription is required for every switch that is managed in a stack. HPE Aruba Networking Central On-Premises device subscriptions do not include hardware support. Hardware support for server appliances and all managed network devices requires an HPE Aruba Networking Foundation Care contract.

Note

For more information on licenses, please refer to the ordering guides.

HPE Aruba Networking Central Subscriptions

Look briefly at the HPE Aruba Networking Central AP, switch, and gateway licensing subscription model. You will review licensing in more detail in later modules.

HPE Aruba Networking Central subscriptions for AOS-10 gateways

HPE Aruba Networking Central SaaS subscriptions for wireless gateways

In HPE Aruba Networking Wireless AOS-10, all HPE Aruba Networking Edge appliances are referred to as gateways rather than controllers since they act as a cloud gateway for wireless services that are deployed within HPE Aruba Networking Central. For wireless use cases, gateways are typically deployed in networks with thousands or tens of thousands of APs for improved scalability, seamless roaming between buildings, or for environments that require enhanced security services. Refer to the list of supported gateways.

HPE Aruba Networking Central subscriptions for HPE Aruba Networking EdgeConnect SD-Branch gateways

The HPE Aruba Networking EdgeConnect SD-Branch solution comprises an HPE Aruba Networking Central SD-WAN software subscription license for each 70xx, 90xx, or 72xx branch gateway, as well as the headend gateway/concentrator. The HPE Aruba Networking Central SD WAN Foundational, Foundational Base, and Advanced subscription licenses are described as follows. The Foundational Base license is applicable for branch gateways and not for VPN concentrators (VPNC). For all 70xx and 72xx gateways, the Foundational license does not have a capacity limit (that is client device limit). However, for the 7005, 7008, 9004, and 9012 gateways, a Foundational Base Capacity license is available. The 7005, 7008, 9004, and 9012 Foundational Base license is limited to 75 client devices in the branch.

Virtual Gateway (VGW) License (VPNC only)

Virtual gateways have bandwidth based licenses. These licenses are valid on public cloud providers like AWS, Azure, and Google Cloud, including on-premises Hypervisors like VMware ESXi. Currently, the bandwidth licenses available are 500 Mbps, 2 Gbps, and 4 Gbps.

HPE Aruba Networking Virtual Gateway is a virtual instance of the headend gateway for ArubaSD-Branch. HPE Aruba Networking Central supports licenses based on the bandwidth capacity for virtual gateways. All license assignments are undertaken by the virtual gateway orchestration app.

The following are the options available for Virtual Gateway Licenses:

- License duration—1 year, 3 years, and 5 years
- Available bandwidths—500 Mbps, 2 Gbps, and 4 Gbps
- Available Aruba Virtual Gateways based on the bandwidth—VGW-500M for 500 Mbps, VGW-2G for 2Gbps, and VGW-4G for 4Gbps

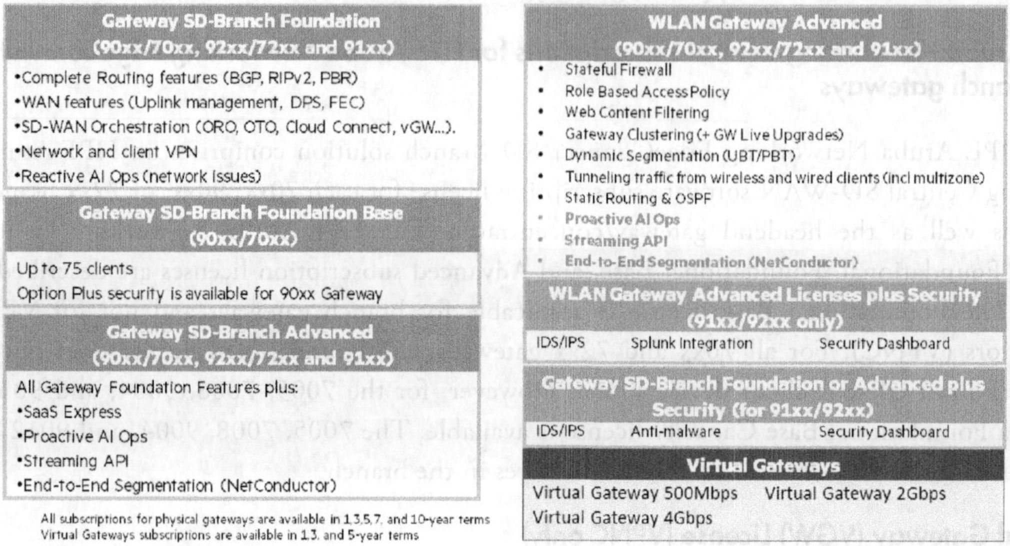

You can find more information at the following URLs:

https://arubanetworking.hpe.com/techdocs/central/latest/content/nms/subscriptions/lic- ovr- sup-ftrs-gtw.htm

https://arubanetworking.hpe.com/techdocs/central/pdfs/2.5.8/licensing-guide.pdf

HPE Aruba Networking Central Foundation and Advanced licenses for APs and SWs

HPE Aruba Networking subscriptions for Access Points are available in two options: Fundamentals and Advanced where the advanced subscription provides all the features included in the Foundation subscription, with additional features related to AI Insights and WLAN services.

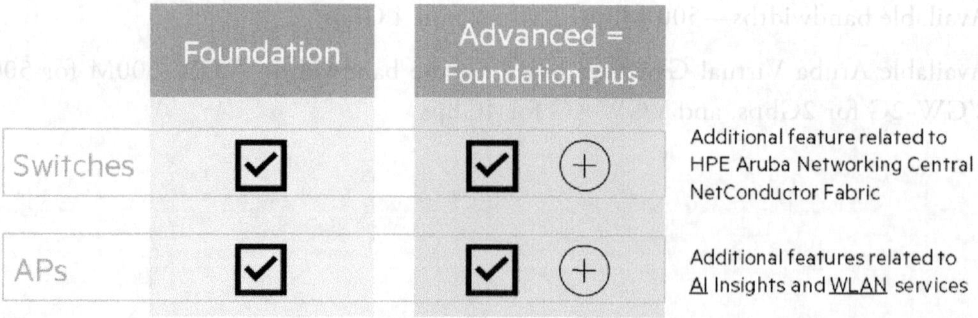

HPE Aruba Networking subscriptions for switches are available in two options: Fundamentals and Advanced where the advanced subscription provides all the features included in the Foundation subscription, with additional feature related to HPE Aruba Networking Central NetConductor Fabric.

For more information, see the following URLs:

https://arubanetworking.hpe.com/techdocs/central/latest/content/nms/subscriptions/overview-licensing.htm

https://arubanetworking.hpe.com/techdocs/central/latest/content/nms/subscriptions/lic- ovr- lic-typs.htm

https://arubanetworking.hpe.com/techdocs/central/pdfs/2.5.8/licensing-guide.pdf

AOS-10 Architecture

The final section examines the HPE Aruba Networking Mobility products' AOS-10 software version release.

Evolution of HPE Aruba Networking APs and gateways with AOS-10

HPE Networking offers APs and gateways running AOS-10. You can now deploy and manage your WLANs from a single and unified cloud-based network management system: HPE Aruba Networking Central.

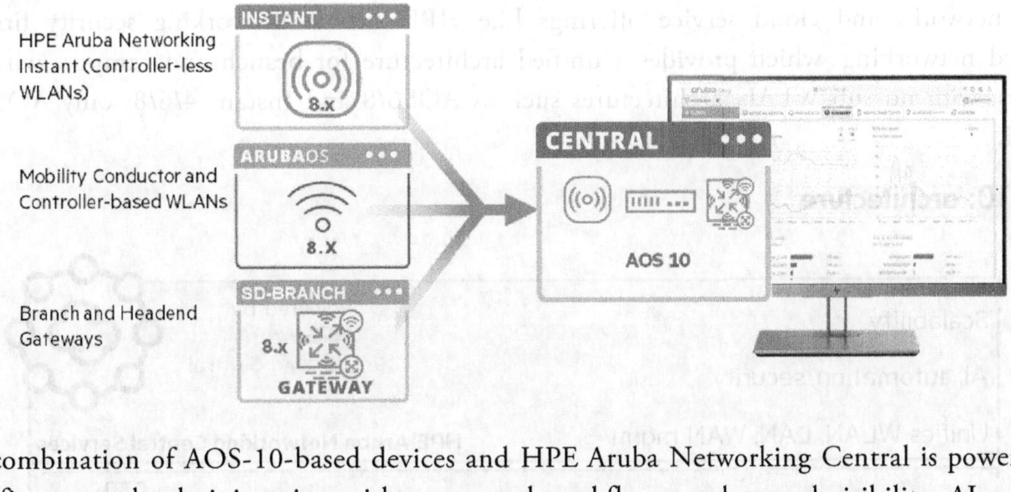

This combination of AOS-10-based devices and HPE Aruba Networking Central is powerful! It simplifies network administration with automated workflows, end-to-end visibility, AI-powered insights, and analytics to enhance and optimize the wireless experience for users.

SD-Branch 8.x is now consolidated into AOS-10 architecture, you simple need to select SD-Branch as the GW type in HPE Aruba Networking Central Group along with the appropriate HPE Aruba Networking Central Subscription, it combines the features of a mobility Gateway and SD-Branch.

Let's explore AOS-10 a bit more.

AOS-10: overview

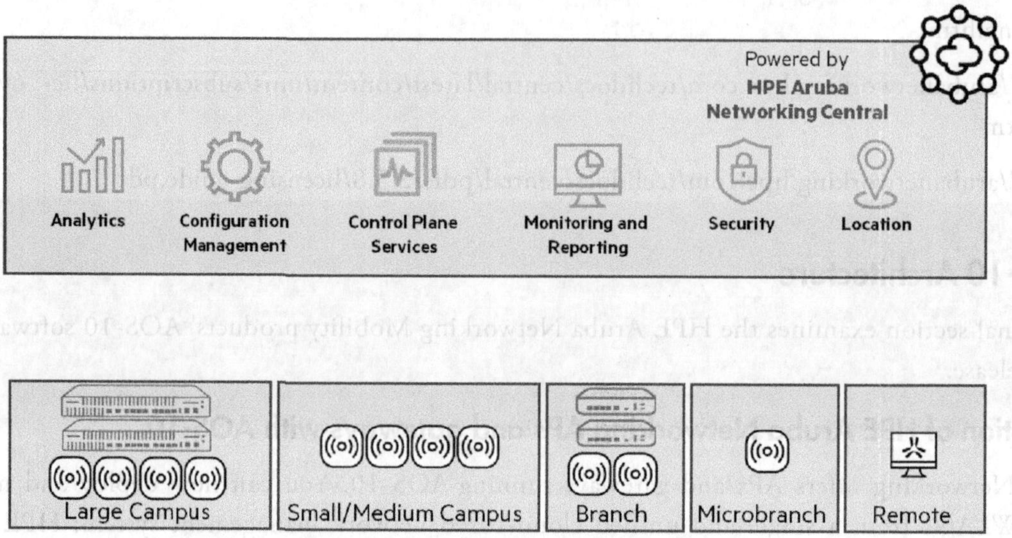

AOS-10 provides a single architecture for customer WLAN access deployments. It integrates access networks and cloud service offerings like HPE Aruba Networking security-first, AI-powered networking, which provides a unified architecture for branch and campus, wired and wireless, with no sub-WLAN architectures such as AOS6/8 and Instant 4/6/8: only AOS-10 is needed.

AOS-10: architecture

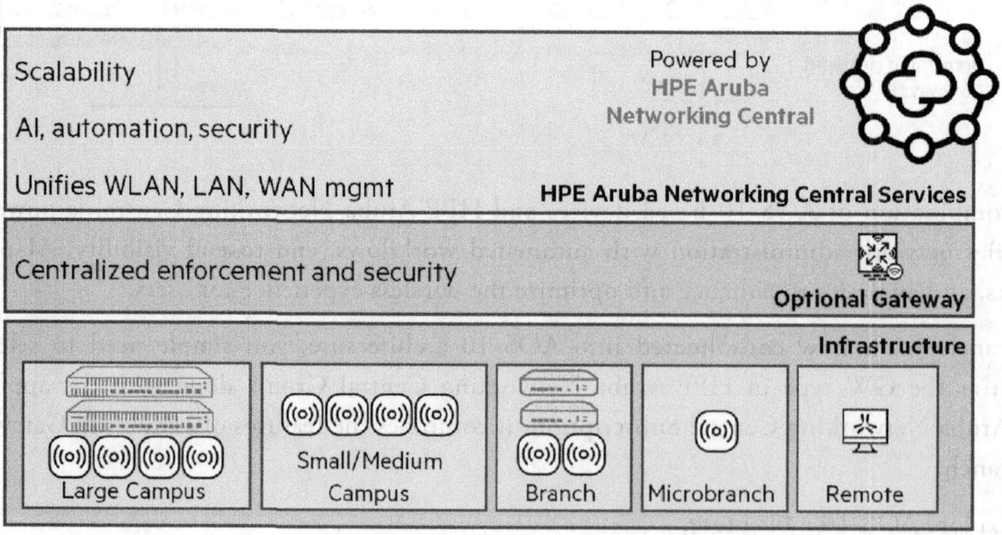

AOS-10 is comprised of three layers.

- The connectivity Layer of on-premises wired and wireless infrastructure with WLAN APs and switches. Gateway could be an optional component but very important for the security
- The services layer for services, applications, and NMS/CMRT
- The policy layer for the campus is implemented using overlay technology and traffic filtering mechanisms to isolate user and application traffic

The advantages of this architecture are that it:

- Scales beyond the Mobility Conductor 8.x domain of 10k APs or Instant Clusters of 2K devices
- Enables cloud services, including AI-powered insights, workflow automation, and robust security
- Unifies WLAN, wired LAN, and WAN management and visibility with consistent UI and workflows

Let's zoom in on the optional gateway component of the Connectivity Layer.

AOS-10: scalability

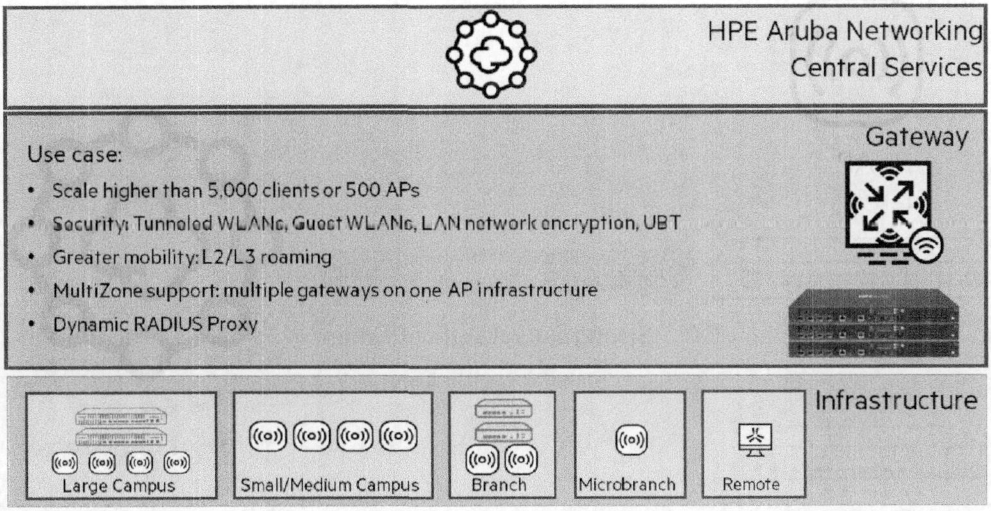

AOS-10 provides additional scalability and security for enterprise customers. Customers should consider a gateway if they have one or more of the following requirements:

- Scale higher than 5,000 clients or 500 APs (Note that this is due to management complexity and not a hard technical limit)

- Security: Tunneled WLANs, segmentation with guest WLANs, network encryption over their LAN (the tunneled user traffic will be encrypted from the AP to the GW by IPsec when Data Encryption is enabled), or user-based tunneling (UBT)
- Greater mobility: scalable and robust L2/L3 roaming
- MultiZone support to allow different gateways to work with the same AP infrastructure in a secure, segmented manner
- Dynamic RADIUS Proxy (DRP): Ensures RADIUS traffic is sent to the correct server. Useful if corporate employee traffic should use a local RADIUS server and guest traffic should use a centralized captive portal server.

Note

The gateway is an optional component. HPE Aruba Networking Virtual Gateways are not currently supported for wireless use cases.

Device to HPE Aruba Networking Central communication

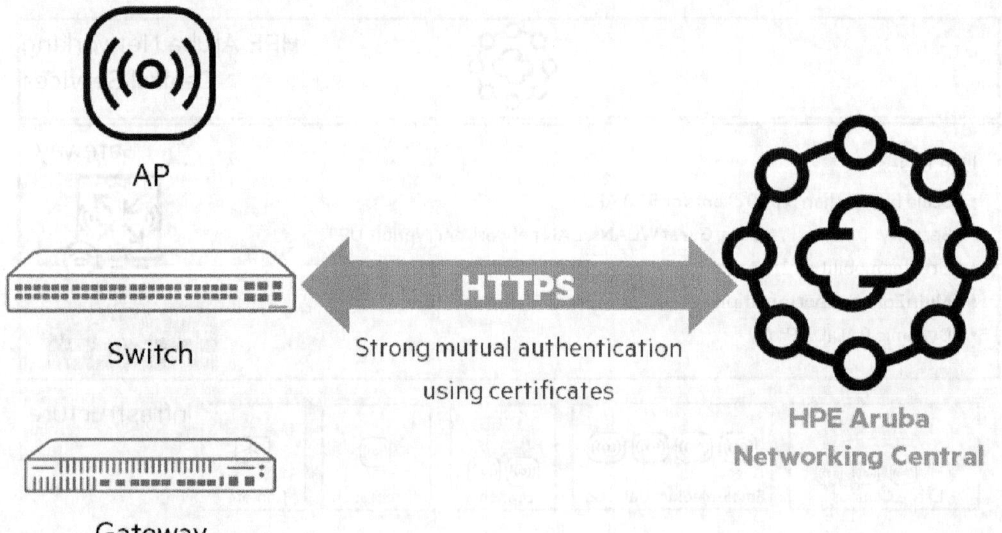

HPE Aruba Networking Central uses a secure HTTPS connection and provides a strong mutual authentication mechanism based on certificates for all communication with HPE Aruba Networking devices. These certificates ensure the highest level of protection.

HPE Aruba Networking Central groups and sites

Three main control mechanisms exist for management and viewing functions: groups, sites, and labels. These were covered in the HPE Aruba Networking Campus Access Fundamentals course and the HPE Press book. Here is a quick review of groups and sites.

Groups

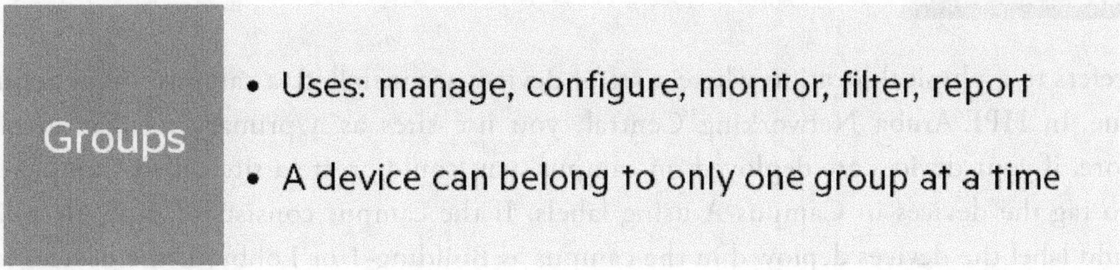

In HHPE Aruba Networking Central, groups are a primary management and configuration element: a device management, monitoring, and maintenance container. Groups enable you to combine devices with identical configuration requirements and manage these devices efficiently by using either a UI-based configuration workflow or a CLI-based configuration template or editor. Device groups enable you to:

- Provision of multiple devices in a single group. For example, a group can consist of various APs. These APs can share standard configuration settings. You can also apply a common security policy for the devices deployed in a specific location.
- Provision of different types of devices in a group. For example, a group can consist of APs, gateways, and switches.
- Create a configuration base and add devices as necessary. When you assign a new device to a group, it inherits the configuration currently applied to the group.
- Create a clone of an existing group. If you want to build a new group based on a current group, you can create a clone of the group and customize it as per your network requirements.

Sites

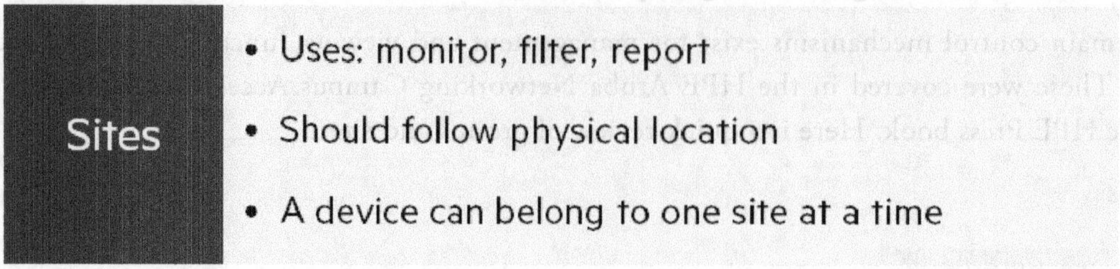

A site refers to a physical location where a set of devices are installed: a campus, branch, building, or venue. In HPE Aruba Networking Central, you use sites as a primary navigation element. Therefore, if your devices are deployed on campus, you could create a site called Campus-A. You can also tag the devices in Campus-A using labels. If the campus consists of multiple buildings, you could label the devices deployed in the campus as Building-1 or Lobby. If the devices in a specific location or an area within a particular location must have similar configurations, you can group them.

A site represents a specific building in HPE Aruba Networking Central Floor Plan and used for an auto-site cluster for AOS-10 Gateways.

Note

A site is first required in HPE Aruba Networking Central to use floor plan.

Learning check

Check your understanding of HPE Aruba Networking security-first, AI-powered networking solutions by answering the following questions.

Question #1

Which option shows the four core attributes on which HPE Aruba Networking security-first, AI-powered networking is based?

- ○ HPE Aruba Networking Central, AIOps, Unified Infrastructure, and Zero Trust Networking
- ○ HPE Aruba Networking Central, AI Clients, Unified Infrastructure, and Zero Trust Networking
- ○ HPE Aruba Networking Central, AIOps, Disperse Infrastructure, and Zero Trust Networking
- ○ HPE Networking AirWave, AIOps, Unified Infrastructure, and Zero Trust Networking

Question #2

HPE Aruba Networking CX switches have which of the following advantages?

- HPE Aruba Networking CX switches have different operation systems for Campus and a Data Center scenarios.
- HPE Aruba Networking CX switches could be managed in HPE Aruba Networking Central which is a cloud management platform. You can also manage HPE Aruba Networking APs and Gateways in HPE Aruba Networking Central.
- All HPE Aruba Networking CX switch models support Virtual Switching Extension (VSX) and Virtual Switching Fabric (VSF) technologies and provide a flexibility to select the right one for a specific scenario.
- Virtual Switch Extension (VSX) supports up to twelve HPE Aruba Networking CX switches in a single virtual stack.

Question #3

Which is FALSE when it comes to HPE Aruba Networking Central licensing for gateways?

- HPE Aruba Networking gateway normal licensing is subdivided into three categories: Foundation, Advanced, and Foundation Base.
- HPE Aruba Networking WLAN gateway licenses allow normal SD-Branch features within a campus.
- SD-Branch gateway functionality requires security licensing.
- Gold, silver, and bronze licensing are available for all hardware gateways to expand capacity.

3 RF Planning

Welcome to chapter 3, RF Planning.

OBJECTIVES

After completing this chapter, you should be able to:

- ✓ Architect based on wireless deployment models.
- ✓ Select the optimal APs and antennas.
- ✓ Perform a virtual site-survey.
- ✓ Validate RF coverage utilizing HPE Aruba Networking Central Floor Plan.

Deployment Models

Let's start by looking at different RF deployment models.

Review common WLAN types

There are several categories of WLAN types regarding user density and coverage. The first and most common type of WLAN is a basic capacity model in a typical office space or campus deployment. This basic capacity model is simple to design for and allows customers to choose from moderate-level AP models such as the AP-720, AP-610, AP-510, AP-503, or AP-500, depending on other needs. Note that the middle number of the model correlates with density needs. The higher the number, the higher the density that the model is designed to support.

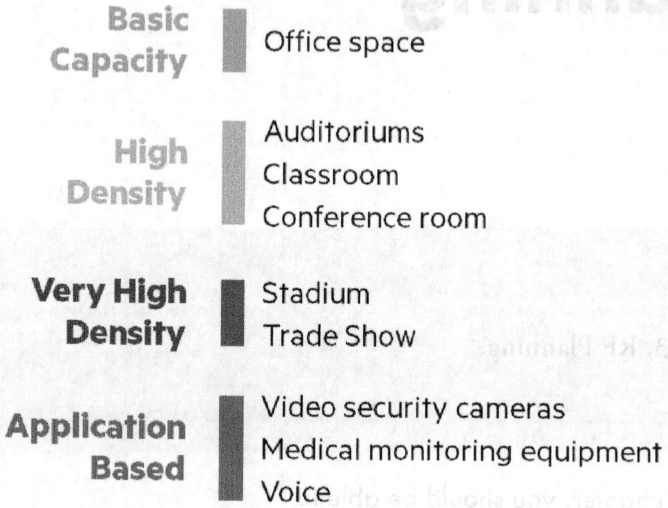

A high-density WLAN is typical for large conference rooms or classroom spaces where users may gather. The AP models here would be the AP-630, AP-610, and AP-510 for Wi-Fi 6E or AP-740 and AP-730 for Wi-Fi 7 deployment.

By comparison, a very-high-density (VHD) or extreme-density design would be users seated side by side in an auditorium or stadium seating environment. Ideal AP models include the AP-750, AP-730, AP-650, AP-630, AP-530, or AP-550 series.

Application-based concerns for real-time traffic, such as live video feeds, collaboration, or VoIP, should be met with high-efficiency standards and a supported minimum signal free from interference. 802.11ax with features such as MU-MIMO, OFDMA, and AirSlice allow the efficiency desired. 802.11be with features such as MLO, 320MHz channels, UTB filtering. Additional choices to use Wi-Fi 6E/7 6 GHz bands, along with layering of applications to specific channels, will further improve the performance of critical applications.

Note that due to the tremendous benefit of Wi-Fi 6E/7 having the additional 6 GHz band, all 600-series and 700-series APs are considered extreme performance APs and an excellent option for any environment where high density and high efficiency are a concern.

Note

Contact HPE in case of wireless network designs with 25K or more access points.

Wireless coverage considerations

Legacy RF designs only required coverage of a floorspace without concern about how much traffic was using the medium. RF coverage is a critical consideration when designing and planning a WLAN deployment. Without RF coverage of the physical space, the clients discover spots where they cannot "hear" an access point.

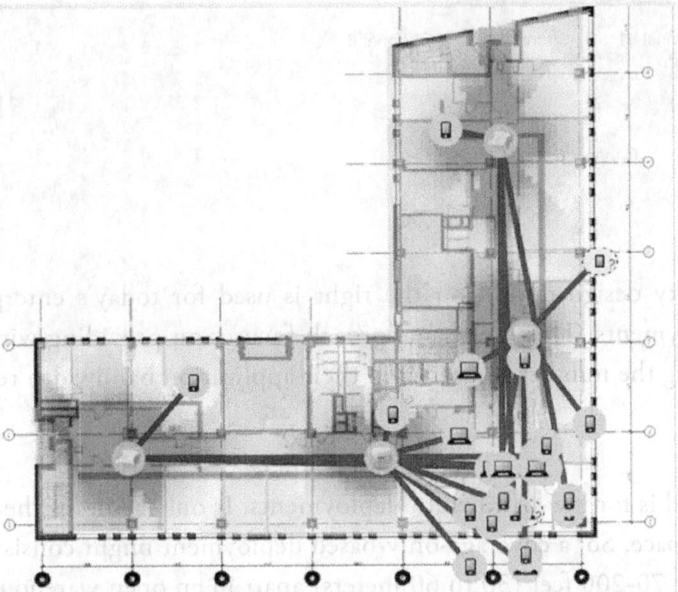

Planning and designing the appropriate overlap between APs' RF coverage areas is necessary. When there is too much overlap, AP coverage areas will be on the same channel in the same physical area, resulting in co-channel interference (CCI). If there is too little coverage, clients can suffer holes in coverage where they might be unable to reliably connect to the WLAN and transmit and receive data to and from the AP.

A coverage approach is necessary at first with any deployment. While this may have been enough when very little reliance on the wireless network existed in legacy designs, today's enterprise wireless networks require coverage and consideration for the wireless capacity users and applications require. Let's compare a coverage design to a capacity design for modern wireless designs.

Review coverage vs. capacity

The two design models provide adequate RF signals for clients to roam without holes in coverage. The first coverage design on the left is suitable for low-bandwidth and sparse client deployments; only the required APs are deployed to provide minimal RF coverage at a basic dBm level.

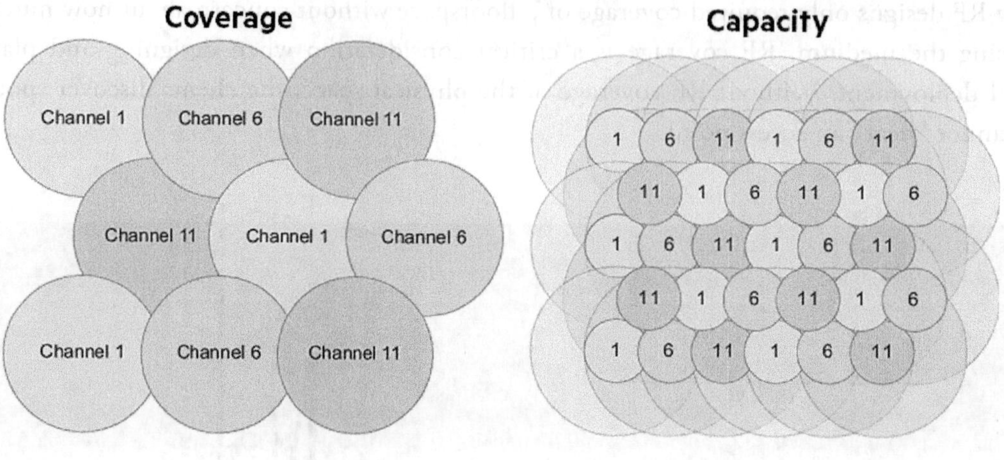

The second capacity design option on the right is used for today's enterprise high-bandwidth, high-density deployments. This modern approach focuses on providing wireless network capacity based on calculating the number of users and their application bandwidth requirements.

Coverage Model

The coverage model is for low-bandwidth deployments. It only achieves the required minimal coverage of the floor space. So, a coverage-only-based deployment might consist of indoor APs placed far apart at roughly 70-200 feet (30 to 60 meters) apart in an open warehouse, running at 50% to 75% power. APs could run at 100% power to only use the bare minimum number of radios, but HPE Networking does not recommend this solution as it removes any redundancy should an AP fail.

A good candidate for a coverage-only deployment is one where minimal data usage and scanning are the only requirements. This should be confirmed based on information gathered from key stakeholders and what a site survey indicates. The deployment could then consist of broadly spread-out APs and clients with a typical ceiling deployment. In practice, this is rarely the case as enterprise networks with even a few clients now see real-time services and high-bandwidth applications critical to business operations.

Capacity Model

The capacity model is the better choice for enterprises, especially high-density, high-bandwidth deployments such as K through 12 school campuses. This might consist of APs roughly 45 to 60 feet (14-18 meters) apart, running at 25-50% or 50-75% power. AirMatch through HPE Aruba Networking Central should be leveraged to ensure proper power levels and roaming coverage. The goal is to provide mobile laptops the same "desktop" experience as when docked with a cable connection. HPE Networking, therefore, recommends capacity-based deployments for all office and education settings.

In a later module, we will cover very high-density (VHD) deployments such as lecture halls, classrooms, and stadiums. Outdoor deployments can cover a range of deployments that include mesh and point-to-point.

6 GHz/5 GHz versus 2.4 GHz coverage

Why 6 GHz and 5 GHz preferred over 2.4 GHz in multiple indoor scenarios even with a smaller coverage area?

- Less noise/interference
- More available bandwidth
- More available channels
- Minimal adjacent channel interference

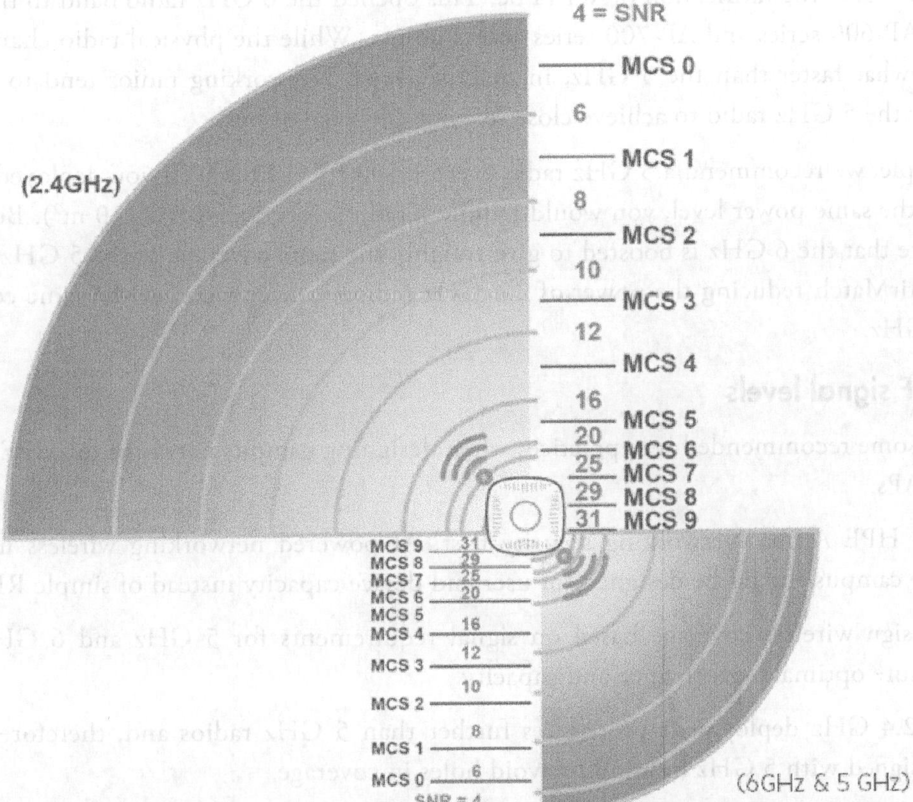

Figure shows data rate and SNR where distance depends on:

- EIRP of the AP
- Rx sensitivity of the client
- Radio frequency being used

In the early 2000s, 802.11 b/g/n WLANs predominantly used the 2.4 GHz range over the 5 GHz bands found in 802.11 a/n standards. With feel clients and low bandwidth applications, the preference largely fell on the desire for the greatest coverage using the least number of radios. This was the classic coverage design model.

The 5 GHz did start to make inroads by providing capacity benefits missing from a coverage design. There was less noise in the 5 GHz band due to fewer devices supporting the band and the huge increase in available channels over the 2.4 GHz. Starting in the 2010s, several factors contributed to a shift towards 5 GHz. More wireless users in the enterprise space and IoT devices made the 2.4 GHz saturated. 2013 saw the ratification of the new 802.11ac 5 Ghz only standard, further cementing the band as the first choice.

With 2020 came the ratification of Wi-Fi 6e. This opened the 6 GHz radio band to the HPE Networking AP-600 series and AP-700 series access points. While the physical radio channel propagates somewhat faster than the 5 GHz, in practice, HPE Networking radios tend to boost the 6 GHz over the 5 GHz radio to achieve close to a 1:1 coverage of both.

For example, we recommend a 5 GHz radio every 1,500 ft^2 (140 m^2). If you deployed the 6 GHz radios at the same power level, you would require a radio every 1,200 ft^2 (110 m^2). But AirMatch will ensure that the 6 GHz is boosted to give roughly the same coverage as the 5 GHz. This is the same as AirMatch reducing the power of 2.4 GHz radios to keep them at the same coverage area as the 5 GHz.

Target RF signal levels

Here are some recommended best practices when designing campus networks for HPE Aruba Networking APs:

- All HPE Aruba Networking security-first, AI-powered networking wireless networks for the campus should be designed for user and device capacity instead of simple RF coverage.
- Design wireless coverage based on signal requirements for 5 GHz and 6 GHz bands to ensure optimal performance and capacity.
- A 2.4 GHz deployment propagates further than 5 GHz radios and, therefore, should be designed with 5 GHz in mind to avoid holes in coverage.

- The 2.4 GHz radio can run at a lower power to avoid co-channel interference or even be disabled among redundant 2.4 GHz APs. This helps to limit the number of channels and provides better propagation of the 2.4 GHz signal. Tools like HPE Aruba Networking Central Floor Plan can validate that coverage gaps do not occur when disabling radios.
- A well-designed 5 GHz deployment can be upgraded to 6 GHz with a 1:1 AP replacement, depending on client density, throughput requirements, and physical environment.
- Wi-Fi 6, 6E and 7 deployments should target a cell size for AP ranges from -55 to -65 dBms, depending on capacity needs. For example, if the customer requirements call for very high throughput with MCS 11 at 40 MHz channel bandwidth for a high-density environment, the target RSSI should be closer to -55 dBms.
- Secondary signal strength is the signal intensity of the second-strongest access point in a particular location. It's critical for the roaming between APs, redundancy, and high-capacity needs.

Types of wireless deployments

A greenfield design is built from scratch with no legacy network existing. All wireless radios should use the latest standard 802.11be. With no existing legacy network to consider, you can design with future technology in mind. An example of a greenfield approach would be the new construction of a campus building.

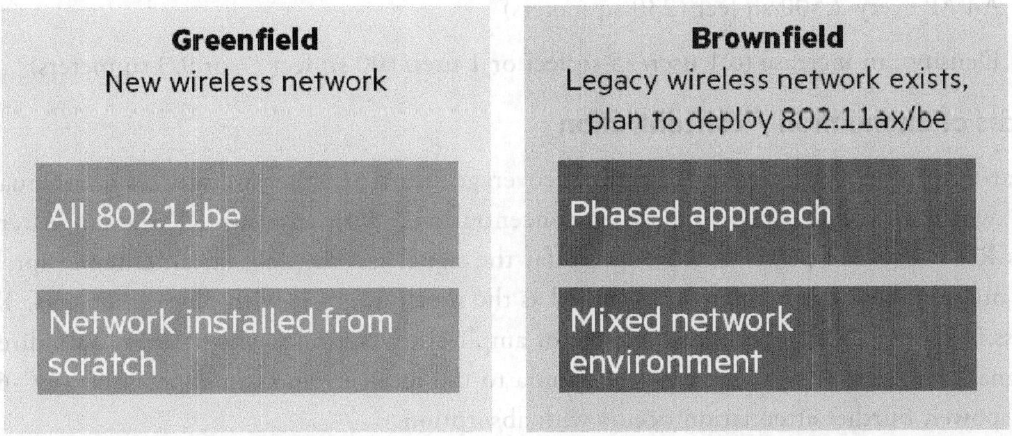

A brownfield deployment describes a scenario where an existing wireless network is in place, and there is a plan to upgrade to the latest technology. This requires considering how to design a network upgrade while minimizing any impact on the existing network. After all, users are likely running production on that existing wireless network, resulting in costs for any downtime experienced.

Additionally, brownfield often has legacy equipment that needs to be supported in a mixed wireless environment. For example, there may be a campus warehouse with legacy IoT devices and scanners that use 2.4 GHz. 802.11be will still be backward compatible, enabling better efficiencies when mixing new and old clients. However, the customer may have chosen AP-600 series APs instead of Wi-Fi 7 AP-700 series unless a plan to update client devices exists.

Typical wireless bandwidth capacity in an office environment

What are some general wireless guidelines for a typical office environment to ensure enough overlap for seamless client roaming?

Typical high-density office

- 2-3 devices per user
- Place APs every 35-50 feet (10-15 meters)
- Each AP provides coverage for 1500 sq feet (140 sq meters)
- APs with 6 GHz radios at the same power level provide coverage for 1,200 sq feet (110 sq meters)

Traditional open office

- 2-3 devices per user
- An AP every 2,500 sq feet (230 sq meters)
- Density can increase to 1 user/75 sq feet or 1 user/100 sq feet (7 or 9.3 sq meters)

Sources of attenuation / concentration

Keep an eye out for locations in the desired coverage area that either are sources of attenuation or places where you will expect high user concentration. With campus RF design, Attenuation affects RF as a loss of power based on how far the signal travels, how far an antenna spreads the signal out, and how much energy is absorbed as the signal interacts with physical objects. This signal loss is known as "free path loss" or loss of amplitude. As a rule, whenever an omnidirectional RF signal doubles in distance from the antenna to the receiver, you can expect a loss of -6 dBms or 1/4 power. Further attenuation occurs with absorption.

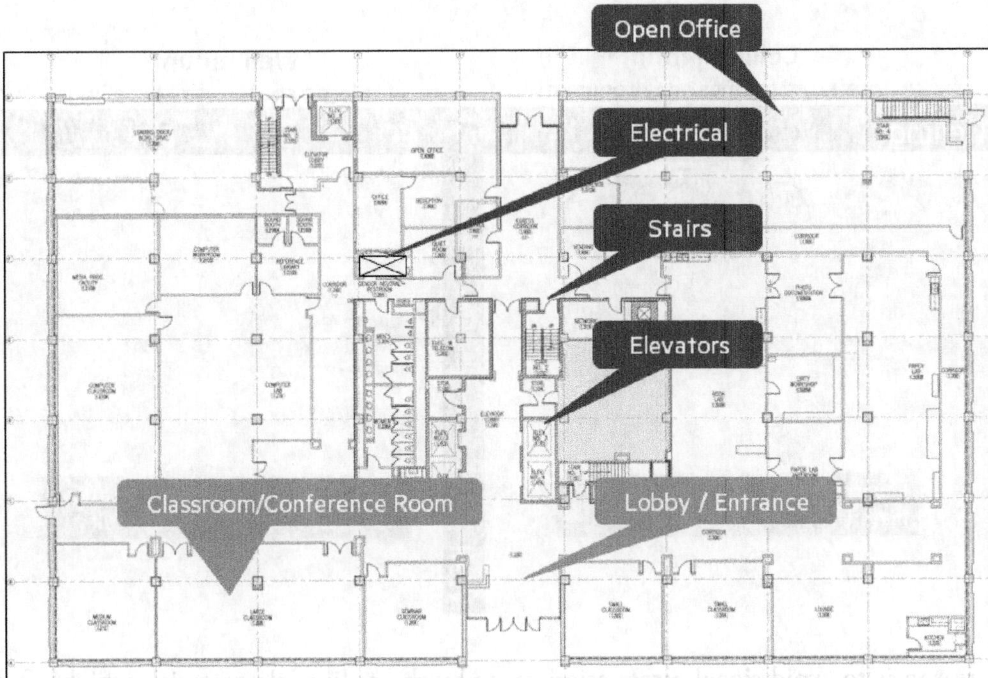

Elevator shafts and stairwell shafts are usually heavily walled and will attenuate radio signals. Electric rooms will also cause radio attenuation. Furthermore, not all open office spaces are necessarily completely open when considering shelves, filing cabinets, desks, and cubicle walls.

User concentration can mainly be predicted based on what the rooms are designed for. A few examples are classrooms and conference rooms; these have a large concentration of users, so HPE Networking recommends locating an AP in these rooms. Entrances and lobbies typically have large amounts of users at specific times, all authenticating to the first AP in the building.

RF design

Place radios where they achieve a line of sight between the AP and wireless devices. This avoids issues with signal loss. As such, most modern WLAN deployments are at the ceiling level. A ceiling deployment can occur at or below the level of the ceiling material. Generally, HPE Networking does not recommend mounting APs above any type of ceiling material, especially suspended or "false" ceilings.

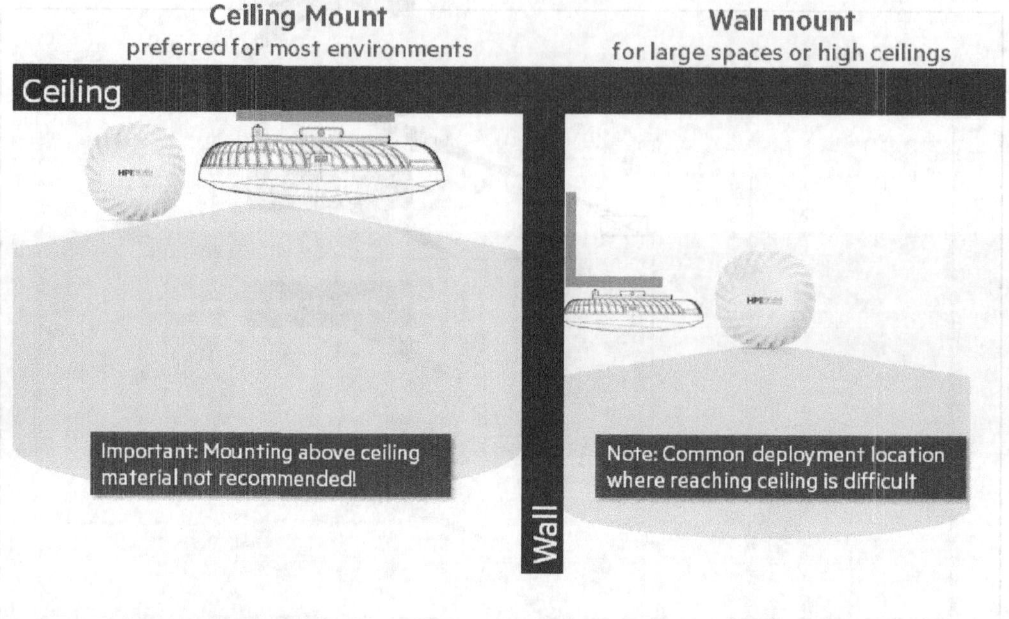

The first reason is to avoid signal attenuation cause by the ceiling tile material, which often has a metallic backing that can significantly reduce signal quality. A second reason is that the plenum area above the ceiling is also used for HVAC, plumbing, lighting, conduits, and other mechanical items. Any of these can directly obstruct RF signals.

Wall deployments are less common indoors than ceiling deployments. They are often used in locations where reaching the ceiling is difficult, such as airports, warehouses, or lecture halls. Wall deployments may also be preferable in areas with a hard ceiling where you cannot run cabling.

AP mounting considerations

AP mounting height, location, and orientation are crucial for wireless coverage.

The following guidelines are for mounting indoor omnidirectional down-tilt antenna APs:

- Avoid mounting APs with internal antennas vertically on a wall; exceptions are hospitality and certain outdoor AP models.
- AP-xx5 series APs should be mounted horizontally with the radome (white surface) facing down.
- Keep the height for AP-xx5 series APs from 12-25 ft (4-8 meters).
- Wall-mounted APs should use a 90° AP mount that supports the AP models; check with third-party vendors to purchase.
- Installing APs above the drop ceiling may introduce attenuation between APs and clients, negatively impacting user experience and HPE Networking AirMatch calculations.
- Avoid mounting APs on building columns, pillars, or I-beams unless the intent is to create RF shadows.
- Avoid installing APs in credenzas, bookcases, or other furniture that could introduce additional attenuation.
- HPE Aruba Networking recommends consulting an experienced and certified WLAN engineer to perform a pre-install survey and review planned AP locations.

Open office environments

How would you plan APs in these two open offices?

Many often consider an open office space an open free space that only needs a few APs to cover the area. However, not all offices are the same. The picture on the left shows low cubicles where Wi-Fi signal can travel a long distance, but the cubicle on the right shows high wall cubicles with metal cabinets that would attenuate the signal.

This requires you to use a denser deployment of APs. Also, consider the number of employees in the area.

Deploy APs in rooms – not in hallways

Hallway only: not recommended

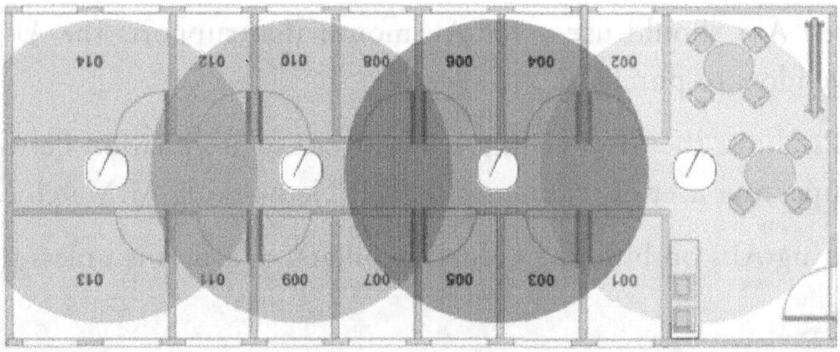

Instead of only using hallway APs at high power to cover users through a wall, HPE Networking recommends using many room APs at low power to cover the user space directly. Hallway APs can assist with covering gaps and roaming as devices move across the floor.

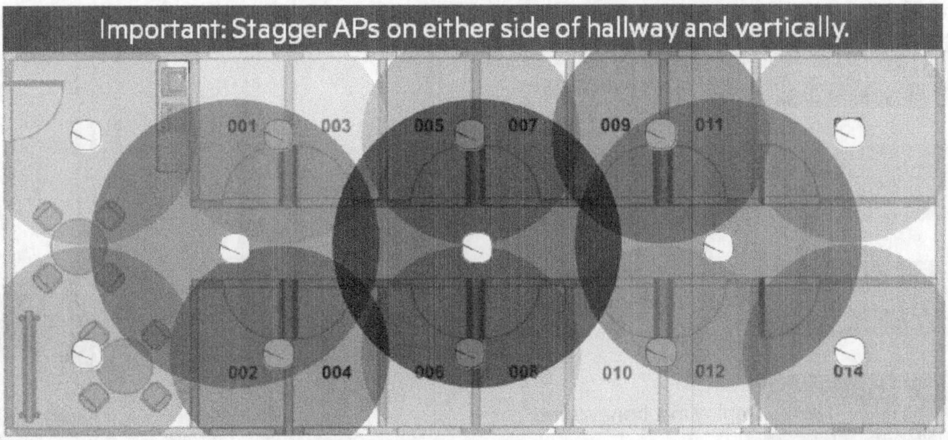

Depending on how much attenuation the building presents, you may place an AP per room or every two or three rooms. Stagger APs inside of rooms on either side of the hallway and vertically. Generally, we should not place APs in rooms directly across the hall from one another but rather stagger them. Change the deployment on adjacent floors so that APs are in different rooms. If the flooring has highly dampening material, such as thick concrete or metal shielding, staggering the APs across floors can be less critical.

AP placement summary

Although HPE Aruba Networking AP-7xx series can support up to 512 client devices per radio, HPE Networking recommends that you plan for a distance of 35-50 feet (10-15m) between APs to provide a good user experience. The minimum RSSI value should be -65dBm for the coverage area from at least two APs. Consider strengthening this to -55 dBms for VoIP or real-time traffic. Also, the SNR (Signal-to-noise ratio) should be greater than 25 dBs.

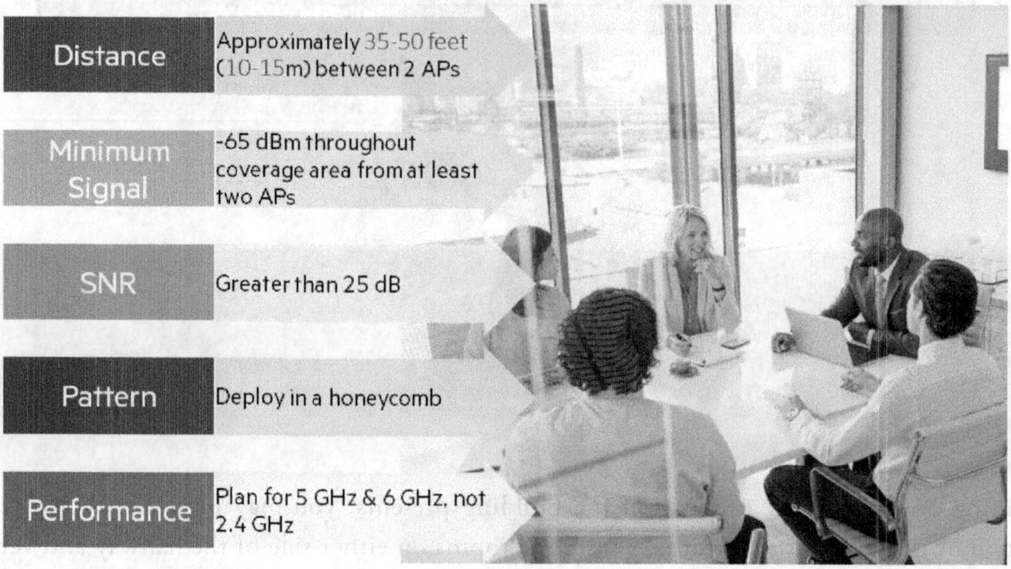

Remember that AP placement should be in a honeycomb pattern, and the network should be planned for 5 GHz instead of 2.4 GHz. Higher frequencies attenuate faster than lower frequencies and, therefore, have smaller cell coverage. Placing APs with 5 GHz in mind will ensure no holes in coverage. All supported frequencies can then be dynamically adjusted with less power using AirMatch to avoid co-channel interference..

Avoid mixing different AP models

The AP-500 and AP-600 series access points support Wi-Fi 6 while AP-600 and AP-700 series access points support 6GHz radios; however, they should not be used within the same contiguous RF service area, particularly when activating the 6 GHz radios in AP-600 and AP-700 series units.

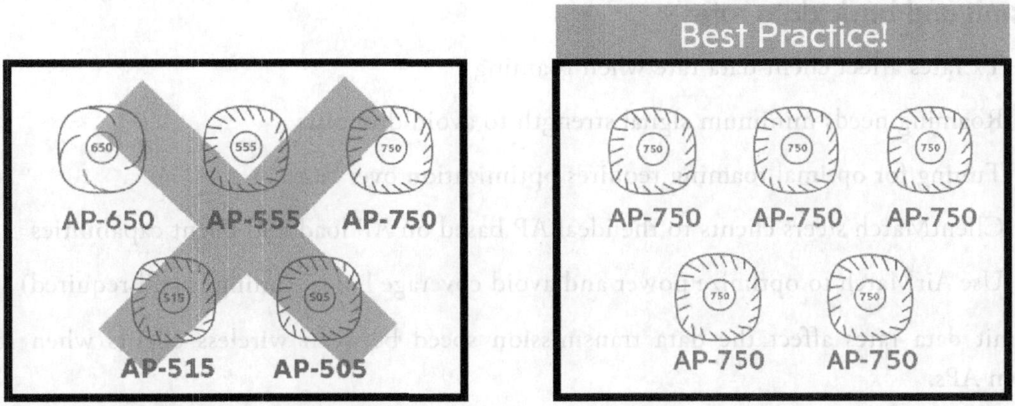

HPE Networking advises using identical AP models within a contiguous RF service area (RF block), which can be perceived as a roaming domain defined by a building or a floor, contingent on the designer's perspective and user requirements.

Transmit power settings

Like channel allocation, HPE Networking recommends using AirMatch to dictate optimal transmit power values for APs.

If the deployment calls for AirMatch not to be used then for a modern office environment with an open floor plan, use the following ARM configuration as a starting point, then monitor and adjust if required:

- 2.4 GHz band, set the min to 6 dBm and max to 9 dBm.
- 5 GHz band, set the min to 15 dBm and the max to 18 dBm.
- The 6 GHz band sets the min to 18 dBm and the max to 21 dBm.
- Do not exceed a power level difference of 6 dBm between the min and max settings on ANY radio bands.
- Tx power for 5 GHz and 6 GHz radios should be a minimum of 6 dBm higher than the 2.4 GHz.

Note that while propagation losses are greater in the 6 GHz than the 5 GHz, the difference is minimal compared to that between the 5 GHz and the 2.4 GHz. Remember that the 6 GHz may be more susceptible to attenuation from walls and other structures, resulting in a significant difference in cell coverage size between 5 GHz and 6 GHz for environments such as closed office spaces, colleges, dorms, hotels, and similar buildings.

Also note that with Advanced License AIOps, you get additional support from HPE Aruba Networking Central AI in deciding which power settings to deploy.

Transmit and basic data rates

- Tx rates affect client data rate when roaming
- Roaming needs minimum signal strength to avoid dropouts
- Tuning for optimal roaming requires optimization over time
- ClientMatch steers clients to the ideal AP based on AP load and client capabilities
- Use AirMatch to optimize power and avoid coverage holes (tuning is still required)

Transmit data rates affect the data transmission speed between wireless clients when roaming between APs.

Roaming requires the client to maintain a certain signal strength and data rate level to avoid dropouts and disruptions. Wireless designers must monitor and tune the environment over time and when deploying a new WLAN to ensure proper performance and optimal client roaming.

ClientMatch can steer clients to the ideal AP based on client capabilities and AP load. Steering clients can take up to 60 seconds, which could still leave the client failing to roam in the meantime. It is important, then, to maintain optimal transmit power settings and avoid coverage holes by utilizing AirMatch.

Estimating AP count for 5 GHz and 6GHz*

Very Low-density, low coverage	• Large open areas with a low user density • Total sq footage divided by **10,000** (929 for sq meters) estimates the number of APs • eg:100,000 sq ft building 100,000 / 10,000=10 APs (sq meters 9,300 / 929 = 10 APs)
Coverage based design	• Basic coverage office/cubicle area with low to medium user density • Total sq footage divided by **5000** (465 for sq meters) estimates the number of APs • eg:100,000 sq ft building 100,000 / 5000 = 20 APs (sq meters 9,300 / 465 = 20 APs)
Capacity based design with low to mid density	• Coverage, office areas with higher user density OR basic coverage office/cubicle with High Demand users • Total sq footage divided by **2500** (232 for sq meters) estimates the number of APs • eg: 100,000 sq ft building 100,000 / 2500= 40 APs (sq meters 9,300 / 232 = 40 APs)

Not all deployments are standard offices

High-capacity design	• High user density indoor space, with high demand users • Total sq footage divided by **1500** (139 for sq meters) estimates the number of APs • eg: 100,000 sq ft building 100,000 /1500 = 67 APs (sq meters 9,300 / 139 = 67 APs) • Recommended for most offices today

Having a rough estimate of AP counts can be a valuable starting point in a design. A simple calculation should consider the square footage of the coverage area and the location type. Let's examine how AP count is decided, comparing coverage versus capacity designs.

A coverage-only design allows for open-area businesses such as offices. Here are two of the most common use cases:

Take the example of low user counts with low bandwidth needs, such as basic web surfing and email. The AP count will be the square footage of the space divided by 10,000 sq ft (930 sq meters).

In a standard open office area with low cubicles and low to medium user density, divide the square footage by 5,000 sq ft (465 sq meters) to find the AP count.

Note that even having a few heavy bandwidth applications may make the coverage design inadequate.

A capacity design is far more practical for modern networks where greater bandwidth and user density would strain a legacy coverage-only approach.

For a capacity-based design with normal to high-density (HD) such as walled offices or high-cubicles, divide by 2,500 sq ft (230 sq meters). Most campus floors would likely fall under this category.

There are unique, very high-density (VHD) requirements for classrooms, auditoriums, and large conference rooms. For a high-capacity design estimate, divide by 1,500 sq ft (140 sq meters). This should also be considered for deployments with a higher bandwidth need, like video training, large file transfers, and very mobile voice applications, which is the standard for most campus designs.

For the very high-density (VHD) with high-throughput wireless network design, divide the square footage by 1,200 sq ft (111 sq meters) to find the quantity of APs.

Modern warehouse trends include automation and AI-driven execution with a higher density of wireless IoT devices and a need in a strong RF signal level. These types of warehouses could not be classified as a very low-density and low coverage scenarios.

AP count: additional factors

Estimating AP count with a simple math equation is not precise. More info is required:

- Data Rates required per user; client standards supported
- Number of devices in an area
- Wi-Fi obstruction
- Wall attenuation
- Future growth ie: Wireless Phones, cameras or IoT devices
- Applications / scenarios

The value of using a quick formula is considering as many factors as possible to predict the needed AP count. Only looking at square footage and office type is essential. However, the square footage per AP also depends on the data rates required (i.e., the bandwidth they want per device) and the number of users in an area.

In today's campus networks, with an expectation of real-time voice and video on mobile devices, the high number of devices per user, and the growth of IoT devices, HPE Aruba Networking always recommends a capacity-based high-density deployment. Benefits of having extra APs include redundancy should an AP go down, better signal overlap for roaming, and the ability to cover the occasional water-cooler effect when users congregate unexpectedly.

Some general best practices for campus design are:

- Use (HD) as the design calculation.
- Remember, with Wi-Fi 6/6E/7, we target a cell size of -55 dBm to -65 dBm.
- Realtime VoIP applications need a minimum of -65db to an AP but ideally should be -55db.
- When using the HPE Aruba Networking Central Floor Plan, use a minimum value of -65db to plan your APs.

Note that Wi-Fi obstructions and wall impedance should be considered along with future growth that the campus anticipates.

Summary of AP/client design

Once the AP counts have been inputted into the HPE Aruba Networking Central Floor Plan, you can use AirMatch to optimize channel and power assignments for radios. While AirMatch will do the work, you should know general RF guidelines regarding campus RF deployments.

Coverage thresholds:

While normal data will perform fine at -75 dBms, a typical voice application always needs a minimum of -65 dBms. To achieve this threshold, plan enough APs to cover areas where Wi-Fi calling is expected.

Jitter, latency, and packet loss thresholds affect real-time collaboration traffic, such as VoIP and conference traffic. Voice traffic has an especially low tolerance for network errors and delays. Using a Wi-Fi network for VoIP requires proper planning to guarantee prioritization, signal, and bandwidth. Make sure to determine busy call hours and total number of calls expected on the wireless network.

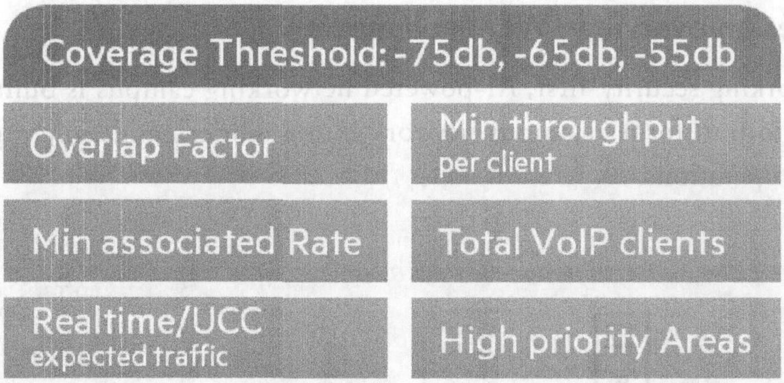

Overlapping coverage is imperative for the performance of mobile Wi-Fi handsets—designed for AP cells with 15% to 20% signal overlap for typical users' walking speed. Consider an expanded overlap policy if the roaming time is shortened due to faster movement, such as a forklift or golf cart moving between cells. Smaller cells or faster roaming speeds require more significant overlap percentages to enable clients to transition between radios.

Minimum throughput per client is the minimum acceptable rate of successful message delivery over a communication channel to an access point.

The minimum association rate is the rate a client can associate with the WLAN. The supported data rates are configured as part of the network policy. Ensure that the clients can adequately decode and receive management frames like beacons and broadcast and multicast traffic, which is essential to successful network connectivity.

Consider the number of voice associations per AP when the network implements VoIP. Voice traffic needs assured bandwidth to avoid interference, as the jitter or loss of packets can affect a voice call.

The number of total voice clients and concurrent calls varies depending on the type of AP and the clients' capabilities.

High-priority areas such as meeting rooms should have their APs to handle higher call density and provide redundancy. The number of users in the room, such as a classroom, may also warrant adding additional APs.

Selecting APs and Antennas

With HPE Networking wireless best practices understood, you are ready to select APs and antennas.

Wireless LAN design overview: WLAN elements

HPE Aruba Networking security-first, AI-powered networking campus is built on three platforms that provide tools to transform a campus network into a modern and agile organization with distributed or central operations.

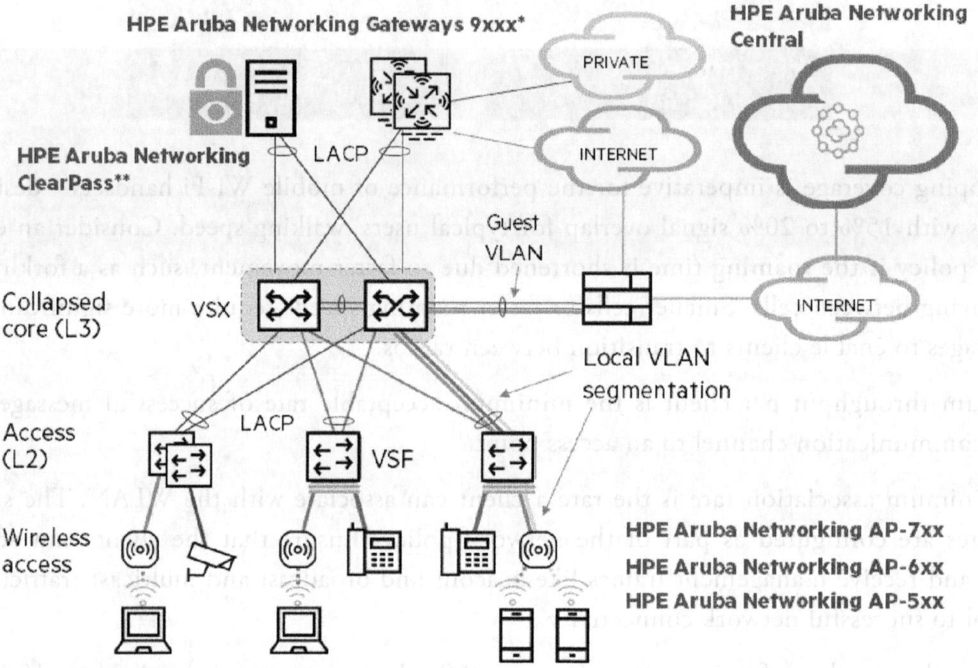

These three platforms are:

- The AOS-CX switch operating system is consistent across campus, branch, and data center.
- AOS-10 cloud-native wireless operating system.
- HPE Aruba Networking Central cloud management SaaS and wired, wireless, and policy infrastructure tools.

This section focuses on the AOS-10 operating system across APs, Gateways, and HPE Aruba Networking Central management. 9xxx series gateways are covered in chapter 8 - Overlay Networks. ClearPassis covered in chapter 7 - Campus Security.

HPE Aruba Networking wireless LAN design overview

HPE Aruba Networking security-first, AI-powered networking campus WLAN elements:

- HPE Aruba Networking Central
- HPE Aruba Networking Gateways 90xx/91xx/92xx: covered in chapter 8 - Overlay Networks
- HPE Aruba Networking Access Points AP-7xx/ AP-6xx / AP-5xx
- HPE Aruba Networking ClearPass*: covered in chapter 7 - Campus Security
- HPE Aruba Networking UXI Sensors: covered in chapter 9 - Network Optimization

*ClearPass Policy Manager (CPPM) is an HPE Aruba Networking server providing AAA services for wired or wireless clients.

HPE Aruba Networking wireless LAN design overview: APs

The connected edge of the HPE Aruba Networking security-first, AI-powered networking solution rests on the strengths of our access points. Wi-Fi 7 and Wi-Fi 6E APs are supported across enterprise, campus, branch, and plug-and-play for remote workers. There is support for indoor and outdoor hazardous environments with the 5xx, 6xx and 7xx series of "EX" APs. The two modern series of APs are the AP-7xx and AP-6xx series.

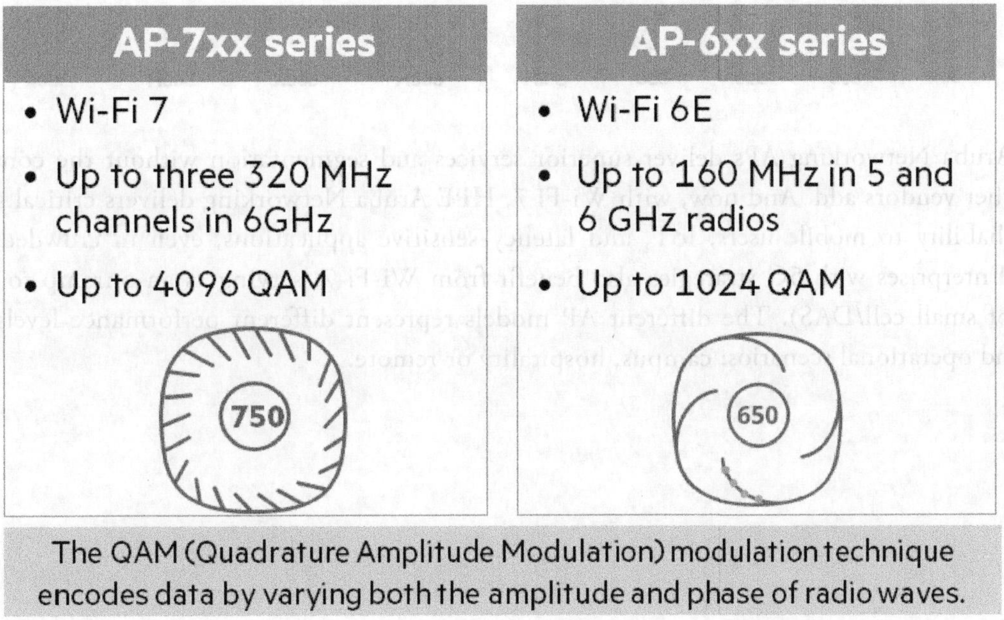

HPE Aruba Networking AP-6xx series access points are designed for Wi-Fi 6E, equipped with 6 GHz radios, and an integrated GPS receiver to self-locate as reference points for accurate indoor location measurements using 802.11mc.

HPE Aruba Networking AP-7xx series access points are the latest offering from HPE Aruba Networking and designed for Wi-Fi 7 deployments to support Extremely High Throughput with up to 320 MHz channels and up to 28 Gbps datarate.

HPE Aruba Networking Indoor Access Points

The HPE Aruba Networking WLAN provides edge connectivity that's intelligent, safe, and fast with AI-powered mobility and enterprise-class design.

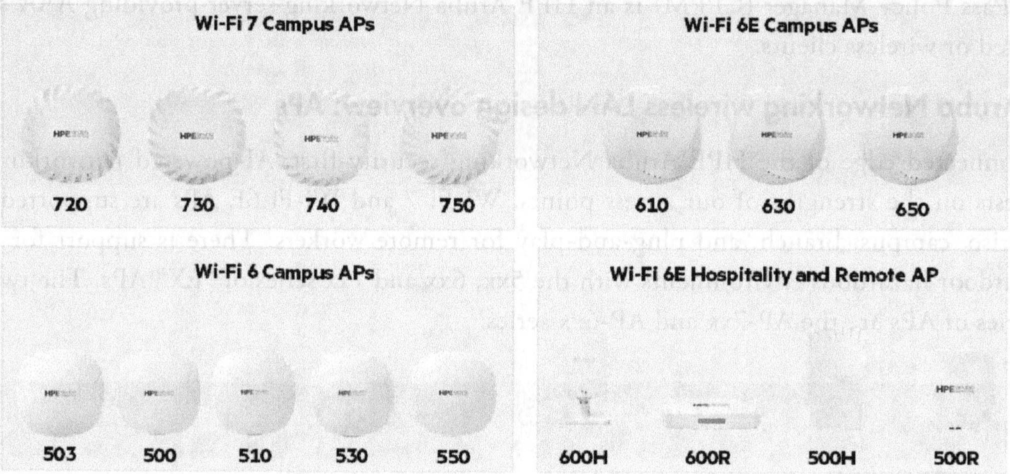

HPE Aruba Networking APs deliver superior services and segmentation without the complexity that other vendors add. And now, with Wi-Fi 7, HPE Aruba Networking delivers critical security and reliability to mobile users, IoT, and latency-sensitive applications, even in crowded Wi-Fi areas. Enterprises with 5G strategies also benefit from Wi-Fi 7, serving as an onramp to 5G (in place of small cell/DAS). The different AP models represent different performance levels, capacity, and operational scenarios: campus, hospitality or remote.

HPE Aruba Networking Outdoor/ruggedized Access Points

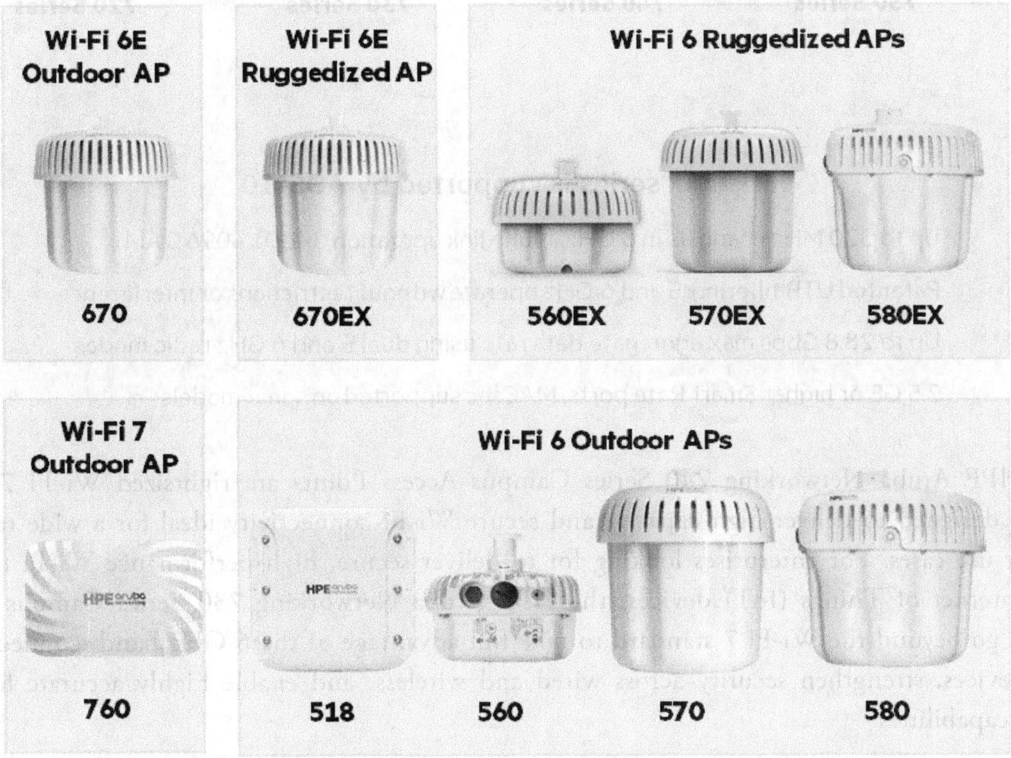

HPE Aruba Networking provides the wide range of outdoor and ruggedized access points: AP-500 for Wi-Fi 6 scenarios, AP-600 for Wi-Fi 6E and AP-700 for Wi-Fi 7 deployments.

Indoor 7xx series Wi-Fi 7

Managing Wi-Fi 7 access points is easier with HPE Aruba Networking Central, which provides intelligent automation, AI insights, and unified infrastructure management. The 7xx series is supported by HPE Aruba Networking AOS-10. HPE Aruba Networking 750 Series Campus Access Points are flagship Wi-Fi 7 access points for high performance, high-capacity, resiliency, and secure wireless connectivity.

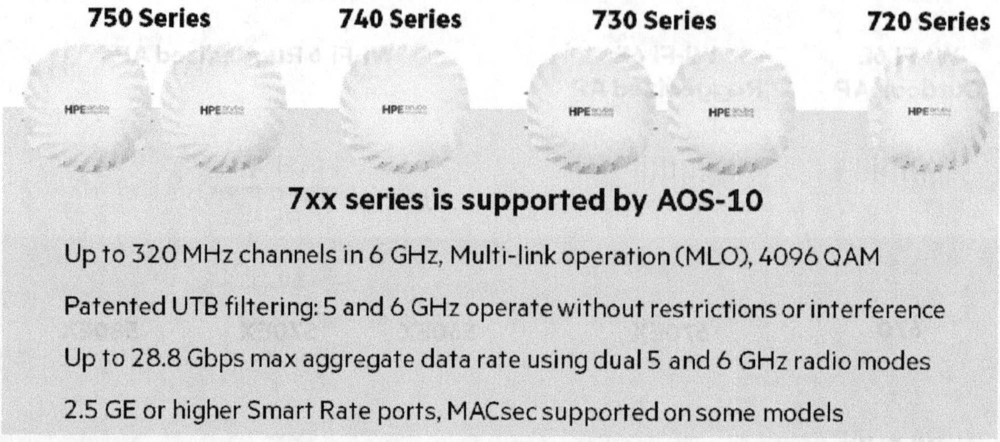

The HPE Aruba Networking 740 Series Campus Access Points are rightsized Wi-Fi 7 access points designed to deliver high-capacity and secure Wi-Fi connectivity ideal for a wide range of indoor use cases. For enterprises looking for to deliver secure, high-performance Wi-Fi to users and Internet of Things (IoT) devices, the HPE Aruba Networking 730 Series Campus Access Points go beyond the Wi-Fi 7 standard to take full advantage of the 6 GHz band, connect more IoT devices, strengthen security across wired and wireless, and enable highly accurate location aware capabilities.

Designed to deliver cutting-edge Wi-Fi performance, the HPE Aruba Networking 720 Series Campus Access Points offer enterprises a cost-effective solution that meets the demand for fast, secure, high-performance connectivity. Leveraging the latest Wi-Fi 7 standard, these access points enhance security across both wired and wireless networks, support Internet of Things (IoT) devices, and provide accurate location-aware capabilities.
The 7xx series Access Points key features:

- Wi-Fi 7 (802.11be) brings multilink operation (MLO) for channel aggregation, 4K QAM for higher throughput and lower latency, and the 6 GHz band for more than double the available capacity
- Up to 28.8 Gbps maximum aggregate data rate using optional dual 5 GHz and 6 GHz radio modes
- Up to three 320 MHz channels in 6 GHz support low-latency, bandwidth-hungry applications like high-definition video and augmented reality/ virtual reality applications
- Patented UTB filtering enables 5 GHz and 6 GHz to operate without restrictions or interference
- AI-powered dynamic power save mode helps reduce energy use

HPE Aruba Networking Wi-Fi 7 AP portfolio comparison

Feature	720 series Indoor AP	730 series Indoor AP	740 series Indoor AP	750 series Indoor AP	760 series Outdoor AP
Wireless radios	triple 2x2	triple 2x2	2x2: 2.4GHz, 4x4: 5 and 6GHz	triple 4x4	dual 2x2 or single 4x4
Radio configuration (GHz)	2.4+5 +6	2.4+5+6 5+5+6 5+6+6	2.4+5+6	2.4+5+6 5+5+6 5+6+6	2.4+5 (2x2) 2.4+6 (2x2) 5+6 (2x2) 5 (4x4) 6 (4x4)
AP peak RF data rate	4.7Gbps	14.4Gbps	8.9Gbps	28.8Gbps	5.8Gbps
Integrated IoT radio	Single	Dual	Single	Dual	Dual
Wired ports (all Smart Rate)	1x2.5GE	2x5GE	1x5GE	2x10GE	1x5GE
Max active PoE consumption	22.8W	36W	40W	51W	37W
PoE requirements	802.3af*/at	802.3at*/bt	802.3at*/bt	802.3at*/bt 2x802.3af*	802.3at*/bt

* With restrictions on functionality

Review HPE Aruba Networking Wi-Fi 7 AP portfolio comparison for the key features. You can find more details in each AP series datasheet.

Key features for 720 series access points:

- Three 2x2 MIMO radios provide triband coverage across 2.4 GHz, 5 GHz, and 6 GHz to deliver 4.7 Gbps maximum tri-band aggregate data rate

- Conveniently powered by PoE (IEEE 802.3at)
- Fast 2.5 Gbps Ethernet connectivity

Key features for 730 series access points:

- Comprehensive tri-band coverage across 2.4 GHz, 5 GHz, and 6 GHz to deliver 9.3 Gbps maximum tri-band aggregate data rate
- Capable of up to 14.4 Gbps maximum aggregate data rate using optional dual 5 GHz and 6 GHz radio modes1
- High availability with dual 5 Gbps ports for redundant Ethernet and power
- MACsec support extends wired Ethernet protection to the access point

Key features for 740 series access points:

- Three MIMO radios (2x2:2.4 GHz + 4x4:5 GHz + 4x4:6 GHz), provide tri-band coverage across 2.4 GHz, 5 GHz, and 6 GHz to deliver 9 Gbps maximum tri-band aggregate data rate
- Fast wired connectivity with 5GbE port

Key features for 750 series access points:

- Three 4x4 MIMO radios provide tri-band coverage across 2.4 GHz, 5 GHz, and 6 GHz to deliver 18.7 Gbps maximum tri-band aggregate data rate
- Capable of up to 28.8 Gbps maximum aggregate data rate using optional dual 5 GHz and 6 GHz radio modes1
- Up to three 320 MHz channels in 6 GHz support low-latency, bandwidth-hungry applications like high-definition video and augmented reality/ virtual reality applications
- High availability with dual HPE Smart Rate ports for redundant Ethernet and power. Configurable to 1, 2.5, 5, or 10 Gbps (or 100 Mbps)
- MACsec support extends wired Ethernet protection to the access point

Indoor 600 series Wi-Fi 6E

The HPE Aruba Networking 6xx series of APs are the first to support Wi-Fi6E, expanding the existing Wi-Fi 6 (802.11ax) standard. Until now, wireless has used two bands: 2.4 GHz and 5 GHz. Wi-Fi 6E takes advantage of the new 6 GHz band. Wi-Fi 6E takes the efficiency features from Wi-Fi 6, like OFDMA, WPA3, and Target Wake Time, and extends them to the 6 GHz band to provide a more contiguous spectrum and less interference. With Wi-Fi 6E, enterprises can support new use cases that demand multi-gigabit speeds like high-definition video.

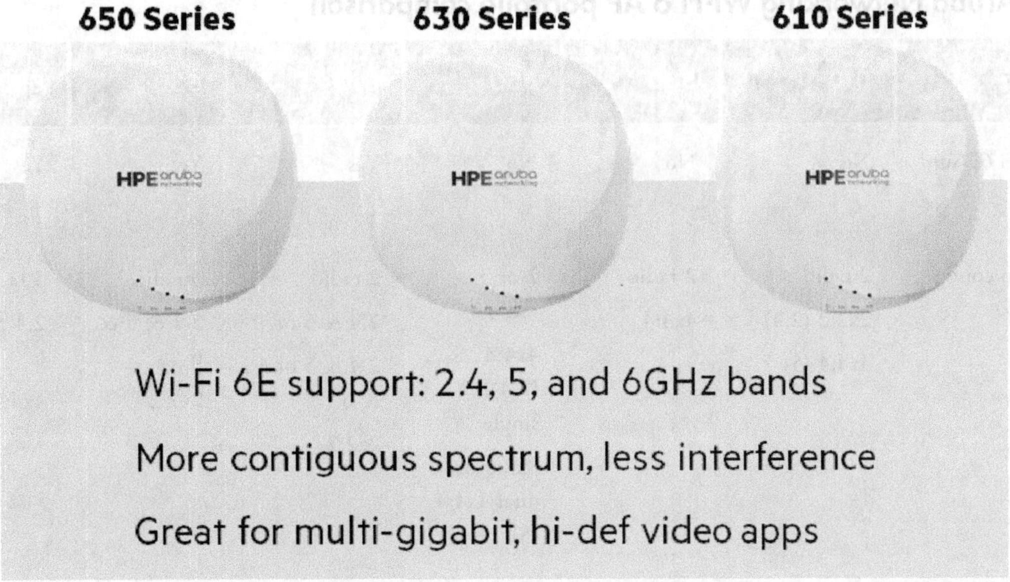

Three models are available in the 6xx series, with all three supporting Wi-Fi 6E and the 6 GHz bands. This includes 1024-QAM modulation, 160 Mhz channels, and OFDMA with up to 37 RUs for the 6 GHz. Another feature is 802.11mc with fine timing measurements (FTM) for precise positioning and location, which are supported on all three models. Peak combined data rates range from 3.6 Gbps to 7.8 Gbps, supported with 2.5GE to 5GE network interfaces. All three support BLE 5.0, 802.15.4 Zigbee radios, and GNSS radios.

Note that HPE Networking currently offers Wi-Fi 6E APs for indoor only. Outdoor and rugged environments are still in the works. This is due to compliance requirements. Before outdoor APs, APs with external antennas, or APs that use standard power (>30 dBm EIRP) can be rolled out, there must be an Automated Frequency Coordination (AFC) service in place that tells the Wi-Fi 6E infrastructure which channels and power levels can be used to avoid interference with incumbents such as ground to satellite transmissions. HPE Networking is partnering with Federated Wireless to develop and deliver an AFC service to support the outdoor use of Wi-Fi 6E. There is no need for AFC indoors since low-power indoor Wi-Fi 6E APs are not subject to interference.

As of this guide, the latest product information can be found at the following link. https://www.arubanetworks.com/products/wireless/access-points/compare/#1297526,298889,338484

HPE Aruba Networking Wi-Fi 6 AP portfolio comparison

	510	530	550	610	630	650
Wi-Fi 6E support	No	No	No	Yes	Yes	Yes
Radio config (GHz)	2 radio 2x2:2 (2.4) 4x4:4 (5)	2 radio 4x4:4	2 or 3 radio 4x4:4 (2.4), single 8x8:8 or dual 4x4:4 (5)	2 radio 2.4 & 5 or 2.4 & 5 or 6 & 6 2x2:2	3 radio 2.4 & 5 & 6 2x2:2	3 radio 2.4 & 5 & 6 4x4:4
AP peak RF rate	2.7Gb/s	3Gb/s	5.4Gb/s	3.6Gb/s	3.9Gb/s	7.8Gb/s
Max clients/radio (recommended)	75	100	100	75	75	100
Ethernet ports	1x2.5GE, 1xGE	2x5GE	2x5GE	1x2.5GE	2x2.5GE	2x5GE
Max active PoE consumption	20.8W	26.4W	38.2W	14.7W	23.8W	32W
Poe requirements	802.3af*/at/-bt	802.3at*/b-t	802.3at*/b-t 2x802.3at	802.3af*/at/-bt	802.3af*/b-t 2x802.3at	802.3af*/b-t 2x802.3at

Wi-Fi 6E with 6 GHz will be supported on any 6xx-series AP, whereas the 5xx-series supports Wi-Fi 6 up to the 5 GHz radios. Both will support the features of the 802.11ax standards. When reviewing which AP would suit a customer environment, take note of the radios configuration, peak data rate, and the recommended max clients per radio. The number of available ethernet ports is essential for resiliency for single AP deployments, but on a campus network, APs are typically deployed using a single ethernet connection.

For example, the AP-650 series has two RJ-45 wired ethernet ports (E0, E1). Their features are:

- Auto-sensing link speed (100/1000/2500/5000BASE-T) and MDI/MDX
- 2.5Gbps and 5Gbps speeds comply with NBase-T and 802.3bz specifications
- POE-PD: 48Vdc (nominal) 802.3af/at/bt POE (class 3 or higher)
- 802.3az Energy Efficient Ethernet (EEE)
- Link aggregation (LACP) support between both network ports for redundancy and increased capacity
- The AP supports direct DC power and Power over Ethernet (POE) on port E0 and/or E1
- When POE power is supplied to both Ethernet ports, the AP combines power from both sources (SmartPOE feature), using E0 as the primary source and E1 as the secondary.

Hospitality and Remote series APs

	500H	600H	500R	600R
Wireless standard	Wi-Fi 6 (802.11ax)	Wi-Fi 6 (802.11ax)	Wi-Fi 6 (802.11ax)	Wi-Fi 6 (802.11ax)
Radio config (GHz)	2 radio 2x2:2, 2.4 & 5	2 radio 3 band 2x2:2, 2.4, 5, 6	2 radio 2x2:2, 2.4 & 5	2 radio 3 bands 2x2:2, 2.4, 5, 6
AP peak RF data rate	1.5Gb/s	3.6Gb/s	1.2Gb/s	3.6Gb/s
Wi-Fi antenna	Two integrated	Integrated omni-	Integrated omni-	Integrated omni-

	500H	600H	500R	600R
	semi-directional	directional	directional	directional
Ethernet ports	4x GE, 1x 2.5GE	4x GE, 1x 2.5GE	3x GE	3x GE, 1x 2.5 GE
PoE Power Source Equipment	1x802.3at or 2x802.3af	1x802.3at/bt	1x802.3af	Power adapter
Max power consumption	20.8W	30.1W	15.4W	40W

All remote APs shown support two radios, with the 600 series supporting 6GHz band. They all support 2.4GHz and 5GHz bands with 2x2:2 MIMO. While only the 600 series support Wi-Fi 6E, the 500H and 500R support Wi-Fi 6. These APs are capable devices in microbranch/home user environments where only a few devices are required.

Key features for 600H:

- Ideal for hospitality, branch, and teleworker deployments
- Flexible coverage across any two bands (2.4 GHz, 5 GHz, and 6 GHz) for up to 3.6 Gbps combined peak data rate
- Up to seven 160 MHz channels in 6 GHz support low-latency, bandwidth-hungry applications such as high-definition video and AR/VR applications
- Combines wireless and wired access in compact desktop or wall mount models that can be PoE powered
- Convenient wired connectivity and support for PoE with a fast 2.5GbE uplink port, two 1GbE ports, and two 1GbE PSE ports capable of supplying up to a total of 30W PoE.
- Internet of Things (IoT)-ready with support for Bluetooth 5 and Zigbee.

Key features for 600R:

- Opens the 6 GHz band for up to triple the capacity of previous generations
- Delivers up to 3.6 Gbps combined peak data rate (when using 5 GHz + 6 GHz operation)
- Cellular module provides backup connectivity and higher resiliency

- Offers low power consumption - virtually unrestricted operation from an 802.3 af (class 3) Power over Ethernet (PoE) source1
- Up to seven 160 MHz channels in 6 GHz support low-latency, bandwidth-hungry applications such as high-definition video and augmented reality/ virtual reality applications
- Multiple uplink/downlink options: 2.5 Gbps uplink/downlink Ethernet port, dedicated 1 Gbps uplink port, and three dedicated downlink ports to help eliminate bottlenecks
- Leverages cloud native HPE Aruba Networking Central to provide zero touch provisioning and a single pane of glass for day 0 to day N management across campus, branch, and remote work environments

Outdoor 700, 600 and 500 series APs

The need for outdoor mounted APs introduces the threat of weather, and therefore, HPE Aruba Networking provides a line of weatherproof and temperature-hardened APs. The AP-5xx series of outdoor APs offer the benefits of 802.11ax Wi-Fi 6 performance while operating in environmentally challenging locations. These are ideal for large public venues (LPVs), pop-up sites, or loading docks.

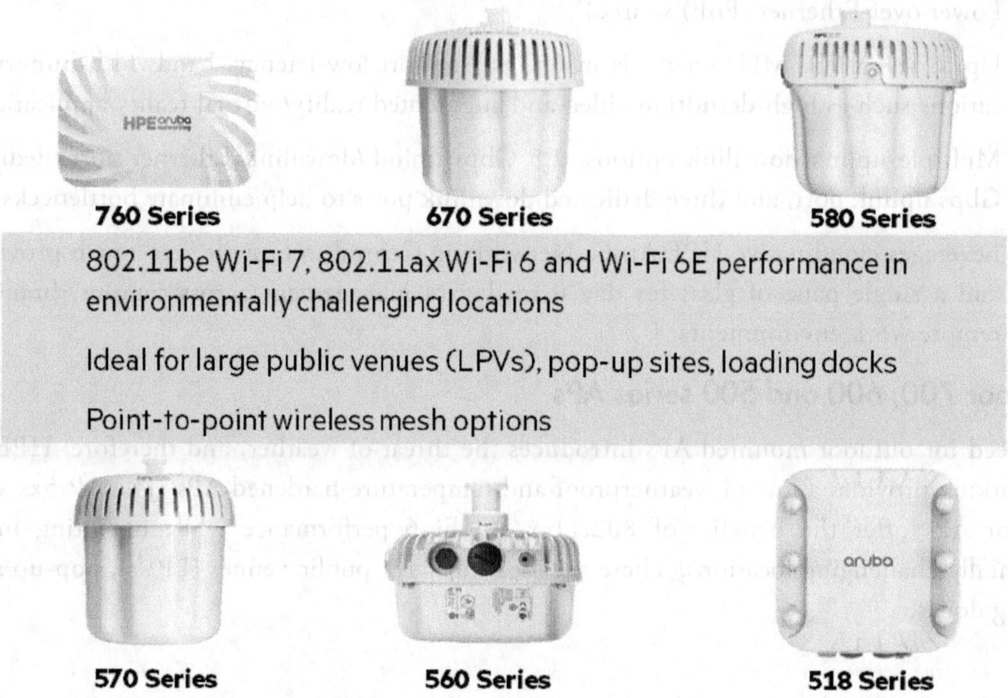

These include two radios (2.4GHz and 5GHz), 802.11ax with 1024-QAM modulation, MU-MIMO, and OFDMA with RUs up to 37 (AP-580 series)—a practical maximum associated client count of up to 100 users per radio. The AP-580 and AP-570 support Smart Rate ethernet, and all models support BLE 5 and 802.15.4/Zigbee.

The HPE Aruba Networking 760 Series Outdoor Access Points bring high performance Wi-Fi 7 to outdoor and environmentally challenging locations such as industrial sites, warehouses, and large public venues. Leveraging the latest Wi-Fi 7 standard, these small form factor access points enhance security across both wired and wireless networks, support IoT devices, and provide accurate location-aware capabilities.

HPE Aruba Networking AP-763 operates as an indoor LPI/SP (composite) device and is ideal for indoor warehouses, LPVs, industrial facilities, and indoor enterprise, where directional-LPI is required (IP50 rating).

HPE Aruba Networking AP-765 operates as an outdoor SP and is ideal for outdoor Wi-Fi 7 coverage, indoor or outdoor warehouses, LPV, industrial facilities (IP66 / IP 67 rating). This model is also available in an EX variant for HazLoc applications.

HPE Aruba Networking AP-764 operates as an outdoor SP device and is ideal for industrial enclosures, LPV enclosures, potential mesh, and for connectorized requirements (IP66/67 rating).

The 670 series delivers more wireless capacity and wider channels taking advantage of Wi-Fi 6E and the 6 GHz band to more than double capacity to enable the speed and reliability needed by enterprise and industrial Internet of Things (IoT) environments. There are three AP 670 models available: AP-674 with external antennas, AP-675 with internal Omni antennas, AP-677 with directional antennas and AP-679 with dynamic directional antennas.

Some models support integrated antennas:

- Tri-band dynamic antennas (software-provisionable omni or directional) (AP-763/AP-765)
- Omni (AP-675), directional 90deg x 90deg (AP-677), dynamic directional 90deg x 30deg or 30deg x 30deg or 90deg x 30deg (AP-679)
- Omni (AP-585), directional 90deg x 90deg (AP-587)
- Omni (AP-575), directional 90deg x 90deg (AP-577)
- Omni (AP-565), directional 90deg x 90deg (AP-567)

They operate at temperature ranges of -40C to +55C. The AP-580 series is purpose-built to survive the harshest outdoor environments and temperatures ranging from -40C to +65C, operation temperatures ranging for AP-670 is -40C to +70C with IP66/IP67 water and dust resistance.

APs for hazardous locations

Outdoor and hazardous indoor environments can be demanding and brutally demanding. HPE Aruba Networking APs deliver high performance and range, no matter the weather.

HPE Aruba Networking Wi-Fi 7 and Wi-Fi 6 ruggedized APs use industrial-rated components and are backed by HPE Aruba Networking's limited lifetime warranty.

HPE Aruba Networking APs feature Wi-Fi 6, Wi-Fi 7, AI capabilities, dynamic segmentation, and more of the latest technological advancements.

Cable category

To provide ethernet connectivity to APs, select the correct category of ethernet cable to meet the AP throughput needs. APs can bridge hundreds of users, and modern standards like WIFI 6A could push multiple gigabits per second. Understanding cabling specs is crucial for a good design.

Category	Type	Mx Tx Speed	Max distance	Signaling rate
Cat. 5	UTP/STP	100Mbps	100m (328 ft)	100MHz
Cat. 5e	UTP/STP	1Gbps	100m 37-55m (121-180ft) at 5Gbps*	100MHz
Cat. 6 (6e)	UTP/STP	1Gbps	100m 37-55m (121-180ft) at	250MHz

Category	Type	Mx Tx Speed	Max distance	Signaling rate
			5Gbps*	
Cat. 6a	UTP/STP	10Gbps	100m	500MHz
Cat. 7	STP	10/40/100Gbps	100m at 10Gbps 50m at 40Gbps* 15m at 100Gbps*	600MHz
Cat. 8	STP	10/25/40Gbps	100m at 10Gbps 30m (98 ft) at 25/40Gbps	2000MHz

*Tested speeds outside the standard

The Cat in Cat5e, Cat6, etc., is short for Category. Network cables are divided into categories based mainly on bandwidth (MHz), maximum data rate (megabits per second), and shielding. The table summarizes commonly used categories. Note that standard speeds and distances are shown along with tested speeds outside the standard, shown with asterisks. Also, Category 7a shares the same specs as Category 7 cabling except for supporting 1000 Mhz signaling and pair sharing for multiple devices on the same cable.

AP mounting kits

There are various options available to install HPE Aruba Networking Indoor or Outdoor Campus Access point depending on the Access Point model and surface type.

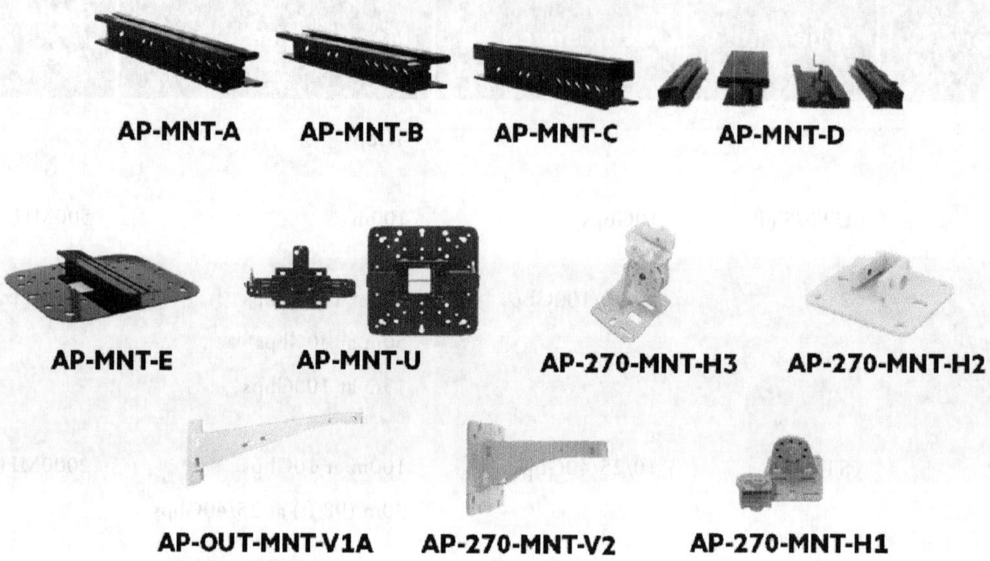

Check additional details in the datasheet and installation guides.

APs mounting kits comparison

Mounting kit	Description
AP-MNT-A	Suspended ceiling rail flat 9/16 for indoor APs
AP-MNT-B	Suspended ceiling rail flat 15/16 for indoor APs
AP-MNT-C	Suspended ceiling rail profile 9/16 for indoor APs
AP-MNT-D	Solid surface for indoor APs
AP-MNT-E	Wall-box for indoor APs
AP-MNT-U	Universal Mount Bracket Kit for indoor APs
AP-270-MNT-H1	Articulating up-down wall or pole mount kit. It's usually for directional antenna Outdoor APs
AP-270-MNT-H2	Non-articulating wall mount for directional APs, or ceiling mount for omni antenna Outdoor APs

Mounting kit	Description
AP-270-MNT-H3	Dual-articulating (up-down, left-right) for wall or pole mounting of directional Outdoor APs, or ceiling mounting APs on uneven ceilings
AP-270-MNT-V2	Short-arm wall or pole mount kit. It is usually for omni antenna Outdoor APs
AP-OUT-MNT-V1A	Long-arm wall or pole mount kit. It is usually for omni antenna Outdoor APs

In this table you can find a brief description for the listed mounting kits. Read the technical documentation to find more details about available options and installations steps.

Ethernet cable length for 1 GbE connections

Implementing ethernet cable in an environment should adhere to recommended cable lengths. With Ethernet, the maximum distance supported is 100 meters, with some exceptions shown in the previous slide. Going beyond the recommended maximum distance invites signal attenuation, where the signal becomes too distorted and weak to be properly interpreted by the receiving side. Unshielded cable (UTP) is also more susceptible to additional noise due to electrical currents and electromagnet interference, given the extended length.

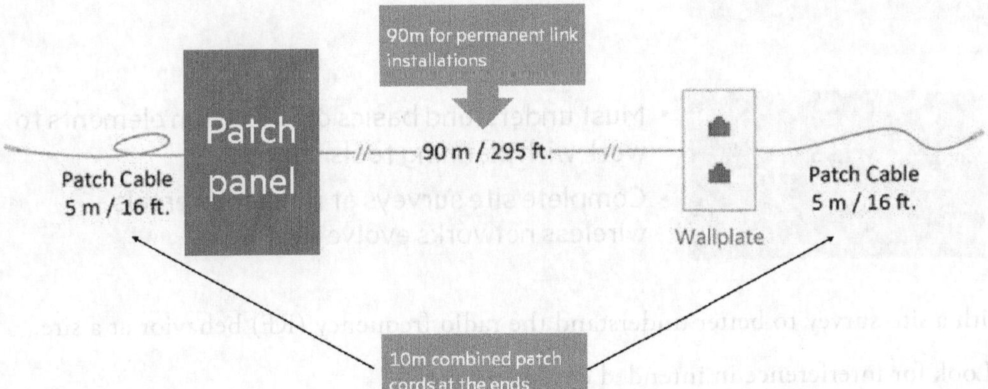

To avoid attenuation, it is vital to understand the total distance of the patch to the plenum cabling. For example, a best practice for 100m maximum standards, like Cat5e, is to use 90m of cabling for the permanent link installations running through the building from patch panel to wallplate. That leaves another 10m combined for patch cabling. Patch cabling is the cable from the switchport to the closet's patch panel and from the room's wall plate to the PC or device NIC.

Vertical fiber between floors and buildings will be covered in the upcoming module "Wired Network Design."

AP Deployment Technologies

The HPE Aruba Networking wireless portfolio includes products for indoor or outdoor coverage. These solutions provide centralized or distributed bridging with policy enforcement to enable wireless capabilities in building 5G cellular roaming, location services, IoT device security, and more.

Where to start?

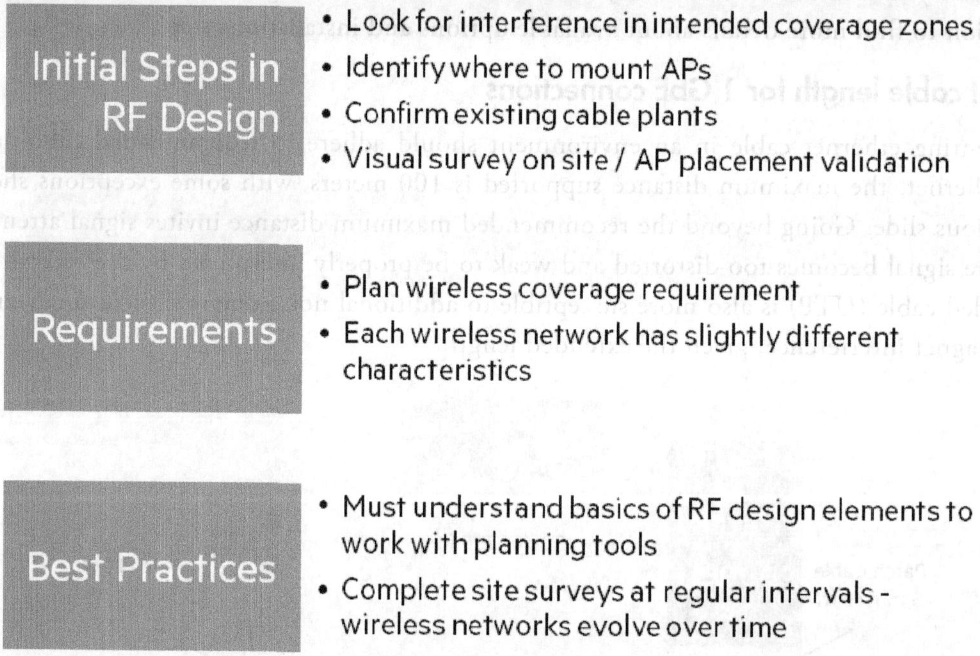

Start with a site-survey to better understand the radio frequency (RF) behavior at a site.

- Look for interference in intended coverage zones.
- Identify where to mount APs.
- Confirm existing cable plants.
- Plan wireless coverage requirement.
- Each wireless network has slightly different characteristics
- Understanding the basics of RF design elements is required to work with planning tools.

Complete site surveys at regular intervals as wireless networks evolve.

Survey methods

A site survey accurately determines how many APs are necessary to provide a target minimum data rate across a given area. This includes identifying the optimal location for AP placement to achieve optimum performance and avoid interference. Modeling AP coverage with a virtual survey for almost any open office environment is trivial.

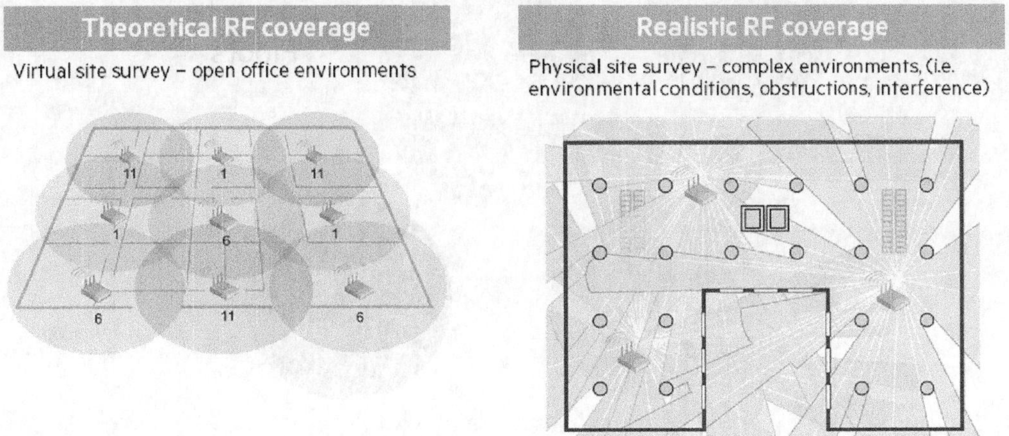

Actual RF coverage for a physical site differs from the theoretical modeling. Environmental conditions, obstructions, and interference must be accounted for at a physical location, as these can impact RF energy propagation. RF behavior is notoriously difficult to predict in any challenging environment accurately. This is why physical site surveys are much better predictors than relying only on the virtual site survey.

Physical building materials

Consider the RF environment that radios must contain within a campus deployment. In every building, the materials will affect the RF signal propagation and consequently may impact the client throughput. Most enterprises have carpeted spaces with materials that may cause reflection or absorption of the RF signals.

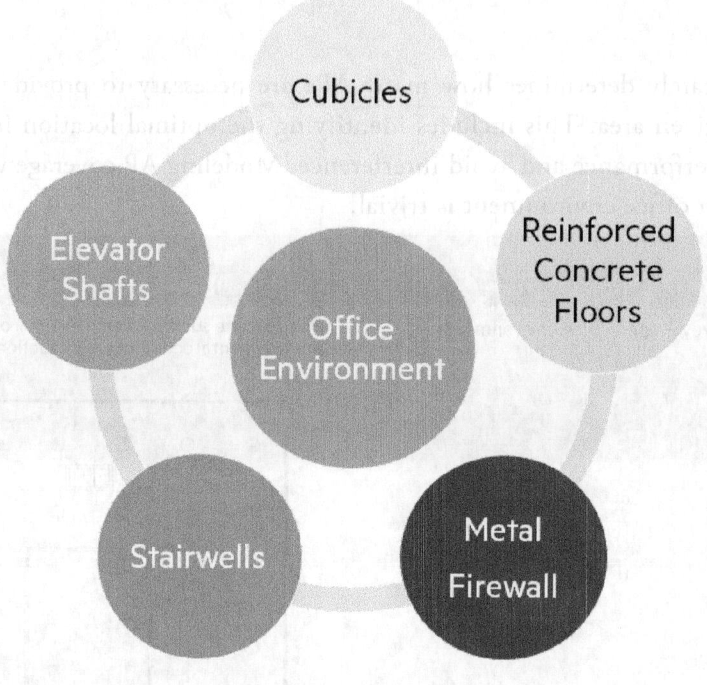

When performing a walkthrough of the environment, you should look for cubicles, stairwells, and elevators, as these may cause a significant loss of RF signal propagation. Concrete walls and floors and metal firewalls between office suites are also a consideration because of their effect on RF signals. These may be difficult to see during a visual walkthrough inspection and require examining a floor plan or blueprint of the building.

Selecting a survey type

Consulting firms and wireless integrators that provide engineering services will generally define a site survey into four categories of an RF survey, as shown.

Predictive surveys are typically performed in low-complexity environments like a relatively new construction with primarily cubicles, as this kind of survey is predictable as to how RF will propagate. You can use HPE Aruba Networking ClearPass's Floor Plan tool to model the deployment and create an initial bill of materials (BOM).

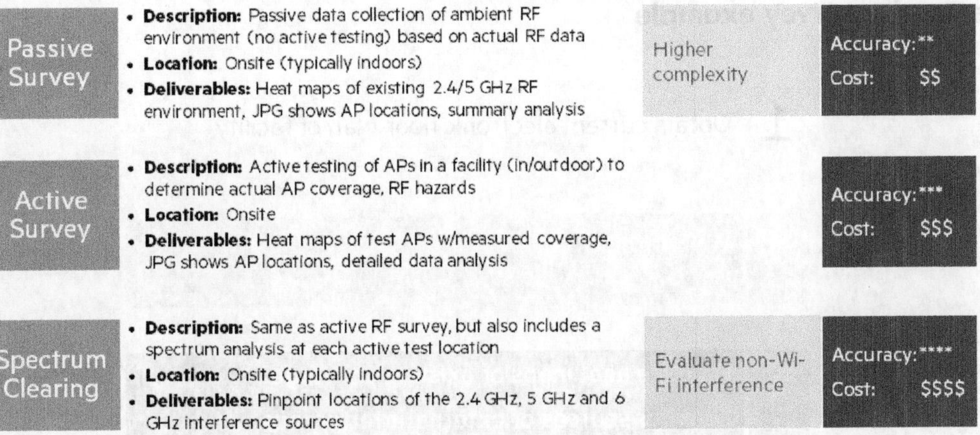

However, a physical site survey may be required in addition to the initial virtual one. On-premises surveys can be passive, active, or spectrum-clearing. A passive survey involves passive ambient RF environment data collection without live testing. Active includes the passive survey with active testing on real APs, such as an AP on a stick, to determine the actual AP coverage footprint and RF hazards. Active surveys also offer greater accuracy in RF design, AP mounting details, cabling requirements, and antenna orientation.

Spectrum clearing includes all the previous steps of a virtual and physical site survey. A spectrum analysis at each active test location is added to evaluate the level of non-Wi-Fi interference you may encounter.

A physical site survey is often necessary to validate the virtual survey for coverage or capacity design.

Note that HPE Networking recommends that wireless professionals perform both predictive and physical surveys.

Onsite passive survey example

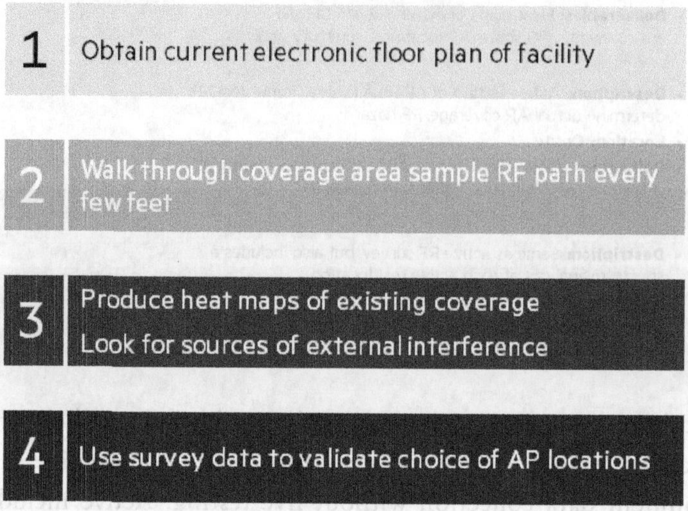

Physical site surveys are the traditional method for planning wireless deployments. Before technology advancements, such as HPE Aruba Networking AirMatch and ARM, APs were planned to use static power and channel settings as determined by the physical site survey.

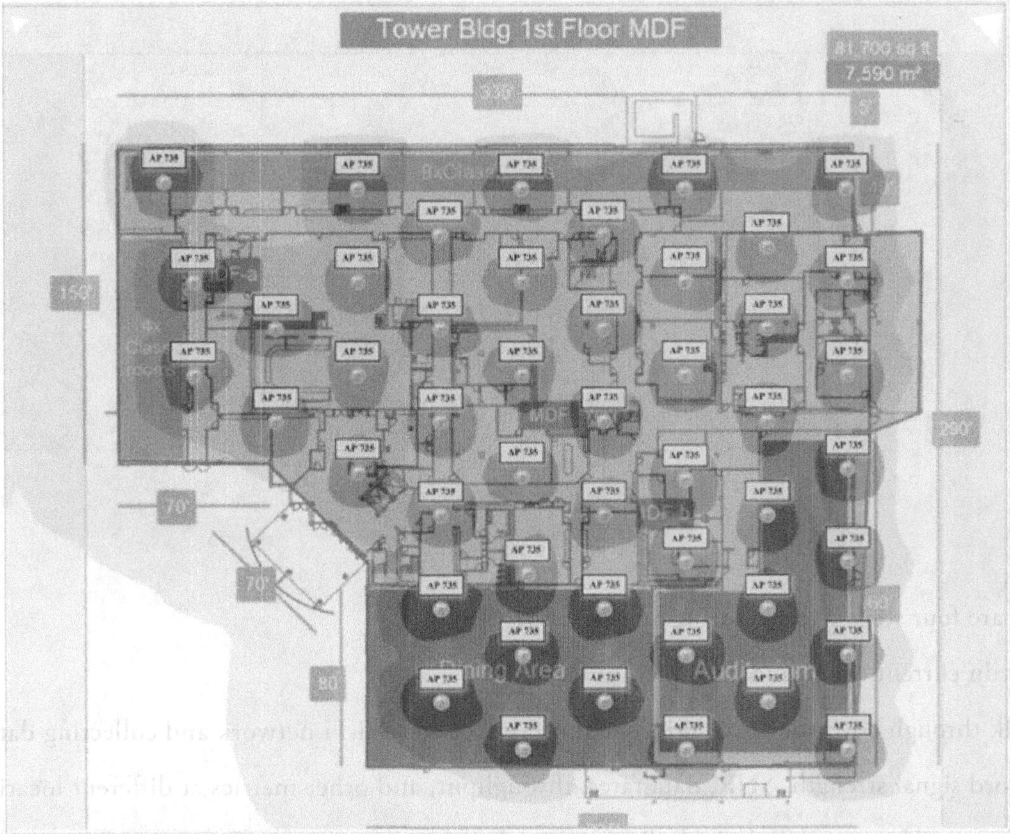

Though many of the same principles in planning apply today, AirMatch and ARM eliminate the need to select exact channel and power settings. The three types of physical surveys are passive, active, and spectrum clearing.

You should use a site survey software application, walk through the coverage area, and sample the RF path every few feet. Analyzing the data will produce heat maps of existing coverage that you can study to find external interference sources.

Onsite active survey example

During an active site survey, the survey equipment is connected to the wireless network at different locations and measures the performance and throughput of the wireless network using various test utilities.

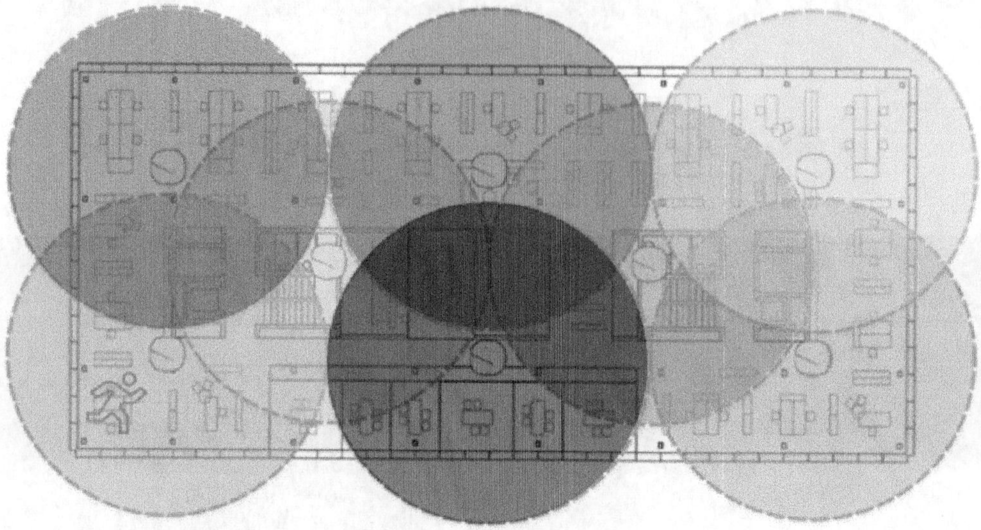

Below are four steps to take during an active survey:

1. Obtain current electronic floor plan of facility
2. Walk through coverage area, actively connecting to the Wi-Fi network and collecting data
3. Record signal strength, SNR, data rates, throughput, and other metrics at different locations
4. Document the survey path, including collected metrics

Onsite survey with temporary placed APs

A physical survey with temporary APs is the most accurate way to determine AP placement and density when considering actual environment variables. The survey should use operational APs to test WLAN signals against actual clients. This process can be very time-consuming but will result in a validated design for the location.

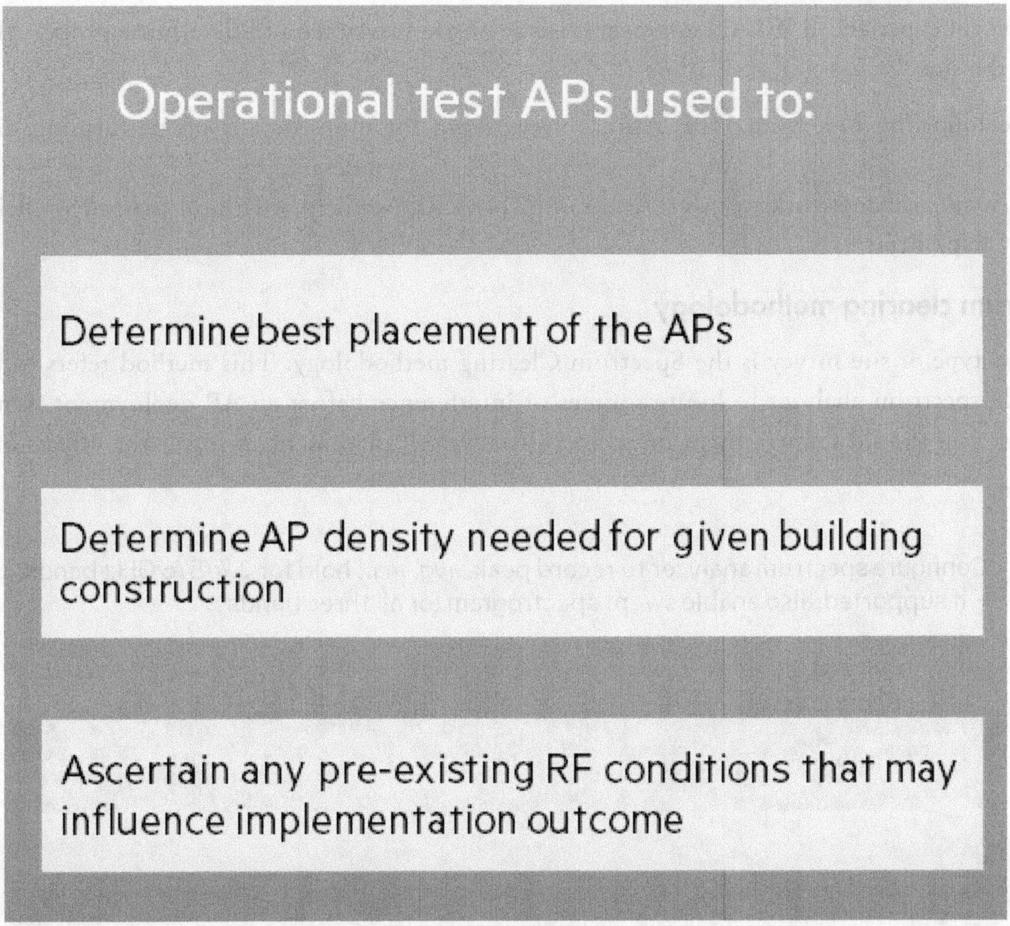

Below are seven steps to take during an active survey:

1 – Obtain the current facility electronic floor plan to mark locations where you will perform active tests.

2 – Using an HPE Aruba Networking Instant AP, provision the AP with ARM minimum power that will be used in the deployment.

3 – Mount AP to a portable tripod, speaker mount, or other stable movable platform – or, if available, mount directly to ceiling rails.

4 – Position AP at a test location and connect to a power source and data link.

5 – Use a professional site survey application to complete a passive survey of the area surrounding the test AP.

6 - Repeat steps 4 and 5 for all identified test locations across the desired coverage area.

7 - Have an experienced WLAN engineer analyze onsite survey data to determine proper AP density for the site.

See the following link from HPE Aruba Networking for more details on performing different types of wireless surveys. https://www.arubanetworks.com/vrd/Indoor80211nVRD/wwhelp/wwhimpl/js/html/wwhelp.htm#href=Chap7.html

Spectrum clearing methodology

The last type of site survey is the Spectrum Clearing methodology. This method refers to using a portable spectrum analyzer to locate sources of interference before an AP deployment. Once discovered, you should remove them or at least distance AP placement to lower the effects of interference.

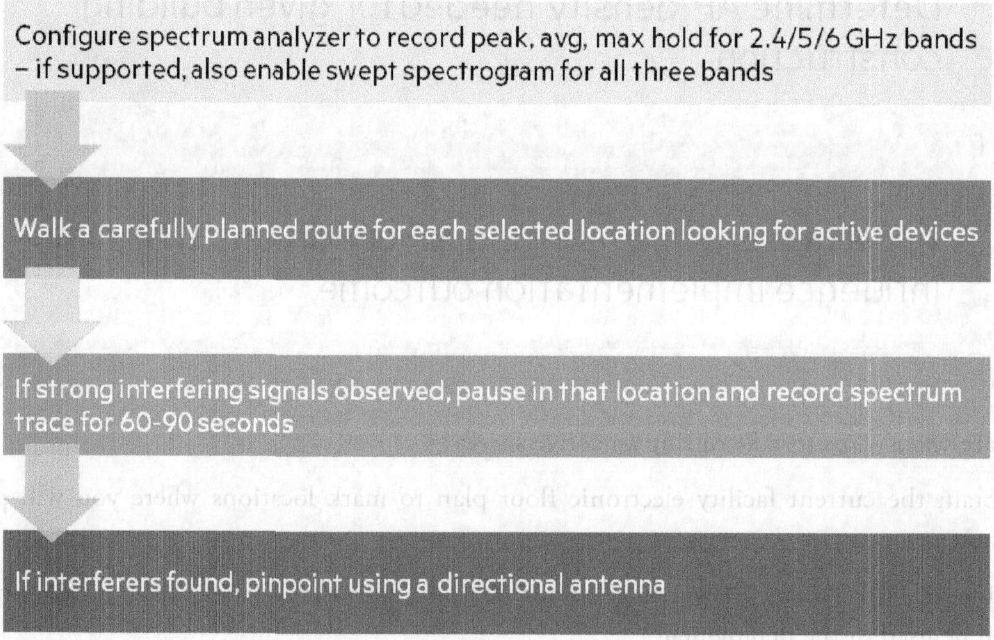

Common examples of devices operating in the same frequency space as the campus WLAN could be cordless phones, analog/digital video cameras, Bluetooth devices, and microwave ovens in break areas. In addition, neighboring APs placed in residential areas like homes or apartments near the campus or any businesses around the campus could all contribute. Consider also airports or weather stations that may remove the use of DFS 5 GHz channels.

WLAN RF planning process

There are five steps in planning a wireless network before deploying APs and antennas.

Step 1: Perform an initial environment evaluation.

Determining the environment type will dictate what design options are called for. Location, building size, age, and materials used (especially shielded areas) should be noted. Pay close attention to the desired coverage areas, user applications, and user roaming; these factors must be addressed. Speaking with key shareholders and completing wired and wireless network questionnaires will help to ensure critical requirements are never missed.

Step 2: AP and antenna selection.

With an understanding of the location and wireless requirements, you can choose which AP(s) and antennas will best meet those requirements. Review the available HPE Aruba Networking AP portfolio to match the right AP with the desired outcome for the customer. For example, a customer wanting the latest standard supported should consider the HPE Aruba Networking 700 or 600 series of APs. If they only need indoor deployment and the benefits of expanded channels, high-density, and future-proofing, then the 700 series is the place to start. Place APs in the HPE Aruba Networking Central Floor Plan to complete the initial predictive plan.

Step 3: Optionally conduct a predictive site survey.

This shows a proof of concept using the HPE Aruba Networking Central Floor Plan with planning mode. HPE Aruba Networking Central Floor Plan is a simple tool showing real-time deployed APs, clients, and heatmaps over existing floorplans and includes a planning mode. This tool is valuable for the pre-deployment of a site as it can calculate the desired coverage and rate provided, given an uploaded floorplan and AP model. This is called a virtual or predictive site survey. An HPE Aruba Networking Central Floor Plan can generate a basic foundation for planning. This should be followed up with a physical visit to the site to verify AP location, signal coverage, and discovery of interference.

Step 4: Optional physical site survey.

Conduct a passive or active physical site survey to properly validate a given facility's RF propagation. The goal is to identify the worst-case challenges in the installation environment when selecting AP locations. Even a passive walk-through is crucial to plan a wireless deployment in a complex environment effectively. Use the walk-through to obtain site images to compare to the floor plan. Note anomalies.

Step 5: Make adjustments to the HPE Aruba Networking Central Floor Plan.

Conducting the physical site survey provides RF propagation assessments that will affect the best choice of the actual AP locations. Look for obstructions such as poles, lights, ventilation, and cable runs. Consider building materials that may affect attenuation or areas of concern, such as mezzanines, atriums, or ceilings above 25ft (8 meters). Update the floor plan to adjust for the findings.

Conclusion

With the HPE Aruba Networking Central Floor Plan completed, you should know what users, devices, and applications will be serviced by the new wireless design. This should consider location, detected interference, roaming requirements, and building obstructions. The customer requirements take into account standards, speeds, and efficiencies used. You should have a clear plan of how many of which AP and antenna models to deploy and be able to demonstrate your deployment plan through a tool like HPE Aruba Networking Central Plan as well as a bill of materials using something like IRIS.

Site survey heat map generator

The HPE Aruba Networking Central Floor Plan, dashboard, and planning tool come with a HPE Aruba Networking Central device license. This integrated tool has been covered earlier to plan and calculate AP counts for new deployments. HPE Aruba Networking Central Floor Plan builds a site location to add a building and upload floorplans, scales the image, and indicates where existing APs are located, their height, and placement direction. This allows a heatmap view of coverage or interference, associated clients and their locations, and filters to narrow to a particular radio band or signal cutoff. HPE Aruba Networking Central Floor Plan can provide live information, discovering a client's location while troubleshooting wireless issues.

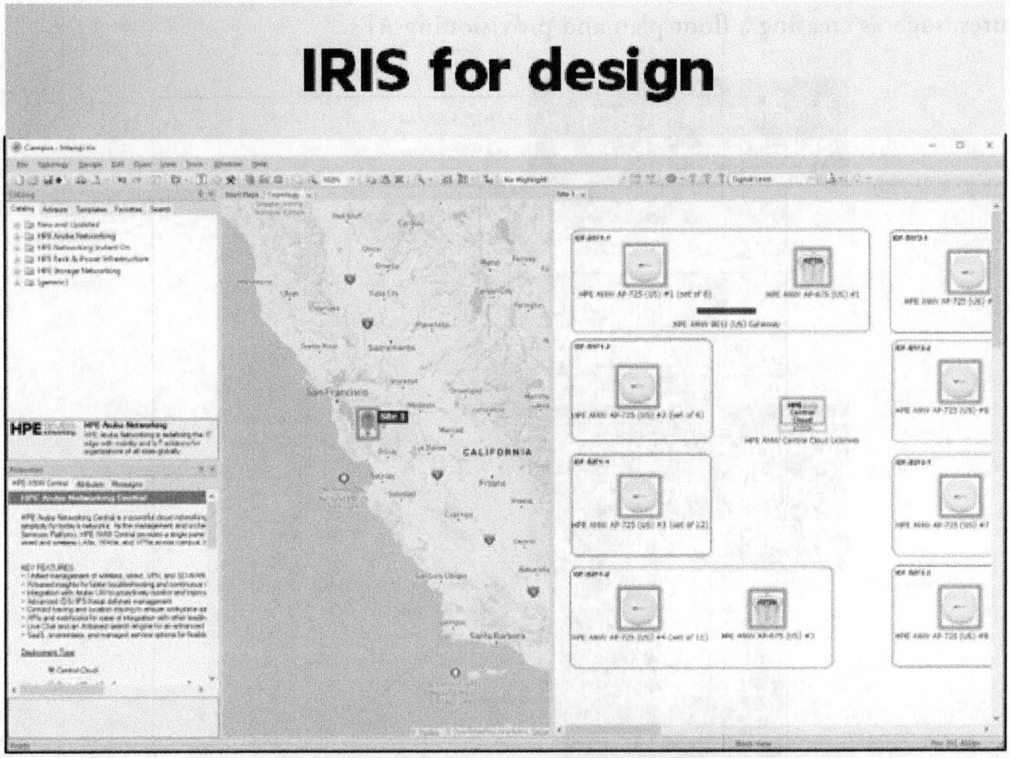

Iris from Intangi is another tool we will use throughout the labs. Iris is a vendor-neutral configuration, design, and quoting platform to efficiently build a bill of materials or BOM. Alternatives like Excel, Draw.io, or HPE Aruba Networking Central Floor Plan lack the scale Iris offers. You can quickly create an express site and organize desired products using Iris Advisors to narrow down which products meet the customer's requirements. Adding switches, APs, Gateways, Transceivers, and appliances can include warranty information, licensing, and organization into Design Groups while showing connection and PoE budgeting. The result of using Iris is a neatly subtotaled bill of materials (BOM) that can be exported as a spreadsheet or Iris file for customers in whatever currency they require.

Internal walls and doors should not be drawn unless they are high dB loss materials. The environment slider bar takes care of drywall style walls and interior doors. External walls should be drawn, often by converting the region into walls.

Floor plan: add calculated APs / removing APs

HPE Aruba Networking Central Floor Plan lets you plan sites, create and manage floor plans, and provision APs. You can use a HPE Aruba Networking Central Floor Plan to do basic planning procedures, such as creating a floor plan and provisioning APs.

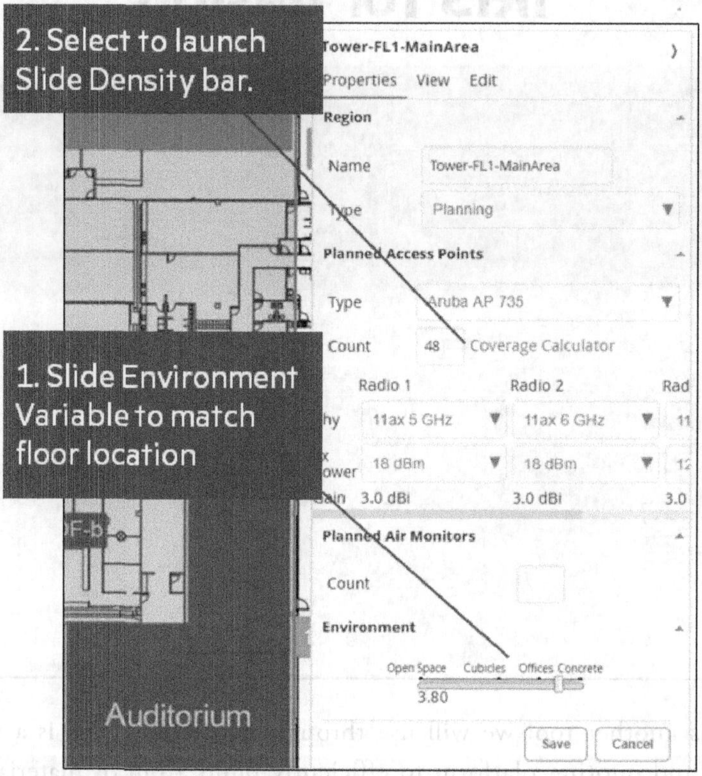

When planning APs, you must slide the density bar to see the AP count. When you want to delete all planned APs, click on Edit and Delete All Planned Devices.

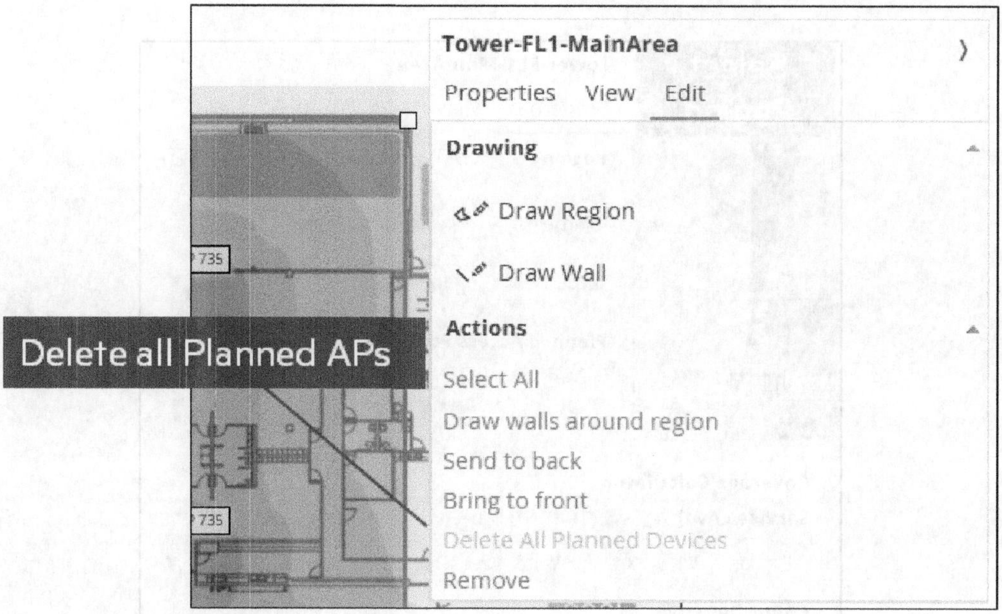

Floor plan: recalculating APs

To recalculate AP, click on Properties, then click on the region area.

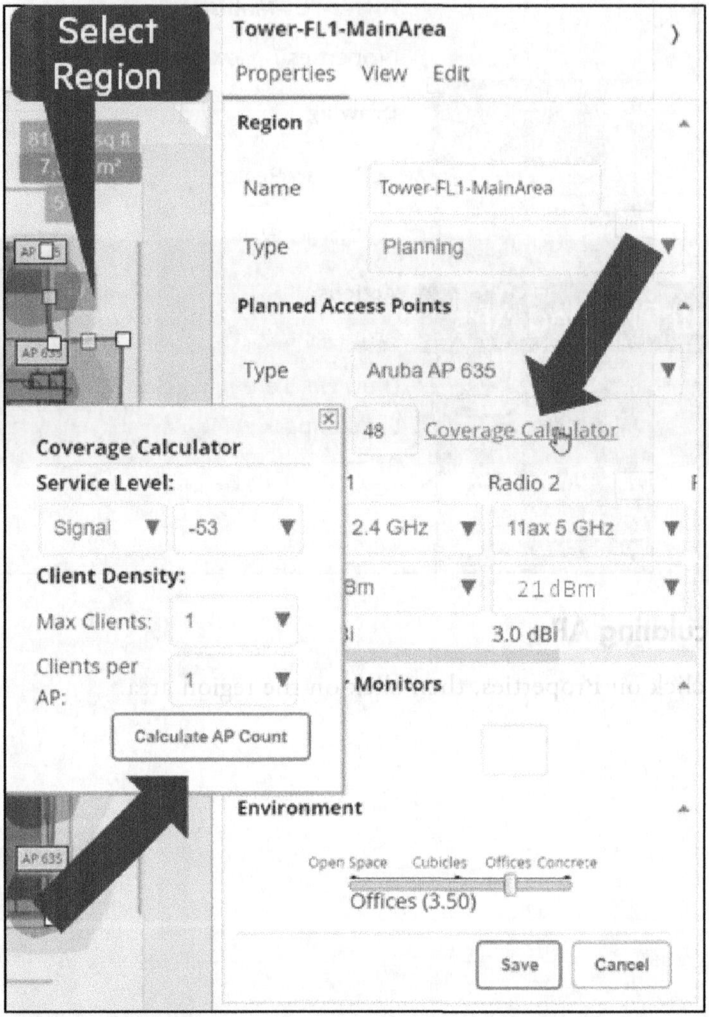

This will change the menu display. Select the type of AP, then click on Coverage Calculator. Include your service level and click on Calculate AP Count.

Creating a Bill of Materials with Iris

Intangi Iris is a tool that helps architects create proposals for customer networks. You begin by creating a site.

You can then select many different HPE and HPE Aruba Networking solutions from a catalog.

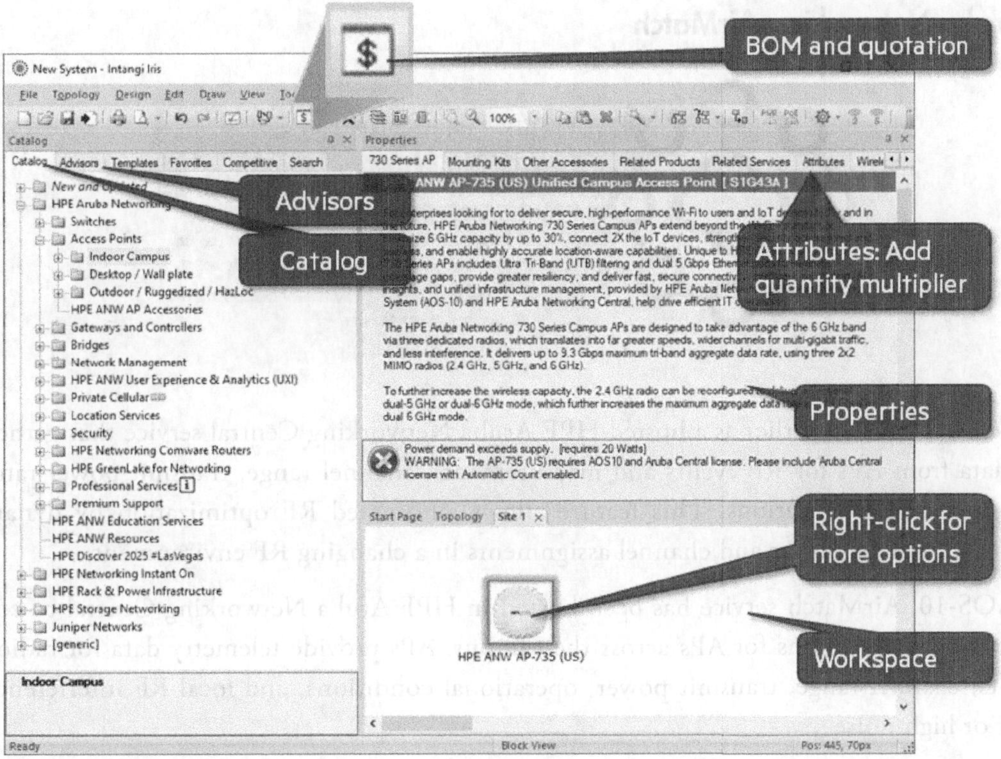

Once you have added a product to the work area, you can click it, and the properties window appears. This window lets you choose a precise model from a series, select modules and transceivers for switches, add licenses for MCs, select accessories, and more.

In this attributes tab, you can set a quantity, which multiplies out the same product configuration, allowing you to fill out a BOM quickly.

To learn more about a product, right-click it and select Documentation. You can then see product information, a datasheet link, or quick specs online.

Click the Advisors tab to get help selecting a product or comparing similar products.

And when you are ready to check out the BOM and price quotation, click here.

—

Iris offers many more capabilities than the ones highlighted here or even those you explore in the lab. You can use it to create complete logical topologies for multi-site networks. It supports design groups and templates to help you work more efficiently. Select Help and open the User Manual to learn more about using Iris. You could watch a training video or visit an online forum.

HPE Aruba Networking AirMatch

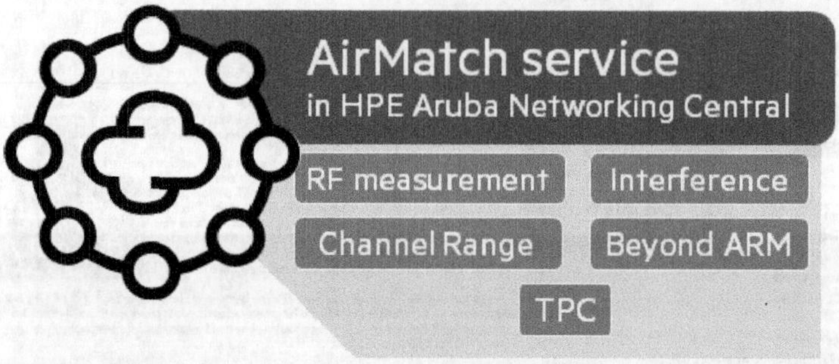

AirMatch was covered earlier as a hosted HPE Aruba Networking Central service that gathers telemetry data from APs for RF events and measurements, channel range, transmit power range and overall operational conditions. This feature allows automated RF optimization by dynamically adjusting AP radios' power and channel assignments in a changing RF environment.

Since AOS-10, AirMatch service has been hosted in HPE Aruba Networking Central to compute and deploy RF allocations for APs across the campus. APs provide telemetry data for radio measurements, channel range, transmit power, operational conditions, and local RF interference such as radar or high noise.

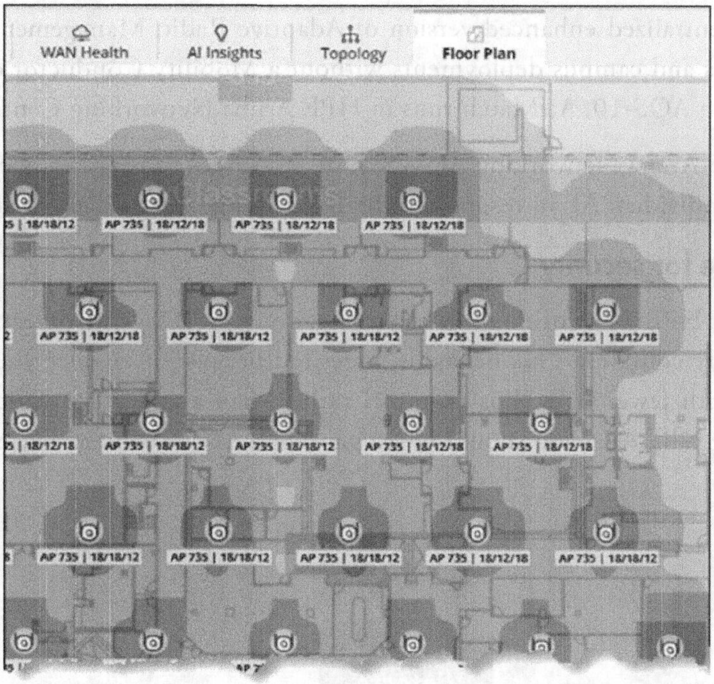

AirMatch considers each AP, the overlapping channels with neighbors, and path loss weight to find the optimal plan to calculate the average radio conflict value. Once a new plan is found, a new conflict value is calculated and compared with the current operating network. This yields an improvement percentage, which, if the improvement is higher or equal to the configured quality threshold, the new plan will be deployed at the configured "Automatically Deploy Optimization" timer.

While a balanced AirMatch wireless coverage approach is recommended, changes to AirMatch should be minimized as any channel change events are disruptive. Note that AirMatch can only optimize the network to the degree that APs are deployed correctly in optimal locations in the environment.

- Computes which channels to assign, the channel width, and the channel transmits power for AP radios.
- Deploy the ideal channel/power plan at configured deployment times.
- Provide neighbor APs list to the Key Management service.
- Provide AP radio channel assignments to the Live Upgrade service.
- Analyzes RF telemetry data to determine configuration changes algorithmically for all HPE Aruba Networking APs on the network. The APs receive regular updates based on environmental conditions.

AirMatch is the centralized enhanced version of Adaptive Radio Management (ARM) used with AOS-8 Instant APs and campus deployments without a Mobility Conductor or HPE Aruba Networking Central. In AOS-10, AirMatch runs in HPE Aruba Networking Central.

Note that the HPE Aruba Networking ARM feature continues to run locally on APs and can alter the power of any individual AP in response to high interference regardless of an existing schedule.

AirMonitor mode for security

When you follow HPE Networking's recommendations, WIDS will detect more threats than WIPS automatically contains. This behavior is desired because it enables you to investigate and mitigate threats with fewer risks. For example, rather than enable automatic protection against APs impersonating valid APs, you should investigate such APs as you are alerted to them.

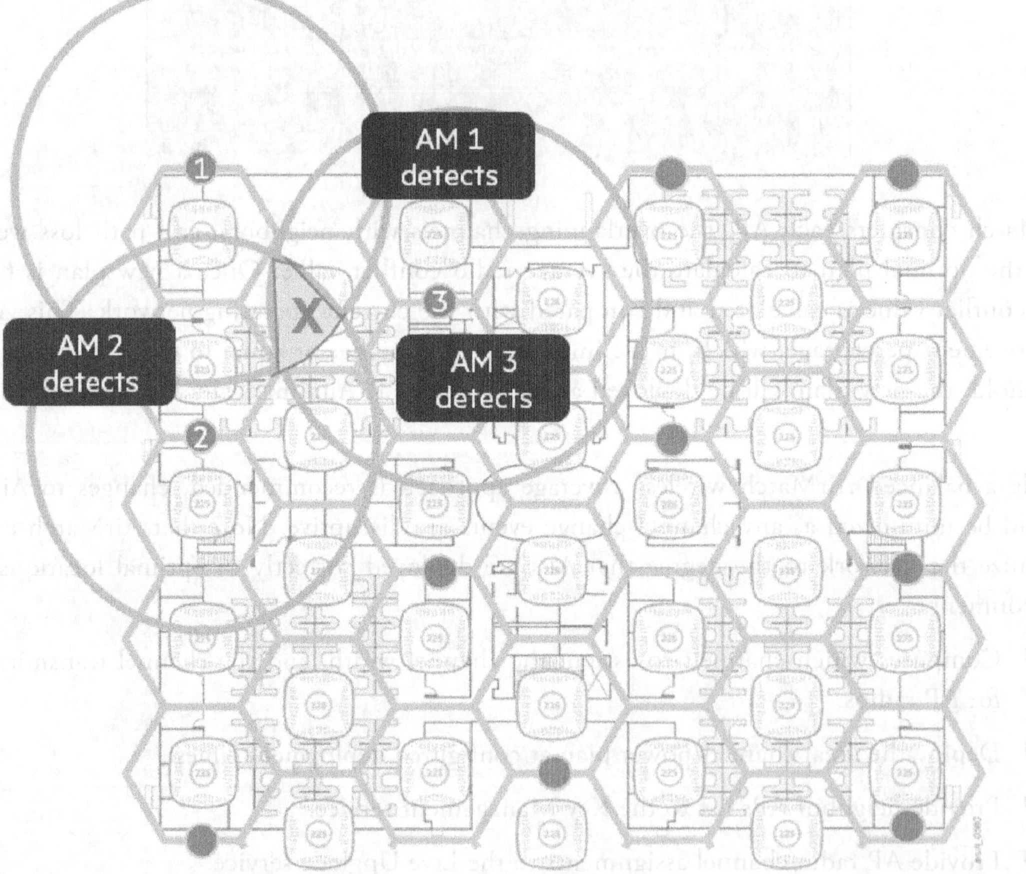

Make a list of the APs/AMs that detected the threat and use trilateration to find the likely area for the threat. When the alert is triggered, search the site for unauthorized devices and question employees about possibly unauthorized people. Is there a legitimate explanation for the threat, or is this a likely instance of a hacker attempting to impersonate an authorized AP?

After a thorough investigation and approval from the security and legal team, you should consider manually marking the AP that triggered the alert as Contained. However, containing is not always the right approach. Consider an AP that is spoofing an authorized AP's BSSID. Attempting to contain that AP could cause a DoS against the legitimate network. Instead, you should remove the interfering AP manually. In addition, you generally want to remove rogue APs rather than contain them over the long term.

Remember that you can implement 802.1X on all Ethernet ports to reduce the risks of unauthorized devices connecting to the network in the future.

Review AM deployment recommendations for rogue AP detection

HPE Networking recommends deploying about one AM per four or five APs to obtain good WIDS/WIPS coverage.

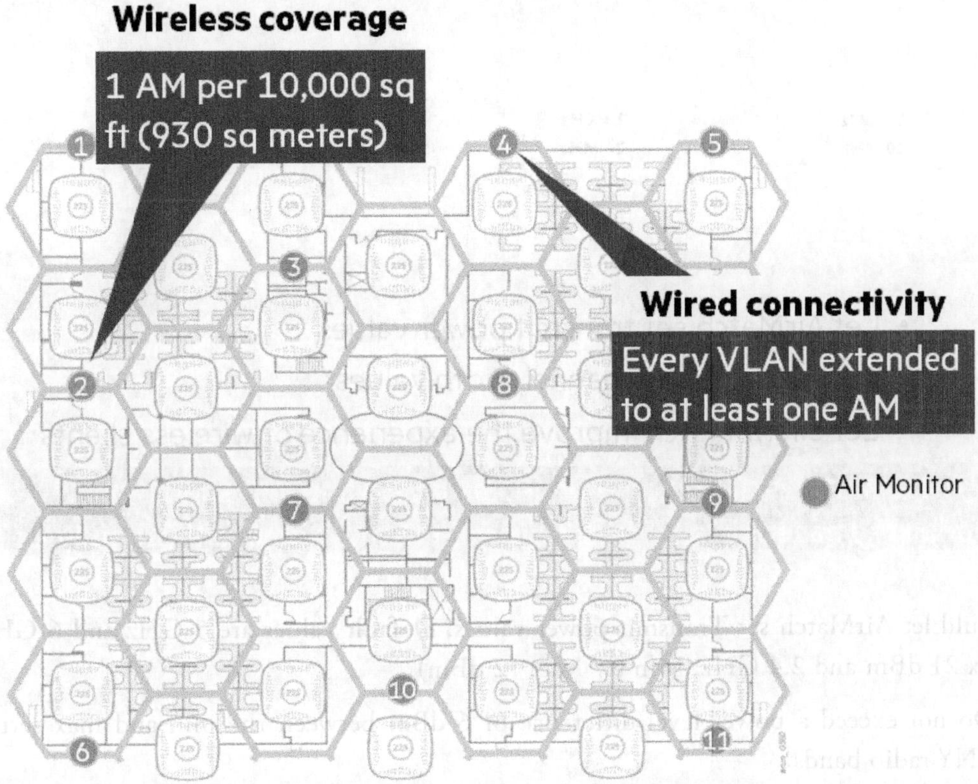

In addition to considering RF coverage for the WIDS/WIPS solution, consider how the solution will detect whether unauthorized APs are connected to the wired network. How WIDS/WIPS detects rogue APs might differ slightly based on the AOS architecture. However, WIDS/WIPS generally builds a list of "wired MACs" detected in the air. These are the source MAC addresses that APs transmit within 802.11 frames, which belong to clients in the wired LAN behind the AP. WIDS/WIPS compares the air-detected wired MACs against MAC addresses it knows to exist in the corporate LAN. If it finds matches or close matches, it knows that the AP is or might be connected to the corporate LAN. WIDS/WIPS can learn the corporate LAN MAC addresses in several ways, including APs/AMs reporting their default gateway MAC addresses and APs/AMs snooping ARP traffic on their VLANs.

To ensure rogue AP detection works correctly, ensure the HPE Aruba Networking solution has at least one device in each VLAN to which rogue APs might connect. One common way to achieve this goal is to extend all the VLANs in each area to AMs in that area. 1 AM per 10,000 sq ft in an open office deployment; more if wall attenuation becomes a concern.

Roaming recommendations

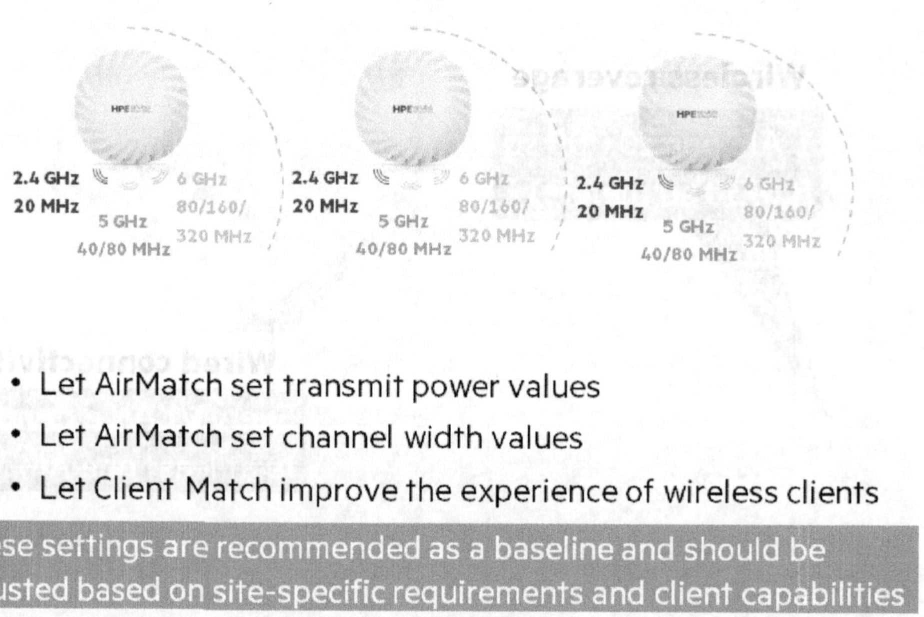

You should let AirMatch set Transmit power values. Default values are 5 GHz and 6 GHz: Min 18 / Max 21 dBm and 2.4 GHz: Min 6 / Max 12 dBm)

- Do not exceed a power level difference of 6 dBm between the min and max settings on ANY radio bands.
- 5 GHz should be 6 dBm more than the 2.4 GHz radio.
- The 6 GHz band sets the min to 18 dBm and the max to 21 dBm.

- Tx power for 5 GHz and 6 GHz radios should be a minimum of 6 dBm higher than the 2.4 GHz.
- Keeps coverage area consistent between radio bands due to lower bands covering more area.
- Using a consistent Tx power policy across all APs in an AP group leads to more predictable roaming behavior.

You should let AirMatch set Channel Width values.

- Set 80 MHz usage with DFS channels only if no radar signal interference is detected.
- Enabling the DFS range for 5 GHz may create holes in RF coverage for clients that only support U-NII-1 and U-NII-3 bands.

ClientMatch

ClientMatch optimizes the client associations by continuously scanning the wireless environment while sharing information about clients and APs. Clients are then steered to the most suitable AP radio based on the dynamic data received.

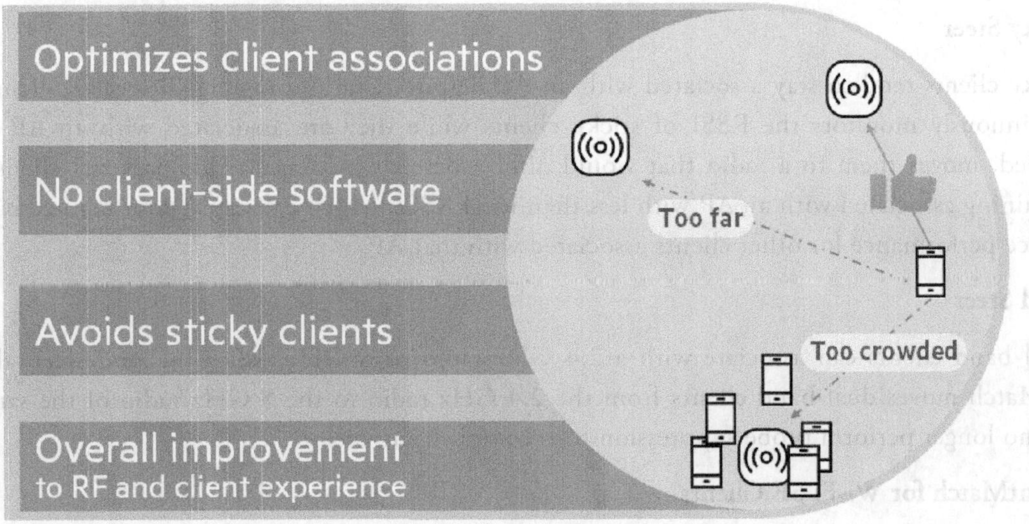

No software is required on the client machine.

ClientMatch avoids sticky clients by continuously monitoring the client's RSSI while it is associated with an AP, and moving the client would significantly improve quality.

ClientMatch monitors the RF neighborhood for each client to provide ongoing client band-steering and load balancing. This can improve the overall throughput for clients unaware of the AP load. Due to the removal of sticky clients and overall improvement to RF, clients experience fewer drops, better-negotiated speeds, and less interference.

HPE Aruba Networking recommends enabling ClientMatch with 802.11k in WLAN wizard as well as on clients. Clients are recommended to also support 802.11v to enable non-legacy ClientMatch.

Note

ClientMatch is not a replacement for proper RF design.

ClientMatch methodology

ClientMatch monitors and identifies client connections with APs and optimizes that connection by steering the client to AP radios that can provide improved service. Below are the different ClientMatch methodologies to accomplish this:

Steer Types

ClientMatch periodically checks the health of the current association of the clients and determines if a sticky steer or band steer should be considered.

Sticky Steer

Sticky clients tend to stay associated with an AP despite deteriorating signal levels. ClientMatch continuously monitors the RSSI of sticky clients while they are associated with an AP and, if needed, moves them to a radio that would offer a better experience. This prevents clients from remaining associated with an AP with less than ideal RSSI, which can cause poor connectivity and reduce performance for other clients associated with that AP.

Band Steer

Dual-band clients can associate with a 2.4 GHz radio or 5 GHz radio. In-band steer, the ClientMatch moves dual-band clients from the 2.4 GHz radio to the 5 GHz radio of the same AP. We no longer perform probe suppression to accomplish this steer.

ClientMatch for Wi-Fi 6E Clients

ClientMatch allows 6 GHz capable clients to participate in sticky and band steers, and in spectrum load balancing.

Band Steer from 2.4 GHz to 6 GHz

The band steer from 2.4 GHz radio to 6 GHz radio feature is enabled by default. ClientMatch attempts band steer only if:

- A client is 6 GHz capable.
- A viable 6 GHz radio is available on the same AP where the client has associated on the 2.4 GHz radio.
- A client is steerable for band steer to 6 GHz radio.

If a client associated on the 2.4 GHz radio is not in good health, then ClientMatch attempts steering the client to the 6 GHz radio on the same AP. If these conditions are not favourable, ClientMatch considers band steer to the 5 GHz radio.

Supported and Non-preferred Channels in 6 GHz

The APs populate the new fields for 6 GHz supported and nonpreferred channels. ClientMatch uses these new fields for the supported and nonpreferred channel checks during sticky steers and load balance steers on the 6 GHz radio.

Steering Methods

After determining the steering type, ClientMatch determines the best neighbor radio to steer the client to and orchestrates the client steer by sending action messages to the APs to carry out the steer. The way ClientMatch steers the clients depends on whether the clients are 802.11v-capable.

Steering for 802.11v-capable Client

To steer 802.11v-capable clients, ClientMatch triggers the AP to send out an 802.11v BSS transition management request to the client and waits for a response.

Steering for Non-802.11v-capable Client

To steer non-802.11v-capable clients, ClientMatch triggers all neighboring AP radios (except the intended destination) to block the client from associating for 5 seconds. Two seconds later, the AP to which the client is currently associated sends an 802.11 de-authentication management frame to the client. When the client tries to re-associate, only the intended AP radio allows the client to associate with it.

HPE Aruba Networking AP professional services

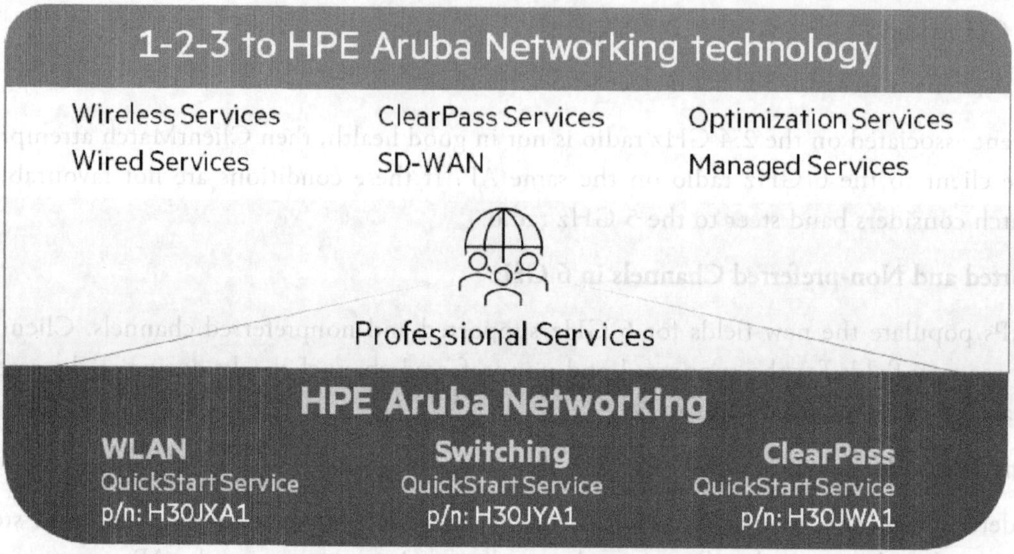

Consider HPE Aruba Networking Professional Services for support with Wireless, Campus Switching, ClearPass, SD-WAN, Optimization, and Managed Services. HPE Aruba Networking advises a skilled engineer to do an RF site survey. Consider speaking with your chosen partner or contacting your HPE Aruba Networking Account manager for professional services.

Learning check

Check your understanding of RF planning by answering the following questions.

Question #1

Which is true regarding HPE Networking deployment models for RF?

- ○ Common WLAN types include basic coverage for office spaces, high density for classrooms, very high density for stadium seating, and coverage for ship-wide area network (SWAN) deployments.
- ○ Both a coverage or capacity design approach will provide adequate RF signal for clients to roam without holes in coverage.
- ○ Even a well-designed 5 GHz deployment should never be upgraded to 6 GHz with a 1:1 AP replacement, regardless of client density, throughput requirements, or physical environment.
- ○ For Wi-Fi 6/6E/7, target a cell size of -55 dBm to -65 dBm.

Question #2

Which of the following is true for HPE Aruba Networking APs?

- O All HPE Aruba Networking 7xx series APs require either dual 802.3at or single 802.3bt power to operate without restrictions.
- O The 7xx series APs are all HPE Aruba Networking Wireless AOS-10.
- O HPE Aruba Networking 6x series supports the new Wi-Fi 7 802.11be standard.
- O The HPE Aruba Networking ruggedized 5xx series, like the AP-518, is specifically designed for hazardous locations.

Question #3

Given plenty of time to perform a site survey, which type generates the most accurate results?

- O Active survey
- O Spectrum clearing including active survey
- O Passive survey
- O Virtual survey

Supplemental Section: Review RF Fundamentals

This section includes optional reading material.

IEEE 802.11 standards

The IEEE original 802.11 standard was introduced in 1997. It was shortly followed by 802.11a/b/g standards supporting 2.4 GHz and 5 GHz bands. Orthogonal frequency-division multiplexing (OFDM) introduced a multi-carrier modulation system that broke channels into sub-carriers for speeds up to 54 Mbps. These standards are more than 20 years old by today's standards and are not recommended for production.

The 802.11n (Wi-Fi 4) amendment was ratified in 2009. This standard supports high throughput (HT) speeds of 600 Mbps with support for up to 450 Mbps on most enterprise radios. In addition, 802.11n supported both the 2.4 GHz and 5 GHz bands. The increase in throughput is largely attributed to the advanced encoding of 64-quadrature Amplitude chapters (QAM), efficient use of antennas with multi-input/multiple-output (MIMO) allowing up to four spatial streams, and bonding two 20 MHz channels together into a wide-band 40 MHz channel.

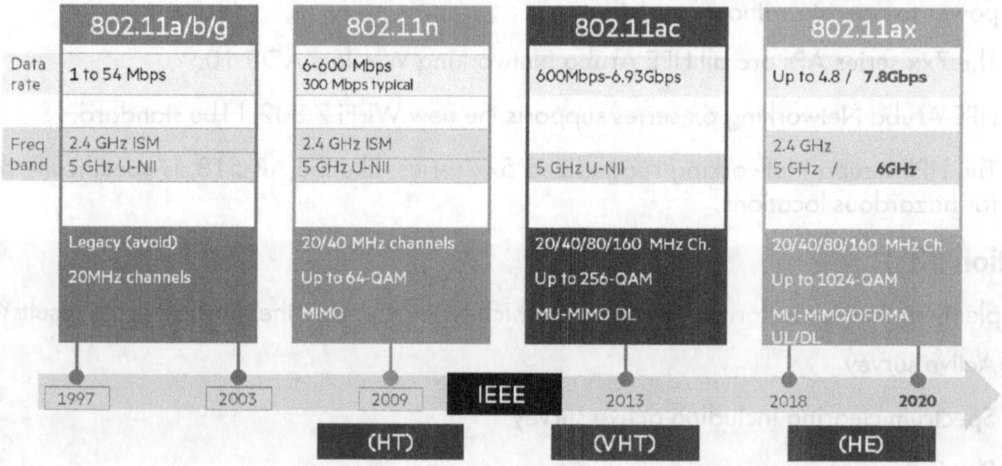

802.11ac (Wi-Fi 5) was introduced in 2013 for the 5 GHz band only. It supports very high throughput (VHT) with higher channel widths of up to 160 MHz. and 256 QAM. The standard allows for up to 4 spatial streams, which, when coupled with the larger channel widths and efficiencies, produced a theoretical 6.9 Gbps. The new multi-user MIMO (MU-MIMO) method allows APs to split eight downlink spatial streams across concurrent client devices.

802.11ax Wi-Fi 6 is the latest standard and operates on both 5 GHz and 2.4 GHz with a max data rate of 4.8 Gbps. Mu-MIMO and up to eight spatial streams are supported on the uplink and downlink, along with better encoding with up to 1024-QAM. OFDMA was also introduced in Wi-Fi 6. This technology allows HPE Aruba Networking's APs to handle multiple 802.11ax capable clients on each channel simultaneously, regardless of device or traffic type. This standard is indicated as high-efficiency (HE).

The 802.11ax Wi-Fi 6E extension was introduced in 2020, with up to 7.8Gbps and a coding scheme (MCS) 10/11. This standard operates in the 6 GHz RF spectrum to enhance throughput in high-density scenarios. Final approval for indoor use was granted by the IEEE Standards Board early in 2021.

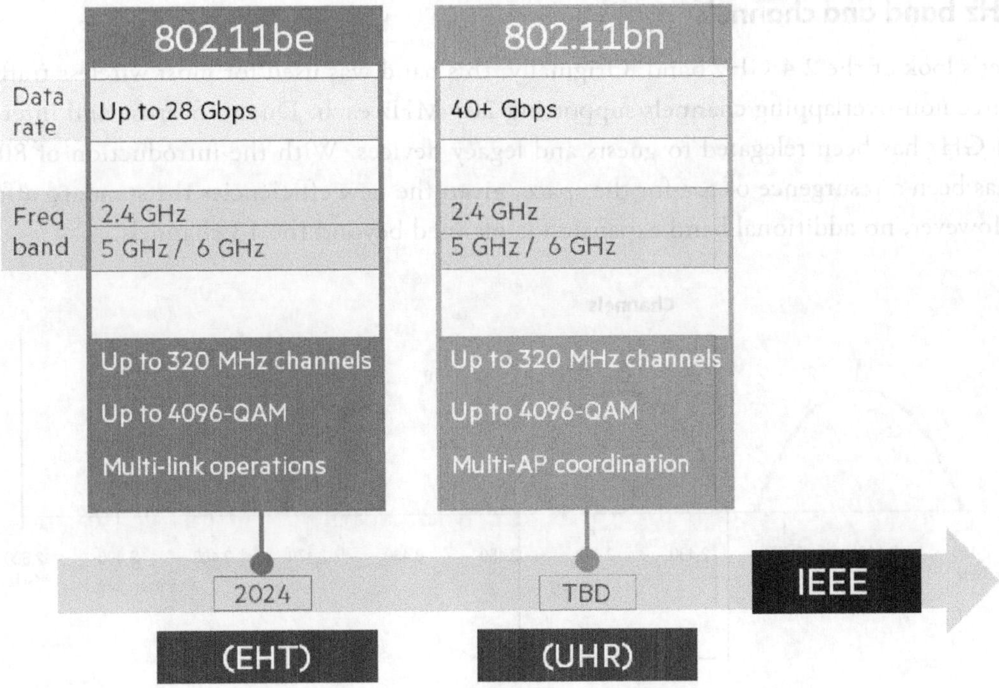

The 802.11be Wi-Fi 7 extension uses the 6 GHz band to greatly increase Wi-Fi capacity by providing access to up to 1200 MHz of additional spectrum, introduces important capabilities that include ultra-wide 320 MHz channels for more data at faster speeds, Multi-link operations (MLO) for greater reliability, and 4K QAM for higher peak data rates. W-Fi 7 is ideal for use cases that demand high bandwidth, low latency, and extremely reliable W-Fi connectivity. Backward compatibility with support for 2.4 GHz, 5 GHz bands in addition to the 6 GHz band. Wi-Fi 7 is also known as Extremely High Throughput (EHT) technology. Find more details with this link: https://www.hpe.com/us/en/what-is/wi-fi-7.html

The 802.11bn Wi-Fi 8 extension aims to provide significant advancements in enterprise networking, IoT, and consumer applications. The focus of this next-generation standard is on enhancing user experience through reliability rather than merely increasing speed. It is designed to complement cellular networks for seamless internet access and aims to improve communication through coordination of multiple access points and optimizing network throughput with better spectrum utilization. Wi-Fi 8 is also known as IEEE 802.11bn Ultra High Reliability (UHR) technology and expected to become available within a few years. Find more details with this link: https://www.hpe.com/us/en/what-is/wi-fi-8.html

2.4 GHz band and channels

First, let's look at the 2.4 GHz band. Originally, this band was used for most wireless traffic, even with three non-overlapping channels supporting 20+ MHz each. Due to overuse and interference, the 2.4 GHz has been relegated to guests and legacy devices. With the introduction of 802.11ax, there has been a resurgence of use for the space, given the new efficiencies the standard affords clients. However, no additional band expansion is planned beyond the 13 channels.

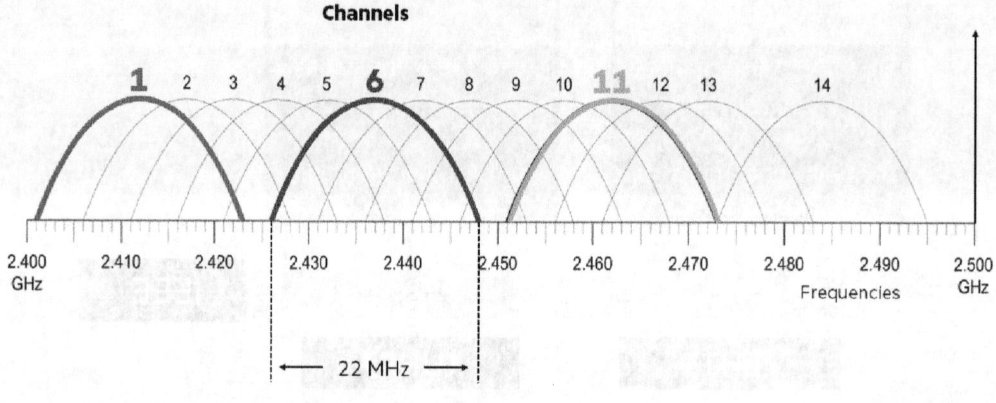

Signal spread over 20+Mhz
Adjacent channels are not usable in same physical RF area

The original standards 802.11 and 802.11b used a direct-sequence spread spectrum (DSSS) signal to reduce overall signal interference. Only 5 Mhz separates each channel, and because any channel will spread over 22 Mhz, channel overlap exists, as shown above. When an AP broadcasts on Channel 1, another transmission on channel 2 can be heard. Both radios will experience adjacent channel interference or ACI. ACI is like talking over the top of another person when having a conversation. Some or all the transmission may be corrupted due to channel overlap.

5 GHz allowed channels in FCC countries

The 5 GHz band is around 500 MHz wide, allowing devices to use up to six larger 80 MHz channels for multi-gigabit speeds. The Unlicensed National Information Infrastructure (U-NII) radio bands designate eight ranges. U-NII-1, U-NII-2, U-NII-2-Extended, and U-NII-3 are for the 5 GHz band. U-NII-5 through U-NII-8 are assigned to the 6 GHz range. IEEE's 802.11h dynamic frequency selection (DFS) restricts which channels you can use in the 5 GHz range when co-located with satellite and radar signals. Two of the six available 80 MHz channels could be unusable if radar interference is detected.

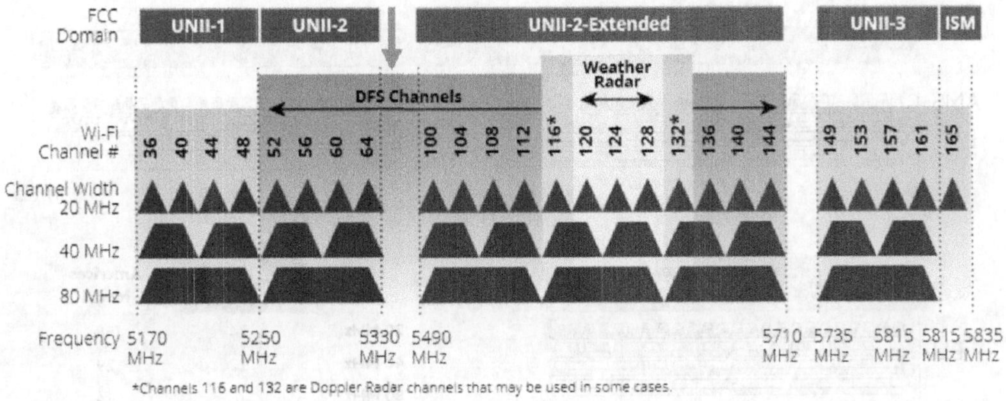

The 5 GHz band has become the predominant radio of choice for enterprise networks since wide-band was introduced in 802.11n. Bonding channels can aid in providing more bandwidth to a channel but becomes a point of diminishing return as fewer channels are available and the noise floor increases. In most high-density enterprise WLANs using 802.11ac or 802.11ax, 20 or 40 MHz channel widths are optimal, with 80 and 160 MHz channels being less practical due to lack of available bandwidth and increased noise floor. Notice that fewer channels become available as the width of channels increases.

Usable 6 GHz channels in Americas & Europe/CEPT

Wi-Fi 6E opens the new 6 GHz band and is available on the new AP-600 series of indoor HPE Aruba Networking access points. Throughout the Americas, regulators are opening 1,200 MHz in the 6 GHz band for unlicensed use. To put this another way, currently, there are more 80 MHz channels in the 6 GHz band than 40 MHz channels in the 5 GHz band. This allows planning 80 or 160MHz wide channels by default when using the 6 GHz band. For the first time, having wide channels and spatial reuse for enterprise deployments will be practical.

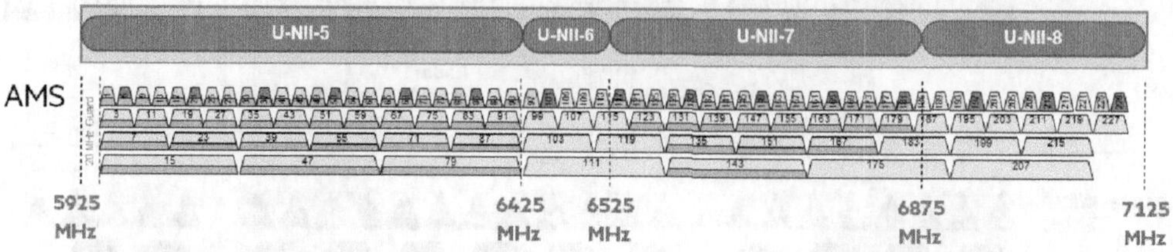

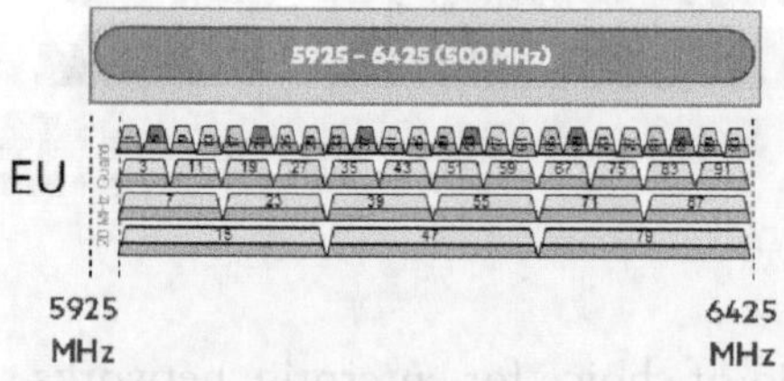

	European Model	Americas Model
20 MHz	24	58
40 MHz	12	29
80 MHz	6	14
160 MHz	3	7

In other regulatory domains, such as Europe and Australia, only the first 500 MHz portion (5925-6425 MHz) has been adopted for unlicensed use. With that said the global organization Dynamic Spectrum Alliance published an article in 2022 urging regulators to consider opening the entire band because of the importance of permitting the unlicensed use of the full 1,200 MHz. The initial product focus is on the low-power indoor (LPI) device class, where access points are limited to indoor operation only. LPI has access to all 1,200 MHz of the frequency, has no DFS requirement, has no spectrum database requirement, and has a massive amount of additional spectrum and capacity.

Note

Technically, channel 233 would be the 59th channel, and you may find references to it in other places, but power will be very low and likely realistically unusable due to band edge limitations.

6 GHz: massive increase in capacity

- More than 2x the existing amount of unlicensed spectrum
- Up to seven 160 MHz wide channels
- Up to three 320 MHz wide channels
- 6 GHz is greenfield - no need for backwards compatibility

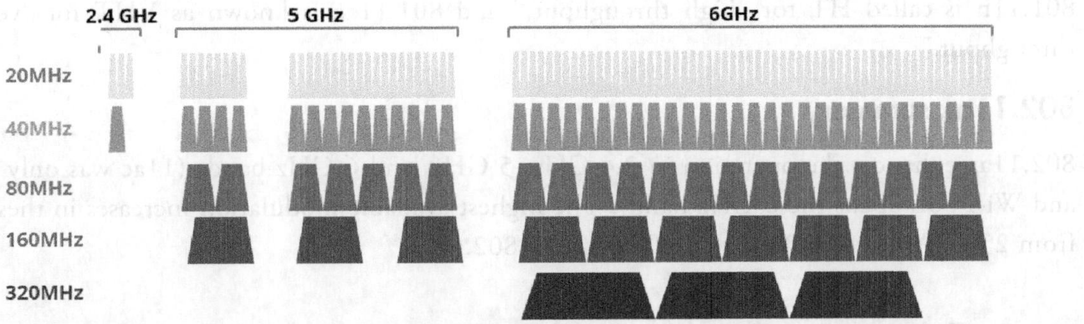

With 6 GHz, we get a massive increase, where some countries are getting more than two times the existing capacity. With seven channels of 160 MHz, spatial re-use becomes more practical, where APs and their connected clients can operate in parallel with other APs and their clients while still having enough channel separation that the same channel can be re-used without causing interference. In most of the EU, Britain, and Australia, spatial re-use of the wide channel 160 MHz is unrealistic. This is because those regions are limited to only about 500 MHz of frequency allocation.

With Wi-Fi 7 up to three 320 MHz channels are available in 6 GHz band.

802.11n / 802.11ac summary

11n supports both the 2.4 and 5 GHz bands and introduces wider 40 MHz channels, better modulation, and an additional four spatial streams. Beam forming has some options, and 802.11.n is also backward compatible with 11. a/b/g.

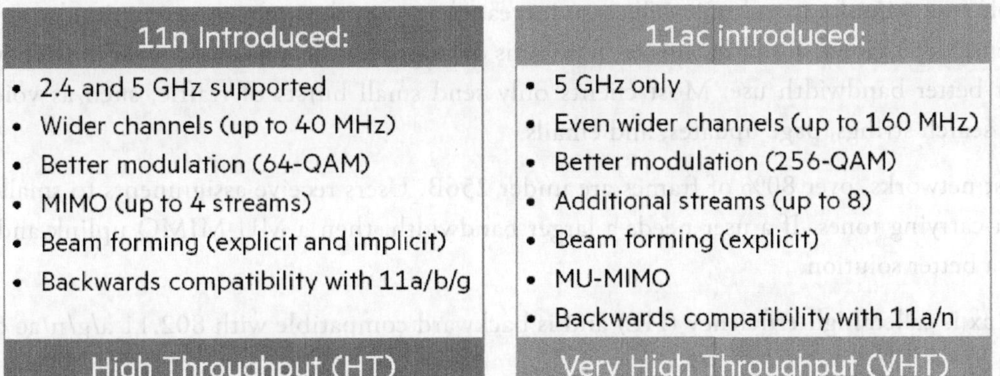

11ac supports only the 5 GHz bands and introduces wider 160 MHz channels, but in enterprise network, HPE Networking does not recommend its use. Better modulation with 256-QAM and up to eight spatial streams. Beam forming is better defined with less options, and it introduces multi-user MIMO. 802.11.ac is also backward compatible with 11. a/b/g/n.

801.11n is called HT for "high throughput," and 801.11ac is known as VHT for "very high throughput."

802.11ax summary

802.11ax enhances the operation in 2.4 GHz, 5 GHz, and 6 GHz bands (11ac was only 5 GHz, and Wi-Fi 6E opens the 6 GHz band). The highest available modulation increases in these bands from 256 QAM in 802.11ac to 1024 QAM in 802.11ax.

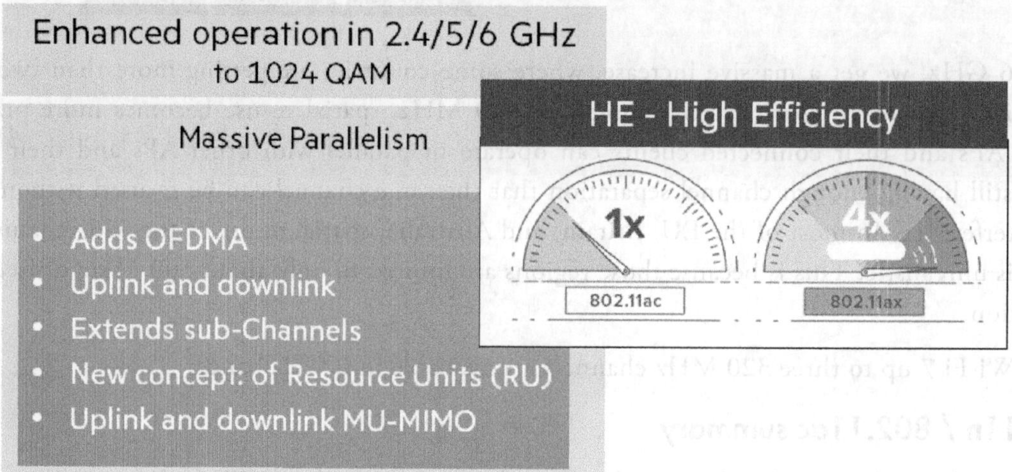

OFDMA introduces uplink and downlink capabilities and extends the subchannels from 64 to 256. OFDMA also introduces the concept of Resource Units (RUs). This allows transmissions to multiple users simultaneously, as well as an increased average throughput per station by at least 4x in a dense deployment scenario. It also maintains or improves the stations' power efficiency, allowing for better bandwidth use. Most clients only send small bursts of traffic, such as voice calls, simple search strings, page updates, and emails.

In most networks, over 80% of frames are under 256B. Users receive assignments to smaller units of data-carrying tones. If a user needs a larger bandwidth, then a MU-MIMO uplink and downlink is a better solution.

802.11ax is called high-efficiency (HE) and is backward compatible with 802.11 a/g/n/ac devices.

802.11be summary

Key takeaway: Wi-Fi 7 standard enhances use of the 6 GHz band

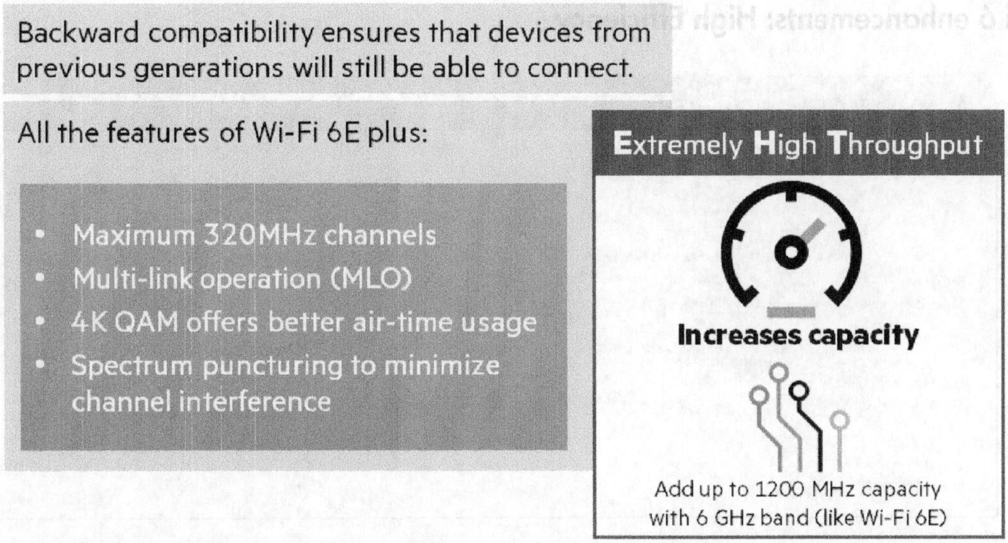

Wi-Fi 7 is here, standard finalized by Wi-Fi alliance in January, and we've seen some product announcements. Some of the cool technology enhancements are

- It brings more speed, capacity and reliability with 320 MHz channels that are double Wi-Fi 6E (and 4 times Wi-Fi 6's 80 MHz) – this means 2x higher throughput (data and speed)
- MLO that aggregates channels across different frequency bands at the same time, for improved load balancing/more efficient, and better reliability.
- And 4K QAM that more densely embeds data into each signal with 20% higher transmission rates
- And like 6E it takes advantage of the 6 GHz band 1200 MHz capacity in the Americas and 500 MHz in Europe and parts of Asia like Japan

Wi-Fi 6 enhancements: High Efficiency

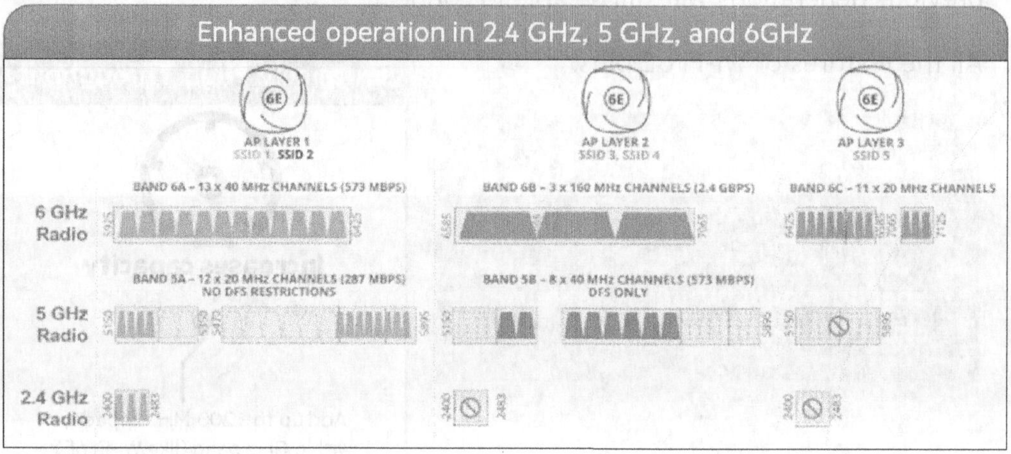

The largest challenge in today's Wi-Fi enterprise is managing the increased density of devices and diversity of applications while still improving performance. 802.11ax brings a four-fold increase in throughput capacity over the previous 802.11ac standard.

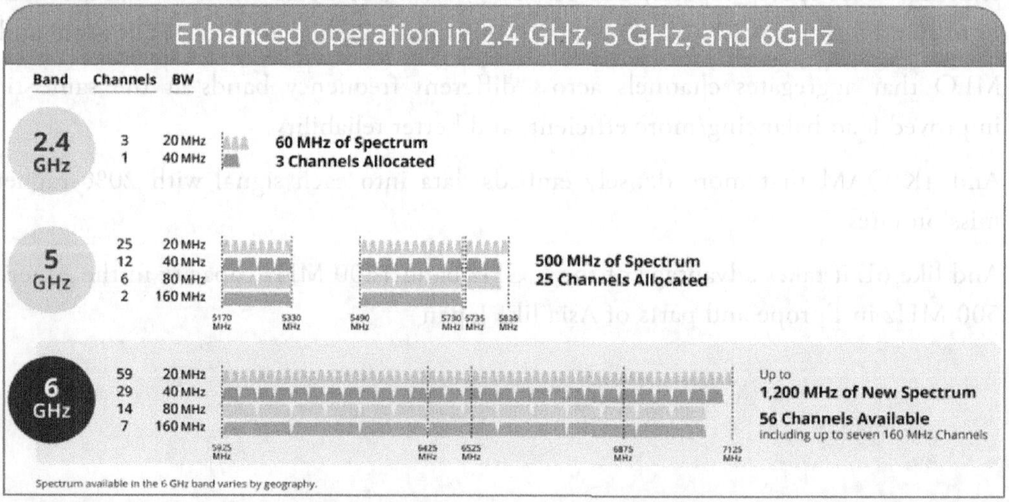

Wi-Fi 6E includes use of the 6 GHz band along with both the 2.4 GHz and 5 GHz bands. When designing customer networks, utilize all three bands where available. Keep in mind that while the 6 GHz band offers a huge wireless expansion of channels, many clients devices will not yet support the band and if they do, the range is currently allocated for indoor use only.

Wi-Fi 6 enhancements: OFDMA

The premiere multi-user feature enabled with 802.11ax is Orthogonal Frequency Division Multiple Access (OFDMA). This allows an AP to send data to multiple clients simultaneously. Channel utilization is enhanced by forwarding each wireless transaction using smaller sub-carriers or resource units called RUs. The result is that clients sharing the same channel are no longer competing for airtime and bandwidth. OFDMA works regardless of device or traffic type. While this feature is enabled automatically, it does require clients to support 802.11ax Wi-Fi 6.

NUMBER OF CONCURRENT CLIENTS PER CHANNEL PER RADIO*	
Prior Wi-Fi generations	**1 at a time**
Wi-Fi 6 in 20 MHz	Up to 9 at a time
Wi-Fi 6 in 40 MHz	Up to 18 at a time
Wi-Fi 6 in 80 MHz	Up to 37 at a time
Wi-Fi 6 in 160 MHz	Up to 74 at a time

* Client density varies based on configured network settings

The table illustrates the number of available RUs per Wi-Fi channel width used—the wider the channel, the more clients can be served. For example, a 20 MHz wide channel can support up to nine clients, while an 80 MHz channel can simultaneously serve up to 37 clients.

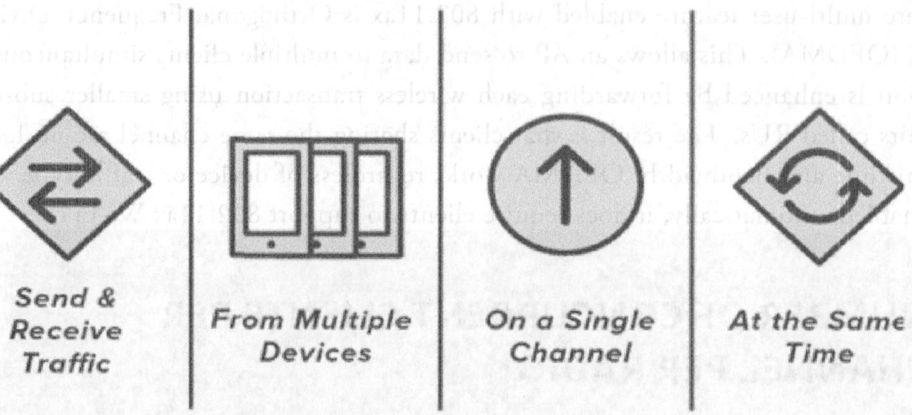

802.11ax provides massive parallelism. It minimizes contention between devices since each is scheduled to transmit data in parallel. OFDMA assigns smaller sections of the channel bandwidth (resource units) to the clients. The RUs can be flexible at each transmission, which allows simultaneous transmission to multiple users with various bandwidth needs. All packet types enjoy improved performance, which is especially critical for latency-sensitive traffic such as voice.

In dense environments, traffic can be pooled into a transport, allowing multiple conversations to occur simultaneously. Another concern with high-density deployments is overlapping BSSs. These dense deployments can block each other from transmitting. Leveraging OFDMA can improve spatial reuse, allowing devices in the same channel to transmit without interrupting other localized devices.

Wi-Fi 6 enhancements: MU-MIMO

Another feature that handles traffic from multiple devices is multi-user MIMO (MU-MIMO). This feature was originally introduced in 802.11ac. For 802.11ax, the standard supports uplink and downlink client traffic across eight spatial streams, allowing up to eight devices to transmit simultaneously using a dedicated stream per device. This allows large packets, such as streaming HD video, to be handled more efficiently, while shorter packets from IoT devices and voice traffic could be better handled using OFDMA.

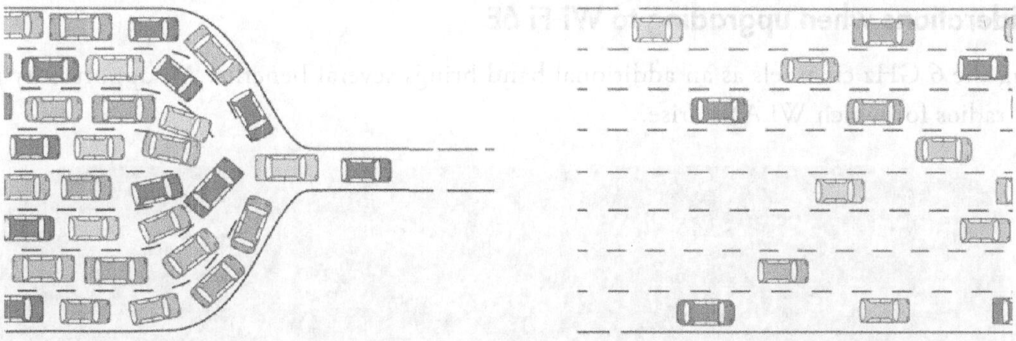

Multi-user MIMO is enabled by default, and recommended to keep it enabled. More special streams result in higher data rates when communicating with a single client and higher aggregate performance with multiple clients simultaneously. Example of single and dual stream clients for MU-MIMO

Wi-Fi 6 enhancements: Transmit Beamforming and TWT

Transmit Beamforming

- Optimizes RF for each of up to eight antennas
- Focuses signal towards each user

Transmit Beamforming can improve data rates and extend the range of signals to specific clients. Wi-Fi 6 increased the number of antennas supported from four to eight. 802.11ax utilizes an explicit beamforming procedure using channel-sounding with a null data packet to compute the antenna weights and optimize RF for each user, similar to the functionality in 802.11ac. Keep this feature enabled.

Target Wake Time (TWT)

- Schedules each client wake time based on expected traffic
- Saves client battery

Target Wake Time (TWT) uses negotiated policies based on expected traffic activity between clients and APs to schedule wake times for each client. Mobile client devices can go into an inactive mode. This extends the battery life of smartphones, tablets, and IoT devices as an underlying benefit. The result is less congestion, energy savings, and an overall better experience.

Leave this feature enabled clients to keep their radios in a sleep state longer and save battery.

Considerations when upgrading to Wi-Fi 6E

Adding the 6 GHz channels as an additional band brings several benefits. An opportunity to plan which radios for which WLANs arise.

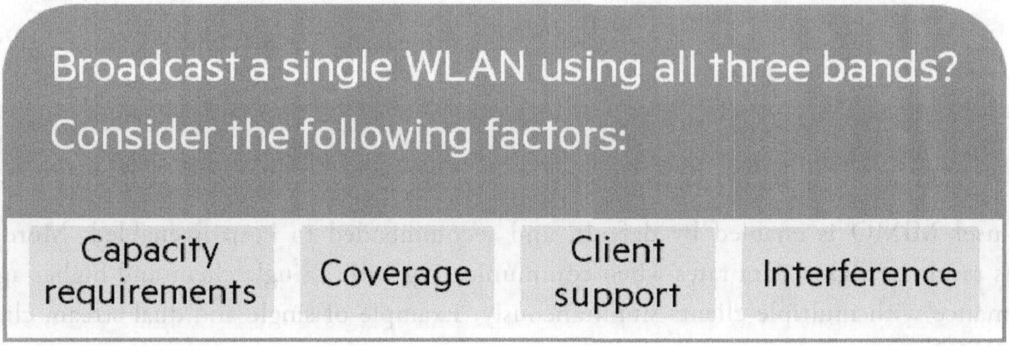

When deciding to broadcast a single WLAN using all three bands, consider the following factors:

Client device support: Not all clients support the 6 GHz band. It may not benefit a location or AP group to enable the 6 GHz band if most devices cannot connect.

Capacity requirements: The 6 GHz band offers the greatest number of non-overlapping channels, resulting in higher data rates, increased capacity, and reduced congestion. If a network is capable, enabling all three bands can help distribute user load effectively.

Interference: The 6 GHz band provides a clean spectrum for better performance than the 5 GHz band. The 2.4 GHz remains susceptible to non-Wi-Fi interferences such as Bluetooth and microwaves. Enabling all three bands or suppressing the 2.4 GHz could help mitigate interference issues.

Coverage: Consider the propagation of the 6 GHz band compared to the 2.4 GHz or 5 GHz bands. The 6 GHz has a slightly shorter range than the 5 GHz and, as a result, is more susceptible to attenuation and absorption issues with obstacles. This should be factored into AP placement strategies and power settings to ensure optimal coverage.

Enabling all three radio bands

Factors that can affect all WLANs when enabling all three bands:

- SSID management
- Band steering
- Application-specific WLANs

Consideration when enabling all three radio bands:

SSID management

For convenience, admins will advertise all WLANs on all bands may lead to increased management overhead and unnecessary SSID clutter that users must navigate. Always limit the number of SSIDs per band to minimize this overhead.

Band steering

Used to move 2.4 GHz clients to a more efficient 5 GHz or 6 GHz bands.

Application-specific WLANs

Consolidate any WLANs designed to host one specific application, such as voice or video. Also, groups of users that can be consolidated based on authentication should be, for example, different types of guests that all use captive portals (guests, volunteers, personal employee devices). HPE Aruba Networking wireless can restrict users by policy regardless of whether they are connected to the same SSID or channel, even in the same VLAN subnet.

Device classes in 6GHz

The 6 GHz band is broken up into device classes by a regulatory framework with the intention of protecting 6 GHz incumbents, specifically fixed-satellite service (FSS).

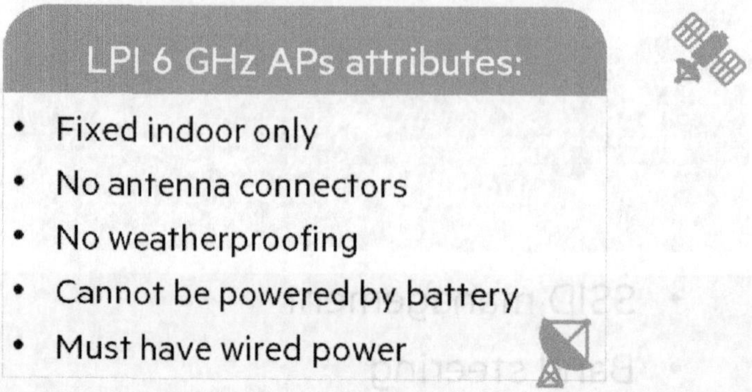

While there is a Very Low Power (VLP) device designation for minimal signal power, HPE Aruba Networking focuses on LIP and Standard power devices.

Low Power Indoor (LPI) APs

Indoor deployments with fixed installation. Regulators prohibit LFI APs from having connectors, weatherproofing, or operating by battery; they must use wired power. This is all to prevent outdoor use and risk interference with existing fixed-satellite service (FSS).

LPI APs

- Fixed indoors only
- No antenna connectors
- No weatherproofing
- Cannot be powered by battery
- Must have wired power

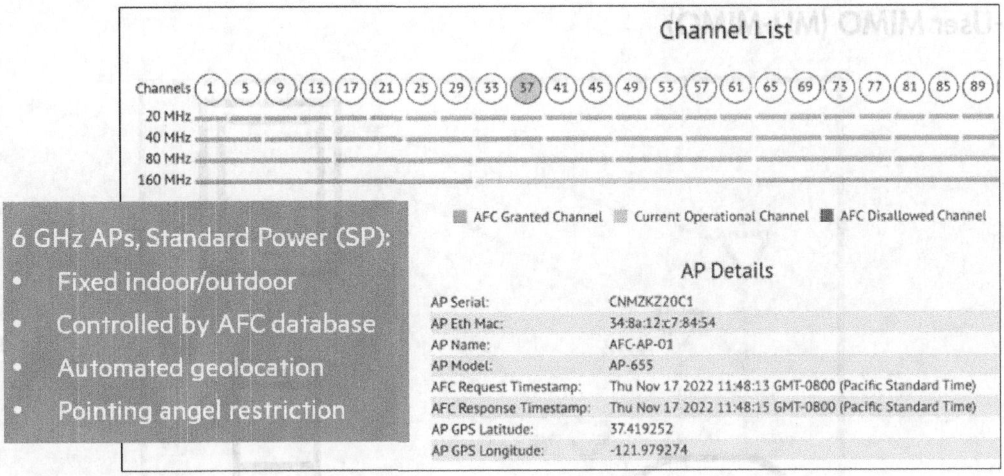

Standard Power (SP) APs

These APs require fixed installation and a permanent mounting to a structure. The Automated Frequency Coordination (AFC) system is used by APs with a CBRS spectrum access system (SAS).

SP APs enable GPS to detect their location automatically. This further avoids interference with incumbents, allowing further utilization of the mid-6 GHz band.

Admins can check AFC granted channels or AFC disallowed channels that may be blocked due to operating incumbent devices that are nearby.

SP APs

- Fixed indoor/outdoor
- Controlled by AFC database
- Automated geolocation
- Pointing angel restriction

Multi-User MIMO (MU-MIMO)

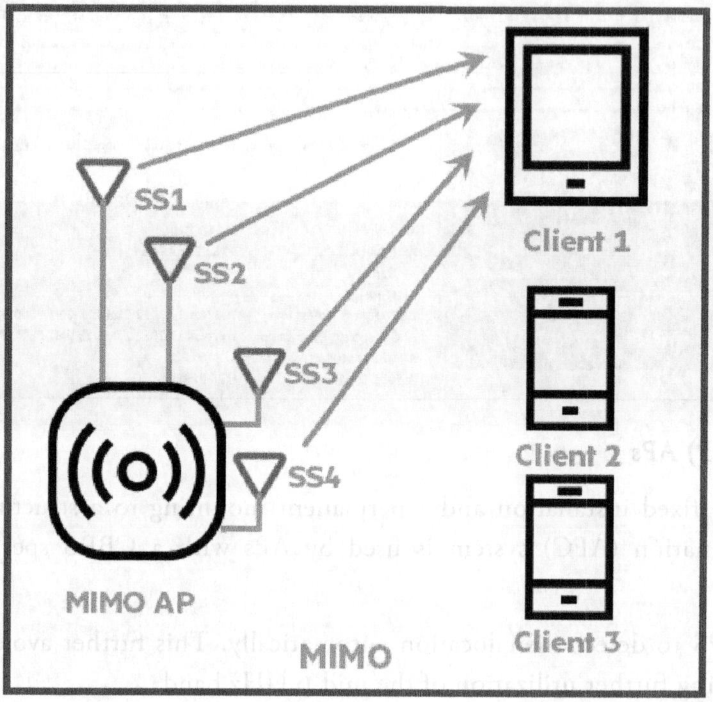

The 802.11ac standard also introduced multi-user MIMO (MU-MIMO). In the 802.11n and 802.11ac Wave 1 standards, all wireless communication is either point-to-point (one-to-one) or broadcast (one-to-all). 802.11ac Wave 2 supported MU-MIMO for downlink traffic only. 802.11ax allows both uplink and downlink MU-MIMO for up to eight spatial streams. Most device chipsets will only do six.

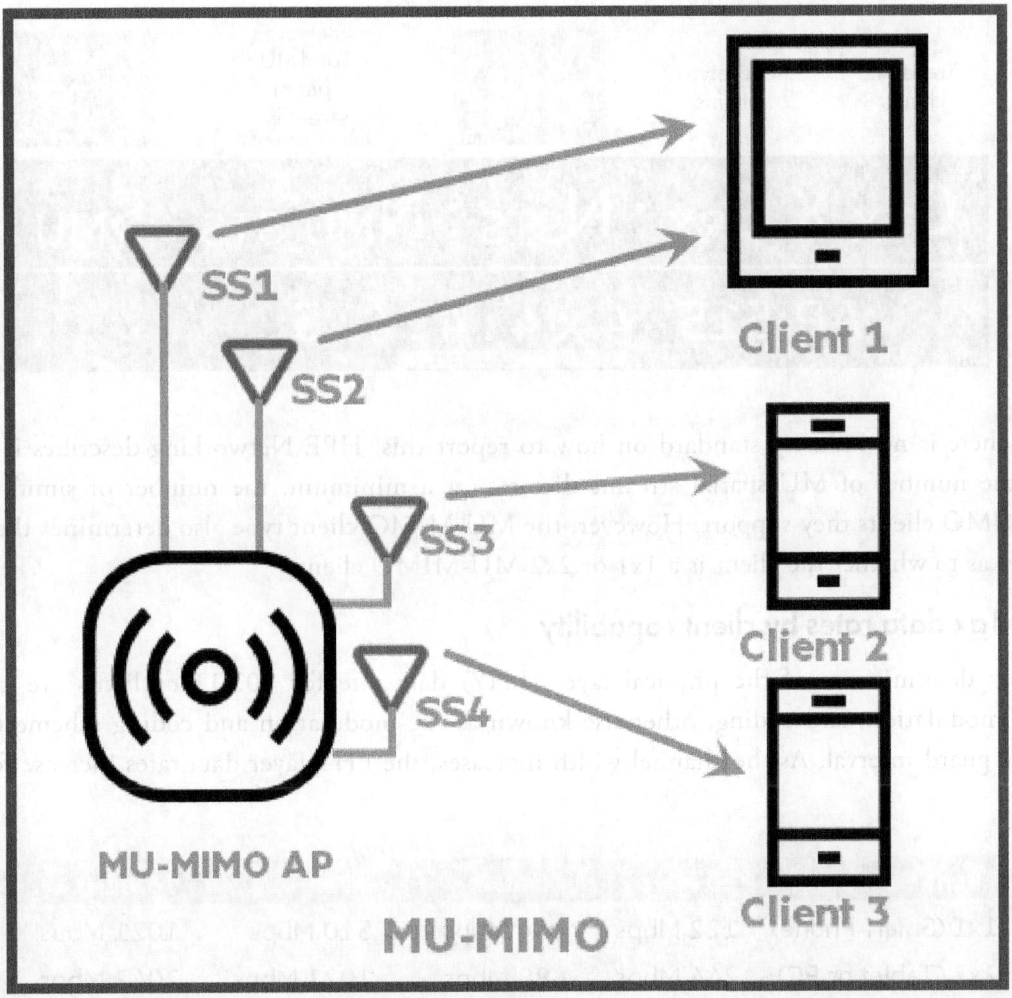

With MU-MIMO, an AP uses its different streams to transmit data to multiple clients at the same time. Instead of using multiple spatial streams between a given pair of devices, spatial diversity can send multiple data streams between several devices at a given instant.

MU-MIMO spatial streams

APs with MU-MIMO have two variables to be aware of when creating a MU-MIMO design. In addition to the number of transmit chains, receive chains, and total spatial streams (the 4x4:4 description), it has the number of multi-user (MU) spatial streams and the total number of multi-user (MU) clients it supports. 802.11ax AP is also capable of uplink and downlink MU-MIMO.

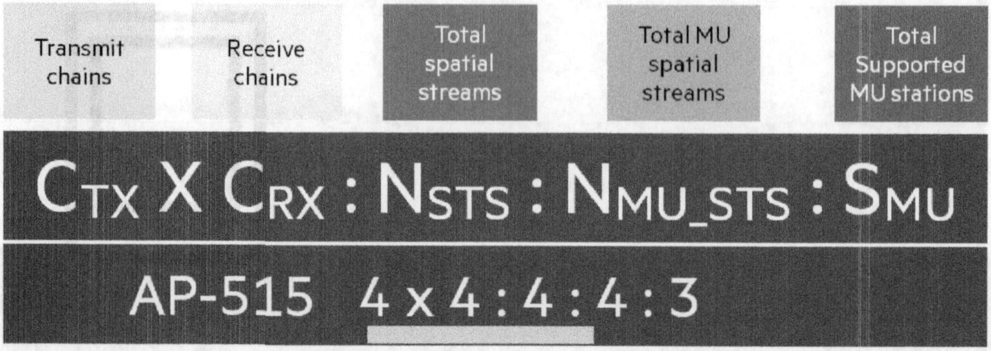

While there is no industry standard on how to report this, HPE Networking describes it as follows: the number of MU spatial streams dictates, at a minimum, the number of simultaneous MU-MIMO clients they support. However, the MU-MIMO client type also determines the client number as to whether the client is a 1x1 or 2x2 MU-MIMO client.

802.11ax data rates by client capability

The key determinants of the physical layer (PHY) data rate for 802.11ac clients are channel width, modulation, and coding, otherwise known as the modulation and coding scheme (MCS) and the guard interval. As the channel width increases, the PHY layer data rates increase for that client.

Client Capability	20 MHz	40 MHz	80 MHz	160MHz
1x1 (Smart Phone)	122 Mbps	244 Mbps	510 Mbps	1021 Mbps
2x2 (Tablet or PC)	244 Mbps	488 Mbps	1021 Mbps	2042 Mbps
3x3 (High end PC)	366 Mbps	731 Mbps	1531 Mbps	3063 Mbps

You can look up details about what features clients support at the site shown here.

- **Channel width:** You can configure an 802.11ax AP to use 20 MHz, 40 MHz, 80 MHz, or 160 MHz channel width.
- **Modulation and coding:** All the earlier options are still available and are used if SNR is too low to sustain the highest rate. However, in the modulation and coding scheme (MCS) table on the next slide, the 802.11n modulation extends to add 256-QAM options with coding of 3/4 and 5/6.
- **Guard interval:** 802.11n and 802.11ac had a long guard interval of 800 ns mandatory, while the short guard interval of 400 ns is an available option. The guard interval is the pause between transmitted RF symbols. It is necessary to avoid multipath reflections of one

symbol from arriving late and interfering with the next symbol. 802.11ax defines three guard intervals of .8us (800 ns), 1.6us, and 3.2us. The longer guard intervals enhance delay spread protection.

Minimum RSSI requirements by data rate

HPE Networking recommends a minimum Received Signal Strength Indicator (RSSI) of -55 dBm throughout a coverage area. RSSI measures how well clients can hear a signal from a given AP. A relatively strong RSSI value of -55 dBm allows the AP and client to negotiate a reliably fast Modulation and Coding Scheme (MCS) of MCS11 data rate on 40 MHz high-density deployments. MCS indicates the transmit and receive rates and the modulation or technology used. Wi-Fi 6 clients that negotiate a poor MCS rate due to interference or free path loss will roll back to older technologies like Wi-Fi 5 and use a lower transmit and receive rate for successful data transmission.

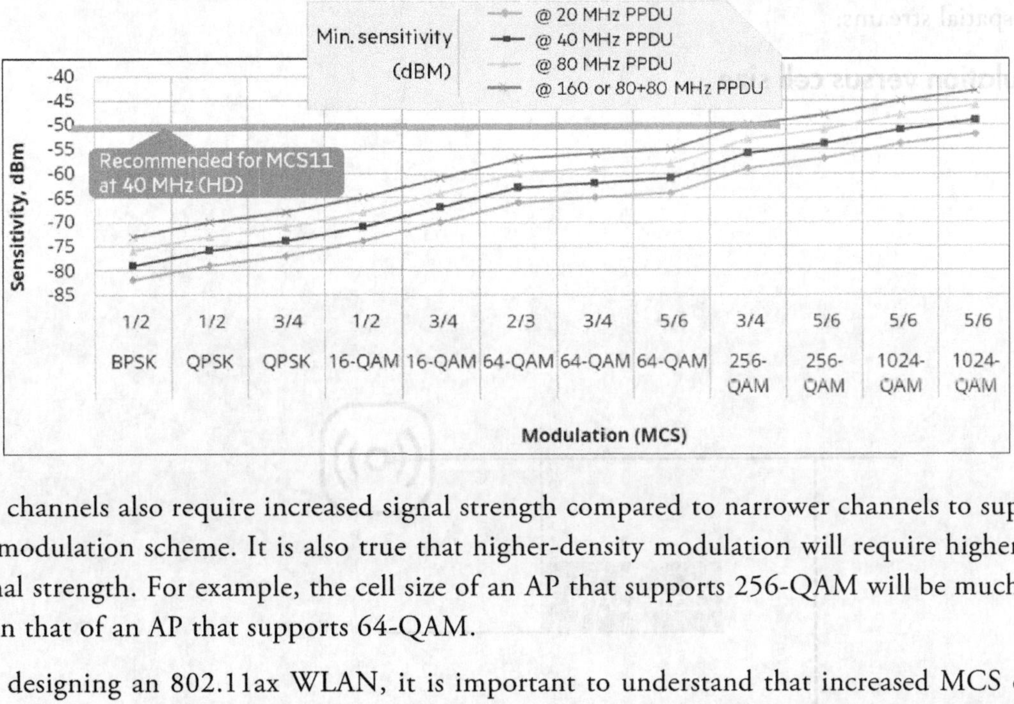

Wider channels also require increased signal strength compared to narrower channels to support a given modulation scheme. It is also true that higher-density modulation will require higher levels of signal strength. For example, the cell size of an AP that supports 256-QAM will be much smaller than that of an AP that supports 64-QAM.

When designing an 802.11ax WLAN, it is important to understand that increased MCS coding comes at a price in terms of bits/sec per hertz spectrum. The required signal level for good reception increases with the complexity of modulation and the channel bandwidth. For instance, this graph shows that a 20 MHz 1xSS MCS 9 (256-QAM 5/6) at a rate of 115 Mbps could be achieved with a minimum RSSI of -57 dBms. If you increased the channel width to 80 MHz using the same modulation rate across 2xSS, you could achieve speeds of 961 Mbps but would require an RSSI of -51 dBms. The difference of 6 dBs translates to 4x the signal strength required. This shows that as the client distance from the AP increases, the RSSI and the data rate will decrease as well.

802.11ax selected rates (Mbps, short GI)

MCS	Modulation & Rate	20 MHz 1x SS	20 MHz 2x SS	20 MHz 4x SS	20 MHz 8x SS	40 MHz 1x SS	40 MHz 2x SS	40 MHz 4x SS	40 MHz 8x SS	80 MHz 1x SS	80 MHz 2x SS	80 MHz 4x SS	80 MHz 8x SS
0	BPSK 1/2	8.6	17.2	34.4	68.8	17.2	34.4	68.8	137.6	36.0	72.1	144.1	288.2
1	QPSK 1/2	17.2	34.4	68.8	137.6	34.4	68.8	137.6	275.3	72.1	144.1	288.2	576.5
2	QPSK 3/4	25.8	51.6	103.2	206.5	51.6	103.2	206.5	412.9	108.1	216.2	432.4	864.7
3	16-QAM 1/2	34.4	68.8	137.6	275.3	68.8	137.6	275.3	550.6	144.1	288.2	576.5	1,152.9
4	16-QAM 3/4	51.6	103.2	206.5	412.9	103.2	206.5	412.9	825.9	216.2	432.4	864.7	1,729.4
5	64-QAM 1/2	68.8	137.6	275.3	550.6	137.6	275.3	550.6	1,101.2	288.2	576.5	1,152.9	2,305.9
6	64-QAM 3/4	77.4	154.9	309.7	619.4	154.9	309.7	619.4	1,238.8	324.3	648.5	1,297.1	2,594.1
7	64 QAM 5/6	86.0	172.1	344.1	688.2	172.1	344.1	688.2	1,376.5	360.3	720.6	1,441.2	2,882.4
8	256-QAM 3/4	103.2	206.5	412.9	825.9	206.5	412.9	825.9	1,651.8	432.4	864.7	1,729.4	3,458.8
9	256-QAM 5/6	114.7	229.4	458.8	917.6	229.4	458.8	917.6	1,835.3	480.4	960.8	1,921.6	3,843.1
10	1024-QAM 3/4	129.0	258.1	516.2	1,032.4	258.1	516.2	1,032.4	2,064.7	540.4	1,080.9	2,161.8	4,323.5
11	1024-QAM 5/6	143.4	286.8	573.5	1,147.1	286.8	573.5	1,147.1	2,294.1	600.5	1,201.0	2,402.0	4,803.9

The table above shows the 802.11ax MCS and modulation rates spread across 20 to 80 MHz up to 8x spatial streams.

Modulation versus cell size

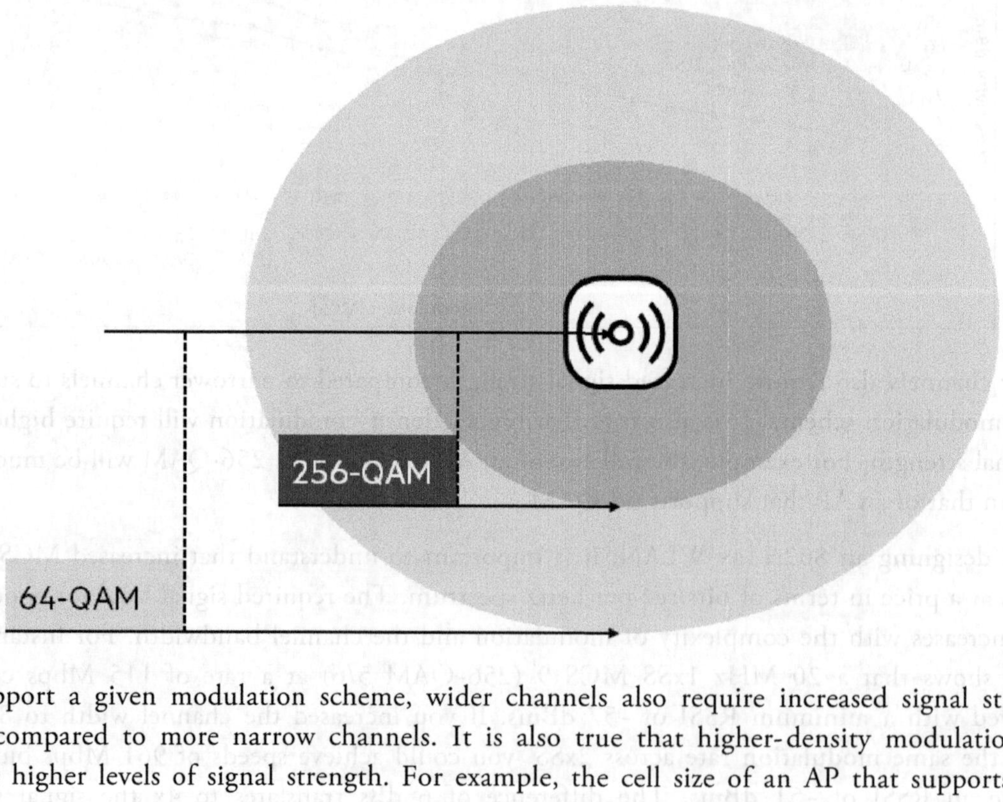

To support a given modulation scheme, wider channels also require increased signal strength when compared to more narrow channels. It is also true that higher-density modulation will require higher levels of signal strength. For example, the cell size of an AP that supports 256-QAM will be much smaller than that of an AP that supports 64-QAM.

Wireless collision domain

Two nearby stations will interfere with one another if they attempt to transmit simultaneously on the same frequency. This concept of a collision domain describes a physical area in which 802.11 devices attempt to send on the same channel and, simultaneously, can decode one another's frame preambles. Transmitting at different times or on different channels results in no collision.

Carrier Sense Multiple Access / Collision Avoidance (CSMA/CA) is a protocol for carrier transmission for devices using 802.11 standards. The goal is to prevent collisions by listening to the broadcasting nodes and informing devices not to transmit data until the channel is free.

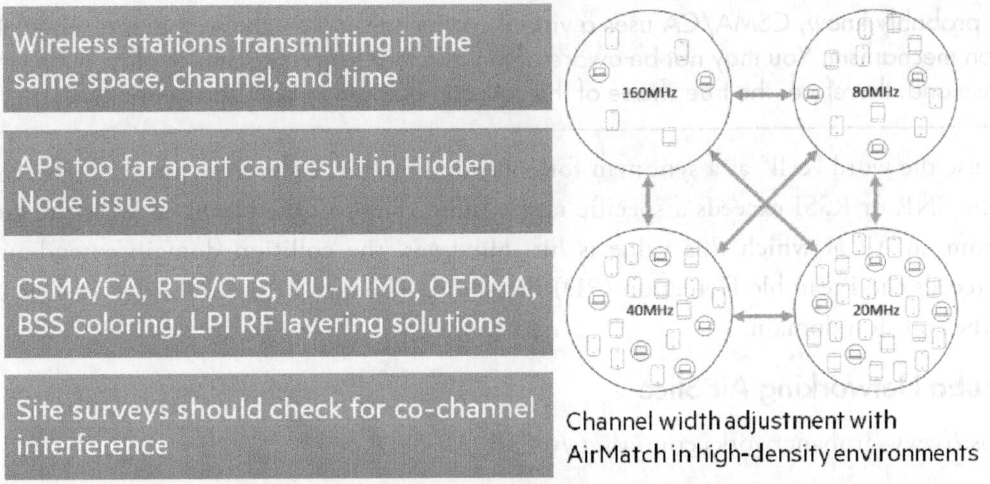

Channel width adjustment with AirMatch in high-density environments

Hidden Node issues can result when APs are too far apart. This situation produces client stations that can reach the AP but are too far from other stations to detect transmissions, resulting in collisions. Request To Send / Clear To Send (RTS/CTS) is a mechanism that can somewhat alleviate hidden node problems. A best practice is to provide enough capacity by avoiding placing APs too far apart.

All wireless stations associate with their AP radio in a contention or collision domain. New efficiency features allow APs and clients to avoid using the entire transmission channel and minimize collisions. This allows for separate spatial streams between multiple clients, as introduced in 802.11ac with MU-MIMO. The most recent standard 802.11ax, allows for sub-carriers within a channel (RUs) with OFDMA and other features like BSS coloring and LPI RF layering solutions (covered later).

HPE Aruba Networking AirMatch in HPE Aruba Networking Central automates channel and EIRP optimization of a deployment. This includes expanding channel widths between 20 MHz, 40 MHz, 80 MHz, and 160 MHz to maximize bandwidth capacity or narrowing towards 40

MHz and 20 MHz should device density increase. EIRP is evenly distributed across all APs to optimize coverage and roaming performance. Finally, channels are automatically assigned to mitigate co-channel interference (CCI) or avoid neighboring clients or radar conditions.

Manual adjustments to AirMatch can be made to ensure that only channels supported by clients are used or to avoid known sources of interference. When doing an 'AP-on-a-stick' site survey, always check for co-channel interference.

Note

As you probably know, CSMA/CA uses a virtual carrier sensing mechanism and a physical energy detection mechanism. You may not be aware of the role that frame preambles play in the virtual carrier sense and, therefore, the true shape of the collision domain in both space and time.

Do not use the word "cell" as a synonym for collision domain. Cells are typically engineered areas where the SNR or RSSI exceeds a specific target value. The so-called "cell-edge" is the radial distance from an AP at which this value is hit. However, the collision domain extends until the SINR exceeds the Preamble Detection (PD) threshold. The area of the cell is far smaller than the area of the collision domain.

HPE Aruba Networking Air Slice

ref: https://www.arubanetworks.com/assets/tg/TB_Air-Slice.pdf

Air Slice is a RF technology providing SLA_grade application assurance. It does so by leveraging HPE Aruba Networking's Policy Enforcement Firewall to match latency-sensitive applications such as AR/VR, Zoom, and Slack. Once matched, Air Slice allocates radio resources, such as time, frequency, and spatial streams, to guarantee performance, increase throughput for high-bandwidth applications, and provide an optimized experience for IoT devices. This is especially effective in mixed-traffic environments where sensitive applications contend for airtime with other services such as basic Internet traffic, messaging, and non-critical services.

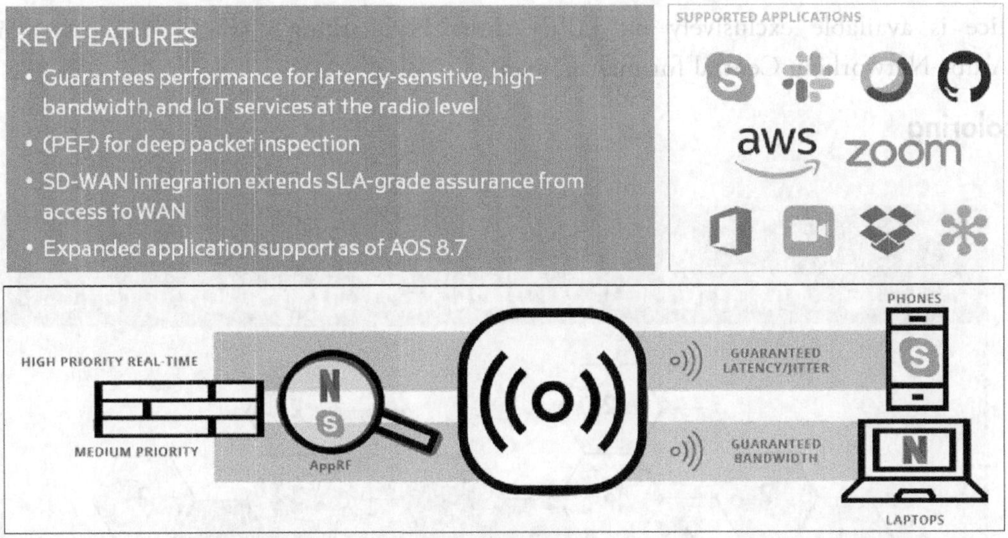

RF Orchestration

Air Slice and all HPE Aruba Networking RF technologies can be configured through HPE Aruba Networking Central cloud management. This allows customization based on network location.

Context Awareness

HPE Aruba Networking PEF enables role-based context for policy customization based on user role, device type, and granular application visibility.

Guaranteed Radio Resources

802.11ax introduces modern network scheduling capabilities like OFDMA RUs, layering with 6 GHz, MU-MIMO upstream and down, and BSS coloring. Air Slice uses these radio resources to assign priority to applications.

Custom Application Support

Air Slice supports custom flow definitions for user applications or industry-specific services.

Wi-Fi 6 and Non-Wi-Fi 6 Client Support

Air Slice ties in with HPE Aruba Networking PEF and ClientMatch to improve the client experience, whether that client is Wi-Fi 6 or earlier. To ensure minimal contention, clients are grouped based on capability and the assigned radio resources.

Integration with HPE Aruba Networking SD-WAN

Air Slice works in tandem with HPE Aruba Networking SD-WAN configurations to enable SLA-grade application assurance across the network and QoS.

Air Slice is available exclusively on HPE Aruba Networking's Wi-Fi 6 access points via HPE Aruba Networking Central for management.

BSS coloring

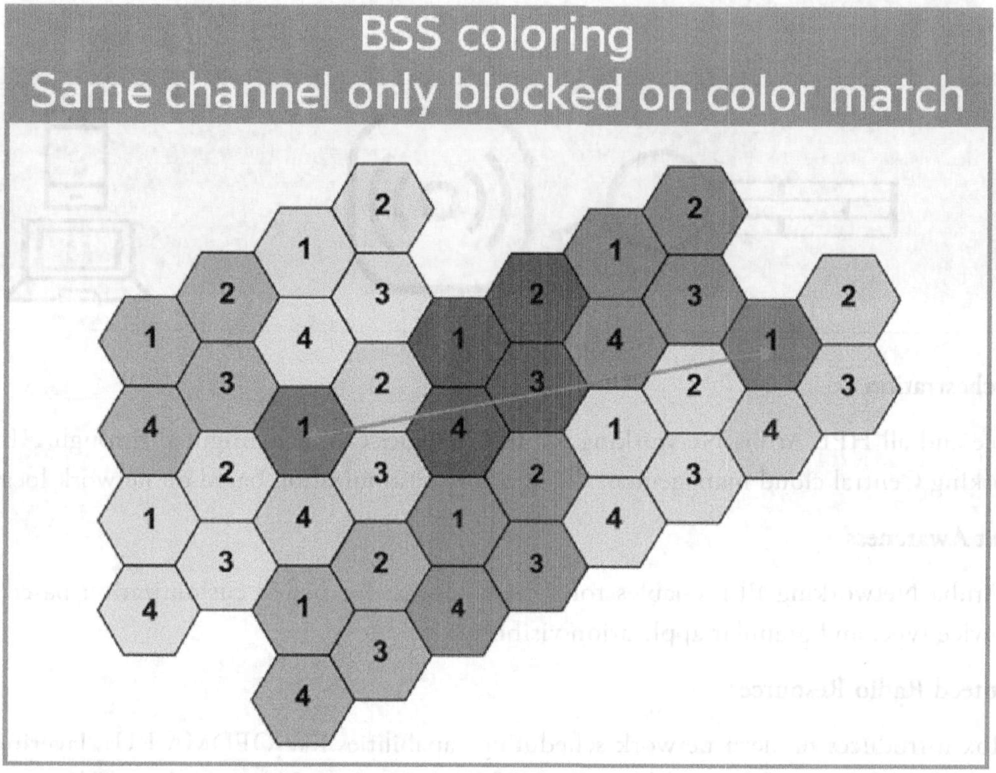

BSS coloring allows the network to assign a "color" tag to a channel and reduce the threshold for interference. Network performance improves because APs on the same channel can be closer together and still transmit at the same time if they are different colors. The field is 6-bits, so there are 63 different colors available.

Outdoor APs: link budget

The link budget calculation is a predicted maximum value of the power generated at the transmitter, lightning arrestor attenuation, antenna gain, Free Space Path Loss (FSPL), and client antenna gain. The actual signal level may be lower due to the unconsidered outside variables. The link budget differs in the opposite direction.

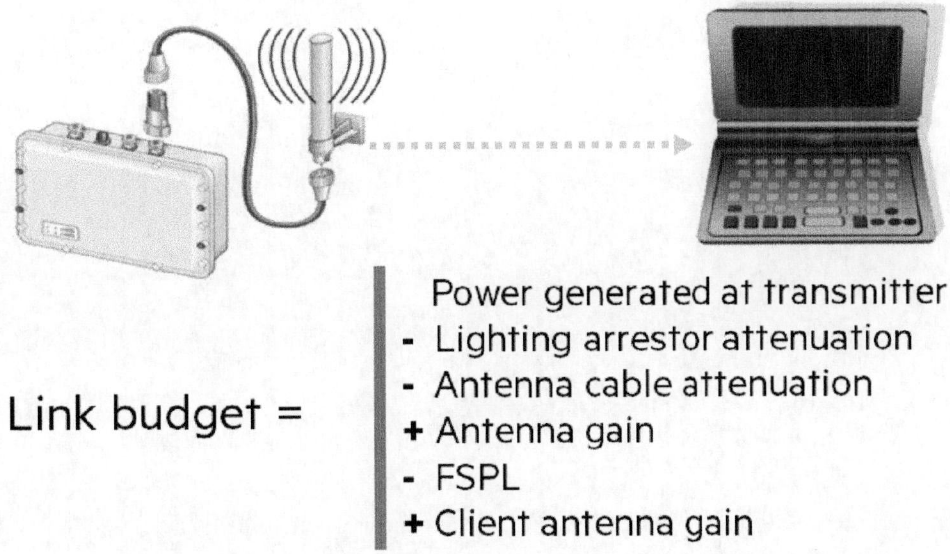

You may need to calculate a link budget for long-distance outdoor Wi-Fi deployments, especially point-to-point links. It is useful to understand how the link budget calculation ensures the Wi-Fi receiver connectivity in outdoor environments, especially those with long-distance RF links.

4 Wireless High-Density Design

With general RF coverage for the campus finished, turn your attention now to the unique challenges of very high-density areas.

OBJECTIVES

After completing this chapter, you should be able to:

- ✓ Overcome high-density challenges.
- ✓ Design for HD & VHD RF coverage.
- ✓ Choose optimal APs for VHD.
- ✓ Solve a VHD scenario using examples.

High-Density Challenges

Let's start by looking at the challenges of high-density and very high-density customer WLANs.

High-density (HD) WLANs today

HPE Networking published the industry's first design guide for high-density or HD WLANs in 2010. At the time, IEEE 802.11n high throughput (HT) APs were the subject of that original design guide. Later, in 2015, IEEE 802.11ac Very High Throughput (VHT) access points replaced the previous 802.11n generation. The focus on wider channels and greater throughput was at odds with the growing density of business and office locations.

Historically, most customers considered a "dense" network to have a capacity design of 2,500 ft² (232 m²) per AP. This would use a cell edge RSSI of -65. These small cells were generally designed to support one device per user for standard data. The density was about 25 devices per cell, assuming an average of 100 ft² (9 m²) per person.

By those standards, today's office network is considered "high-density," given the proliferation of BYOD, IoT devices, smartphones, and tablets. Modern wireless networks often support at least three devices per person or 75 devices per cell, yet the cell sizes remain as they were.

Today, architects are designing networks to support five to seven Wi-Fi-enabled devices per user. For example, a 20-seat board room could contain 60 user devices, 5-10 fixed multimedia devices, and half a dozen IoT devices.

802.11ac High Efficiency (HE) introduced several critical technologies to contend with the growing wireless density. These included using the 5 GHz bands, MU-Mimo, 256 QAM, Tx beamforming, and Frame Aggregation.

Today, HPE Networking supports 802.11be Extremely High Throughput (EHT). This new WI-FI 7 standard brings modern enhancements to support high-density deployments, includingultra-wide 320MHz channels, 4096 QAM, Spectrum puncturing, and Multi-Link Operations. But perhaps most impactful is the emerging support for the 6 GHz band to allow expanded frequency through layering.

What is a very high-density WLAN?

Simply put, Very high-density (VHD) WLANs are defined as RF coverage zones with many wireless clients and APs in a single physical space.

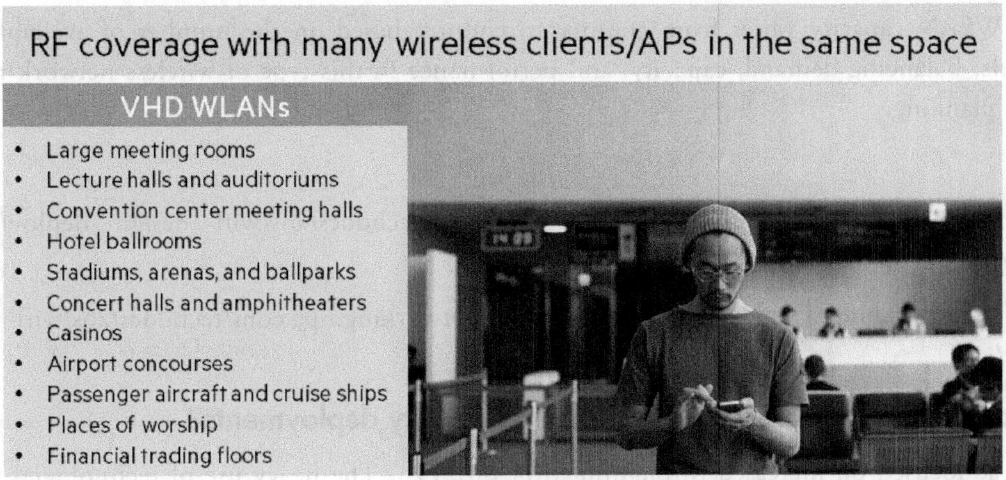

This creates a single RF collision domain designed to serve at least 100 devices per cell but may serve as many as 500. Devices may belong to individuals or be embedded for machine-to-machine or peer-to-peer service (such as streaming devices like Roku, Chromecast, or an Amazon Fire TV stick).

Where Are VHD WLANs Found? With the proliferation of wireless-enabled personal and enterprise mobile devices, a surprisingly diverse range of facilities that need VHD WLAN connectivity:

- Large meeting rooms
- Lecture halls and auditoriums
- Convention center meeting halls
- Hotel ballrooms
- Stadiums, arenas, and ballparks
- Concert halls and amphitheaters
- Casinos
- Airport concourses
- Passenger aircraft and cruise ships
- Places of worship
- Financial trading floors

The high concentration of users in any high-density environment presents challenges for designing and deploying a wireless network. The explosion of Wi-Fi-enabled smartphones means that each person could have three or more 802.11 NICs vying for service. At the same time, maximum

VHD WLAN capacity varies from country to country based on the number of available radio channels. Balancing demand, capacity, and performance in this type of wireless network requires careful planning.

LPV calculator: https://arubanetworking.hpe.com/techdocs/aos/wifi- design- deploy/design-guides/lpv/calculator/

Wi- Fi Design and Deployment: https://arubanetworking.hpe.com/techdocs/aos/wifi- design-deploy/

How do 802.11ax technologies affect high-density deployments?

802.11ac focused on increased throughput over 802.11n. The heavy use of technologies such as 80 MHz and 256 QAM and increased spatial streams resulted in a jump in bandwidth to real-world gigabit speeds over 802.11n, supporting a hundred megabits. The expectation for Wi-Fi 6E and 802.11ax is that it would increase single-user throughput to gigabit speeds. However, the technologies put forth in 802.11ax support high efficiency (HE) over faster single-user throughput.

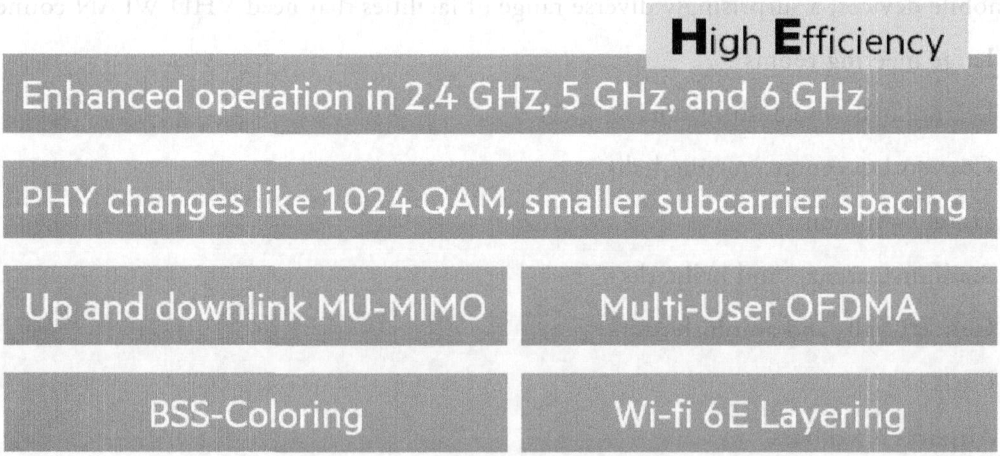

Below is a summary of 802.11ax technologies and their use for improving high-density deployments. Many of these are improvements on older technologies found in 802.11n and 802.11ac.

PHY Changes

The 802.11ax specification introduces significant changes to the physical layer of the standard. In addition to supporting the 6 GHz band, 802.11ax re-opens the use of 2.4 GHz, whereas the previous standard 802.11ac was 5 GHz only. 802.11ax supports 20, 40, 80, and 160 MHz channels, the latter becoming practical with the use of the 6 GHz band for indoor wireless.

Compatibility

802.11ax maintains backward compatibility with 802.11a/b/g/n and 802.11ac devices, such that an 802.11ax STA can send and receive data to legacy STAs. The subcarrier spacing is one-fourth the subcarrier spacing of previous 802.11 revisions. At the same time, the OFDM symbol duration and the cyclic prefix were increased by four times. This improves efficiency and robustness in indoor/outdoor and mixed environments while keeping the raw link data rate the same as the previous 802.11ac standard. Note that 1024 QAM and smaller cyclic prefix ratios for indoor environments increase the maximum data rate.

MU-MIMO

Like 802.11ac, Wi-Fi 6E uses beamforming to direct packets simultaneously to spatially diverse users. 802.11ax supports up to eight MU-MIMO uplink/downlink transmissions at a time from the four downlinks only supported in 802.11ac. Each transmission supports individual MCS and many streams.

Multi-User OFDMA

The OFDMA mechanism serves dedicated subchannels simultaneously to users in dense environments. Otherwise, these users would suffer inefficient contention awaiting their turn to use the channel. Previous 802.11 standards have supported OFDM since 802.11a. 802.11ax builds on this existing technology and further assigns specific subcarriers to individual users. This results in the existing 20, 40, 80, and 160 MHz wide channels being divided into smaller sub-channels with a predefined number of subcarriers. The smallest subchannel is called a Resource Unit (RU) with a minimized size of 26 subcarriers.

BSS-Coloring

The main goal of BSS coloring is to improve wireless efficiency in dense deployments. To accomplish this, BSS coloring allows APs and clients to mitigate collisions by APs assigning 802.11ax frequency a value from 1-63 color value so other APs can avoid broadcasting on the same shared frequency or color marking. This makes networks more efficient when transmitting data to multiple devices in congested areas. BSS coloring allows Wi-Fi 6 APs to precisely adjust Clear Channel Assessment (CCA) paraments.

Wi-fi 6E Layering

Campus networks have long had the option to customize SSID layout through traditional dual-band radios. Wi-Fi 6E and a third 6 GHz band enable further dedicated frequency use for dense, overlapping environments. A simple example would be reserving 2.4 GHz for IoT only, guests and corporate to the 5 GHz, and a dedicated 6 GHz only corporate SSID. With 6 GHz, it becomes possible to further subdivide users into groups of partitioned channels. For example, divide the 5 GHz and 6 GHz bands into three separate blocks: 6A, 6B, 6C, 5A, and 5B. The frequency plans could split channels as follows:

- Band 6A (Clinical Data Network): 6 x 80 MHz, or 12 x 40 MHz channels
- Band 6B (Clinical Imaging): 3 x 160 MHz channels
- Band 6C (Clinical Voice/IOT): 11 x 20 MHz channels
- Band 5A (Clinical Voice/IOT): 12 x 20 MHz channels
- Band 5B (Guest/Public): 8 x 40 MHz channels

Layering would allow the positioning of dedicated new APs where they will be most effective.

Note

Customers may have increased their AP densities as high as 1 per every 1,000 ft^2 (92 m^2). While stadiums or extreme density locations may have special use cases for this density level, it is generally considered overly dense for general campus enterprise. Layering and partitioning would allow these customers to resolve contention issues and improve utilization without removing APs.

For more information: https://www.ni.com/en/solutions/semiconductor/wireless- connectivity-test/introduction-to-802-11ax-high-efficiency-wireless.html.

How do 802.11be technologies affect high-density deployments?

Wi-Fi 7, also known as IEEE 802.11be, extends the capabilities of Wi-Fi 6E with significant new capabilities:

Ultra-wide 320 MHz bandwidth channels double the transmission capacity of the 160 MHz channels supported by Wi-Fi 6. This increase in bandwidth helps to reduce delays and improve overall transmission rates.

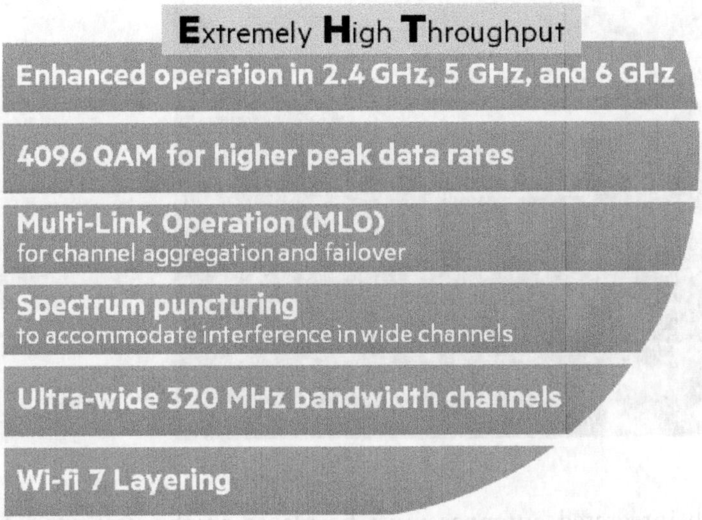

Multi-Link Operation (MLO) for channel aggregation that enables devices connected to a Wi-Fi 7 access point to combine different channels across frequency bands, allowing concurrent transmission and reception of data over multiple links. Prior to Wi-Fi 7, devices typically used a single band to transmit data.

4K QAM (Quadrature Amplitude Modulation) provides higher peak data rates by enabling each signal to more densely embed greater amounts of data. Data transmission efficiency and performance is significantly improved compared to the only 1K QAM with Wi-Fi 6 and Wi-Fi 6E

Spectrum puncturing helps accommodate interference in wide channels by allowing subchannels to operate within wide channels by opening up 20 MHz increments. This helps work around interference or other requirements while still enabling 320 MHz channels to function.

Aesthetics

In many VHD areas, aesthetics requirements can significantly limit where and how APs are placed. The availability of suitable mounting locations can significantly impact the performance of the overall RF design. In the auditorium shown above, different ceiling heights, dense users, and tightly controlled aesthetics severely limit the options available to the wireless architect. Expensively Furnished Rooms Pose AP Placement Challenges. However, HPE Networking cannot advocate strongly enough the importance of proper placement. Wi-Fi is fundamentally a line-of-sight technology. Concealing APs behind almost any building materials will negatively affect the performance of a VHD system. Sometimes, the impact is severe.

Generally, APs with integrated antennas must be placed in the user space, with a clear line of sight to the user seating. HPE Networking has spent significant sums on the industrial design of every one of its APs. Great care has been taken to make them look attractive. Remember that high up on the ceiling, APs tend to blend in with all the other ceiling furniture (such as smoke detectors, occupancy sensors, and public address speakers). The status lights on the AP can be disabled so they cannot be seen from the ground. Very High-Density 802.11ac Networks Planning Guide Validated Reference Design. It is our experience that if you work with a facilities team or building architects, a suitable compromise can be worked out. But in such meetings, you must take a firm stand that the system's performance cannot be compromised. Otherwise, the investment is not justified. If the APs cannot be placed in the user space, then it is necessary to switch to external antennas, which must be surface mounted in the user space. This means penetrations and

pigtails, with suitable fixed mounting for the AP above the ceiling surface. Often, the cost of this step will settle the argument over placing APs on the ceiling. Check the HPE Networking URL for available antennas. https://www.arubanetworks.com/assets/matrix/matrix-antennas.pdf.

Your HPE Networking systems engineer can help you position the performance impact of undesirable mounting locations with architects and facilities teams. Remember that APs should never be hidden inside walls as they require ventilation and a means to dissipate heat. Mounting inside the plenum with surface-mounted antennas is acceptable.

Note

Do not mount Access Points with omni-directional antennas to the wall facing the room. Access Points should be facing the floor or use external antennas.

HD & VHD RF Coverage

Now you look at best practices for designing RF coverage in high-density (HD) and very high-density (VHD) areas.

Higher-density, open-office layouts

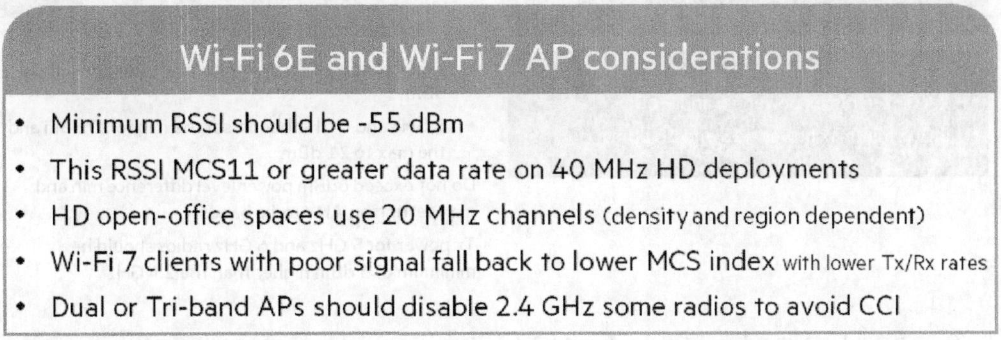

Wi-Fi 6E and Wi-Fi 7 AP considerations:

- Minimum RSSI should be -55 dBm or better throughout the coverage area.
- This RSSI allows for a Modulation and Coding Scheme (MCS) with at least an MCS11 data rate on 40 MHz high-density (HD) deployments.
- Wi-Fi 7 clients with a poor MCS rate will fall back to a lower MCS index with a lower Tx/Rx rates to improve successful data transmission.
- Dual or Tri-band APs should consider disabling 2.4 GHz radios to avoid co-channel interference. AirMatch may not disable these radios dynamically.
- High-density open-office environments may want to use 20 MHz channels.

Power settings

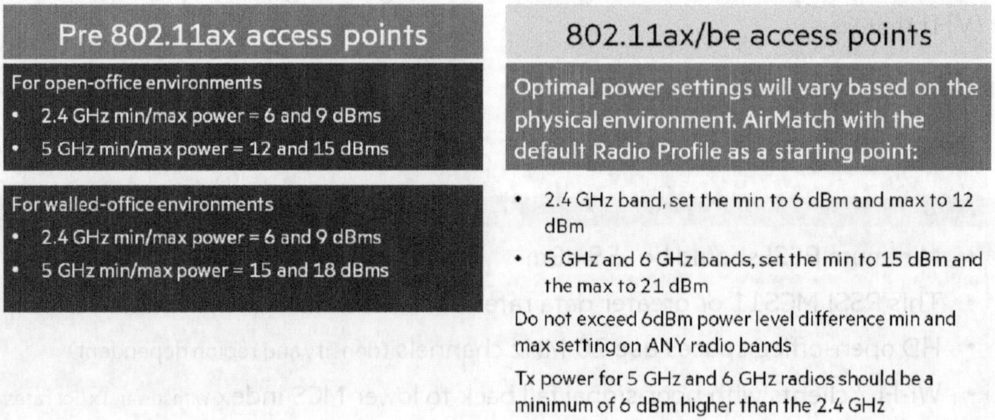

After configuring the initial settings, the AirMatch software automatically adjusts power settings continuously as needed. Remember, you can utilize the following as a general guideline for a typical wireless design.

AirMatch with the default Radio Profile as a starting point:

- 2.4 GHz band, set the min to 6 dBm and max to 12 dBm
- 5 GHz band, set the min to 15 dBm and the max to 21 dBm
- 6 GHz band, set the min to 15 dBm and the max to 21 dBm

Do not exceed a power level difference of 6 dBm in all environments between the min/max on all radio bands. The minimum power level difference between equal coverage level 2.4 GHz and 5 GHz radios should be 6 dBms to support desired band steering by clients.

Note also that the 802.11ax and 802.11be EIRP min/max considerations have been included from module 3 on the right hand of the slide.

Note that optimum power settings vary based on your physical environment, and you should always follow the recommendations from your professional site survey to finalize your initial settings.

RF coverage strategies

You can use radio coverage in three ways, regardless of the type of area you want to serve.

Overhead Coverage: APs are placed on a ceiling, catwalk, roof, or other mounting surface directly above the users to be served.

Side Coverage: APs are mounted to walls, beams, columns, or other structural supports in the RF area.

Floor Coverage: This design creates picocells using APs mounted in, under, or just above the floor of the coverage area.

External directional antennas give additional coverage options!

For Overhead Coverage, you can place APs on a ceiling, catwalk, roof, or other mounting surface directly above the users. You should always use the integrated antenna APs for ceilings of 10m (33 feet) or less.

With Side Coverage, you mount APs to walls, beams, columns, or other existing structural supports in the space to cover.

In a Floor Coverage design, you will create pico-cells using APs mounted in, under, or just above the floor of the coverage area. You should use APs with integrated antennas for any VHD area of under 5,000 seats.

Overhead coverage

- Uniform signal is desired
- No RF spatial reuse is possible (with omnidirectional antennas)
- 20-MHz channel width
- Requires access to the ceiling

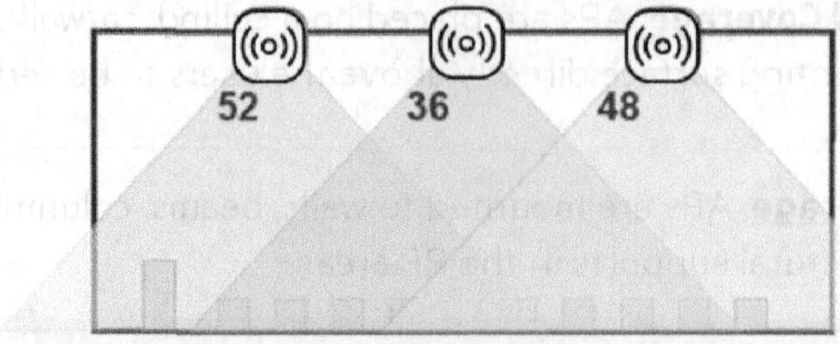

Side View

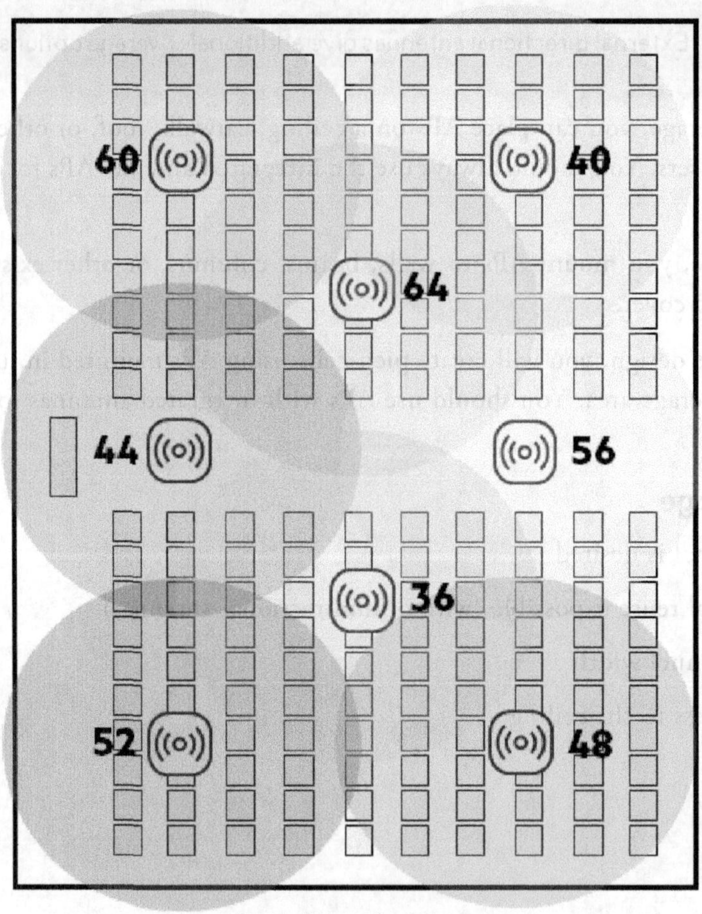

Overhead view

Overhead coverage is a good choice when the customer desires a uniform signal everywhere in the room. Overhead coverage prevents RF spatial reuse because of the wide antenna pattern and multipath reflections. We'll look more into this in the next slide.

Note the 20-MHz channel width for Wi-Fi 6 deployments. This strategy avoids using the same channel number more than once due to no RF re-use.

The design goal is to provide an optimal number of seats per AP. Utilizing 500 users per AP is a goal for indoor Auditoriums and stadium seating with less than 10,000 users. Examples of pushing this further down to 105/AP or even 75/AP can be found in real-world stadium deployments. These are examples of a static, non-repeating channel plan a wireless architect intentionally chose. Overhead coverage requires access to the ceiling with minimal difficulty or expense to pull cables and install equipment.

No RF spatial reuse with omnidirectional antennas

RF spatial reuse occurs when every AP can be heard everywhere in the room. It could happen in high-density environments, such as the interior of the stadium bowl, due to the RF propagation. This is an issue using Wi-Fi 5 with a Large Public Venue (LPV) design since it requires all 2.4 GHz and 5 GHz channels (3 in 2.4 GHz and 23 in the 5 GHz for a total of 26 times 20 Mhz channels) to meet the density requirements. The WLANs cannot reuse the channels due to the RF propagation in the stadium bowl. You'll also need DFS channels to support the channel bandwidths for a high density of users.

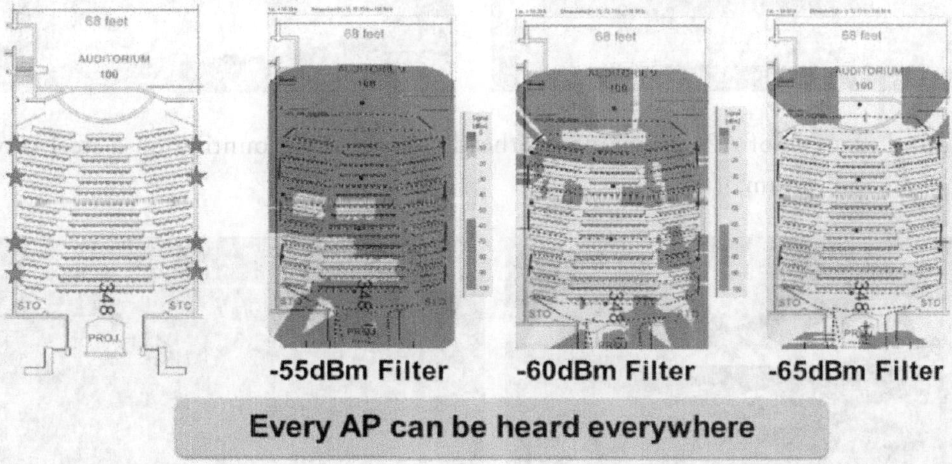

By expanding into the 6 GHz, you open another 1,200 MHz in the US (500 MHz in the EU and Australia at the time of this guide).

The first advantage is using an additional 58 available 20 Mhz (+24 channels). If you choose, you could just add the additional 58 APs without causing CCI in a non-RF spatial reuse environment. Another strategy may be to leave the 2.4 GHz for IoT and devices that do not support other frequencies. Then leverage 40 MHz channels across the 5 GHz and 6 GHz bands for 8 channels in the 5 GHz (avoiding DFS), and the 6 GHz would add 29 for a total of 37 channels (20 channels in Europe/CEPT).

Other benefits are higher efficiencies with 802.11ax designed specifically for VHD environments that typically result in 4x the throughput over 802.11ac. These include OFDMA, MU-MIMO up/down at 8 SS and BSS Coloring. The strategy of layering different user roles and Air Slice would further improve QoS for sensitive real-time applications.

Examples – overhead coverage

Here is an example of overhead coverage APs provide in an auditorium. Note the overhead mounting placement of the APs.

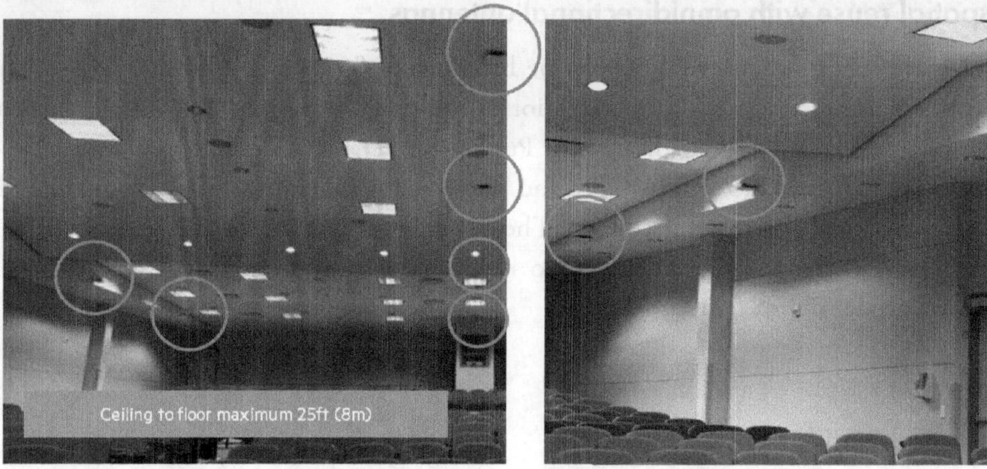

Here is another example of APs providing overhead coverage by mounting on a steel catwalk on the ceiling of an auditorium.

Side coverage

- Wall, beam, and column installations
- Difficult ceilings to reach

- No RF spatial reuse in indoor environments
- 50% of the wall-mounted AP signals
- Adjacent APs on same wall: always skip at least one channel number

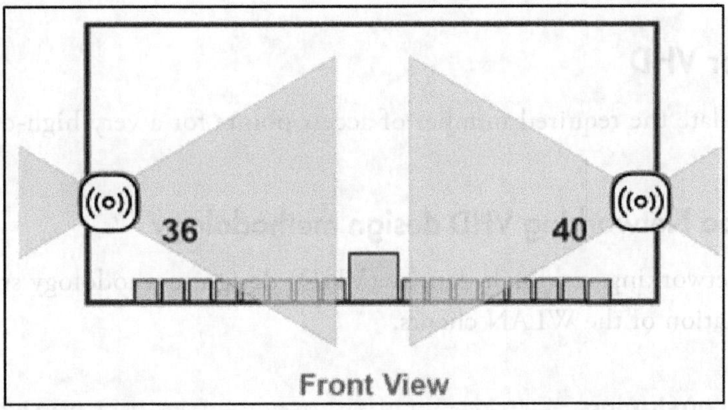

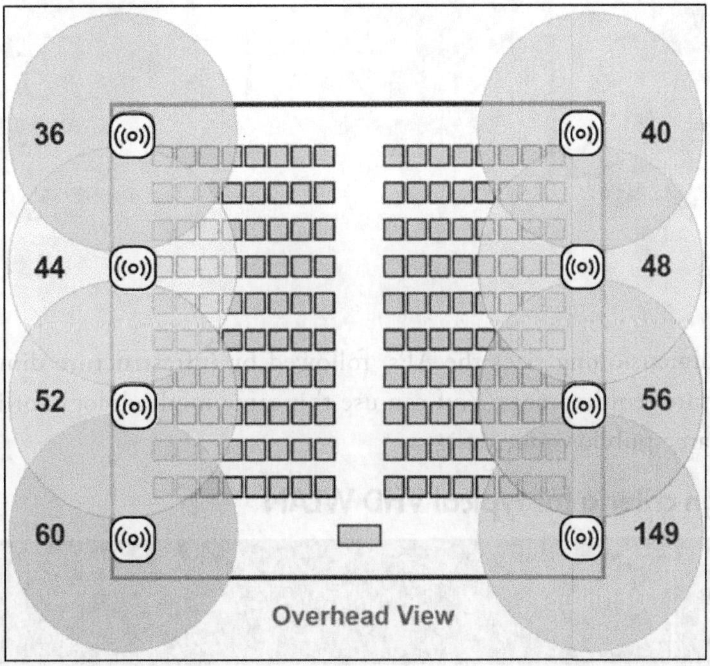

Side coverage is common in VHD areas with wall, beam, and column installations. Some ceilings are too difficult to reach, while others have costly finishings, or there may be no ceiling, such as open-air atriums. Side coverage does not allow RF spatial reuse in indoor environments when mounting to walls or pillars. Wall-mounted AP signals lose 50% of their signal to the next room (and 75% of the signal in the corners).

Note

Adjacent APs on the same wall always skip at least one channel number.

Selecting APs for VHD

Now, you can calculate the required number of access points for a very high-density (VHD) location.

Review HPE Aruba Networking VHD design methodology

The HPE Aruba Networking very high-density (VHD) design methodology starts with the device dimensioning evaluation of the WLAN clients.

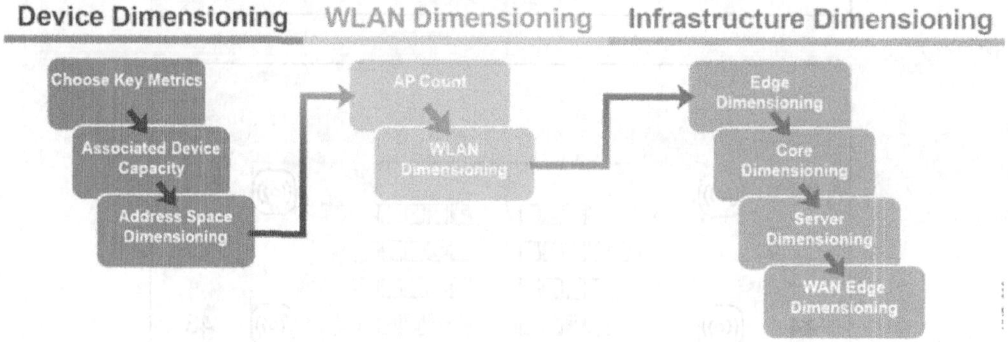

Next, the WLAN dimensioning sizes the APs, followed by infrastructure dimensioning to determine the infrastructure requirements. You can use this same methodology for indoor high-density areas and outdoor large public venue designs.

Step 1- key design criteria for typical VHD WLAN

Metric	Definition	Typical Value
Seating capacity	Number of people the facility can hold	Varies
Average devices per person	Typical number of discreet Wi-Fi enabled devices carried by a persion visiting the VHD facility	1 to 5
Take rate	Percentage of seating capacity with an active Wi-Fi device	50% - 100%

Metric	Definition	Typical Value
Associated device capacity (ADC)	Take rate multiplied by the average number of Wi-Fi enabled devices per person.	Varies
Seats or area covered per AP	How many square meters (or feet) or seats each AP must serve - essentially the physical size of a radio cell	Varies
Associated devices per radio	The design target of how many associated devices should be served by each radio	512
Average single-user goodput	The minimum allowable per-user bandwidth when multiple users use the same AP	512Kbps - 2Mbps
6 GHz vs 5 GHz vs 2.4 GHz divide	Distribution of clients across two or three bands	6 GHz: 30% 5 GHz: 50% 2.4 GHz: 20%

To determine the device dimensioning of a VHD WLAN, determine the following metrics:

- Seating capacity
- Average devices per person
- Take rate (which means what percentage of users will have active wireless devices)

From these basic metrics, you can calculate additional values such as seats or area covered per AP and associated devices per radio.

Note that client device support for 6 GHz has evolved rapidly in the last few years since the ratification of 802.11ax in 2020. Even as recently as 2021, designs for VHD regarded the 6 GHz as a supplemental use case and stayed with traditional distributions of 75% for 5 GHz and 25% for 2.4 GHz. Given the wide-spread support of modern devices in 2023 for 6 GHz, you may choose instead to design for the 6 GHz band with a distribution shown in the figure of a ratio of 30%, 50%, and 20% (6, 5, and 2.4 GHz) accordingly. Any distribution you choose should be validated with real-world statistics.

Step 2 – estimate associated device capacity

The goal is to get the associated device capacity (ADC). This starts with determining the seating/standing capacity of the VHD area you want to cover. Second, estimate the take rate (50% is a common minimum). Third, choose the number of devices you expect per person. This varies by venue type – it might be lower in a stadium and higher in a university lecture hall or convention center salon.

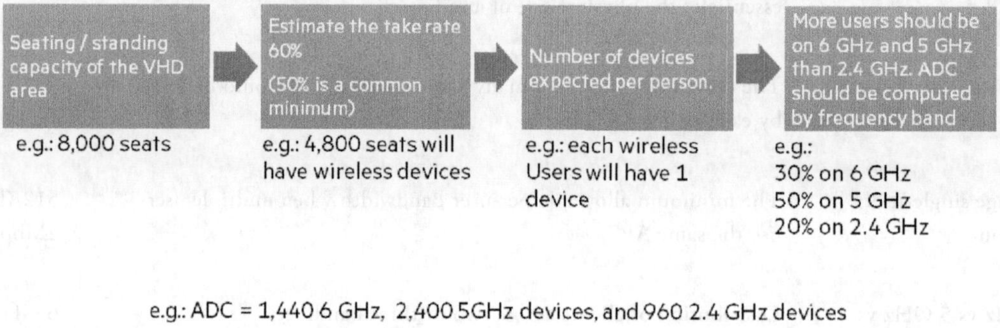

For example, 60% of an 8,000-seat stadium would be 4,800 devices, assuming each user has a single device. Or 100% of a 1,000-seat lecture hall where every student has an average of 2.5 devices would have an associated device capacity (ADC) equal to 2,500. More users should be on 6 GHz and 5 GHz than 2.4 GHz. You should compute ADC by frequency band. You could target a ratio of 30%, 50%, and 20% (6, 5, and 2.4 GHz), and you should assume association demand to be evenly distributed throughout the coverage space.

Step 3 – ADC address design examples

These are a few sample ADC and address space estimates for an outdoor 4,000-seat school stadium. Notice the plan is for guest users to have a take rate of 35% now and 50% planned, with the majority of Guests still on the 6 GHz and 5 GHz over the 2.4 GHz.

User Group	Devices (Current)	Devices (Planned)	6 GHz %	5 GHz %	2.4 GHz %	Subnet Min. size
Guest / Fans	1,400 (35% take rate)	2,000 (50% take rate)	30% (600)	50% (1000)	20% (400)	/21
Staff	50	150	25% (38)	75% (112)	0%	/24

User Group	Devices (Current)	Devices (Planned)	6 GHz %	5 GHz %	2.4 GHz %	Subnet Min. size
Ticketing	25	50	25% (12)	75% (38)	0%	/26
POS	25	100	25% (25)	75% (75)	0%	/25
Team	15	100	25% (25)	75% (75)	0%	/25
ADC Total	**1,515**	**2,400**	**700**	**1300**	**400**	-

The table below shows the average associated device capacity (ADC) and address space estimates for a high school auditorium to support 6,000 students and 370 faculty. Take rates are shown in the table, growing from 35% to 75% as planned. Students are expected to average 2.5 devices each, while faculty use 1.5.

To find the student ADC, multiply 6,000 students by the average device count of 2.5. Multiple the result by a 35% (current) and 60% take rate (planned). This results in the current student ADC of 5,250 and planned 9,000 devices. Next, break down the devices per radio band for the planned count. You can do the same for the 370 faculty and verify your results against the table above. The final ADC is calculated by adding the resulting students' and faculty's ADCs per band.

User Group	Devices (Current)	Devices (Planned)	6 GHz %	5 GHz %	2.4 GHz %	Subnet Min. size
Students	5,250 (35% take rate)	9,000 (60% take rate)	30% (2700)	50% (4500)	20% (1,800)	/18
Faculty	194	330	30% (100)	55% (180)	15% (50)	/23
ADC Total	**5,444**	**9,330**	**2800**	**4,680**	**1,850**	-

Step 4 – estimate AP count

Recall that in an open office area, you would design for one AP every 2000 ft^2 to 2,500 ft^2 (186m^2 to 232 m^2), with an average of three devices per user. Practical campus deployments should use a maximum client count per radio for the 5xx-AP series, which is 75 to 100.

150 associations per 5 GHz or 6 GHz radios
150 associations for backwards compatibility with 802.11ac

AOS supports up to 512 associations per radio on 700 series APs; plan for 150 devices to remain backwards compatibly with 802.11ac and 802.11ax clients.

All VHD areas experience inrush/outrush

Remember to increase max users in the SSID profile

$$\text{AP Count (5 GHz Radio Count)} = \frac{\text{Associated Device Capacity (5 GHz)}}{\text{Max Associations per radio}}$$

In a VHD design, with 802.11be clients, you could plan for the maximum associations supported, with 512 associations per radio. However, in mixed client environments with 75% using 802.11ac, you would design only 150 associations per radio to remain backward compatible. Remember that 802.11ac does not support 2.4 GHz or 6GHz.

For an all 802.11ax or 802.11be client design, consider the following example. The HPE Aruba Networking 700 series APs can support 512 users per radio. You could plan for 512 802.11be client associations per radio maximum. Planning for extra headroom allows for user "breathing" when initially connecting. Also, remember to increase the maximum number of users in the SSID profile.

Note that stadium examples may push these down to 110 or 75 users per AP using engineered piconet design to overcome the issue with non-spatial reuse; this, however, is beyond the scope of this training.

Step 5 – dimension gateways

For high-density designs up to 32,000 users, the platform size must be greater than or equal to the ADC. For this number of users, it is critical that you use redundant PSU, so you should not use 9004 or 9012 series gateways. Also, active/active redundancy is critical for this large number of users.

Model	9004	9012	9106	9114	9240 base	9240 silver	9240 gold
Maximum number of devices (access points)	**128**	**256**	**2K**	**4K**	**4K**	**8K**	**16K**
Maximum number of clients	**2K**	**2K**	**8K**	**10K**	**32K**	**48K**	**64K**
Cluster size	**4**	**4**	**6**	**6**	**6**	**6**	**6**
Firewall throughput (Gbps)	4	6	10	20	20	30	40
AES encrypted firewall throughput AES-GCM-256 (Gbps)	4	6	10	20	20	30	40
Maximum bridge table	64K	64K	256K	512K	1M	1.5M	2M
Maximum datapath / firewall sessions	128K	128K	2M	2M	4M	4M	4M
Concurrent IPsec tunnels	2K	2K	16K	32K	32K	64K	128K

Model	9004	9012	9106	9114	9240 base	9240 silver	9240 gold
ARP	2K	2K	16K	20K	64K	96K	128K
Form factor	Desktop/ 1RU	1RU	Desktop/ 1RU	1RU	1RU	1RU	1RU
1000Base-T ports	4	12	4	4	0	0	0
10GbE ports (SFP+)	0	0	2	4	0	0	0
25GbE ports (SFP28)	0	0	0	0	4	4	4
Redundant PSU	**No**	**No**	**No**	**Yes**	**Yes**	**Yes**	**Yes**

Note

Performance metrics could change with new AOS versions.

Note

9xxx series gateways replaced 7xxx series

Step 6 – edge design

For the edge switch design in a VHD environment, HPE Networking recommends having full, non-blocking 2.5GE/5GE/10GE ports downstream to APs, full 802.3bt uPoE (Class 6) with 60W on all ports, and Cat-6A cabling.

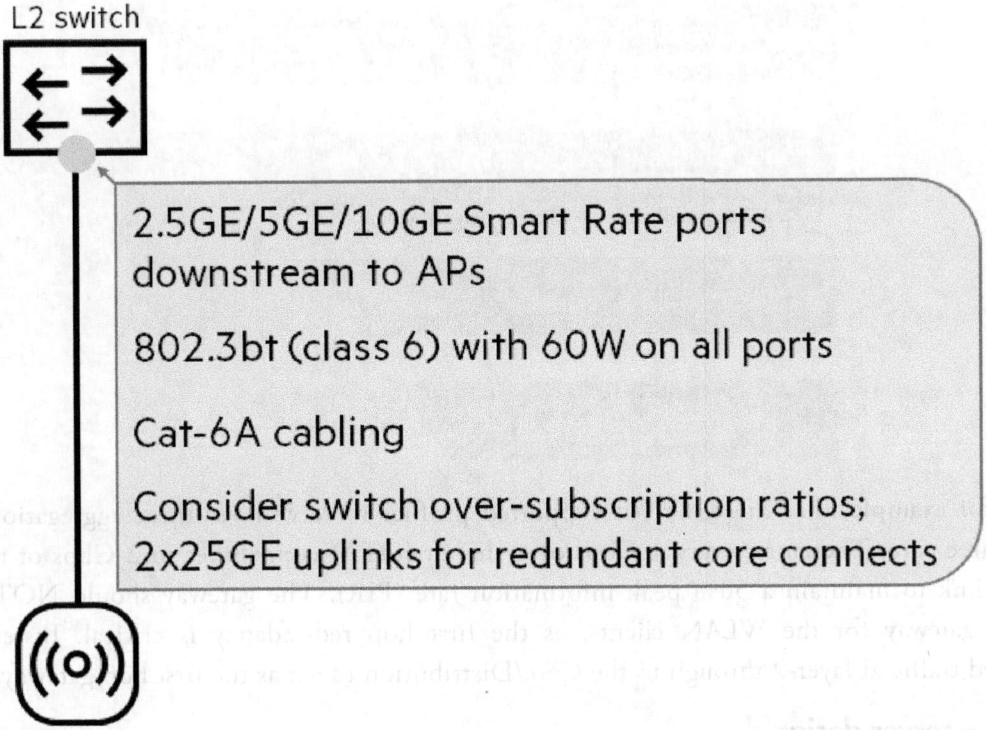

For example, the flagship HPE Aruba Networking 750 series AP can handle up to 28.8 Gbps of wireless throughput. The AP supports 2x10Gbps ethernet ports using 802.3bz (Smart Rate). These can be bundled using LACP for combined capacity. In the worst case, maximum power consumption will require 51W POE with a USB attached (46W on DC power). Therefore, 802.3bt uPoE should be provided on a single port, or 802.3at (PoE+) can be combined if both ethernet ports are utilized. 802.3af class 3 or lower will not power on the AP. Note that a mounting bracket is pre-installed on the back of the AP to secure the AP to any mount kits (sold separately).

Step 7 – core design

For the switch core design, verify that the ARP cache and forwarding tables in core switches are large enough to handle a big flat user VLAN and that the gateway-to-core uplinks are twice the size of the WLAN throughput computed in the capacity plan.

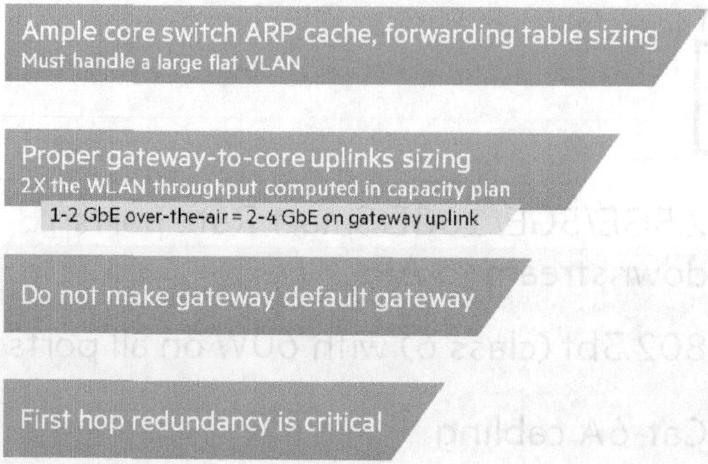

Note, for example, that the 8325 has supported profiles to serve layer three aggregation versus layer three core. Therefore, every 1 Gbps over-the-air (OTA) contributes to 2 Gbps of the gateway uplink to maintain a 50% peak information rate (PIR). The gateway should NOT be the default gateway for the WLAN clients, as the first hop redundancy is critical. Instead, pass tunneled traffic at layer-2 through to the Core/Distribution to act as the first hop gateway.

Step 8 – server design

- **DHCP/DNS: key metric is transaction time**
  - Should be <= 5ms
  - This is MUCH more critical than transaction rate
  - Carrier-grade DHCP/DNS servers strongly recommended (Infoblox)
  - Lease times should be 2X duration of event (8 hours suggested)
  - Model DNS at 1 request/device/second
- **Captive portal rate = DHCP arrival rate**
- **RADIUS loads depend on whether devices are using 802.1X**

The key metric for the DHCP/DNS servers in a high-density design is transaction time, which should be <= 5ms. This is much more critical than the transaction rate and should be modeled at 5% of seating capacity over 5 minutes.

For example, in an 18,000-seat arena, * 5% / 300 seconds = 3 discovers per second. HPE Networking strongly recommends carrier-grade DHCP/DNS servers, such as Infoblox, and lease times should be 2X the event duration (8 hours suggested). Model DNS at one request/device/second, and the captive portal rate = DHCP arrival rate. The RADIUS loads depend on whether devices are using 802.1X.

Step 9 – WAN edge design

You can estimate the WLAN uplink bandwidth using the HPE Aruba Networking total system throughput process.

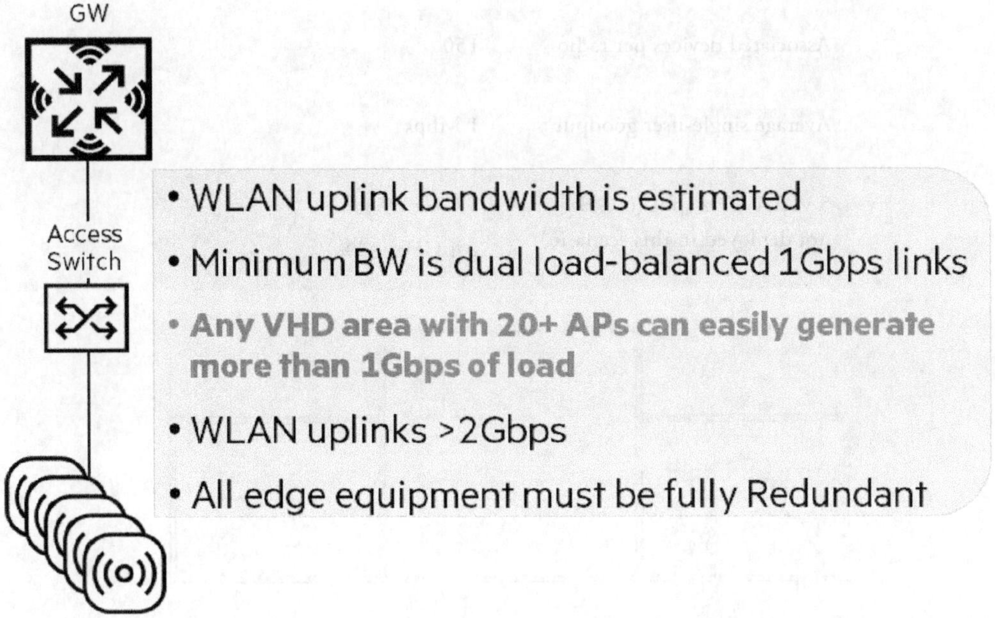

Minimum BW is dual load-balanced 1Gbps links for a country with 20+ channels in 5GHz using all/most DFS channels. Any VHD area with 20+ APs should easily generate 1Gbps of load. You may require WLAN uplinks greater than 2Gbps if attempting RF spatial reuse. Also important is that all edge equipment must support full high-availability (HA).

Deployment Examples

Let's look at a VHD deployment example.

Example scenario: typical multi-auditorium

In this example, you have a Hotel conference center or university building with multiple adjacent auditoriums.

Metric	Target
Take Rate	100%
Average devices per person	Work/study - 3 - 5 (future)

Metric	Target
	Fan/gues - 1-2 (future
Associated devices per radio	150
Average single-user goodput	1 Mbps
5 vs 2.4 GHz split (6 GHz not deployed in this scenario)	5 GHz: 75% 2.4 GHz: 25%

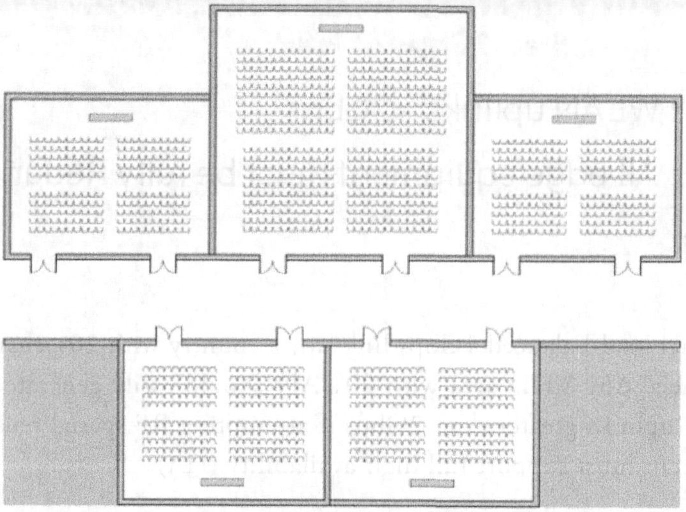

Physical Layout

Step 1: understanding load in auditoriums

You can use the VHD Spatial Stream Blend Lookup Table and Network Characteristics of Common Auditorium Applications to understand the device load in the auditorium.

The table will show information concerning:

- Common apps for web browsing, email, and office collaboration.
- Class presentation and exam software that is bursting with high concurrent usage.
- Cloud service latency that is not visible to users.

VHD Usage Profile	Devices / Persion (Now)	Devices / Person (Future)	1SS (%)	2SS (%)	3SS (%)
Work / Study	3	5	30%	60%	10%
Fan / Guest	1	2	50%	50%	0 %

Now you relate the Work / Study and Fan / Guest rows above to the corresponding rows in the table below.

User category	Application	Bandwidth	Latency	Duty Cycle
Work / Study	Play courseware (non-video)	500 Kbps	Medium	Medium
	Play courseware (video streaming)	1 Mbps+	Low	High
	Test / exam / quiz	Under 250 Kbps	Real-time	Synchronized bursts
Fan / Guest	General internet usage	500 Kbps	Medium	Low
	Email	Under 250 Kbps	High	Low
	Social media (text)	500 Kbps	Medium	Low
	Photo / video cloud sync	1 Mbps+	High	Low

Step 2/3 - estimate associated device capacity

Use the provided formulas to calculate ADC. Note that we have not included the 6 GHz in this example for backward compatibility.

Start with seating capacity | Use per-user device to estimate ADC | Break out by frequency band | Determine address space

Room Number	Seats	ADC (Now)	ADC (Future)	5-GHz ADC (Future)	2.4-GHz ADC (Future)	Minimum Subnet Size
Room A	200	600	1,000	750	250	/22
Room B	200	600	1,000	750	250	/22
Room C	450	1,350	2,200	1,650	550	/20
Room D	200	600	1,000	750	250	/22
Room E	200	600	1,000	750	250	/22
Staff / House	--	25	75	75	0	/24
GUEST ADC	**1,250**	**3,750**	**6,200**	**4,650**	**1,550**	**/19**
STAFF ADC	**--**	**25**	**75**	**75**	**0**	**/24**
TOTAL ADC	**1,250**	**3,775**	**6,275**	**4,725**	**1,550**	

Estimate staff / facility devices separately

Step 4 - estimate the AP count using a AP-730

Next, estimate the AP count. Assume using an HPE Aruba Networking AP-730 because the frequencies of the 2.4 GHz, 5 GHz, and 6 GHz can be used simultaneously. Note that we only look at the 5 GHz when planning capacity to remain backward compatible with the high number of expected 802.1ac devices. The 2.4GHz will be considered in the next slide.

$$\text{AP Count (5 GHz Radio Count)} = \frac{\text{Associated Device Capacity (5 GHz)}}{\text{Max Associations per radio}}$$

Room Number	5-GHz Guest	5-GHz Staff	Total 5-GHz Devices	Devices per Radio	AP Count
Room A	750	15	765	150	5
Room B	750	15	765	150	5
Room C	1,650	15	1,665	150	11
Room D	750	15	765	150	5
Room E	750	15	765	150	5
Hallway	500	15	515	150	3
TOTAL	**5,150**	**90**	**5,240**		**34**

Take 5-GHz ADC | Divide by per-radio metric

Step 5 – calculate system throughput

Finally, determine the system throughput calculations, excluding co-channel interference (CCI). Convert the number of APs to the number of channels that they offer. In this example, the environment permits the use of DFS channels. Within each VHD, each AP represents one 5GHz channel because the 5GHz band offers many non-overlapping channels. However, each area only has three non-overlapping 2.4 GHz channels. Therefore, all APs will advertise a 5 Ghz channel, but some APs may not advertise a 2.4 Ghz channel due to the density of the deployment. As shown in the slide, the three APs broadcasting in the 2.4 GHz would do so at 20 MHz and lower throughput.

Room Number	AP Count	Channels (USA DFS)		Avg. Channel Bandwidth		formula	Aggregate Bandwidth
		2.4 GHz	5 GHz	20 MHz (2.4 GHz)	40 MHz (5 GHz)		
Room A	5	3	5	162.5 Mbps	325 Mbps	(3 * 162.5) + (5 * 325)	2,113 Mbps
Room B	5	3	5	162.5 Mbps	325 Mbps	(3 * 162.5) + (5 * 325)	2,113 Mbps
Room C	11	3	11	162.5 Mbps	325 Mbps	(3 * 162.5) + (11 * 325)	4,063 Mbps
Room D	5	3	5	162.5 Mbps	325 Mbps	(3 * 162.5) + (5 * 325)	2,113 Mbps
Room E	5	3	5	162.5 Mbps	325 Mbps	(3 * 162.5) + (5 * 325)	2,113 Mbps
Hallway	3	3	3	162.5 Mbps	325 Mbps	(3 * 162.5) + (3 * 325)	1,463 Mbps
TOTAL	34						13,978 Mbps

AP count

Convert to channels
- 1 for each AP with 5 GHz radio at 40 GHz
- 1 max for each room with 2.4 GHz radios at 40 MHz
- 3 max for each room with 2.4 GHz radios at 20 MHz

Multiply by estimated channel capacity

Total maximum load if zero CCI

The average channel bandwidth uses HE (high-efficiency) 802.11ax utilizing two spatial streams (2SS). Generally, the maximum MCS supported for VHT or HE is 7, given an overhead deployment. You should be able to support a 40 MHz channel width, given that the maximum number of channels is less than 7 in the largest room (Room C). This would average about 325 Mbps channel bandwidth at 802.11ax MCS 7 (use 150 Mbps for clients only supporting 802.11ac). Note that spatial reuse is possible given the small number of APs; therefore, we can leverage the greater efficiency with wider channels. Smaller channels would maximize the number of AP radios in a large auditorium with no spatial reuse.

For the full MCS Table, check this URL: https://mcsindex.com/

You can then multiply the number of channels (7 in Room C, for example) by the estimated channel capacity (325 Mbps) to calculate an aggregate bandwidth. This results in 2.275 Gbps for Room C.

Note

- The overwhelming wireless client devices will support 2SS or two spatial streams.
- MU-MIMO may not be as prevalent in VHD environments, so you may choose to plan around the feature.

- 6 GHz will greatly boost the available spectrum and can be designed similarly to 5 GHz indoors. However, be prepared for slow adoption by some campus environments. This example demonstrated calculating aggregate bandwidth without the benefit of the 6 GHz band.

Learning check

Check your understanding of wireless high-density design by answering the following questions.

Question #1

Which of the following accurately describes today's very high-density challenges?

- ○ With the proliferation of Bluetooth devices, fewer Wi-Fi clients are expected today than in years past.
- ○ A very high-density WLAN is a high-density area of clients and an AP count of 1 AP per 1,500 ft² (140m²), but due to obstructions, it allows for RF re-use of channels.
- ○ A very high-density (VHD) WLAN is an area with so many users and APs that RF channel reuse cannot be expected.
- ○ While 802.11be does bring 10 gigabit speeds to wireless, 802.11ax is still defined as Extremely High Throughput (EHT).

Question #2

Which three are recommended best practices or things to keep in mind for very high-density deployments in open offices (select three)?

- ☐ Dual-band APs should always enable 2.4 GHz and run them with 10 MHz wide channels.
- ☐ Minimum RSSI should be -55 dBm to allow for a Modulation and Coding Scheme (MCS) with at least an MCS11 data rate on 40 MHz high-density deployments.
- ☐ HD open-office environments may want to use 20 MHz channels to avoid CCI.
- ☐ Wi-Fi 7 clients with a poor signal will fall back to a lower MCS index and lower Tx/Rx rates.
- ☐ Tx power for 5 GHz and 6 GHz radios should be a minimum of 12 dBm higher than the 2.4 GHz.

Question #3

Which is true for selecting APs for very high-density deployments?

- ○ Include into the associated device capacity (ADC) calculation only 802.11be capable clients.
- ○ Key criteria when designing for a very high-density location start with seating capacity, average devices per person, and the take rate (percentage of seats with Bluetooth users).

- A best practice since the ratification of 802.11ax in 2020 has been to plan for 75% of clients to connect using a 6 GHz radio.
- You should compute the associated device capacity (ADC) by frequency band.

Question #4

A new auditorium has 4,000 seats with an 85% take rate and an average of two and a half devices per user. What is the associated device capacity per frequency band if you expect a target ratio of 30% 6GHz, 50% 5GHz, and 20% 2.4GHz?

- 2400 6GHz, 4000 5GHz, and 1600 2.4GHz
- 2550 6GHz, 4250 5GHz, and 1700 2.4GHz
- 500 6GHz, 7500 5GHz, and 2000 2.4GHz
- 3400 6GHz, 3400 5GHz, and 1700 2.4GHz

5 Wired Network Design

Welcome back! Let's shift gears and move from Wi-Fi to wired campus design.

OBJECTIVES

After completing this chapter, you should be able to:

- ✓ Design 2-tier and 3-tier wired architectures.
- ✓ Layout closets and cables.
- ✓ Plan the access layer for HPE Aruba Networking security-first, AI-powered networking.
- ✓ Plan the core and aggregation layers for HPE Aruba Networking security-first, AI-powered networking.

2-Tier versus 3-Tier Design

Let's start with the benefits of 2-tier versus 3-tier campus design. We will also review individual switch models and where they should be placed in the campus (or data center) design.

2-tier versus 3-tier topology

When you plan a wired network, you must select products and design how they will fit together. You can classify the architecture based on the number of tiers or layers.

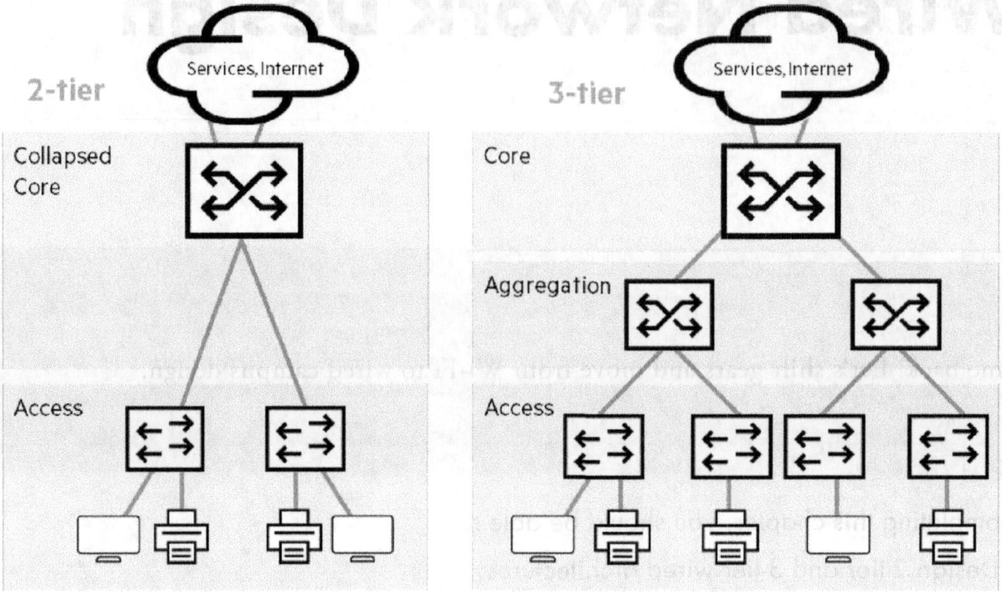

Here are a few points to review:

- A 2-tier topology connects endpoints to an access layer and from there to a core. The core provides external resources or other connections for centralized servers.
- A 3-tier topology often divides the access layer into groups or zones.
- In a 3-tier, each group of access layer switches connects to an aggregation layer switch, which aggregates links and often applies various policies.
- Often the aggregation layer switch supports ports of different speeds, the aggregation switch might connect to access layer switches on 1GbE or 10GbE links and to the core on 10GbE or 40GbE links.

Benefits of a 2-tier topology

HPE Networking recommends a 2-tier topology for most small and medium campuses because this topology provides simplicity and the best performance. In a 2-tier topology, traffic flows directly from the edge to the core, reducing latency. The customer also avoids buying several aggregation layer switches and fewer transceivers.

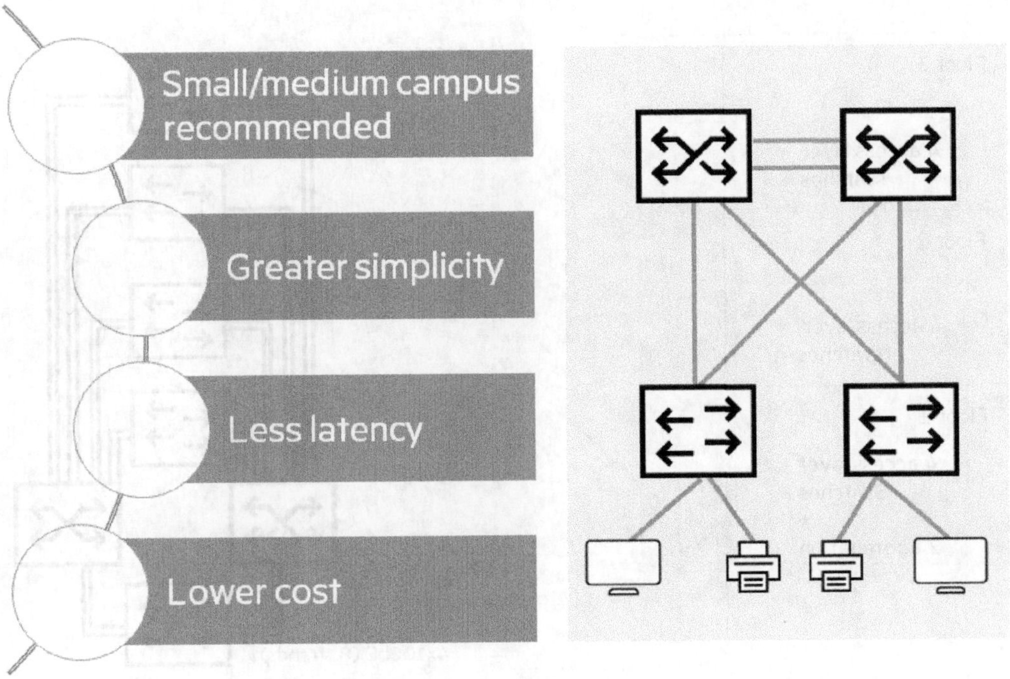

A 3-tier topology typically routes between the aggregation layer and the core, whereas a 2-tier topology can extend the layer-2 network across the campus, permitting greater flexibility in VLAN design. For example, the customer may want to extend a security camera VLAN across the campus. A flat topology that permits the extension of VLANs can be particularly important for virtualized data centers. The lower latency of a 2-tier topology might also offer benefits for a campus where users run many UC applications.

When 3-tier may be required

A customer's physical site and cabling restraints are the primary factors determining whether you can use a 2-tier topology, which requires the access layer to connect directly to the core. Access layer switches typically reside in distributed closets, while core switches reside in a central location. You will need to examine the distances between the switches and the types of cabling available for each connection.

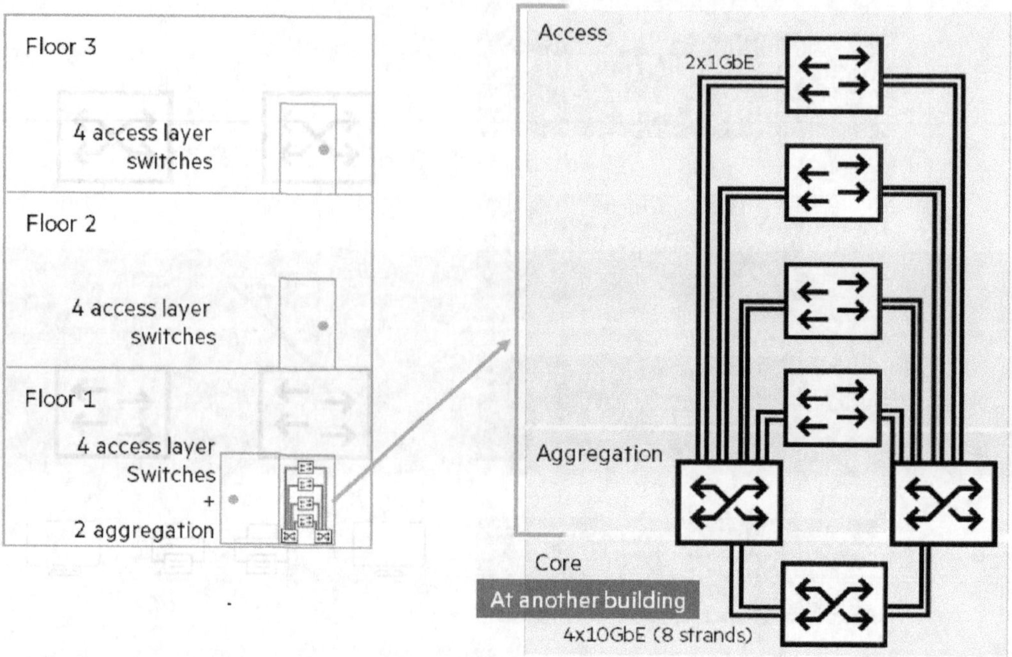

Single-building campuses can often use a 2-tier topology having sufficient cable to connect all access layer switches to the core. However, multi-building campuses may pose a challenge. Large sites may not have enough building-to-building fiber links to connect each access layer switch to the core, particularly when each switch needs a redundant link.

Using an aggregation layer and 3-tier topology has traditionally addressed these cabling issues. In each building, access layer switches connect to one or two aggregation layer switches. The aggregation layer switches then connect the entire building to the core.

For example, a building has 12 access-layer switches but limited fiber to connect the access switches to the core of another building. You can use an aggregation layer to aggregate the many 1GbE connections from the access layer into just four 10GbE connections, which fit on the limited cabling. VSF can also help solve cabling limitation issues by stacking access switches in each IDF.

Options for implementing routing

Once you select the architecture you want to use, you must choose where to implement routing.

- Traditionally, you implemented routing between the aggregation and core layer in a 3-tier topology to prevent issues from broadcasting domains that extend too far. However, because you can now use VSF stacking to eliminate spanning tree, you do not need to worry as much about larger layer 2 domains.

- Therefore, you can alternatively implement routing at the core.
- In a 2-tier topology, you normally only route at the core. However, many HPE Aruba Networking access layer switches support routing, which allows you to implement routing there as well.

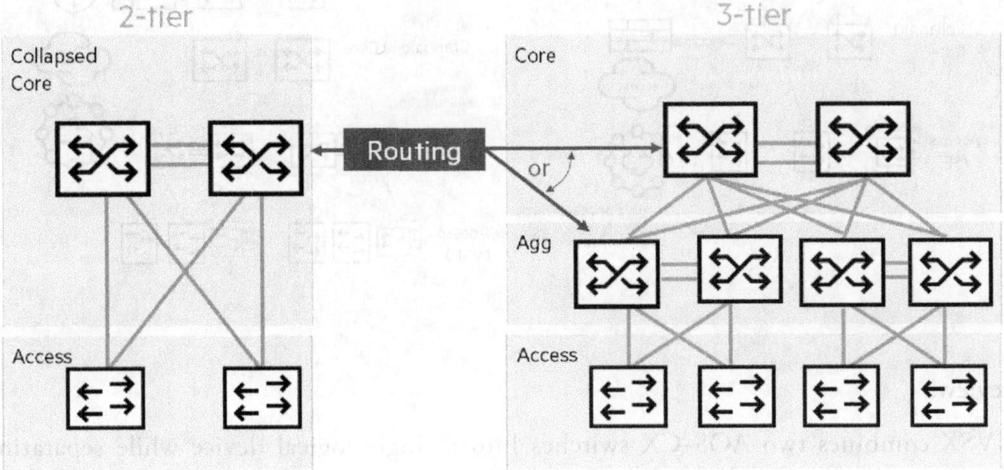

In the past, some architects preferred routing traffic at the access layer to avoid implementing a spanning tree. When properly designed, routed links load balance traffic better and failover faster than links using a spanning tree.

VSF stacking can provide many of the same benefits without requiring routing at the access layer.

VSX pair placement

The topology shows where VSX should be implemented across a 2-tier collapsed core and a 3-tier standalone core design. Note again that layer 3 routing is recommended between a standalone core and aggregation switches. This allows ECMP without the need for VSX to run on the core.

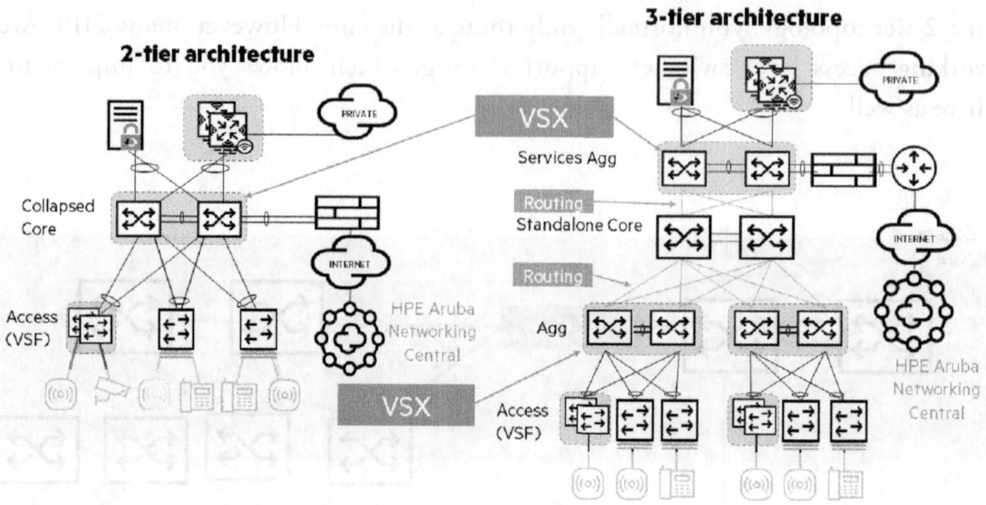

VSX review:

- VSX combines two AOS-CX switches into a single logical device while separating management and control planes. The goal of VSX is to enable a shared data plane for allowing VSX-LAGs (VSX-LAGs).
- Supported VSX models:
- 6400, 5420, 8320, 8325, 8360, 8100, 8400, 9300, and 10000 models. 6300 supports VSX starting with AOS-CX 10.16.
- You cannot use different models in a VSX pair. For example, 8320s and 8325s cannot be combined into a VSX pair.
- The Inter-switch link (ISL) synchronizes software and hardware tables shared between VSX pairs. The ISL is a standard link aggregation group (LAG) that carries the ISL protocol (ISLP).
- Active Gateway feature is a first-hop redundancy protocol (FHRP) that allows VSX pairs to share a virtual IP (VIP) address and virtual MAC (vMAC). Allows active-active forwarding on both VSX members.

VSX pair placement in campus fabric

In HPE Aruba Networking Central NetConductor distributed campus fabric VSX is typically deployed on Stub and Border switches.

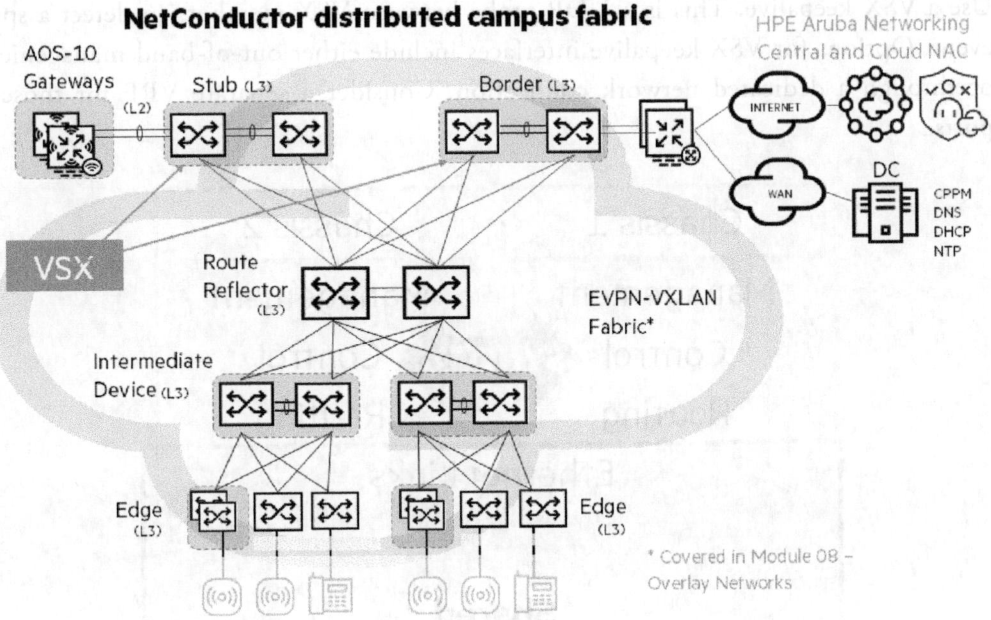

VSX design considerations

Consider the following best practices when designing VSX for a campus network:

- Live-Upgrade feature with zero downtime, given that each device has its own independent control plane and link to downstream access devices.
- With VSX-LAGs enabled, STP is not required but is recommended as an additional enhanced loop protection security mechanism. The link aggregation control protocol (LACP) enabled 2,4,6, or 8 physical ports into a single trunk interface for redundancy and capacity. LACP Fallback allows an active LACP interface to connect with its peer before it receives LACP protocol data units (PDUs). Use this feature for access witches using Zero Touch Provisioning (ZTP) to connect to aggregation switches across an LACP connection to receive provisioning.
- HPE Networking recommends running spanning-tree with MSTP. This is enabled by default on the 6xxx series switches. With point-to-point layer 3 routed links between aggregation and core, there is no requirement for MSTP. L2 Loops are prevented by L3 boundaries between layers.
- For VSX-LAGs, use LACP active mode. This required enabling needed VLANs. The best practice is to avoid allowing all VLANs.

- Use a VSX keepalive. This is a UDP probe between VSX members to detect a split-brain event. Options for VSX keepalive interfaces include either out-of-band management ports or through a dedicated network connection. Consider a separate VRF for those layer 3 ports.

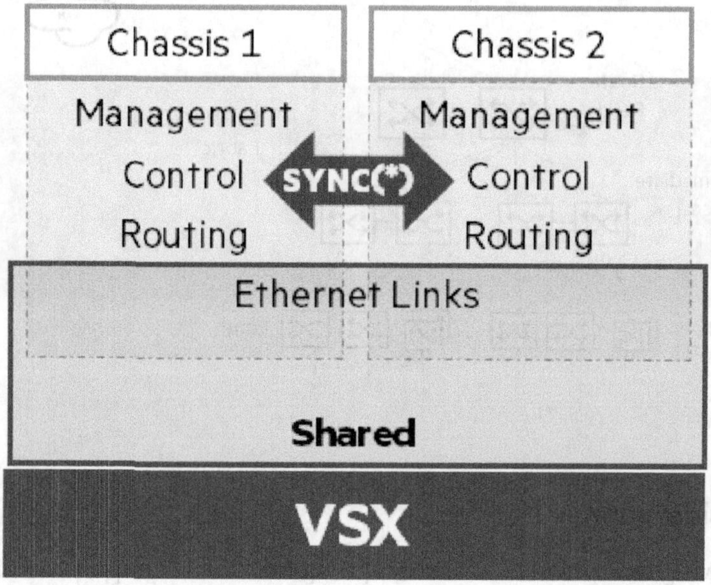

Active Gateway (AG) is an HPE Networking FHRP proprietary protocol. AG configures a shared virtual IP (VIP) within a subnet on both VSX members to be used as the default gateway for end-points. Both members must set the same VIP and vMAC addresses. There are four ranges reserved for private use:

- X2-xx-xx-xx-xx-xx
- X6-xx-xx-xx-xx-xx
- xA-xx-xx-xx-xx-xx
- xE-xx-xx-xx-xx-xx

Multicast considerations

When a multicast-enabled network, the default PIM Designated Router (DR) is the VSX primary switch, the VSX secondary can also establish PIM peering to avoid long convergence times if a VSX primary switch fails.

Configure PIM as active-active for each VSX pair.

2-tier and 3-tier wired

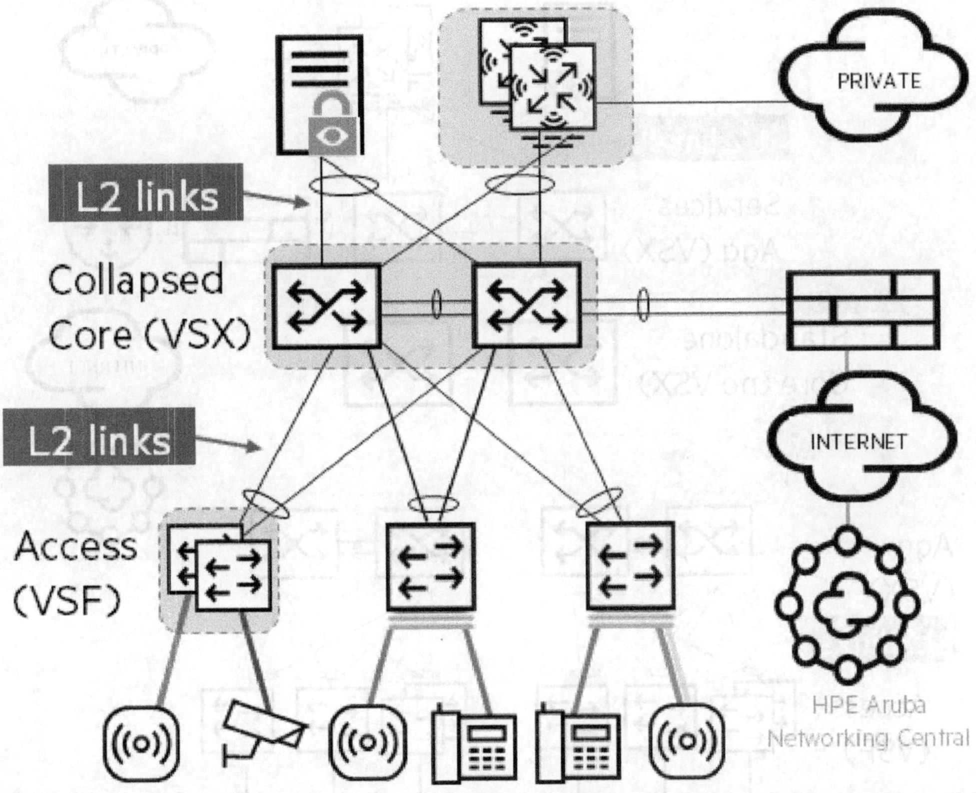

HPE Networking recommends either a 2-tier LAN with a collapsed core or a 3-tier LAN with a routed core. You will explore both a 2-tier collapsed core and the standalone core 3-tier design options.

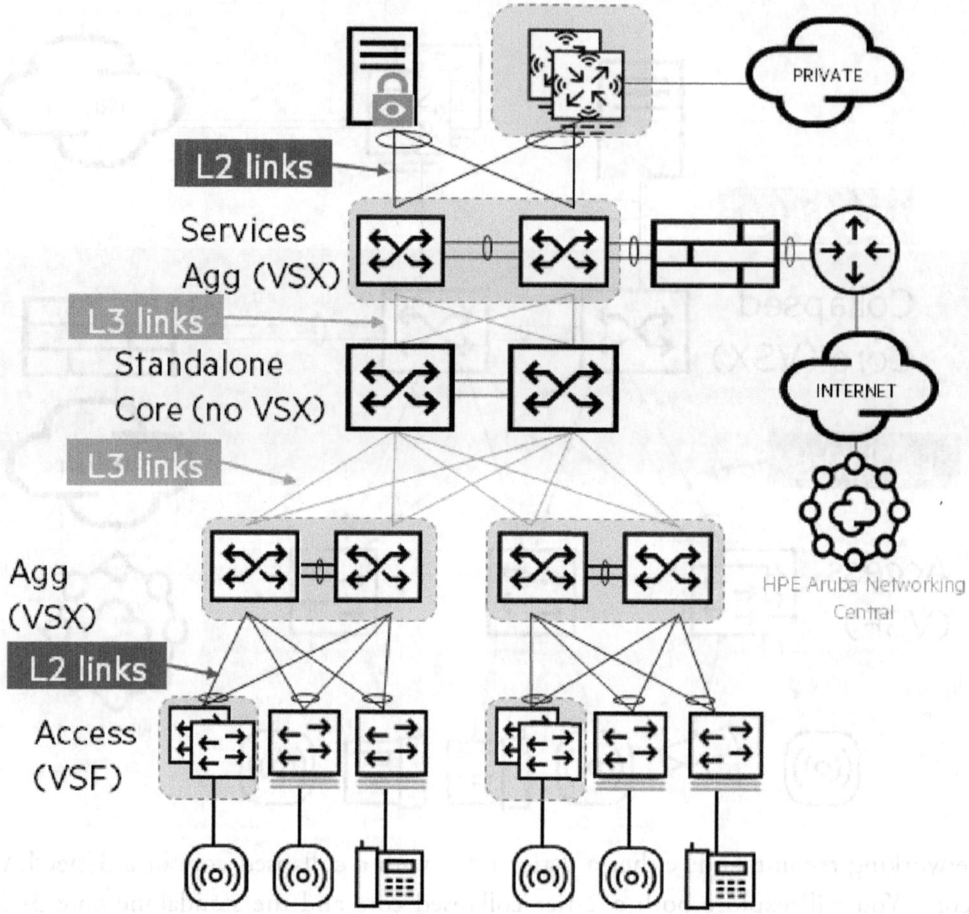

2-tier campus LAN

The 2-tier campus design is defined by a single "collapsed" core in place of both a core and aggregation layer switches. The fundamental purpose of the core is to interconnect access switches to avoid daisy-chaining IDF closets together and, instead, produce a hub-spoke topology where the hub is the core. This design is well suited for small buildings with few wiring closets so long as the core switches have enough high-speed ports to uplink all the access layer closets (IDFs). A 2-tier can be scaled to larger environments when the fiber cabling from each wiring closet is homed in a single location.

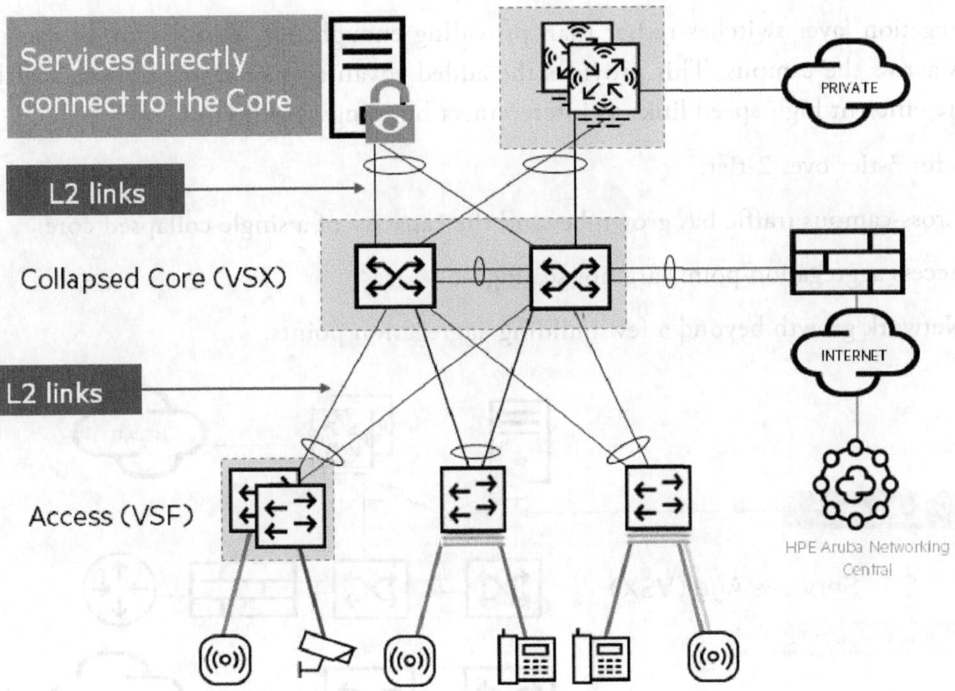

Build the 2-tier underlay utilizing stacking with VSX at the core to provide VSX-LAG layer 2 connections. Since services and access layer switches all connect to the core, redundant layer 2 VSX lags are recommended. For a simple design, use the core switches to provide layer 3 services and routing for traffic connected to layer 2 access switches.

Another option is to route between access and core layers to establish a resilient network over which a network overlay can run. Route using OSPF peering between each access and core switch, adding a loopback. Note that even in such a design, VSX should still be used in the core to provide service connectivity, such as SD-WAN gateways or ClearPass server clusters (shown in the topology).

The access layer switches in a 2-tier campus should also stack using VSF for easy deployment. Use uplink ports on different VSF stack members with one into each VSX-LAG configured core switch when running a layer-2 design. Enable the port-based authentication to apply HPE Aruba Networking Colorless Ports through port policies. Consider enabling layer 2 protection mechanisms along with STP, such as Loop Protection, BPDU Filter, Root Guard, and BPDU Protection.

3-tier redundancy core

A 3-tier campus is one where the network has grown to warrant using aggregation switches. Often a building with several IDF closets would localize the traffic to a pair of aggregation switches in the basement. Therefore, the core of a 3-tier campus is primarily responsible for interconnecting the aggregation layer switches rather than providing enough port density for all the access layer switches across the campus. This provides the added advantages of segmentation, scalable design, and more efficient high-speed links to interconnect buildings together.

Reasons for 3-tier over 2-tier:

- Cross-campus traffic has grown beyond the capacity of a single collapsed core.
- Access aggregation points are not scaling.
- Network growth beyond a few building aggregation points.

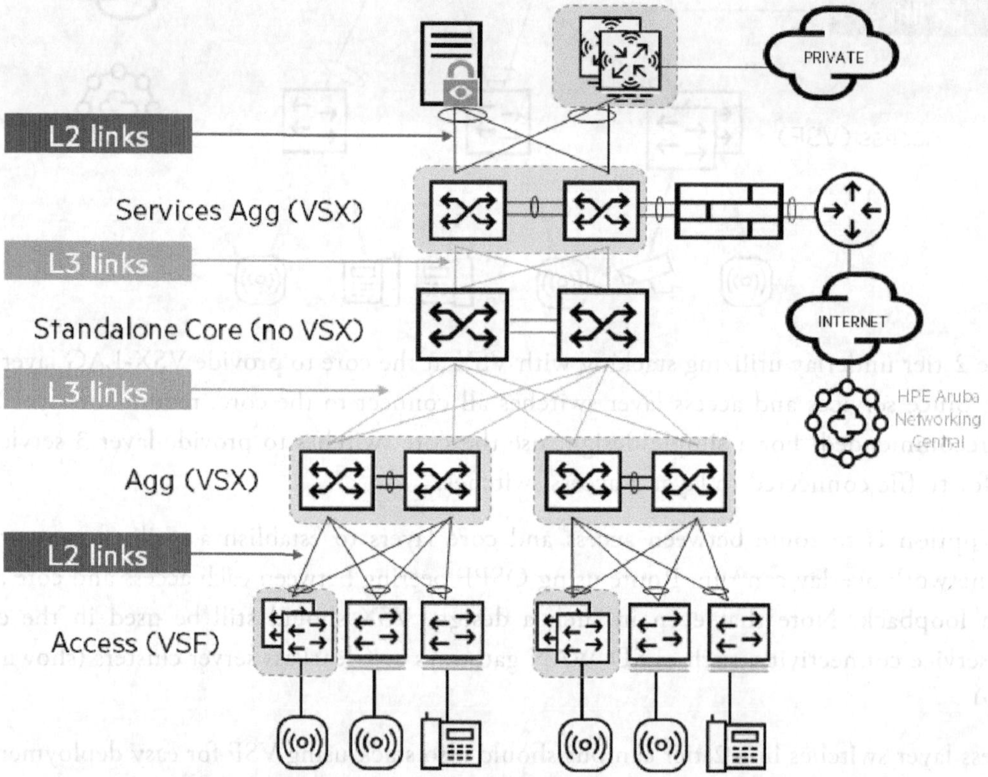

A 3-tier campus design is recommended for larger networks with thousands of users or where growth is limited due to wiring restrictions in a 2-tier campus. If a 2-tier is the core of a single building, a 3-tier connects multiple instances of those buildings together in a central core. The building "core" is then designated as an aggregation switch.

A pair of core switches can join the pairs of aggregation switches together with user high-speed layer 3 links. This should be designed to take advantage of equal-cost multipath (ECMP) to dynamic load sharing. Do not use VSX-LAGs to layer 3 aggregation switches. Use point-to-point routed links.

The access switches would remain in the same role as in the 2-tier design. Any connections from a layer-2 device, such as a gateway cluster or servers like ClearPass, should connect using VSX-LAG from the core. Use VSX-LAG for these devices but not for the routed aggregation switches. Gateways connected without an aggregation block make sense in environments with lower wireless tunneling.

When to use VSX in the core

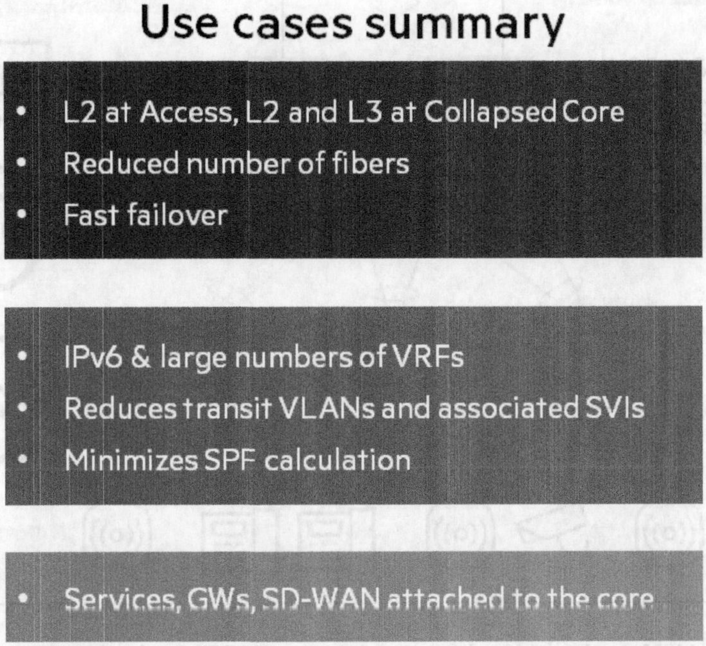

You should implement VSX at the core when only the aggregation layer operates at layer 2. In this case, VSX LAGs will help eliminate the need for spanning-tree, and VSX will enable the core switches to provide active-active default gateway services for access VLANs. This approach simplifies the topology. Traffic flows at layer 2 to the core, routing all the traffic. The simpler topology could lead to a reduced failover time.

While the VSX often operates at the aggregation layer, HPE Aruba Networking sometimes recommends using VSX at the core. These cases include when the network uses IPv6 and must support a large number of VRFs. Each VRF requires its transit VLAN and associated SVI on each VSX LAG. By combining the core switches in a VSX pair, you half the number of VSX LAGs, thus halving the number of transit VLANs and SVIs. This will simplify and minimize the burden of SPF calculations when you use OSPF as the routing protocol.

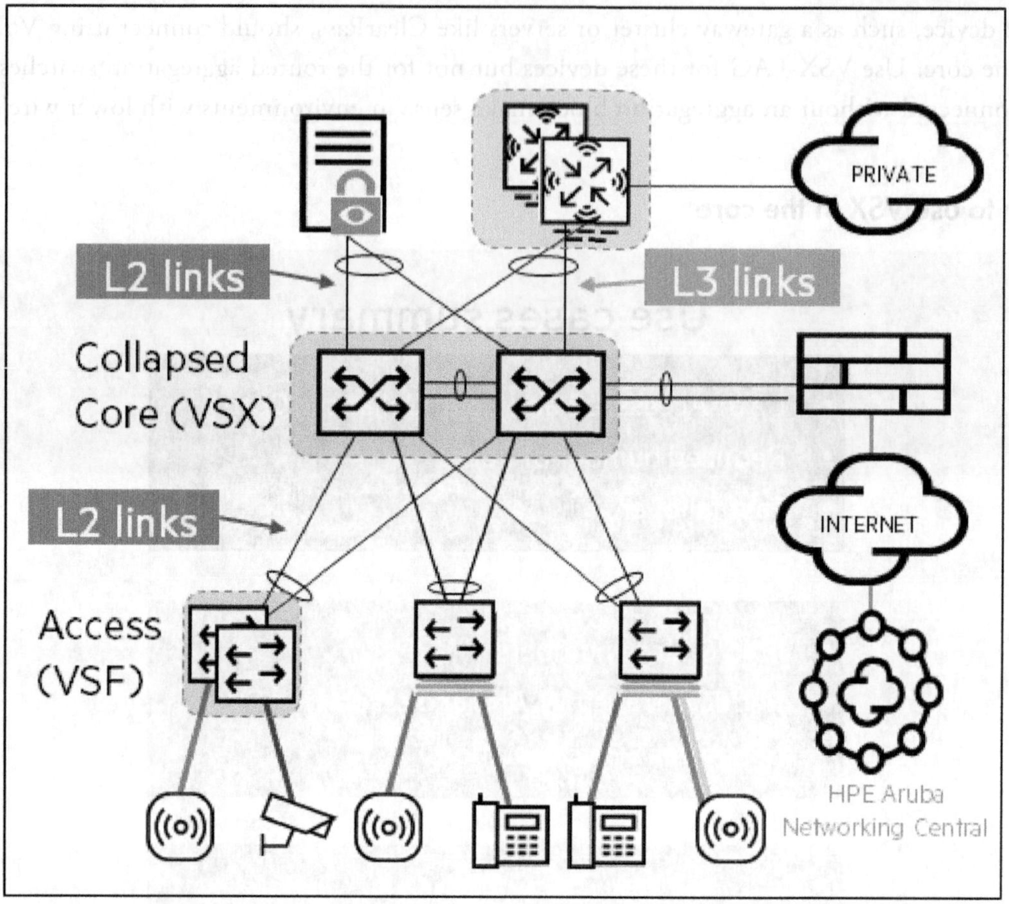

You must also use VSX at the core when you connect HPE Aruba Networking Gateways to the core. Then, the core switches can use VSX LAGs to connect to the GWs and provide active-active default gateway services for wireless device traffic. Because the GWs could forward traffic from a large number of wireless devices, you will want to ensure the core VSX pair has large enough MAC and ARP tables to accommodate all of the devices.

Note that VSX could be used at the core for required L2 services that can only connect to that core, EVEN if a 3-tier design.

Core layer

Here is a quick summary of what functionality each layer should provide.

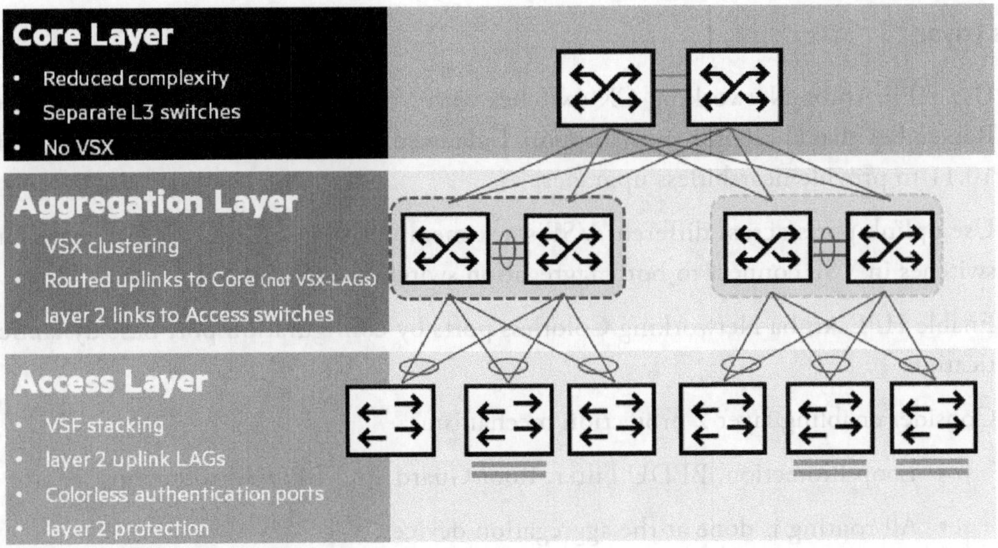

Core Layer

- Reduced complexity by only forwarding routed traffic.
- The standalone core uses separate switches instead of VSX, with dual ECMP connections to all aggregation switches. This allows for sub-second failover as soon as a physical link between a core and aggregation switch fails. Design with point-to-point links and ECMP in mind, as these are the most deterministic, and convergence is measured in milliseconds.

Aggregation Layer

- Use HPE Aruba Networking CX switches with VSX redundancy.
- Consider port density and bandwidth requirements to meet the demands of the campus access layer switches.
- Use VSX-LAGs for efficient layer 2 connectivity to servers and devices.
- Design with OSPF loopbacks with reachability on all switches, including layer 2 access switches. This will lay the foundation for future scalability in network automation and the orchestration of tunnels.

Access Layer

- Use HPE Aruba Networking CX switches with VSF support, such as the 6200s or 6300s. Remember that the 6300s now support Enhanced Software Upgrade (ESU) introduced in 10.11 to provide near-hitless upgrades.
- Use uplink ports across different VSF stack members in layer 2 access switches. The access switches in VSF connect to both aggregation switches across VSX-LAGs.
- Enable HPE Aruba Networking Colorless ports by configuration port base dynamic authentication.
- Consider enabling layer 2 protection mechanisms:
  - Loop Protection, BPDU Filter, Root Guard, and BPDU Protection.
  - All routing is done at the aggregation devices.

HPE Aruba Networking CX product portfolio

The HPE Aruba Networking CX portfolio includes modular access, aggregation, and core switches. At the access and aggregation layers, you get power over Ethernet (PoE), multi-gigabit speeds, secure segmentation, and stacking capabilities. You get high availability, bandwidth, and speed at the campus core and data center with 40 and 100 GbE scalability.

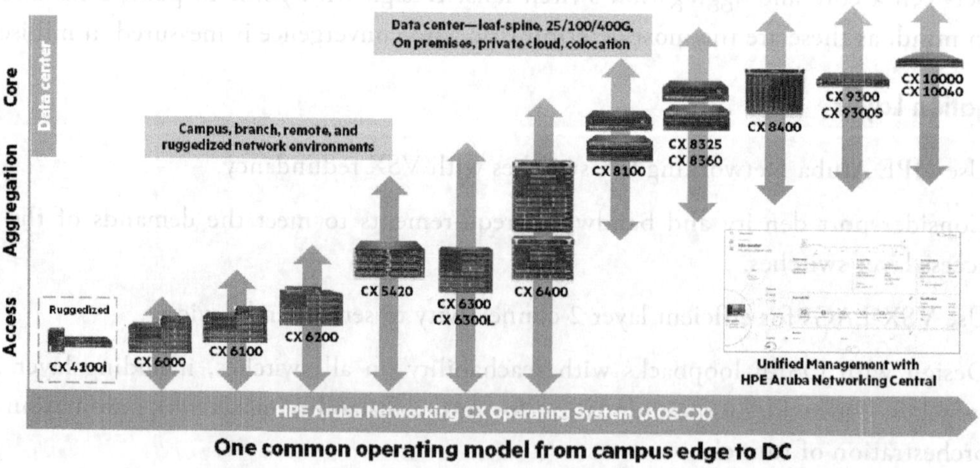

An overview and product positioning for the HPE Aruba Networking CX switching product portfolio are shown.

HPE Aruba Networking CX 10000 Series and 10400 switch series

The HPE Aruba Networking CX 10000 Series Switch Series helps network and security administrators distribute intelligence to the data center network-server edge, overcoming legacy network and security design limitations that often require overly complex, highly inefficient, and costly architecture designed to address decades old requirement. It combines the best-of-breed network operating system, HPE Aruba Networking CX, for data center, campus, and edge in addition to the fully programmable data processing unit (DPU). This allows the DPU-enabled CX 10000 series and CX 10040 switches to deliver stateful software-defined services inline, at scale, with wire-rate performance and orders of magnitude scale and performance improvements over traditional L2/L3 switches at a fraction of their total cost of ownership (TCO).

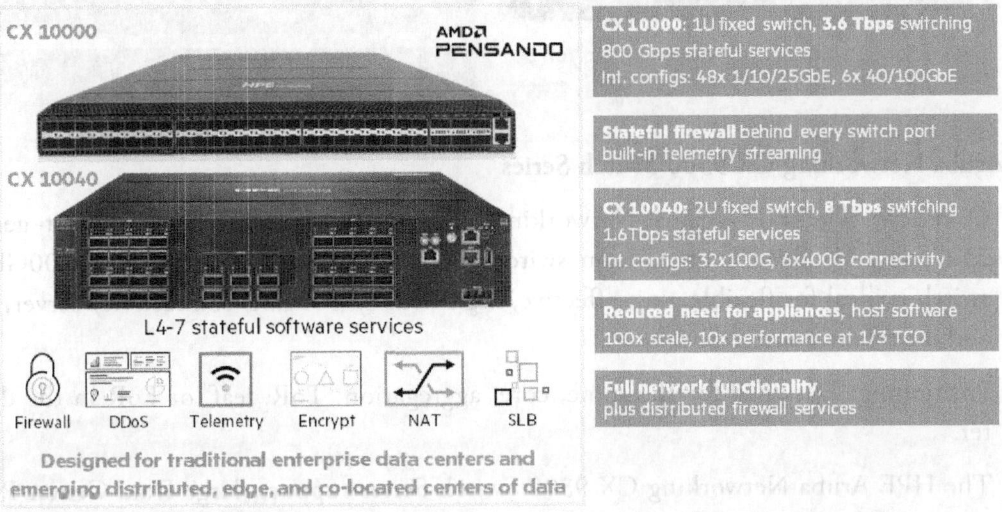

The switches allows operators to extend industry standard leaf-spine networking with distributed stateful segmentation, east–west firewalling, NAT, encryption, and telemetry services — all delivered inline, all the time, on every access port, closer to where critical enterprise applications run. The switches' distributed services architecture is agnostic of network deployment architecture. This allows the flexibility to enable stateful services delivery, when deployed as access, leaf top of rack (ToR), or end of row (EoR) in a data center, and potentially in the aggregation layer in campus or edge data center designs.

HPE Aruba Networking CX 9300 Switch Series

High performance 400GbE data center switches

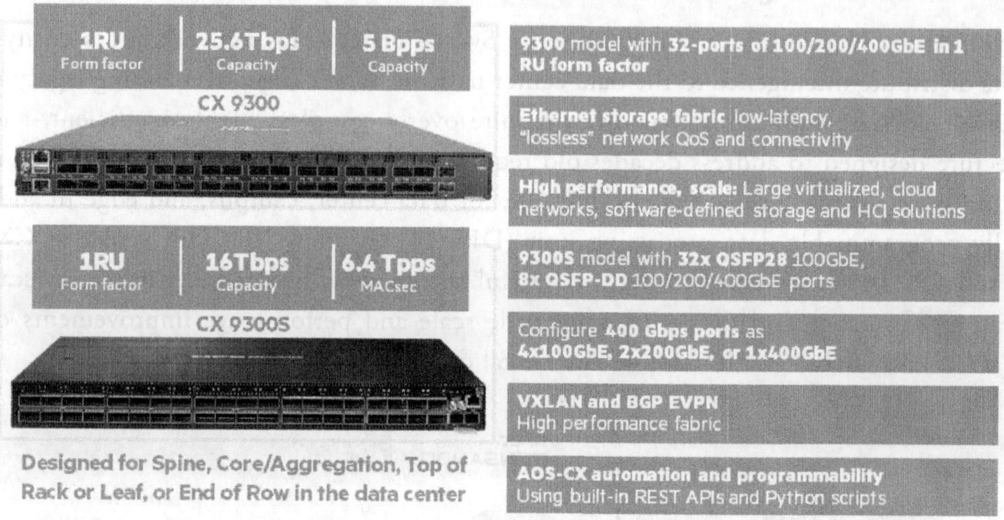

HPE Aruba Networking CX 9300 Switch Series

- Overview: The HPE Aruba Networking CX 9300 Switch Series is a next-generation 25.6Tbps, 1U fixed configuration switch supporting 32 ports of 100/200/400GbE. The switch is ideal for flexible, cost-effective, high-density 400GbE connectivity server, storage, and intra-fabric connectivity.
- Positioning: Designed for the spine, core, aggregation, ToR, leaf, or EoR in the data center.
- The HPE Aruba Networking CX 9300 Switch Series supports large Data Center PODS of up to 6,000 x 25GbE servers or up to 2,000 x 100GbE servers.
- Supports front-to-back or back-to-front airflow.

9300s series port-splitting considerations for two different port splitting profiles

- Profile 1: Maximizing 100G ports with 8xQSFP-DD (400G). Still supports up to 8xQSFP-DD (400G) or up to an additional 32x100G ports by using 400GbE QSFP-DD to 4xQSFP56 100G AOC splitters.
- Profile 2: Maximizing 25G while allow 4xQSFP-DD ports (400G). Still supports up to 8xQSFP-DD (400G) or up to 40x25G ports by using 100GbE QSFP28 to 4x25GbE SFP28 DAC splitters on ports 13,15,16,25,26,27 plus the DD ports 17,19,21,23.

Port-grouping considerations

- Groups 1,2,7,8 (ports 1-4,5-8,33-36,37-40) are PHY ports with speeds up to 100G.
- Groups 3,4,5,6 (ports 9-12,13-16,25-28,29-32) are PHYless ports with speeds up to 200G.
- Ports 17-24 are QSFP-DD ports not in a port group with speeds up to 400G.

HPE Aruba Networking CX 8400 Switch Series

Resiliency and performance built for the core

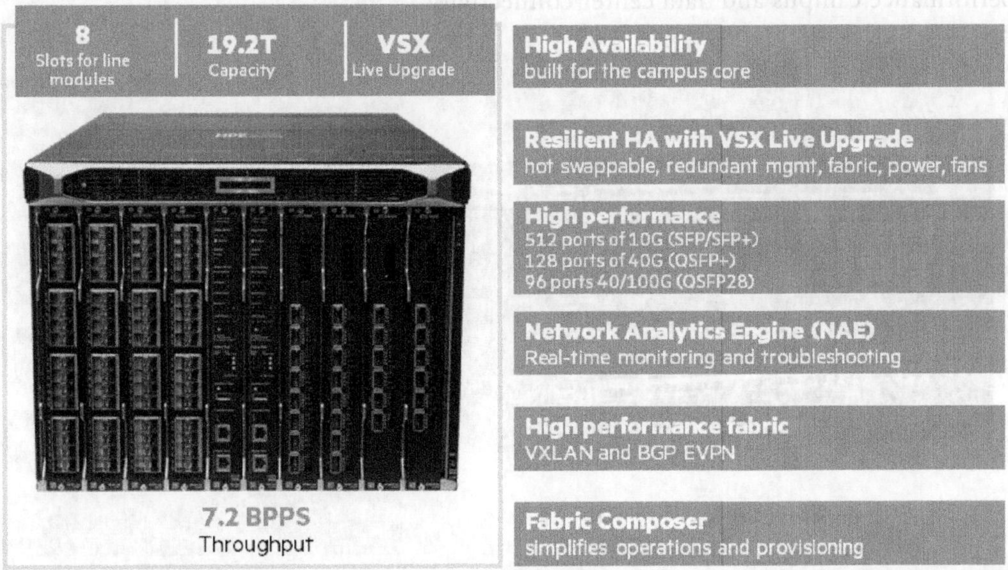

- Overview: A carrier-class, 8-slot chassis providing up to 19.2 Tbps of capacity with high density 10/40/100GbE connectivity with fully redundant fabric, management, power, and fans.
- Positioning: Ideal for large enterprise campus core and data center networks where high availability is a must, and the largest scale (table sizes and buffers) is required.
- Supports a maximum of 256 10GbE (SFP/SFP+) or 25GbE (SFP28) ports, or 64 40GbE (QSFP+) ports, or 48 ports 40/100GbE (QSFP28) combination.
- Empty configuration weight: 76 lbs (34 kg), full configuration weight: 241 lbs (109 kg).
- Can be mounted on a 2-post rack.
- Available modules:
  - HPE Aruba Networking 8400X 32-port 10GbE SFP/SFP+ with MACsec Advanced Module (JL363A)
  - HPE Aruba Networking 8400X 8-port 40GbE QSFP+ Advanced Module (JL365A)

- HPE Aruba Networking 8400X 6-port 40GbE/100GbE QSFP28 Advanced Module (JL366A)
- HPE Aruba Networking 8400X-32Y 32-port 1/10/25G SFP/SFP+/SFP28 Module (JL687A)

HPE Aruba Networking CX 8360 Switch Series

High performance campus and data center connectivity

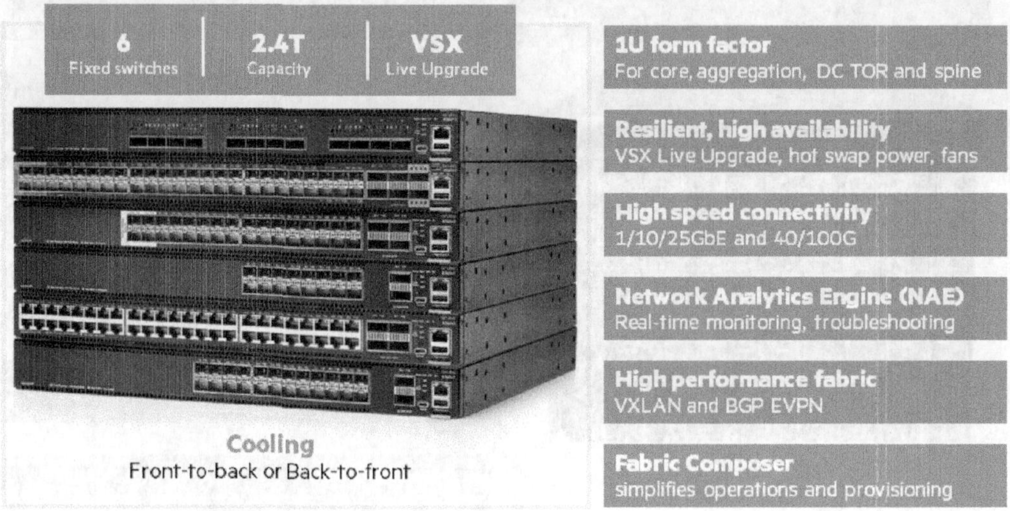

The HPE Aruba Networking CX 8360 Switch Series offers a flexible and innovative approach to addressing the application, security, and scalability demands of the mobile, cloud, and IoT era. These switches serve the needs of the next-generation core and aggregation layer of campuses, as well as virtual and highly dynamic data center environments. They provide up to 4.8 Tbps of capacity, with line-rate Gigabit Ethernet interfaces including support for Smart Rate (1/2.5, 5 Gbps), 10 Gbps, 25 Gbps, 40 Gbps, 50 Gbps, and 100 Gbps. The CX 8360 series includes industry- leading line rate ports with 1/10/25/50GbE (SFP/SFP+/SFP28) and 40/100GbE (QSFP+/QSFP28) connectivity in a compact 1U form factor.

Key features:

- High performance 4.8 Tbps and 2678 Mpps
- Intelligent monitoring and visibility with the HPE Aruba Networking Network Analytics Engine
- High availability with industry leading VSX redundancy, and redundant power supplies and fans

- Designed for core/aggregation in the campus or Top of Rack or End of Row in data center environments
- MACsec secured connectivity over untrusted domains
- HPE Aruba Networking OS-CX automation and programmability using built-in REST APIs and Python scripts
- Advanced Layer 2/3 feature set includes BGP, OSPF, VRF, and IPv6
- 50G support on 8360-48Y6C, 32Y4C, and 16Y2C switches
- Addition of Smart Rate (1/2.5/5/10G) ports on the 8360-48XT4C

HPE Aruba Networking CX 8325 Switch Series

High performance campus and data center connectivity

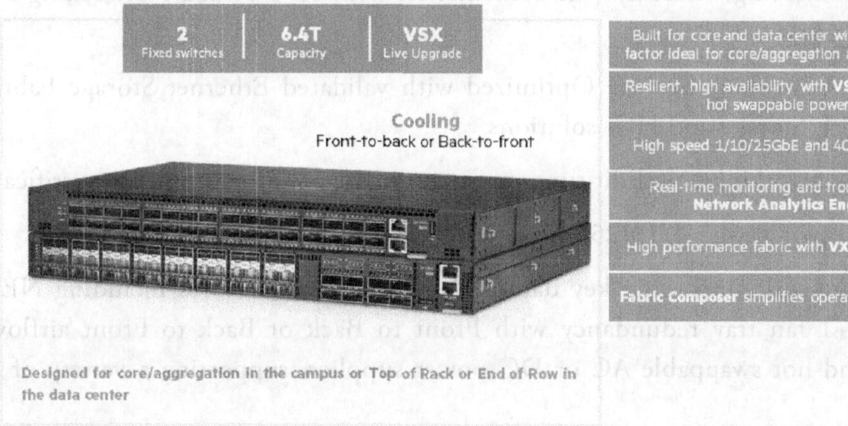

The HPE Aruba Networking CX 8325 Switch Series offers a flexible and innovative approach to addressing the application, security, and scalability demands of the mobile, cloud and IoT era. These modern family of stackable switches serve the needs of the next generation core and aggregation layer, as well as emerging data center requirements at the Top of Rack (ToR) and End of Row (EoR). They provide over 6.4 Tbps of capacity, with line-rate Gigabit Ethernet interfaces including 1 Gbps, 10 Gbps, 25 Gbps, 40 Gbps, and 100 Gbps. Additionally, the new CX 8325H Switch Series are designed to provide high-performance connectivity (10/25/40/100G) with fixed, redundant power and cooling in a half-width, single RU form factor, while still meeting the requirements of a wide array of compute and storage, HCI networking use cases.

CX 8325 series key features:

- High performance 6.4 Tbps of bidirectional switching capacity with 2,000 Mpps
- High availability with industry-leading VSX redundancy, and redundant power supplies and fans
- Designed for core/aggregation in the campus or Top of Rack or End of Row in the data center
- HPE Aruba Networking CX Switch Operating System automation and programmability using built-in REST APIs and Python scripts
- Advanced Layer 2/3 feature set includes BGP, OSPF, VRF-Lite, and IPv6
- Dynamic VXLAN* with BGP-EVPN for deep segmentation in data center and campus networks
- Intelligent monitoring, visibility, and remediation with HPE Aruba Networking Network Analytics Engine
- HPE Aruba Networking Storage Optimized with validated Ethernet Storage Fabric solutions with HPE Storage and HCI solutions
- HPE Aruba Networkingg NetEdit support for automated configuration and verification
- Compact 1 U switch with 1/10/25GbE and 40/100GbE connectivity
- High hardware scale, including key data center and DCBx features including NEBS certification, N+1 fan tray redundancy with Front to Back or Back to Front airflow, N+1 redundant and hot swappable AC or DC power supplies, supporting a variety of deployment models

CX 8325P-32C Key features:

- Supports PTP (1588 v2), SMPTE, Telecom Profiles G.8275.1 and G.8275.2 with Class C performance, Synchronous Ethernet (SyncE, G.8262.1)
- PTP timing interfaces: 1 PPS, 10 MHz, and ToD+1 PPS input and output clocks GNSS module with antenna input interface
- Supports Global Navigation Satellite System (GNSS) to support Telco Boundary Clock and, in the future, Telco Grand Master Clock

CX 8325H Key features:

- High-performance, high density 1/10/25/40/100G connectivity with fixed, redundant power (1+1) and fans (3+1).
- Enables low-latency and lossless network QoS and connectivity characteristics that storage requires.

- CX 8325H-18Y4C is designed to support any combination of 1/10/25G interfaces on its 18 ports - no port grouping necessary. The 4 ports of 100G can be split using breakout cabling depending on your needs.
- CX 8325H-16C allows for different port speeds with breakout cabling, allowing for a single 100G to be split into 4x25G.

HPE Aruba Networking CX 8325H switch series

Cost-effective, half-width Ethernet network switches for Storage, HCI and DC applications

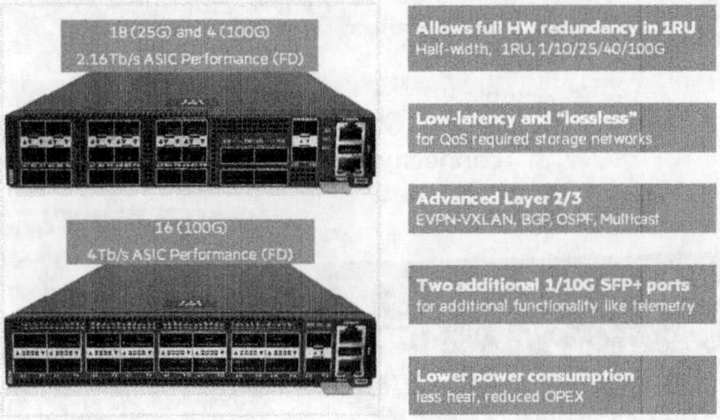

The HPE Aruba Networking CX 8325H switches are designed to provide a high-performance 1/10/25/40/100G connectivity with fixed, redundant power and cooling in a half-width, single RU form factor while meeting the requirements of a wide array of compute and storage lossy and lossless networking use cases. These half-width switches help organizations balance density, high availability, and affordability which helps to save space and costs on server, storage, and HCI connectivity.

HPE Aruba Networking CX 8325H is also ideal for smaller edge or colocation data centers that require limited port density or have limited rack space. The CX 8325H switches can be used in small campus and branch locations where space is at a premium (e.g., retail locations, healthcare facilities, media, financial, and enterprise branches) and which don't always have IT racks, but often have storage closets being used to store servers and IT equipment.

Key benefits:

- High-performance, high density 1/10/25/40/100G connectivity with fixed, redundant power (1+1) and fans (3+1).
- Enables low-latency and lossless network QoS and connectivity characteristics that storage requires.
- CX 8325H-18Y4C is designed to support any combination of 1/10/25G interfaces on its 18 ports - no port grouping necessary. The 4 ports of 100G can be split using breakout cabling depending on your needs.

- CX 8325H-16C allows for different port speeds with breakout cabling, allowing for a single 100G to be split into 4x25G.
- Supports VXLAN-EVPN, BGP, OSPF, Multicast.

HPE Aruba Networking CX 8320 Switch Series

High performance campus and data center connectivity

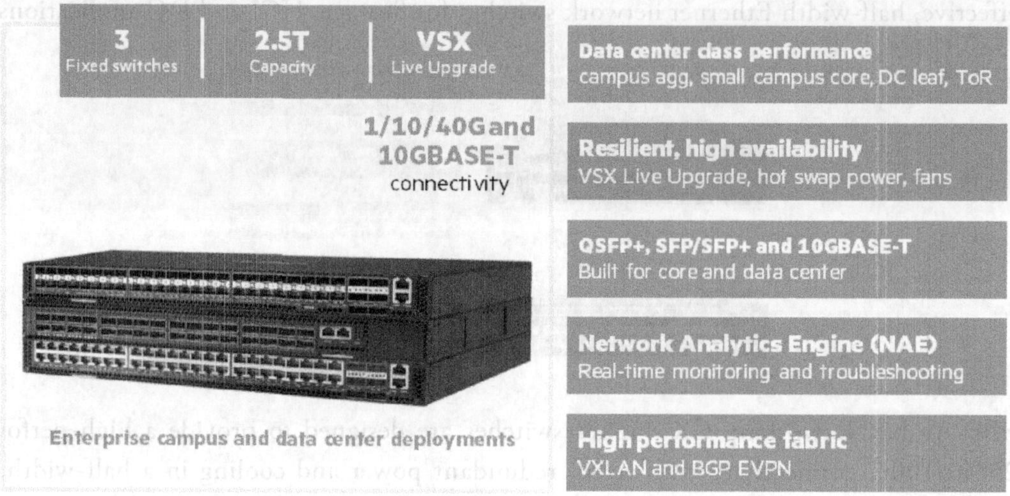

The past several decades in networking have been defined by static, closed networking solutions designed for the client-server era. The HPE Aruba Networking CX 8320 Switch Series is a game-changing campus core and aggregation solution offering a flexible and innovative approach to dealing with the demands of the mobile, cloud, and Internet of Things (IoT) era. The CX 8320 series also serves as a top-of-rack (ToR) switch for data centers needing 10GbE connectivity to servers and 40GbE to the spine. The CX 8320 series provides industry-leading line rate 1/10GbE (SFP/SFP+ and 10GBASE-T) and 40GbE connectivity in a compact 1U form factor. It provides up to 2.5 Tbps of switching capacity with redundant power and fans combined with HPE Aruba Networking Virtual Switching Extension for a high availability solution ideal for enterprise campus and data center deployments.

Key features:

- High performance 2.5 Tbps with 1,905 MPPS
- High availability with HPE Aruba Networking Virtual Switching Extension, and redundant, hot-swappable power supplies and fans
- HPE Aruba Networking CX Switch Operating System enables automation and programmability using built-in REST APIs and Python scripts

- Intelligent monitoring, visibility, and remediation with HPE Aruba Networking Network Analytics Engine (NAE)
- One touch deployment with the HPE Aruba Networking CX mobile application
- HPE Aruba Networking Switch Multi-Edit Software support for automated configuration and verification
- Advanced Layer 2/3 feature set includes BGP, OSPF, VRF, and IPv6
- Compact 1U switches with 1/10GbE (SFP/SFP+ and 10GBASE-T) and 40GbE connectivity

HPE Aruba Networking CX 8100 Switch Series

Enterprise campus aggregation and data center top-of-rack server connectivity

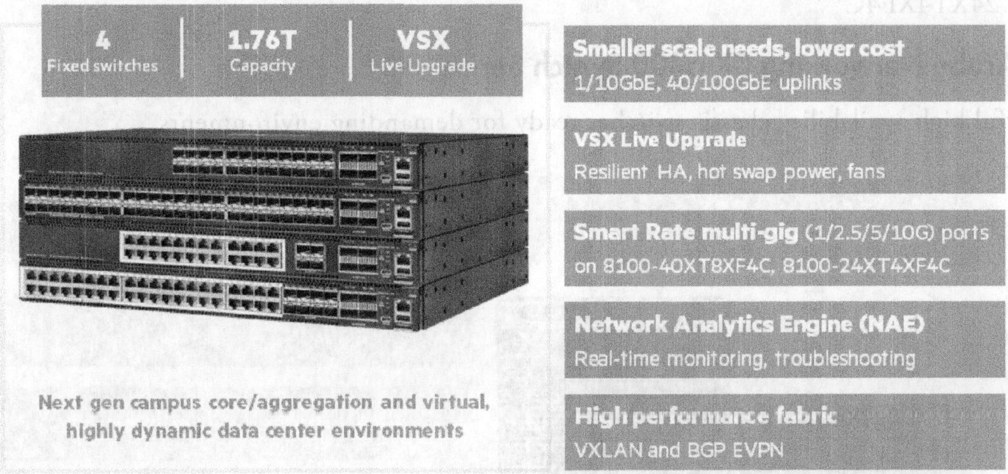

The HPE Aruba Networking CX 8100 Switch Series offers a flexible and innovative approach to addressing the application, security, and scalability demands of the mobile, cloud, and IoT era. These switches serve the needs of the nextgeneration core and aggregation layer of campuses, as well as virtual and highly dynamic data center environments. They provide up to 1.76 Tbps of capacity, with line-rate Gigabit Ethernet interfaces including support for Smart Rate (1/2.5/5 Gbps), 10 Gbps, 25 Gbps, 40 Gbps, and 100 Gbps. The CX 8100 series includes industry-leading line rate ports with 1/10GbE (SFP/SFP+) and 40/100GbE (QSFP+/QSFP28) connectivity in a compact 1U form factor. 4x10 Gbps and 4x25 Gbps break out from 40/100G ports offer advanced flexibility in connectivity and aggregation. These switches deliver a fantastic investment for customers requiring 1GbE/10GbE with a mix of fiber and copper ports, and 40/100GbE uplink ports.

Key features:

- High performance 1.76 Tbps and 1309 Mpps
- Intelligent monitoring and visibility with HPE Aruba Networking Network Analytics Engine
- High availability with industry leading VSX redundancy, and redundant power supplies and fans
- Designed for core/aggregation in the campus or top of rack or end of row in data center environments
- AOS-CX automation and programmability using built-in REST APIs and Python scripts
- Advanced Layer 2/3 feature set includes BGP, OSPF, VRF, and IPv6
- Addition of Smart Rate (1/2.5/5/10G) ports on the 8100- 40XT8XF4C and 8100-24XT4XF4C

HPE Aruba Networking CX 6400 Switch Series

Powerful high availability chassis switches ready for demanding environments

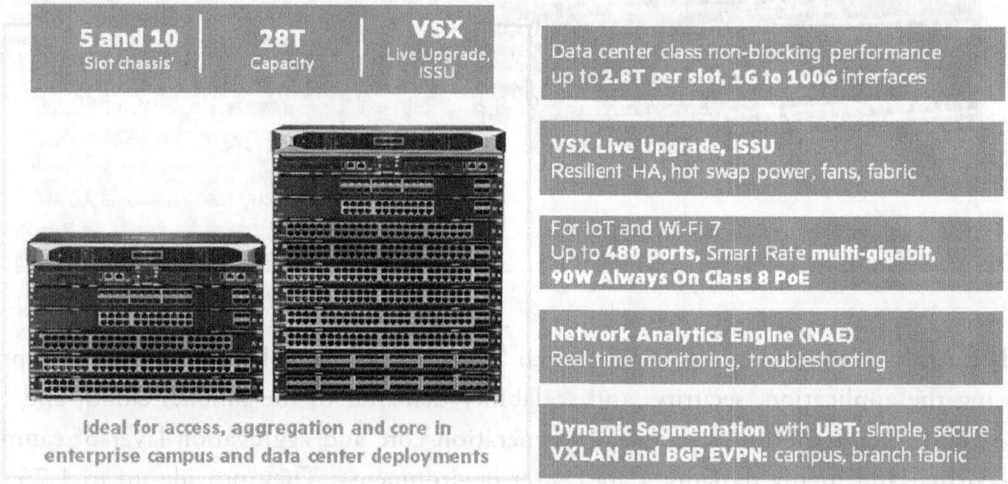

The HPE Aruba Networking CX 6400 Switch Series is a modern, flexible, and intelligent family of modular switches ideal for access, aggregation and core in enterprise campus and data center deployments. Created for game- changing operational efficiency with built- in security and resiliency, the 6400 switches provide the foundation for high performance networks supporting IoT, mobile and cloud applications. Built from the ground up with a combination of cutting-edge hardware, software and analytics and automation tools, the 6400 switches are part of the HPE Aruba Networking CX Switch Operating System portfolio, designed for today's enterprise campus, branch and data center networks. By combining a modern, fully programmable OS with the HPE Aruba Networking Network Analytics Engine, the 6400 switches provide industry leading monitoring and troubleshooting capabilities across the network.

Key features:

- Powerful, modular Layer 3 switches with BGP, MPLS, EVPN, VXLAN, VRF, and OSPF with robust security and QoS
- High performance switching with up to 28 Tbps with 11.4 Bpps - High availability with VSX live upgrades, redundant power supplies and fans
- High capacity HPE Smart Rate1 (1/2.5/5/10GbE) multi-gigabit, 60W to 90W PoE, and SFP+ modules - High speed, non-blocking 1/10/25/40/50/100GbE ports
- Intelligent monitoring, visibility, and remediation with HPE Aruba Networking Network Analytics Engine (NAE)
- Manage via single pane of glass with HPE Aruba Networking Central across wired, wireless, and WAN
- HPE Aruba Networking NetEdit support for automated configuration and verification
- HPE Aruba Networking dynamic segmentation enables secure and simple access for users and IoT devices

HPE Aruba Networking CX 5420 switch

Powerful, secure, Layer 3 modular chassis

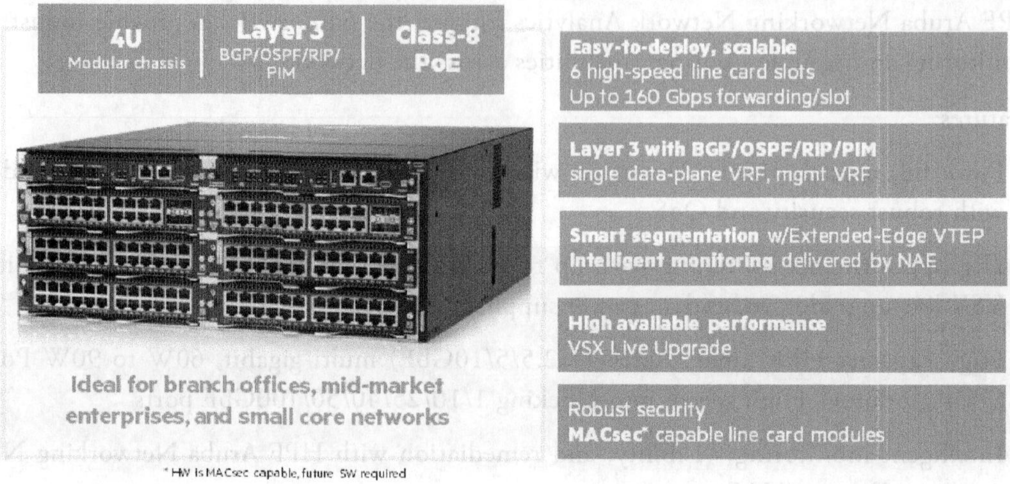

The HPE Aruba Networking CX 5420 Switch Series is an intelligent, flexible and modular Layer 3 switch chassis created for today's enterprise small and medium campus, branch enterprise networks with built-in MACsec hardware capability, analytics and automation. CX 5420 switch chassis is part of the CX switching portfolio that provides an enterpriseclass access layer solution that is easy-to-deploy, simple, secure and resilient. By combining a modern, fully programmable OS with HPE Aruba Networking Network Analytics Engine, the CX 5420 Series provides industry leading monitoring and troubleshooting capabilities across the network. A powerful Gen7 ASIC architecture delivers performance and extensive feature support with flexible programmability for the tomorrow applications. This advanced modular 4U chassis switch offers up to 960Gbps with low latency, 6 half-width line card slots, VSX Live Upgrade, and HPE Smart Rate 10GbE up to 90W Class-8 PoE with BGP/OSFP/PIM support.

Key features:

- Powerful, modular, intelligent Layer 3 switch chassis with BGP, VSX, and OSPF with robust quality of service (QoS).
- Flexibility with 6 half-width modules that allow growth when needed and support for port diversity in a condensed 4RU chassis with redundant MM and PSU.
- High performance switching with up to 960 Gbps using 714 Mpps, built-in high availability with HPE Aruba Networking Virtual Switching Extension (VSX) Live Upgrade and Always-On PoE.
- Simplified operations with the common HPE Aruba Networking CX Switch Operating System (AOS-CX) used across the CX portfolio delivering a more consistent, more efficient operator experience.

- Easy to manage and simplify configurations with HPE Aruba Networking Central, HPE Aruba Networking Switch Multi-Edit Software, with easy-to-use web GUI or industry-standard CLI.
- Automated, unified role-based policies for users, Internet of Things (IoT) and devices with HPE Aruba Networking Dynamic Segmentation and intelligent monitoring, visibility, and remediation with NAE.

HPE Aruba Networking CX 6300 Switch Series

Powerful Layer 3 stackable switches ready for demanding environments

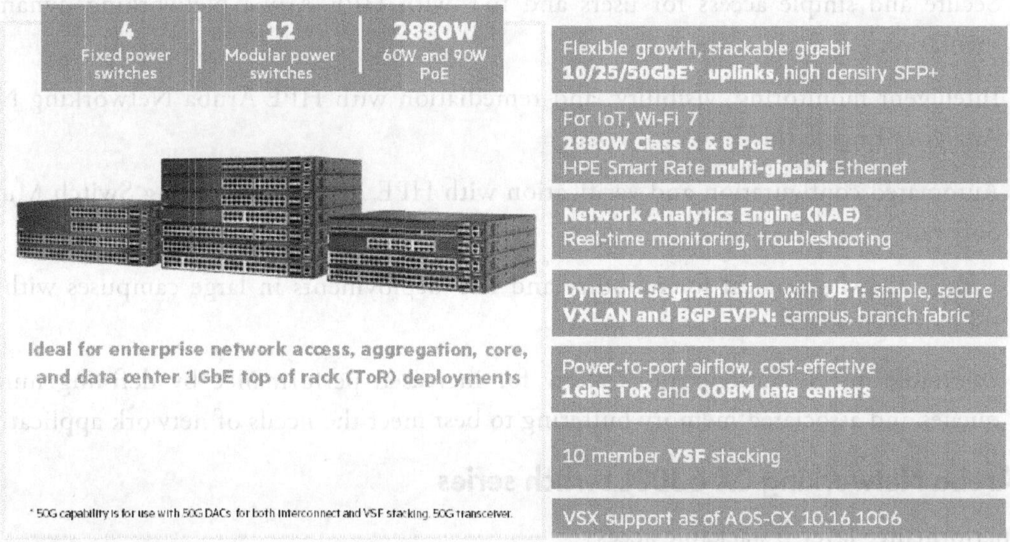

The HPE Aruba Networking CX 6300 Switch Series is a modern, flexible, and intelligent family of stackable switches ideal for enterprise network access, aggregation, core, and data center top of rack (ToR) deployments. Created for gamechanging operational efficiency with built-in security and resiliency, the 6300 switches provide the foundation for high-performance networks supporting IoT, mobile and cloud applications.

Key features:

- Stackable Layer 3 switches with BGP, EVPN, VXLAN, VRF, and OSPF with robust security and QoS
- Non-blocking performance with up to 1760 Gbps switching capacity, up to 1310 Mpps of forwarding throughput and up to 400 Gbps stacking bandwidth.

- Compact 1U switches with full density HPE Smart Rate (1G/2.5G/5G/10GbE) multi-gigabit, up to 90W PoE (Class 8), 10G LRM SFP+, and built- in high speed 1/10/25/40/50/100GbE uplinks available on select models1
- High availability and scale with 2 to 10-member Virtual Switching Framework (VSF) stacking. VSF in-service software upgrades delivers ISSU orchestration with no downtime or restart when upgrading within the same major release
- Management via single point of visibility and control with HPE Aruba Networking Central across wired, wireless, and WAN
- Secure and simple access for users and IoT with HPE Aruba Networking dynamic segmentation
- Intelligent monitoring, visibility, and remediation with HPE Aruba Networking Network Analytics Engine (NAE)
- Automated configuration and verification with HPE Aruba Networking Switch Multi-Edit Software
- Large TCAM size ideal for mobility and IoT deployments in large campuses with several thousand clients
- Selectable queue configurations allow for increased performance by defining number of queues and associated memory buffering to best meet the needs of network applications

HPE Aruba Networking CX 6300L switch series

High performance layer 2 stackable access

- High performance, power, and connectivity with up to 10G SR downlinks, 25G/50G uplinks, and Class 8 (90W) PoE.
- Stronger data link layer security with standards-based MACsec 256 encryption and authentication on uplinks and downlinks.
- Time synchronization support with precision time protocol (PTP) boundary clock (BC) and transparent clock (TC).
- Always-on reliability with VSF stacking and modular hot-swappable fans and power supplies

HPE Aruba Networking CX 6200 Switch Series

Lite Layer 3 stackable access switches for branch offices and SMBs

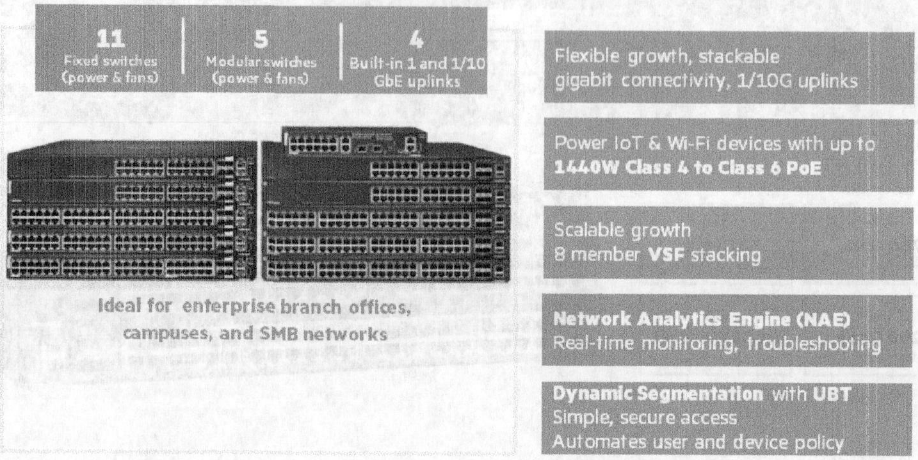

The HPE Aruba Networking CX 6200 Switch Series is a next-gen family of stackable access switches ideal for enterprise branch offices, campuses, and SMB networks. Created for game-changing operational efficiency with built-in analytics and automation, CX 6200 series switches provide an enterprise-class access layer solution that's simple and secure. The stackable CX 6200 switches are built from the ground up with a combination of cutting-edge hardware, software, and analytics and automation tools. By combining a modern, fully programmable operating system with the HPE Aruba Networking Network Analytics Engine (NAE), the CX 6200 series brings industry leading monitoring and troubleshooting capabilities to the access layer. A powerful HPE Aruba Networking Gen7 ASIC architecture delivers reliable performance and enterpriseclass feature support with flexible programmability for tomorrow's applications. The CX 6200 series is designed for simple deployment using the intuitive HPE Aruba Networking CX Mobile App that speeds install, configuration and enables stacking of up to 8 switches. The CX 6200 series includes fixed (CX 6200F series) and modular (CX 6200M series) switches with built-in high-speed uplinks. Fixed switches support up to 30W of PoE per port. Flexible, modular switches offer hot-swappable power supplies and fans for enhanced resiliency and redundancy, and up to 60W of PoE per port to power the latest IoT devices, security cameras, and wireless APs.

Key features:

- Enterprise-class connectivity with support for ACLs, robust QoS, and common protocols such as static and Access OSPF routing
- Scalability with 8-member switch VSF stacking for up to 384 downlink ports
- Versatile 1G/10G uplinks with LRM and MACsec 256 support on modular switches

- Convenient built-in 1G/10G uplinks on fixed power switches and additional cost-efficient 1G uplink switch models
- Industry standard Power over Ethernet with up to 30W PoE (Class 4) (PoE) per port on fixed power switches and up to 60W PoE (Class 6) per port on modular power switches
- Intelligent monitoring, visibility, and troubleshooting with HPE Aruba Networking Network Analytics Engine
- Manage through single pane of glass with HPE Aruba Networking Central across wired, wireless, and WAN
- Simple, one touch deployment with the HPE Aruba Networking CX Mobile App
- Automated configuration and verification with HPE Aruba Networking Switch Multi-Edit Software
- Secure and simple access for users and IoT with HPE Aruba Networking dynamic segmentation

HPE Aruba Networking CX 6200 Switch Series modular and fixed power models

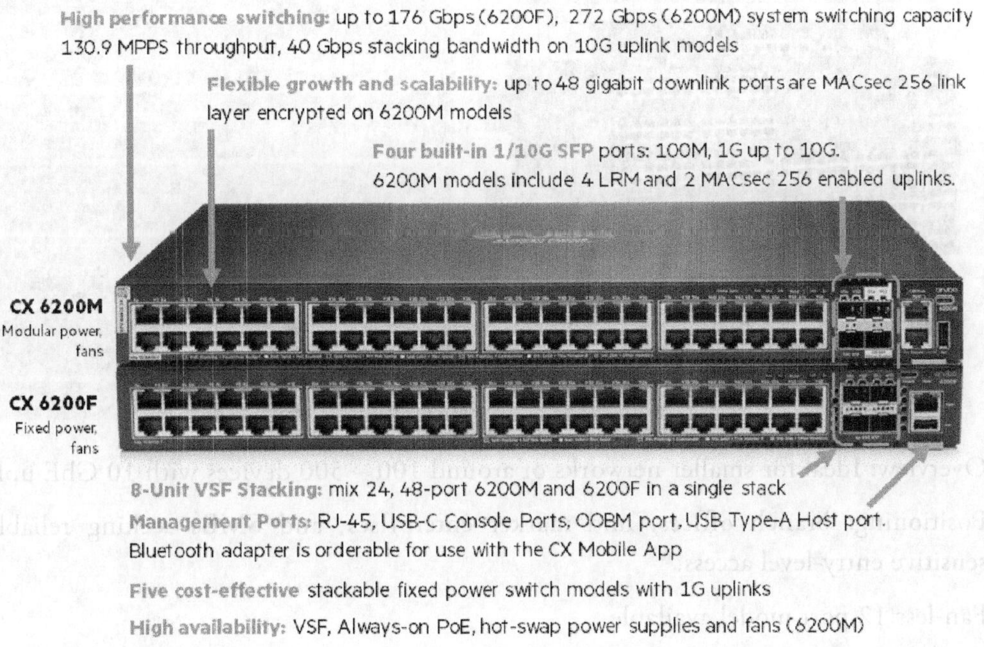

- Overview: These fixed-port access switches support 8-member stacking with built-in 1/10GbE uplinks and PoE.

- Positioning: Ideal for branch offices, mid-market enterprises, and SMBs looking for a reliable, simple-to-deploy, cost-sensitive access solution.
- Supports module fans/power supplies on the 6200M series versus the 6200F (fixed) versions.
- Up to 30W PoE (Class 4) per port on 6200F series and up to 60W PoE (Class 6) per port on modular power switches.
- Built-in 1G/10G uplinks on fixed power switches.
- Supports a mix of 24, and 48-port modular and fixed CX 6200 series models within a single VSF stack.
- Note that the CX 6200F series 12G supports an 8-member stack with other 12p 6200F switches only.

HPE Aruba Networking CX 6100 Switch Series and CX 6000 series

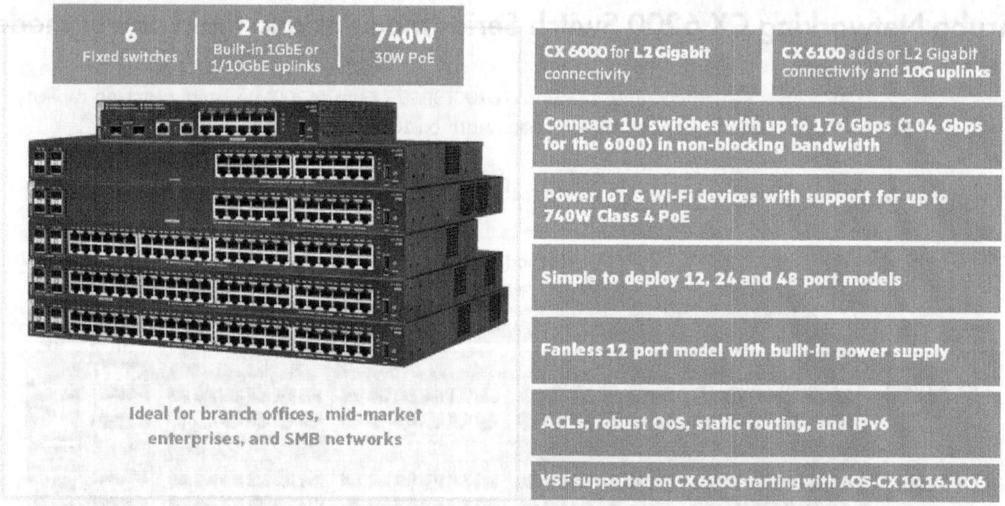

- Overview: Ideal for smaller networks of around 100 – 500 devices with 10 GbE uplinks.
- Positioning: branch offices, mid-market enterprises, and SMBs seeking reliable, cost-sensitive entry-level access.
- Fan-less 12 port model available.
- Positioned for branch offices, mid-market enterprises, and SMBs.
- Compact 1U switches with up to 176 Gbps (104 Gbps for the 6000) in non-blocking bandwidth.
- Up to 740W of Class 4 PoE+.

- 6100 supports 1/10GbE uplinks (6000 supports 1GbE uplinks only).
- layer 2 with static routing support.
- Fixed 12 (fanless),24, and 48 port models.

Important

Important! CX 6100 series support maximum of 6 members in a stack with AOS-CX version 10.16.1006. You can stack 6100 24/48 port models together. You can stack 12-port model switches only with 12-port model switches. VSF stacking between the 6100 JL679A 12-port SKU and other 6100 24/48P models is not supported. VSF stacking 4100i models with 6100 models is not supported.

HPE Aruba Networking CX 4100i Switch Series

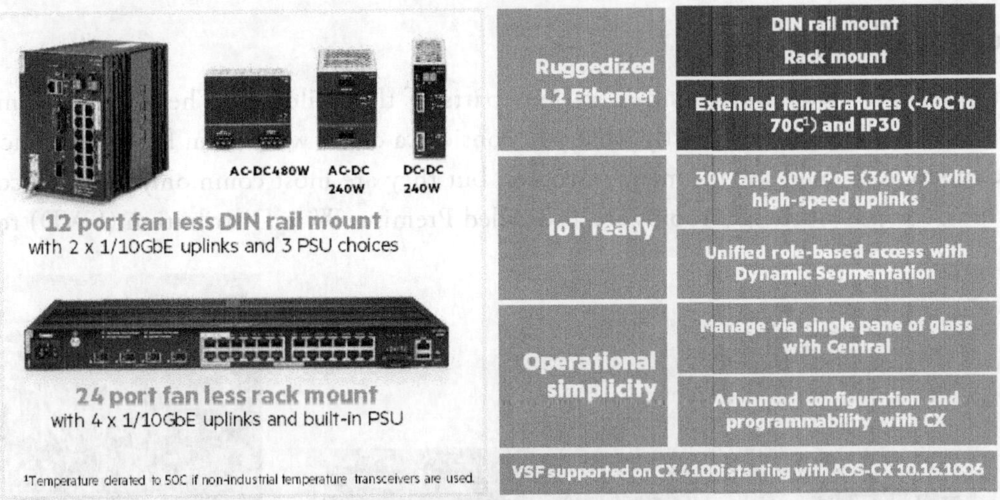

HPE Aruba Networking CX 4100i Switch Series is a layer-2 switch with a ruggedized design. It is a special use case for challenging environments. Two fanless models can survive extended temperatures, minus 40 to 70 degrees Celsius.

- Mount switch and 1U model; both fan-less.
- Ruggedized for substation applications and high tolerances for constant vibrations, shocks, and electronic immunity of railway applications.
- Operational temperatures are from -40 °C to 70°C.
- System Switching Capacity: Up to 128 Gbps, System Throughput Capacity: Up to 95 Mpps.

- 12-port 1GbE (8-port Class 4 POE and 4-port Class 6 POE) 2-port SFP+ DIN Mount Switch.
- 24-port 1GbE (20-port Class 4 POE and 4-port Class 6 POE) 4-port SFP+ Switch.

Important

Important! CX 4100i series support maximum of 4 members in a stack with AOS-CX version 10.16.1006. You can stack 24-port model switches together and with 12-port model switches. VSF stacking CX 4100i series models with CX 6100 series models is not supported.

Closets and Cables

Next, this section reviews closet and cabling designs.

Network distribution

A wiring closet is commonly found in various parts of the buildings. These closets come in all shapes and sizes; they may be exactly what you consider a closet with room for a single rack. They may be rooms you could use for many purposes, but they are most commonly used for computer networking equipment. Larger rooms are also called Premises Wire Distribution (PWD) rooms.

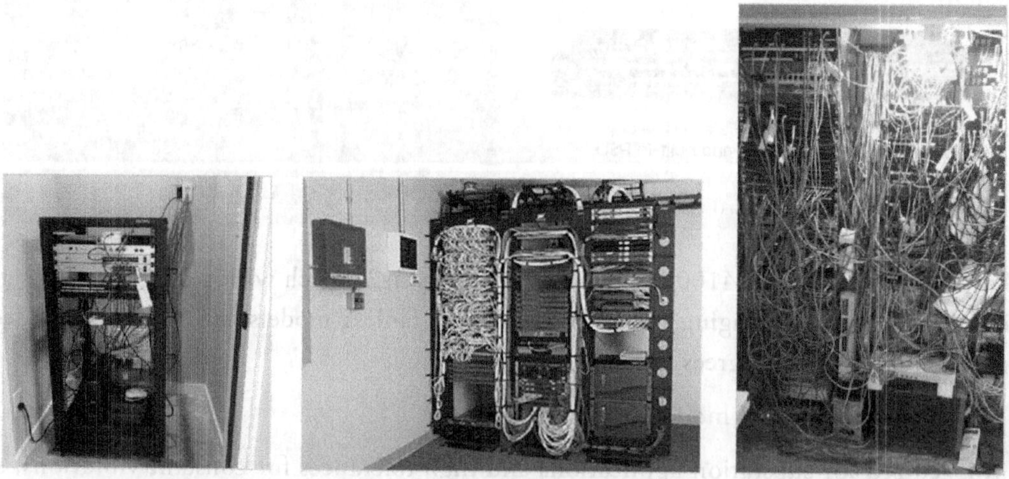

Note that not all closets are kept in immaculate condition.

- MDF/IDF = Main Distribution Framework / Intermediate Distribution Framework, used for Aggregation/Core. This is the point where you are egressing the campus network.
- MCR/SCR = Main CrossConnect Room / Secondary* Other locations outside of the US

may use this term. This main hold the "MDF/IDF."

- MER/SER = Main Equipment Room / Secondary

IDF / MDF closets

MDF is where traffic egresses the campus network, and IDFs distribute that connectivity throughout the building. The Main Distribution Frame (MDF) typically houses devices that interconnect and manages the wiring between itself and any number of IDFs. An MDF closet can also have cabling for user devices or network devices.

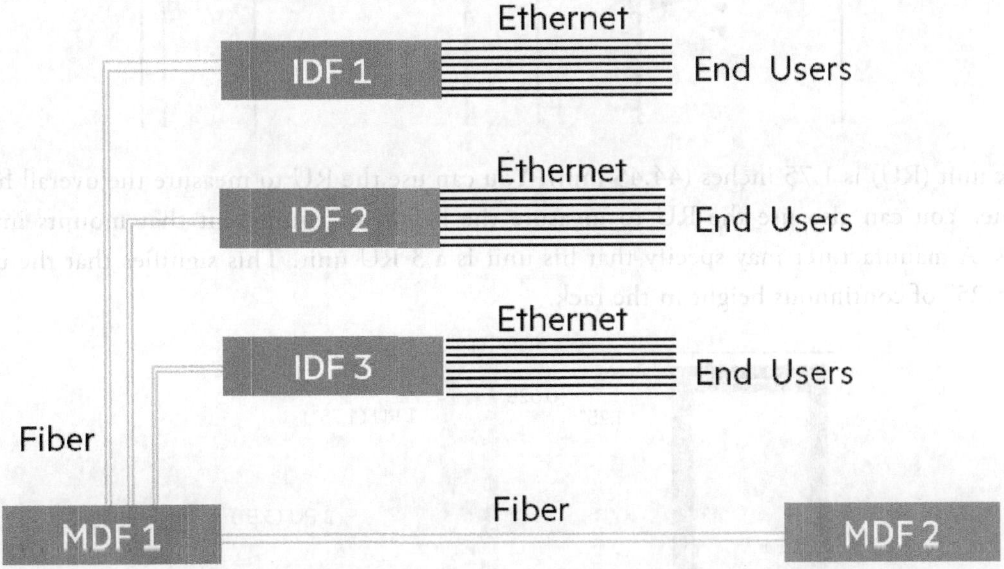

An Intermediate Distribution Frame (IDF) is on the customers' site. IDFs typically have cabling for individual user devices or network devices.

Rack unit (RU)

- The standard rack configuration is from 42U to 52U in height
- The standard rack depth can vary, typically from 600 mm to 800 mm (23.6" - 31,5")
- The deep rack depth can be from 1000 mm to 1200 mm (39" - 47")

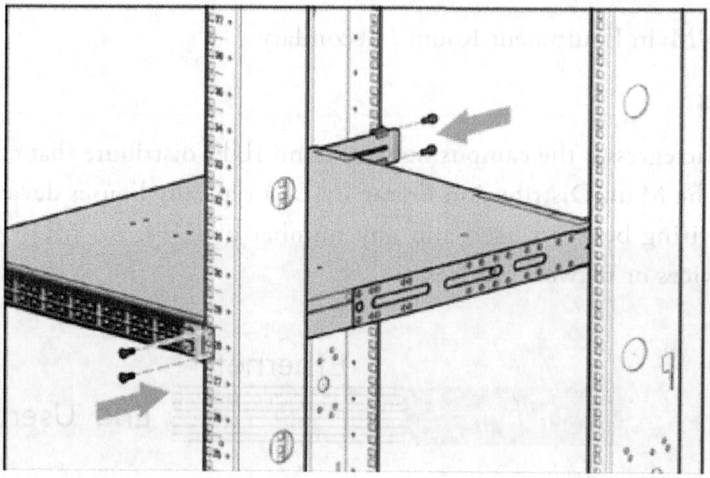

A rack unit (RU) is 1.75 inches (44.45 mm). You can use the RU to measure the overall height of a frame. You can also use the RU to measure the height of equipment that mounts into these frames. A manufacturer may specify that his unit is a 3 RU unit. This signifies that the unit will need 5.25" of continuous height in the rack.

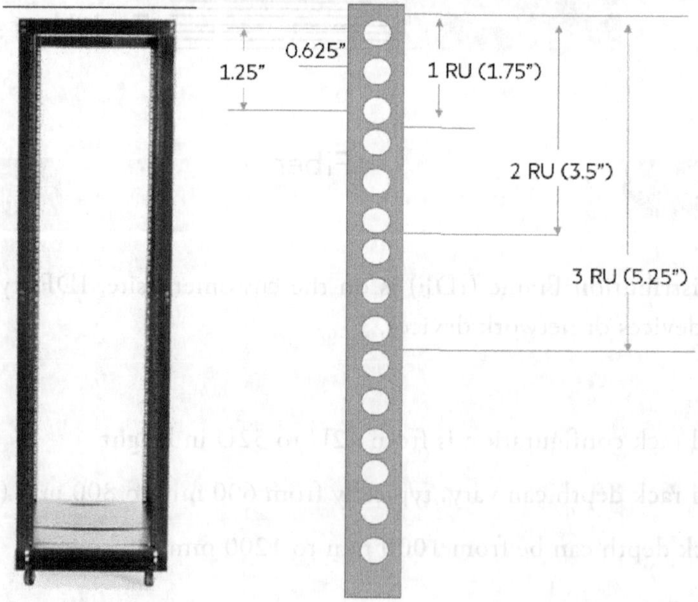

You can find the per-RU boundary by looking at the spacing of the screw holes or rack nut openings. Where two are close together, that's the boundary. No rack mount device of any size should bridge or overlap unevenly with these smaller gaps.

Patch panels and cable organizers

Companies commonly use patch panels and cable organizers placed inside the frame. It is important to know that this may limit the RU availability.

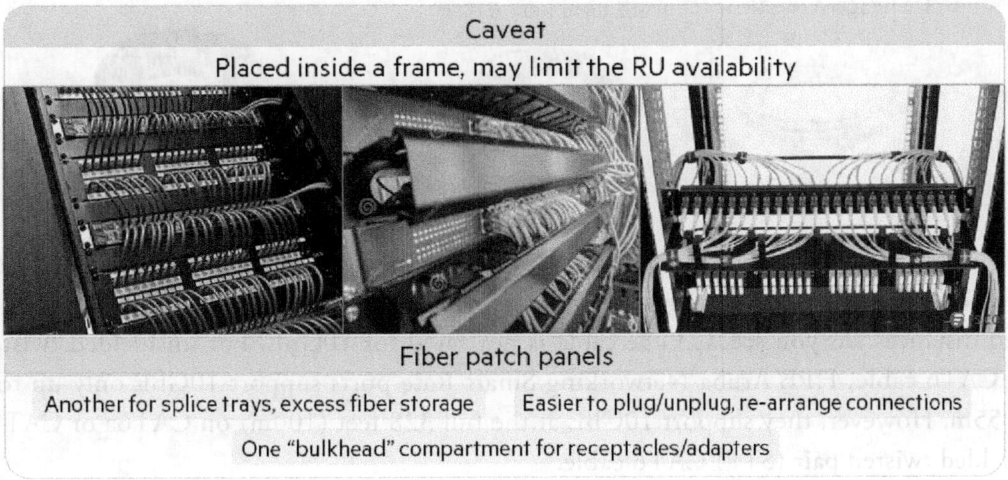

Fiber patch panels are common because they allow you to plug and unplug cables frequently and arrange connections between devices. A fiber patch has one compartment that is the bulkhead for receptacles/adapters and another compartment that you can use for splice trays and excess fiber storage.

Consider the media: copper

Consider the media type. Horizontal cabling usually uses BASE-T copper cable. In some smaller buildings, vertical cabling might also be copper. However, copper only supports high enough speeds for links over limited distances. Here you see the speeds supported at various distances by CAT5e, CAT6, CAT6a, and CAT7 cables. Also, be aware that most BASE-T ports on HPE Aruba Networking switches are 10/100/1000Mbps ports. Use switches with Smart Rate ports if you need to connect devices or switches at 10GbE. Replacing cabling, in general, is very expensive. For that reason, you should design cabling with the future in mind. For existing cabling, you sometimes have to design around the fact that you cannot easily replace the cabling.

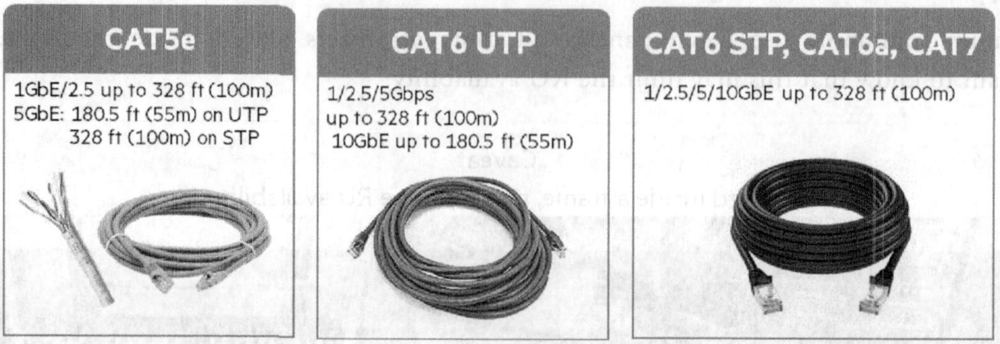

Because access layer uplinks in modern networks often require at least 10GbE, if your customer site uses copper cable between closets, make sure that it will support 10GbE speeds at the required distances. As you see, CAT5e cable is not rated for 10GbE. For unshielded twisted pair (UTP) CAT6 cable, HPE Aruba Networking Smart Rate ports support 10GbE only up to 180.5 feet or 55m. However, they support 10GbE at the full 328 feet (100m) on CAT6a or CAT7 cable and shielded twisted pair (STP) CAT6 cable.

Smart Rate ports also support 2.5GbE operation or 5GbE operation on CAT5e cable. On unshielded CAT5e, 2.5GbE is supported up to 328 feet (100m), and 5GbE is supported up to 180.5 feet (55m). On CAT6 cable or better, and shielded CAT5e, 2.5GbE and 5GbE are supported up to 328 feet (100m).

Smart Rate

Today's cabling systems were designed for 1 Gigabit Ethernet, and moving up to 10 Gigabit Ethernet may require expensive wiring upgrades.

Mode	Cablet type	Cable length
1 Gbps (1000 BASE-T), 100 Mbps (IOOBASE-T)	CAT5e/Class D or better	Up to 100 m (328 ft)
2.5 Gbps - Available when link partner is either 802.3bz compliant, HPE Smart Rate or NBASE-T compliant	CAT5e/Class D or better	Up to 100 m
5 Gbps - Available when link partner is either 802.3bz compliant, HPE Smart Rate or NBASE-T compliant	CAT5e/Class D	Up to 55 m (180 ft) in high noise environments

Mode	Cablet type	Cable length
		up to 100 m
	Shielded CAT5e/Class D	up to 100 m
	CAT6/Class E or better	up to 55 m
10 Gbps (IOGBASE-T)	CAT6/Class E	up to 100 m
	CAT6A/Class E	up to 100 m
	CAT7/Class F	

Smart Rate is a multi-gigabit twisted-pair network interface that is interoperable with the NBASE-T 2.5/5Gbps products and with existing industry standard 1GbE/10GbE devices.

It allows most existing cable installations in campus LAN environments to provide higher bandwidth connectivity and distribute PoE/PoE+ power to connected devices.

Alien noise environment (ANEXT) is electrical "crosstalk" between adjacent cables and connecting hardware. This is not a problem at the lower frequencies for CAT5e and CAT6 cabling, but the higher frequency signals for CAT6A make ANEXT the limiting noise source for these systems.

The IEEE 802.3bz standard for multi-gigabit Ethernet gives high speed and power for APs such as 802.11ax and 802.11ac while using existing CAT5e and CAT6 twisted pair wiring. This saves you from having to rip and replace the expense and complexity of new cabling infrastructure. At HPE Networking, we call this multi-gigabit Ethernet technology HPE Smart Rate.

- Class A: 100 kHz using Category 1 cable/connectors
- Class B: 1 MHz using Category 2 cable/connectors
- Class C: 16 MHz using Category 3 cable/connectors
- Class D: 100 MHz using Category 5e cable/connectors
- Class E: 250 MHz using Category 6 cable/connectors
- Class EA: 500 MHz using Category 6A cable/connectors
- Class F: 600 MHz using Category 7 cable/connectors

- Class FA: 1000 MHz using Category 7A cable/connectors
- Class I: link/channel up to 2000 MHz using Category 8.1 cable/connectors (specification under development)
- Class II: link/channel up to 2000 MHz using Category 8.2 cable/connectors (specification under development)

Additional resource: https://www.hpe.com/psnow/doc/a00058568enw

Quick terms

Cabling

- Active Optical Cable (AOC)
- Multi-fiber push-on connectors (MPOs)
- Direct attach copper (DAC) cable

Trans. Distance

- Max 100 meters for UTP or STP, due to attenuation
- As signal traverses' cable, it weakens due to resistance
- This weakening is measured as attenuation

Data Rate

- Number of transmitted **b**its **p**er **s**econd
- Measured in Mbps or Gbps
- Products include transceiver modules, optical, DAC, and AOC

Here is a summary of quick terms related to networking cabling.

Active Optical Cable (AOC): AOC uses electrical-to-optical conversion. A cabling technology that uses the same electrical inputs as a traditional copper cable but then carries the signal using optical fiber between connectors.

Multi-fiber push-on connectors (MPOs): Fiber connectors comprise 8, 12, or 24 fibers.

Direct attach copper (DAC) cable: Shielded twinax copper cable with pluggable connectors designed for short transmission distances of 7-15 meters.

Data Rate: Number of bits transmitted per second. Measured in Megabits per second (Mbps) or Gigabits per second (Gbps). Products include transceiver modules, optical and Direct Attach over Copper (DAC), and Active Optical Cables (AOC).

- 400 Gbps (Optical, AOC at this time)
- 200 Gbps (AOC at this time)
- 100 Gbps (optical, DAC, and AOC)
- 50 Gbps (DAC and SR only at this time)
- 40 Gbps (optical, DAC, and AOC)
- 25 Gbps (optical, DAC, and AOC)
- 10 Gbps (optical, DAC, and RJ45 10GBASE-T)
- 1000 Mbps (also known as Gigabit) (optical and RJ45 1GbT)
- 100 Mbps (also known as Fast Ethernet) (optical only)

Transmission Distance: A limit of 100 meters exists for UTP or STP cables due to attenuation. As the signal travels along these cables, the transmission weakens due to resistance. This weakening is measured as attenuation. Longer distances equal more attenuation, causing the signal-to-noise ratio to drop below an accepted level, making it impossible to distinguish between signal and background noise. The result is data loss.

Review the following URL for more information. https://www.arubanetworks.com/techdocs/Switches/xcvrs/PDF/AOS- S%20and%20AOS-CX%20Transceiver%20Guide.pdf

Switch to switch - direct attach cable (DAC)

Direct attached copper (DAC) cables are twinax cables factory-terminated to a transceiver. You use them for relatively short connections, often from a switch in one equipment rack to another switch, server, or other device in the same rack. You might connect from one modular switchblade to another blade in the same switch.

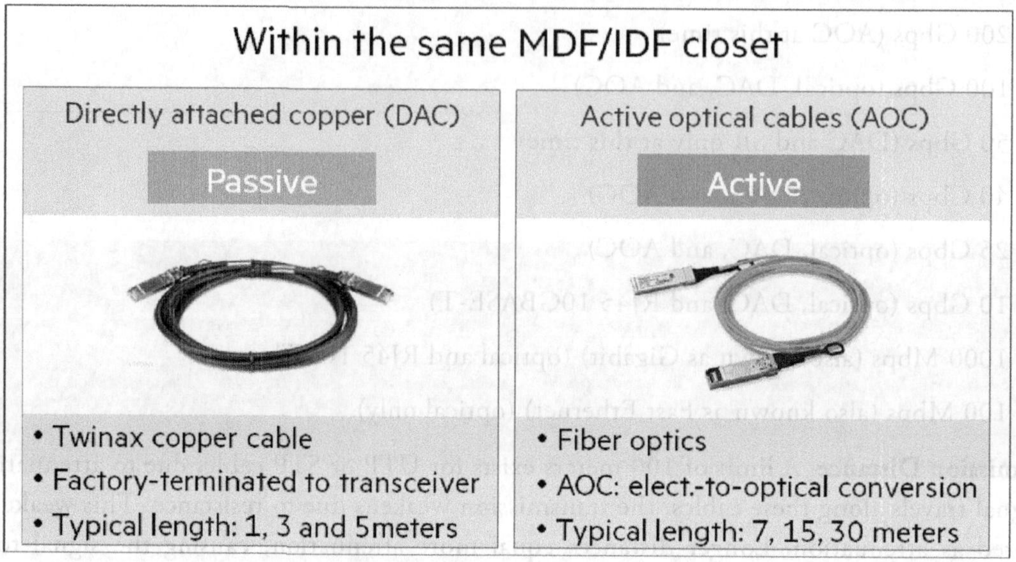

High-speed direct attach cables fall into two categories:

Directly Attached Cables (DAC). DAC can be passive without any electrical components to ensure minimal power consumption and is limited to 7 meters. Alternatively, they can contain electrical components to boost the signal and reach greater distances (over 5 meters or more) while ensuring signal quality.

Active optical cables or (AOC). These use electrical-to-optical conversion on the cable ends to improve the cable's speed, distance, and performance while remaining compatible with standard SFP+ interfaces. They typically reach up to 15 meters.

HPE Networking breakout cables for high-speed ports

A QSFP+ breakout cable allows you to establish four separate SFP+ 10GbE connections to one QSFP+ switch port. You might use this option when a switch needs to support a mix of 40GbE and 10GbE connections to servers. You can propose a switch with many QSFP+ ports and use some of the ports for 40GbE connections and others for 10GbE connections.

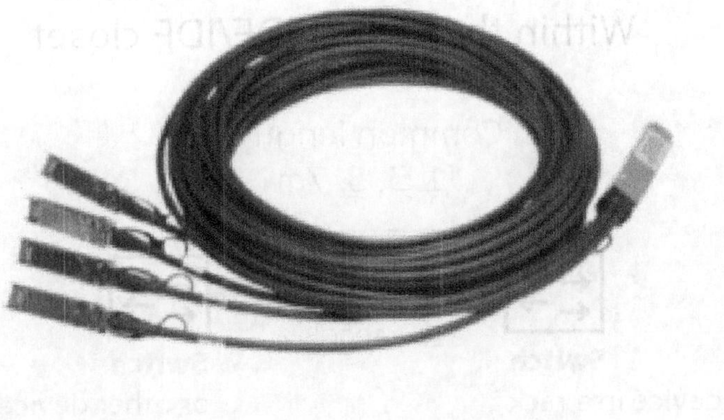

Other breakout cables provide similar options, dividing out:

- QSFP28 into four 25 GbE links (QSFP28 to 4x SFP28)
- QSFP56 into two 100 GbE links (QSFP56 to 2x QSFP28)
- QSFP-DD into four 100 GbE links (QSFP-DD to 4x QSFP56)It is important that you understand that HPE Aruba Networking AOCs are QSFP-DD to QSFP56, not QSFP28. You must connect these breakout cables to QSFP56 ports, not the typical QSFP28 ports for 100GbE. For a list of HPE servers and systems that support QSFP56 ports, refer to the AOS-S and AOS-CX Transceiver Guide.
- QSFP-DD into two 200 GbE links (QSFP-DD to 2x QSFP56)

HPE Networking provides DAC and AOC breakout cables; sometimes these are called splitter cables. You can refer to the AOS-S and AOS-CX Transceiver Guide for a list of the available cables and their lengths.

Note that you can only breakout bandwidth in this way for the standards listed above. The 50 (SFP56), 25 (SFP28), 10 (SFP+), and 1 (SFP) GbE standards do not use multiple channels and cannot be split into multiple streams.

Switch to switch - direct attach cable (DAC)

Direct attached copper (DAC) cables are twinax cables factory-terminated to a transceiver. You use them for relatively short connections, often from a switch in one equipment rack to another switch, server, or other device in the same rack. You might connect from one modular switchblade to another blade in the same switch.

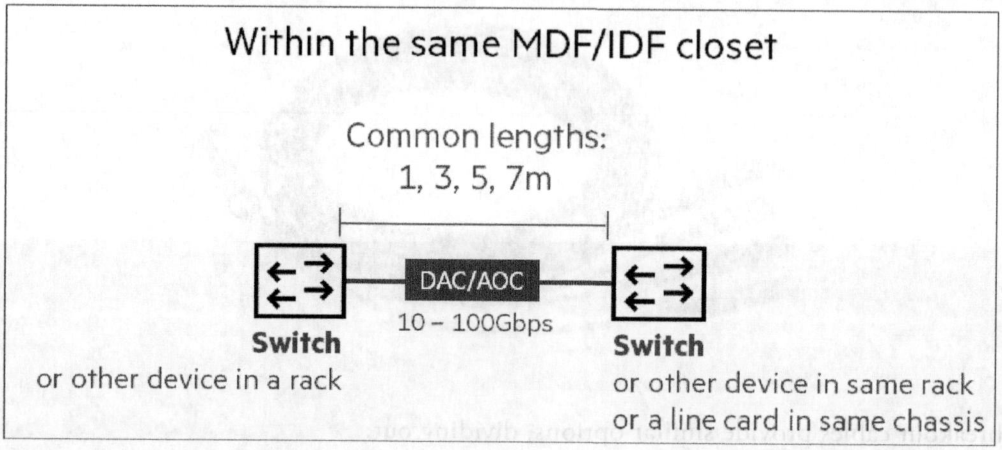

As shown in the figure, DAC cables connect from one line card to another. There are also break-out DACs that connect from one device and up to four other devices. A common option from HPE Networking is a DAC cable with SFP+ transceivers.

For more information, search for "AOS-Switch and AOS-CX Transceiver Guide" currently at this URL: https://www.arubanetworks.com/techdocs/Switches/xcvrs/5200- 3362/index.html#GUID-D7147C7F-2016-0901-0A24-0000000004A4.html

Wired connection using fiber optics

Use fiber optic cables to connect devices over 328 feet or 100 meters apart in your LAN. This technology uses light impulses to transmit data using fiber. Maximum length can vary widely depending on cable and transceiver type, from a few meters to around 500 meters, perhaps up to 40 kilometers.

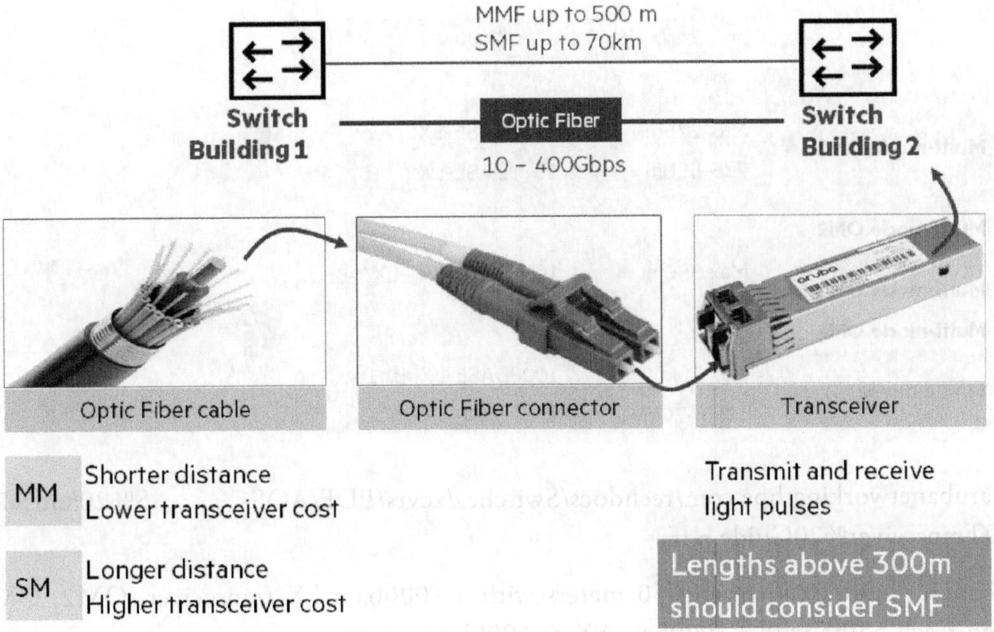

Optic fiber achieves faster data rates. A less expensive fiber transfers 10 Gigabits per second, while a higher-quality fiber provides 100 Gigabits per second in data rates.

We organize fiber optic cables into two main groups: Multimode (MM) is less expensive for shorter distances. Single Mode (SM) is more expensive but supports longer distances.

There are several standard fiber connectors, but you typically use a Lucent Connector (LC), as shown here. You connect each end of the cable to a transceiver, which you have installed into a network device or server. Fiber optic transceivers are the active components that transmit and receive data through light pulses.

Cables and transceivers: SFP

- SFP56 = 50G
- SFP28 = 25G (also supported by SFP56 ports)
- SFP+ = 10G (also supported by SFP28 and SFP56 ports)
- SFP= 1G, 100Mbps (may also be supported by SFP+, SFP25, SFP56 ports

	1GbE	
Fiber type	Distance	Transceiver
Multi-mode OM1	Max 275m	1000BASE-SX
	275-550m	1000BASE-LX
Multi-mode OM2	Max 550m	1000BASE-SX, -LX
Multi-mode OM3		
Multi-mode OM4		
Single-mode	Max 10km	1000BASE-LX, BX10
	10km-70km	1000BASE-LH70

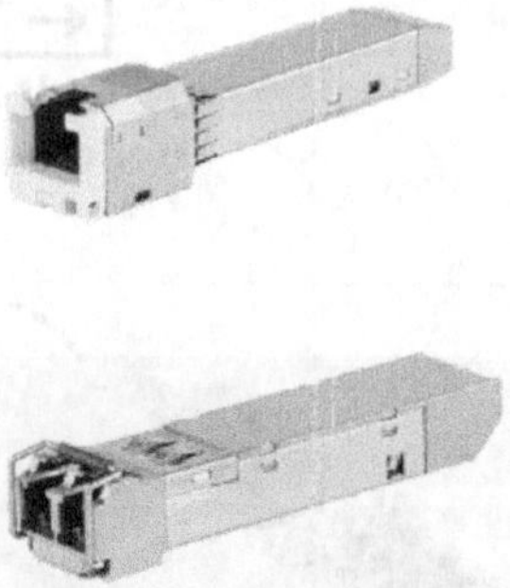

https://arubanetworking.hpe.com/techdocs/Switches/xcvrs/PDF/AOS-S%20and%20AOS-CX%20Transceiver%20Guide.pdf

For 1G speeds, OM1 can reach 550 meters with a 1000base-LX transceiver. OM2, OM3, and OM4 can reach 550m with a 1000base-SX or 1000base-LX type transceiver.

Here is a list of OM fiber types, their power, and thickness:

- OM1, for fiber with 200/500 MHz*km overfilled launch (OFL) bandwidth at 850/1300nm (typically 62.5/125um fiber)
- OM2, for fiber with 500/500 MHz*km OFL bandwidth at 850/1300nm (typically 50/125um fiber)
- OM3, for laser-optimized 50um fiber having 2000 MHz*km effective modal bandwidth (EMB, also known as laser bandwidth), is designed for 10 Gb/s transmission.
- OM4, for laser-optimized 50um fiber having 4700 MHz*km EMB bandwidth designed for 10 Gb/s, 40 Gb/s, and 100 Gb/s transmission.

Cables and transceivers: SFP+

- SFP+ = 10G
- Also supported by SFP28 and SFP56 ports

	10GbE	
Fiber type	Distance	Transceiver
Multi-mode OM1	Max 33m	10GBASE-SR
	33-220m	10GBASE-LRM
Multi-mode OM2	Max 82m	10GBASE-SR
	82-220m	10GBASE-LRM
Multi-mode OM3	Max 300m	10GBASE-SR
Multi-mode OM4	Max 300m	10GBASE-SR
Single-mode	Max 10km	10GBASE-LR
	10km-30km	10GBASE-ER

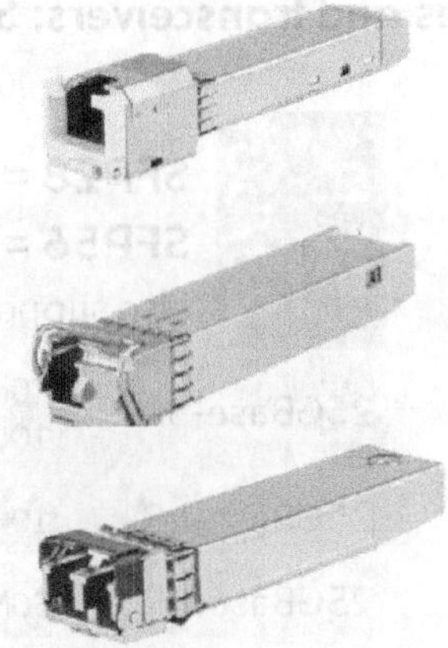

For 10G speeds, Multi-Mode OM1 reaches up to 33 meters with a 10Gbase-SR transceiver but can reach 220 meters with a 10Gbase-LRM transceiver. OM2 can reach 82 meters with a 10Gbase-SR transceiver and 220 meters with a 10Gbase-LRM transceiver. OM3 and OM4 can reach 300 meters with 10Gbase-SR transceivers. Single-mode fiber can achieve distances between 10 km and 30 km with the correct transceivers for greater distances. Remember that the transceiver to the fiber patch panel connects at both ends.

If a client has a 30-meter OM1 fiber between two closets, but each closet has a fiber patch panel, your equipment will have a five-meter fiber cable between the switch and the fiber path panel. The entire length of this connection is 40 meters. You would need a 10Gbase-LRM transceiver to make this a 10G connection.

Select CX models now support LRM transceivers to run 220m of 10 GbE over OM1 and OM2 MMF. This is a major feature in enabling customers to migrate to CX.

Cables and transceivers: SFP28 and SFP56

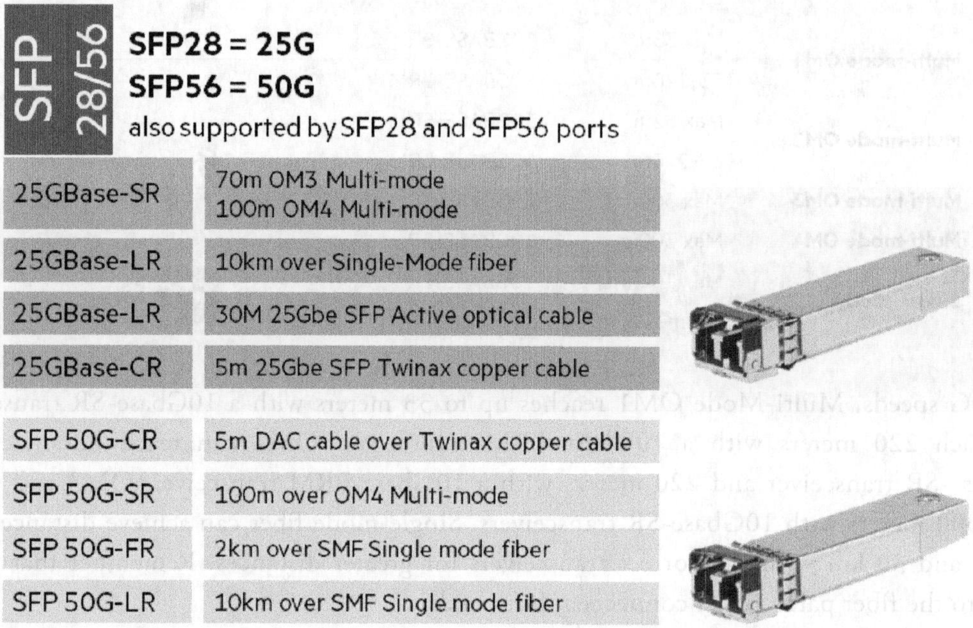

Transceiver solution overview:

- SFP56 = 50G
- SFP28 = 25G (also supported by SFP56 ports)
- SFP+ = 10G (also supported by SFP28 and SFP56 ports)
- SFP= 1G, 100Mbps (may also be supported by SFP+, SFP25, SFP56 ports

This chart provides the supported distance and type of fiber to support SPF28 at 25G. For multimode OM3, the 25G distance is 70 meters, but with OM4, you can use a 100-meter distance. Therefore, with single mode, you can support 10km.

Active Optical Cable (AOC) is a cabling technology that accepts the same electrical inputs as a traditional copper cable but uses optical fiber between the connectors. The distance with this cable is 30 meters. Twinax is similar to a coax cable but has two inner conductors and a 5-meter limit. DAC Twinax 5-meter cable supports the SFP56 at 50G. By using OM4, you can attain 100 meters. A single-mode fiber can go 2km and 10km depending on the transceiver.

For more information, check the following link for HPE Aruba Networking's transceiver guide:

https://arubanetworking.hpe.com/techdocs/Switches/xcvrs/PDF/AOS-S%20and%20AOS-CX%20Transceiver%20Guide.pdf

Cables and transceivers: QSFP

Quad SFP transceiver solution overview:

- Q-DD or QSFP-DD = 400G (or quad small form-factor pluggable double density)
- QSFP56 = 200G (also supported by Q-DD ports)
- QSFP28 = 100G (also supported by Q-DD ports)
- QSFP+= 40G (also supported by Q-DD ports)

QSFP

Q-DD or QSFP-DD* = 400G
QSFP56 = 200G
QSFP28 = 100G
QSFP+= 40G

*QSFP-DD
Quad **S**mall **F**orm-factor
Pluggable **D**ouble **D**ensity

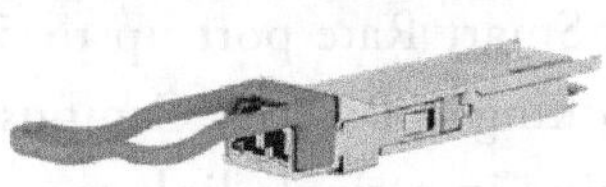

40GbE			
Fiber type	**Distance**	**Transceiver**	**Strands**
Multi-mode OM3	Max 100m	QSFP+ MPO SR4	12
		QSFP+ LC BiDi	2
	Max 300m	QSFP+ MPO eSR4	12
Multi-mode OM4	Max 150m	QSFP+ MPO SR4	12
		QSFP+ LC BiDi	2
	Max 400m	QSFP+ MPO eSR4	12
Single-mode	Max 10km	QSFP+ LC LR4	2

100GbE			
Fiber type	**Distance**	**Transceiver**	**Strands**
Multi-mode OM3	Max 70m	QSFP28 MPO SR4	12
Multi-mode OM4	Max 200m	QSFP28 MPO SR4	12
Single-mode	Max 10km	QSFP28 LC SR4	2

Check the AOS- S and AOS- CX Transceiver Guide for more information. https://www.arubanetworks.com/techdocs/Switches/xcvrs/xcvr_guide/Content/fir-int.htm

Access layer

Let's turn our attention now to the access layer design.

Identify AP-7xx series access requirements

Each HPE Aruba Networking AP model delivers a maximum data rate. An AP radio's actual throughput is typically under 50 percent of the theoretical data rate. However, some APs could still require over 1Gbps from their wired connections to ensure that the wired backend does not limit the wireless-side bandwidth the customer is investing in.

720 Series

- 4.7 Gbps (2.4GHz + 5GHz + 6GHz)
- 1x 2.5 GbE Smart Rate
- 802.3at (class 4): no restrictions
- Single 802.3af (class 3): some restrictions

730 Series

- 14.4 Gbps (2.4GHz + 5GHz + 6GHz)
- 2x 5 Gbps Smart Rate
- 802.3bt (class 5): no restrictions
- Single 802.3at (class 4): some restrictions

740 Series

- 8.9 Gbps (2.4GHz + 5GHz + 6GHz)
- 1x 5 GbE Smart Rate
- 802.3bt (class 6): no restrictions
- 802.3at (class 4): some restrictions

750 Series

- 28.8 Gbps (2.4GHz + 5GHz + 6GHz)
- 2x 10 Gbps Smart Rate
- 802.3bt (class 6) or dual 802.3at (class 4): no restrictions
- Single 802.3bt (class 5): some restriction

Different models require different approaches to ensure adequate bandwidth on the AP's side. The AP 720 series has a single 2.5 GbE Smart Rate port, but HPE Aruba Networking has designed this AP for low-density deployments. The 730 has two Smart Rate port up to 5 Gbps. This AP is for high-density deployments. The HPE Aruba Networking 750 Series Campus Access Points have two Smart Rate 10 GbE ports, which you can combine in a single link aggregation. HPE Aruba Networking has designed these APs for higher-density deployments.

The 720 series will function with 802.3at (class 4) without restriction. The 730 series can function with 802.3bt (class 5) without restriction.

The 740 and 750 series need the extra power of an 802.3bt (class 6) to operate without restriction. The 750 series APs can support different combinations of PoE from both ports. Search for additional information in the datasheet.

Identify AP-6xx series access requirements

The AP 610 series is designed for low-density deployments and has a single Smart Rate port. The HPE Aruba Networking 630 and 650 series APs have two Smart Rate GbE ports, which you can combine in a single link aggregation. HPE Aruba Networking has designed these APs for higher-density deployments.

610 Series

- 3,603Mbps (2.4GHz + 5GHz + 6GHz)
- 1 x 2.5 Gbps Smart Rate Ethernet port
- 1 x 802.3af*/at/bt
- Dual radios

630 Series

- 3,890Mbps (2.4GHz + 5GHz + 6GHz)
- 2 x 2.5GE Smart Rate Ethernet ports
- 2 x 802.3at/bt
- Single 802.3at*/bt
- Three dedicated radios

650 Series

- 7,779Mbps (2.4GHz + 5GHz + 6GHz)
- 2 x 5GE Smart Rate Ethernet ports
- 2 x 802.3at/bt
- Single 802.3at*/bt
- Three dedicated radios

*Indicated power draw with restrictions. Check spec sheet for *worst-case* power consumption

The 610 series will function with 802.3at (class 4) without restriction. The 630 and 650 series can function with two 802.3at (class 4) or a single 802.3bt (class 5) connection without restriction.

Identify AP-5xx series access requirements

Each HPE Aruba Networking AP model delivers a maximum data rate. An AP radio's actual throughput is typically under 50 percent of the theoretical data rate. However, some APs could still require over 1Gbps from their wired connections to ensure that the wired backend does not limit the wireless-side bandwidth the customer is investing in.

500 Series

- 2.4Ghz 574Mbps / 5Ghz 1.2 Gbps
- max aggregate data rate 1.77Gbps
- 40-70% of total: 740 Mbps-1.23 Gbps
- E0: 1GbE
- 802.3at (class 4): no restrictions
- Single 802.3af (class 3): some restrictions

530 Series

- 2.4Ghz 1.15 Gbps / 5Ghz 2.4 Gbps
- max aggregate data rate 3.55Gps
- 40%-70% of total: 1.42 Gps-2.48 Gbps
- E0, E1: HPE SmartRate ports 5 Gbps
- 802.3bt (class 5) or dual 802.3at (class 4): no restrictions
- Single 802.3at (class 4): some restrictions

510 Series

- 2.4Ghz 575Mbps / 5Ghz 4.8 Gbps
- max aggregate data rate 5.37Gps
- 40-70% of total: 2.1-3.76 Gbps
- E0: HPE SmartRate port 2.5Gbps
- 802.3bt (class 5) or 802.3at (class 4): no restrictions
- Single 802.3af (class 3): some restrictions

550 Series

- 2.4Ghz 1.5 Gbps / 2x 5Ghz 2.4 Gbps
- Max aggregate data rate 5.95 Gps
- 40%-70% of total: 2.38 Gbps-4.16 Gbps
- E0, E1: HPE SmartRate ports 5 Gbps
- 802.3bt (class 5) or dual 802.3at (class 4): no restrictions
- Single 802.3at (class 4): some restriction

Different models require different approaches to ensure adequate bandwidth on the AP's side. The AP 500 series has a single GbE port, but HPE Aruba Networking has designed this AP for low-density deployments. The 510 has one Smart Rate port up to 2.5 Gbps. This AP is for high-density deployments. The HPE Aruba Networking 530 and 550 series APs have two Smart Rate GbE ports, which you can combine in a single link aggregation. HPE Aruba Networking has designed these APs for higher-density deployments.

The 500 series will function with 802.3at (class 4) without restriction. The 510 series can function with 802.3at (class 4) or 802.3bt (class 5). without restriction.

The 530 and 550 series need the extra power of an 802.3bt (class 5) or two 802.3at (class 4) to operate without restriction.

Uplink requirements and oversubscription

The concept of adequate capacity is critical to good quality of service (QoS) design. If all edge devices were operating at near-full capacity, you would need to plan the uplink bandwidth to match the bandwidth provided to the edge ports. However, you can oversubscribe the uplinks because most wired devices do not require nearly all their 1Gbps port.

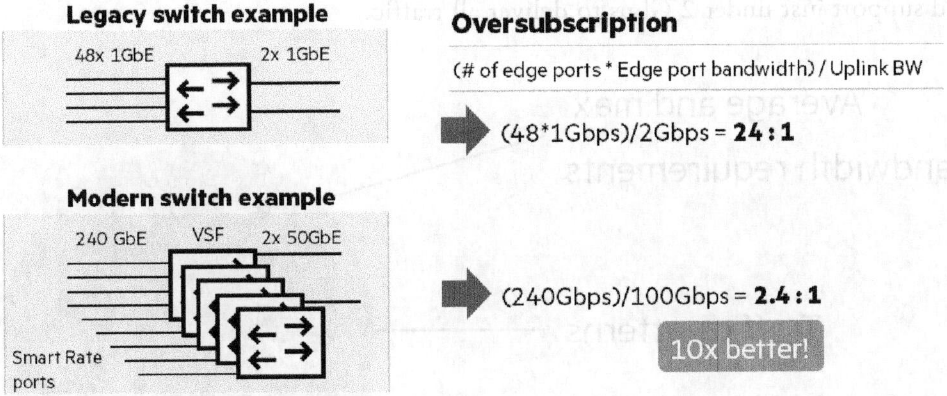

This formula describes the oversubscription ratio. This ratio equals the number of edge ports times edge port bandwidth divided by uplink bandwidth.

Legacy access switches would provide 2 or 4-gigabit uplinks for 24 to 48 access ports. For this example, an access layer switch has 48 Gigabit edge ports and 2 Gigabit uplinks. The oversubscription ratio is 48 times 1Gbps divided by 2Gbps, or 24. This ratio is usually reported in this format, 24:1.

Note

This ratio assumes that the switch can use most of the bandwidth on the uplinks, which it can if the uplinks are combined in a link aggregation.

Modern switches are often stacked and support 10, 25, 40, 50, or 100 Gbps uplinks. For example, the figure shows an IDF closet with five 48-port 6300Ms. 6300Ms support 4xSFP56 uplinks with two uplinks on each 6300M connected to DAC cables for VSF stacking. That still leaves ten total SFP56 at 50Gbps available. Using just two uplinks for redundancy provides 100Gbps combined uplink bandwidth for 240Gbps access ports. The resulting oversubscription is 2.4 to 1. Even with five times the switch ports, the jump in bandwidth still results in the VSF stack having a ten times improved oversubscription ratio over the legacy switch.

Using VSF stacking at the aggregation layer and core makes combining all access layer uplinks in a link aggregation possible. If you were using an MSTP-based architecture, traffic would tend not to be distributed as evenly over the uplinks.

Factors that affect appropriate oversubscription

Several factors affect an appropriate oversubscription ratio, including traffic patterns and connected devices' average and maximum bandwidth requirements. For example, a typical office employee rarely needs more than 40 Mbps. If the switch connects to 48 such endpoints, the uplink could support just under 2 Gbps to deliver all traffic.

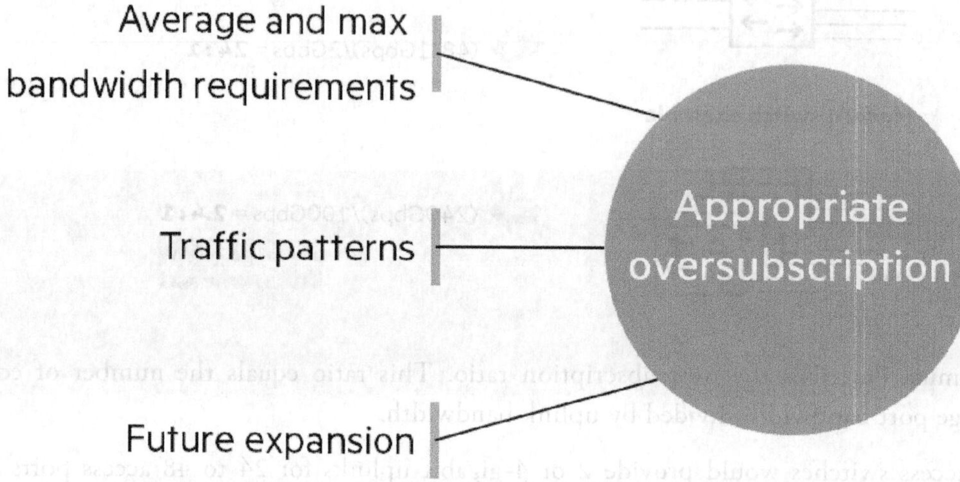

You should also consider traffic patterns. For example, are the endpoint devices IP security cameras that send a constant stream of data, or are they user devices that send and receive bursts of traffic as users download an attachment or browse the internet? If you thought that all 48 devices would require 40 Mbps at all times, you would probably want to plan more than 2 Gbps to give yourself a buffer. For this example, though, 40 Mbps per device defines the peak requirements the uplink can meet.

In more normal circumstances, some devices will have lower demands while others have higher, and the uplink will only operate at 50 percent utilization or less. This gives you a buffer for future increases in demand.

Customers will have the best experience when you plan to meet peak requirements on all devices and provide a buffer for expansion. Make sure the solution will continue to meet customers' needs for at least three years.

Example designs for fixed-port switch ports

In the past, the most oversubscription you would plan was 24:1, which a 48-port switch with two 1GbE links could provide. This worked when most devices had peak requirements under 30-40Mbps and lower average requirements. Examples included desktops typical employees use, such as users sending emails or browsing the Internet. Even users running compressed video conferencing applications could fall below these levels if they were not running several other applications simultaneously. VoIP phones, printers, low-resolution security cameras, and many IoT devices may fall within these ranges.

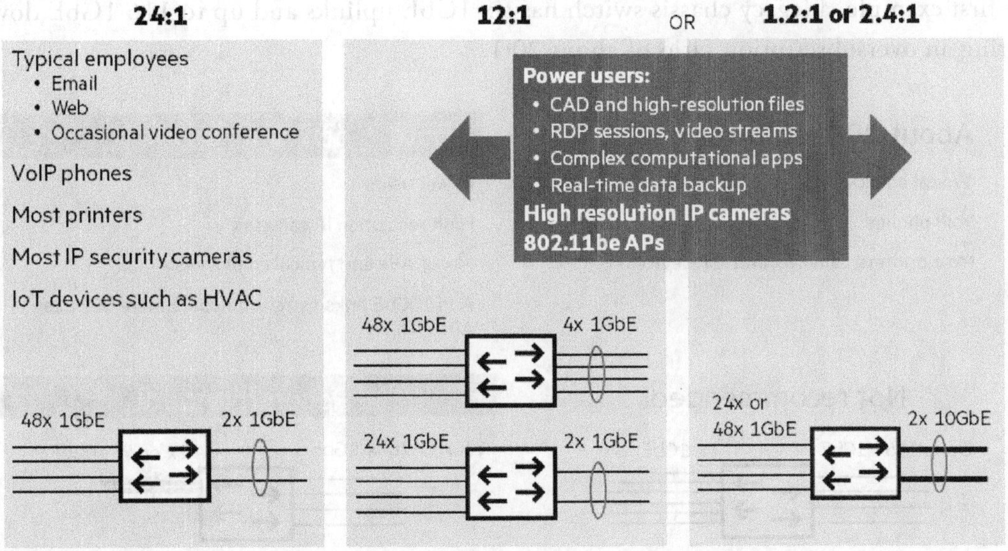

Modern users often require more bandwidth. Examples of high bandwidth users include users who work with large CAD files or high-resolution medical graphics. Power users might also run many RDP sessions or multiple streaming videos. Very high-resolution cameras also require as much as 70 or 80 Mbps.

But a critical piece to remember is bridged wireless traffic. Each 500, 600 and 700 series APs can easily push Smart Rate speeds of 1 to 2.5 Gbps or more consistent bandwidth. You can think of access points as another access layer hosting up to 75-100 users (recommended maximums). Most HPE Aruba Networking access switches limit the number of Smart Rate ports; there are examples like the 6300M 24SR10 with 24 ports of Smart Rate provided connectivity. Plan accordingly by leveraging multiple SFP28 or SFP56 uplinks to keep the oversubscription lower.

The leap to 10GbE can more than satisfy requirements now and in the future with a generous 2.4:1 oversubscription ratio. Depending on the environment, the customer might prefer two 10GbE links to multiple 1GbE links to reduce the required cabling. Typically, you spread APs across several switches to prevent a single failure from wiping out the wireless coverage in an area. However, even if a switch only connects to eight APs, the uplink bandwidth requirements might push past the ability of the switch to deliver with 1GbE, so you would decide to use 10GbE connections.

Example designs for 6300/6400 switches at the access layer

If you deploy the HPE Aruba Networking CX 6405 Switches at the access layer, a single switch can support a large number of ports with a great deal of flexibility. These two examples show the high-end edge ports that a single chassis switch might support.

In the first example, a legacy chassis switch has six 1GbE uplinks and up to 116 1GbE downlinks, delivering an oversubscription ratio of about 20:1.

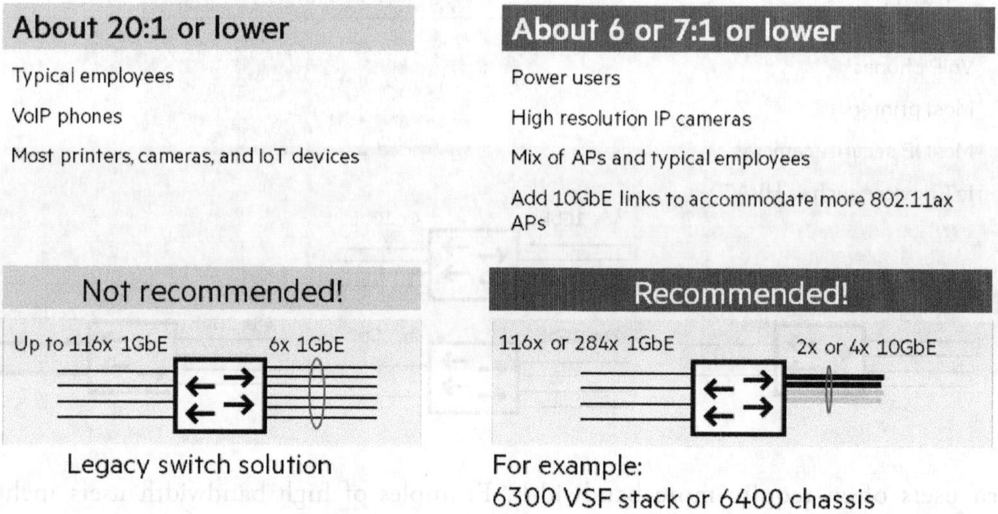

In the second example, the uplinks upgrade to 10GbE on a 6300 or 6400 series, supporting more power users and APs. A fixed port switch with, at most, 48 ports will often underuse the two 10GbE links. With the modular 6405, you can increase the number of edge ports to around 116. Even with 116 edge ports, the switch will provide about a 6:1 oversubscription ratio, which can meet the needs of power users, high-resolution IP cameras, and a mix of typical employees with a few 802.11ac APs. Another example with 284 1GbE edge ports and four 10GbE uplink ports for an oversubscription ratio of about 7:1.

You might choose a design like the one in the first example when providing high-density wired connections to devices with relatively low bandwidth needs. You can create this design with two modules, each providing 20 10/100/1000 ports and four SFP ports, and four other modules, each providing up to 24 10/100/1000 ports.

The design in the second example could also meet the needs of a mixed environment with several APs that support the bulk of the traffic, a few printers, a few IoT devices, and some desktops. You can create this design with two modules, each providing 20 10/100/1000 ports and four SFP+ ports, and four other modules, each providing up to 24 10/100/1000 ports. This design provides plenty of extra SFP+ ports for additional uplink bandwidth.

Also note that the 6405 series supports 40GbE ports, so you can deploy 40GbE to the edge if needed.

Uplink plan for VSF fabrics

A VSF fabric allows multiple switches to share highspeed uplinks and to be managed as a single device. This example plan features a four-member VSF fabric. The stack has 192 downlink ports and two 10GbE uplinks distributed on two members that are one hop apart. This plan provides an oversubscription ratio of under 10:1, which is suitable for all applications and devices except for a high density of high-bandwidth 802.11ax/be APs.

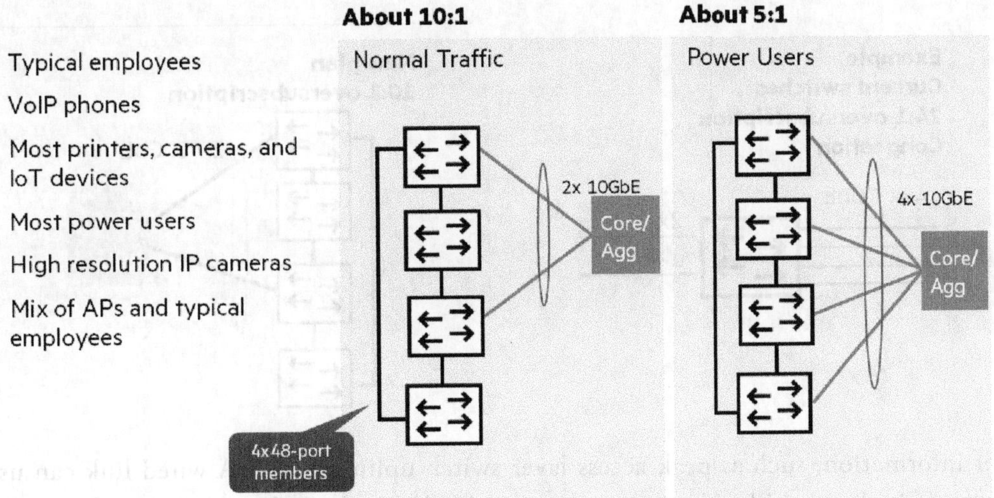

You can add 10GbE links to other members to increase the bandwidth.

You can also add more members to the group and create a higher oversubscription ratio. However, although you can combine some switch models in larger groups, you should usually cap the members at four or five. You should also loop the fabric in a ring topology or full mesh (if supported).

Using current traffic information to assess plan's validity

If your customer has a current solution, you can begin with the customer's current oversubscription ratio as a baseline.

You can then work with network managers to analyze current traffic volumes and patterns to determine how well the oversubscription works. You can use a network management solution like HPE Aruba Networking Central and CLI diagnostic commands like the show interface command.

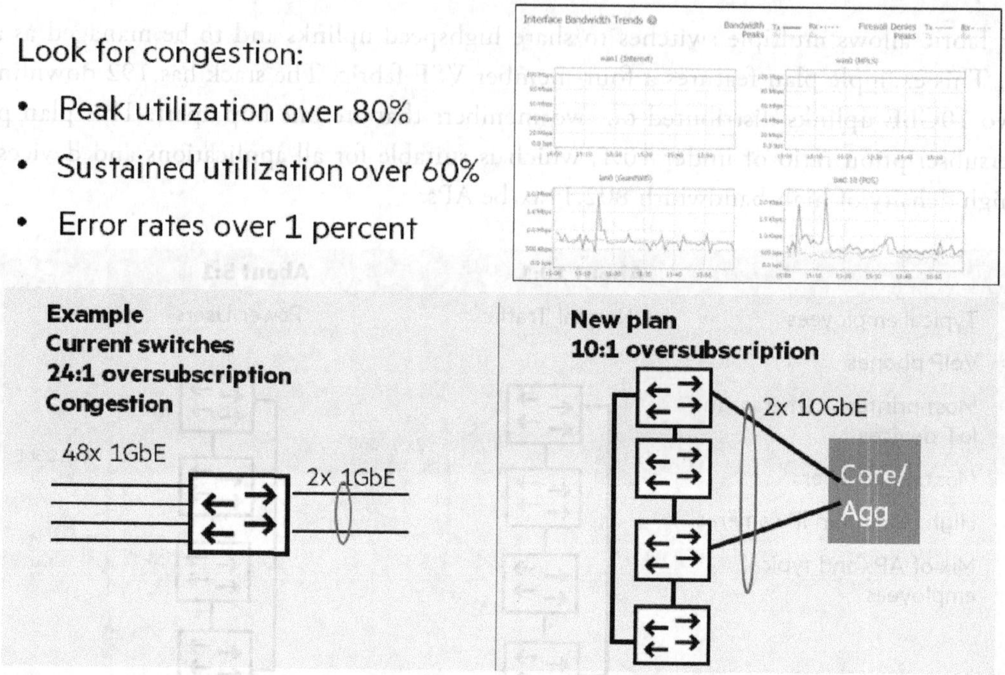

Look for information, such as peak access layer switch uplinks usage. A wired link can use about 80 percent of its bandwidth. If you see usage at this level, the link is probably dropping traffic, and you need to provision more bandwidth.

The link should not sustain over 50 or 60 percent usage for transmitting or receiving information.

You should examine the error rates over a sustained period. High error rates (over one percent) indicate that congestion is occurring.

When you detect congestion, you can make a plan to resolve the issue. For example, a legacy customer has a 48-port Gigabit switch with two 1GbE uplinks. The current oversubscription is 24:1. You might plan to upgrade to a VSF stack with two 10GbE uplinks, achieving less than half the current oversubscription.

Although you can use CLI diagnostic commands to determine how the customer's oversubscription is working, you might remember that network management solutions consolidate information, show trends, and otherwise help you analyze the data.

When considering peak usage, be aware that brief peaks do not matter as much as peaks that sustain over longer periods. Generally, if a link occasionally but rarely rises to 50 or 60 percent usage, you do not need to worry. However, if the link is often used at over 50 percent, you might want to provision more uplink bandwidth.

Note that you are looking at both transmit and receive usage on the access layer edge ports. The receive usage will often be higher because traffic tends to flow from servers toward endpoints.

You might find lower usage on one uplink and unacceptably high usage on another. A spanning tree can introduce this type of inefficient load balancing. If this is the case, you might take steps to improve load balancing across the links rather than add more bandwidth, such as implementing VSF at the aggregation layer or core.

You can also look at average usage. However, peak usage usually gives the best information about the traffic that the switch needs to support.

Finally, consider variations over different switches. You should look at the usage on different switches to determine whether the traffic has been distributed more or less evenly or whether particular locations handle a higher traffic load. You might need to plan different oversubscription ratios for different switches to account for variable usage patterns.

Upstream VSF bandwidth allocation

Special stacking cables provide sufficient bandwidth that you do not need to worry about provisioning if you create the recommended ring or complete mesh. For VSF, you need to make sure to provision enough bandwidth. When a VSF member needs to forward traffic on a link aggregation, it favors the aggregation's local link. This cuts down on the inter-member traffic flowing from endpoints toward the core. In a VSF fabric where half of the members have uplinks, roughly half of the upstream traffic can forward locally, while the other half must cross a VSF link to a member with an uplink one hop away.

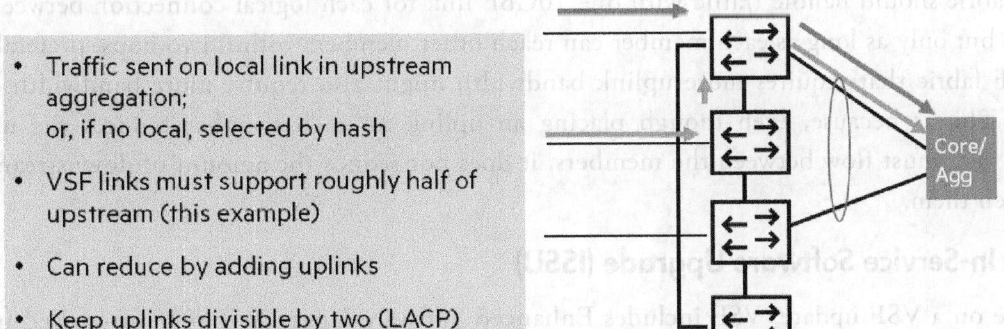

In this example, each member has two 10GbE VSF links, and the uplinks only support 20Gbps. Therefore, the VSF bandwidth should be adequate. Adding uplinks to more members can reduce upstream traffic on VSF links. However, you should keep the total number of uplinks even to balance them across the typical two aggregation or core switches. Do not make your plan until you consider downstream traffic.

Downstream VSF bandwidth allocation

Now, examine the traffic arriving from the core and destined to the endpoints. Downstream traffic, which often forms the bulk of traffic in a campus LAN, might arrive on any of the VSF fabric's uplinks regardless of the destination because the upstream switch selects the link. The member that receives the traffic must forward it to the member that connects to the destination endpoint. In a four-member group, the VSF links will carry roughly three-fourths of the downstream traffic, and you cannot reduce that by adding uplinks.

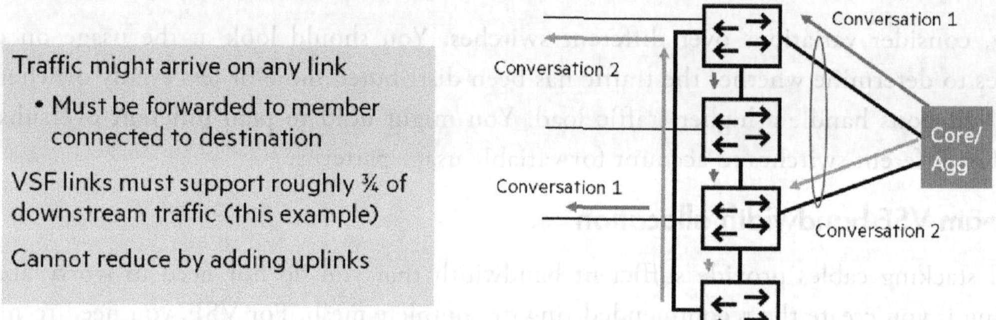

In this example, the VSF fabric has two 10GbE uplinks, and each member has two 10GbE VSF links for four links between them. As long as the 10GbE uplinks are not congested, the VSF links should meet the demands. However, if the VSF fabric requires more uplink bandwidth, such as a 10GbE link on each member, it might require more bandwidth on the VSF links.

VSF fabric should handle traffic with one 10GbE link for each logical connection between members – but only as long as each member can reach other members within two hops, preferably one. A VSF fabric that requires more uplink bandwidth might also require more bandwidth on VSF links. This is because, even though placing an uplink on each member reduces the upstream traffic that must flow between the members, it does not reduce the amount of downstream traffic between them.

VSF - In-Service Software Upgrade (ISSU)

A note on a VSF update. VSF includes Enhanced Software Upgrade (ESU) introduced in 10.11 for the HPE Aruba Networking CX 6300 Switch Series. The new name for ESU is In-Service Software Upgrade (ISSU). The conductor, standby and members of the stack are upgraded dynamically without having the necessity for them to reboot at all starting with AOS-CX 10.13. The topology shows an example of VSF OOBM MAD and links.

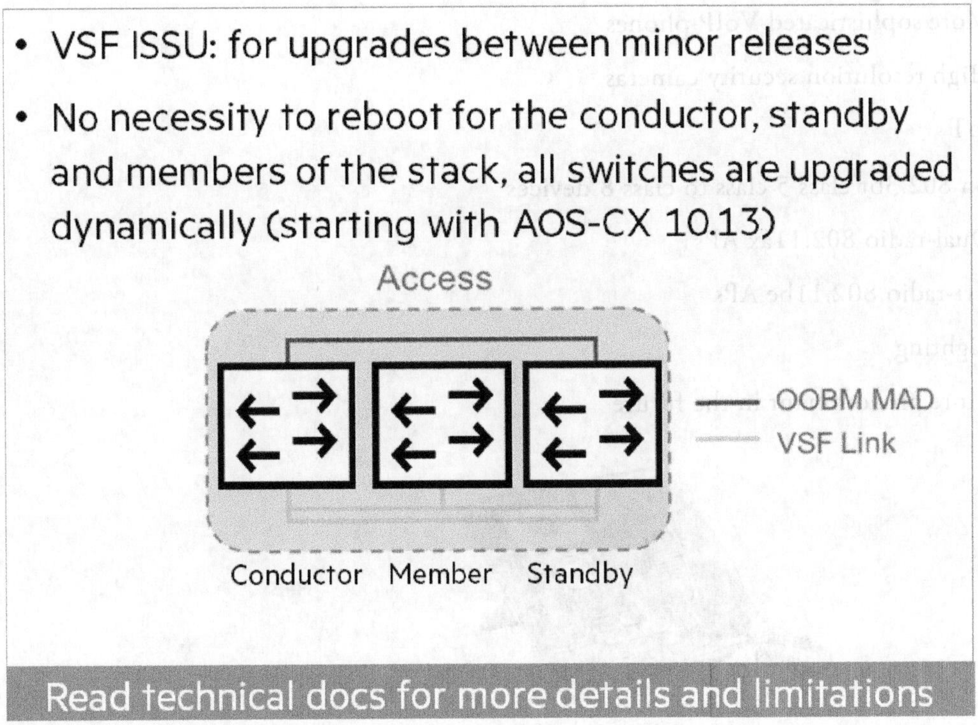

VSF ISSU operation:

Step 1: Image download and validation

Step 2: Install the image in the standby & members

Step 3: Freeze the system

Step 4: Perform the upgrade for the line modules

Step 5: Recover the previous state

Step 6: Failover occurs between the conductor and the standby

Step 7: The previous Conductor will upgrade its image

Edge port PoE

PoE Devices

- Less common in new devices
- Basic IP phones
- Security cameras

Common PoE+ devices

- More sophisticated VoIP phones
- High resolution security cameras
- IoT

Common 802.3bt class 5 class to class 8 devices

- Dual-radio 802.11ax APs
- Tri-radio 802.11be APs
- Lighting
- More predominant in the future

Another important feature for many access layer switches is the customer's need for PoE and PoE+ and 802.3bt class 5 and class 6 power. First, you need to identify the devices that require power so that you can plan which edge ports need to support this feature. The current AOS-switch line supports PoE+ on most switches and 802.3bt class 6 on other switches. Devices requiring power include VoIP phones, security cameras, IoT devices, and enterprise APs. More and more devices are requiring 802.3 bt power.

Quick PoE supplies PoE power to powered devices as soon as the switch is plugged into AC power so device can initialize at same time as switch OS boots up.

High availability with always-on PoE that supplies PoE power even during scheduled reboots and firmware upgrades

Considerations for the PoE budget

Once you determine whether the customer requires PoE class 3, 4, 5, or 6, 7, or 8, you can plan the power budget for each access layer switch.

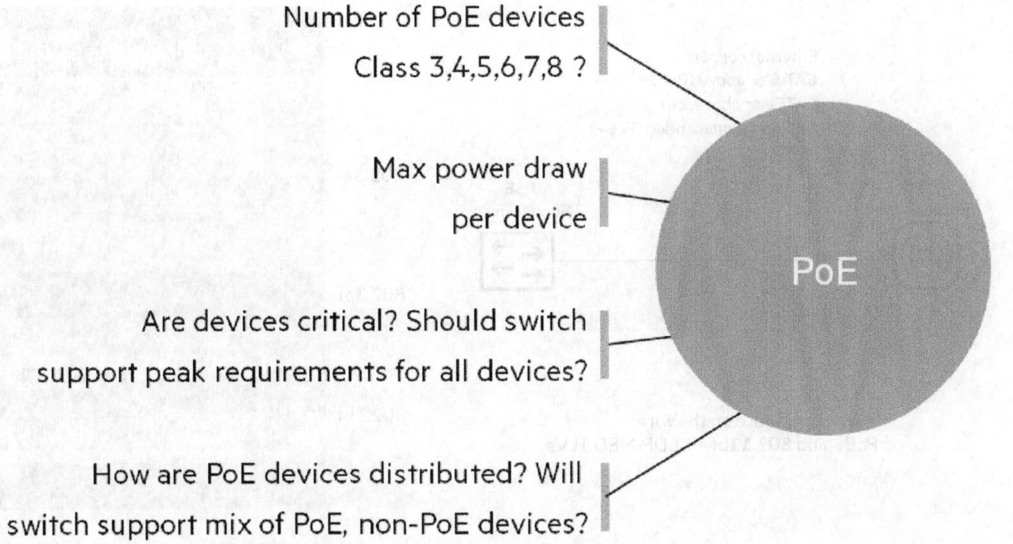

- In addition to clarifying how many total PoE devices the network needs to support and how to distribute these devices, you need to determine how much power each device will draw.
- Do you expect the switch to support all PoE devices drawing power at their peak requirements? When talking about critical devices, such as APs and VoIP phones, the switch should meet peak requirements, even if it is unlikely that all devices will draw the maximum power simultaneously.
- Also consider how you will distribute the devices. Will a switch support a mix of PoE and non-PoE devices, or will it need to deliver PoE on every port?

PoE and PoE+ review

Make sure to plan how switches allocate power to ports. To understand your options, you should understand how devices request power and how switches allocate it.

Powered devices (PDs) receive PoE power while power-sourcing equipment (PSE) provides the power. When devices connect to PSEs, the PSEs detect if the devices are PDs.

Classes indicate the power ranges that PDs require. The power available to the PD is less than the power required at the PSE due to cable loss. PDs can indicate their classes when they connect. Only PoE+ devices (802.3at) can use class 4.

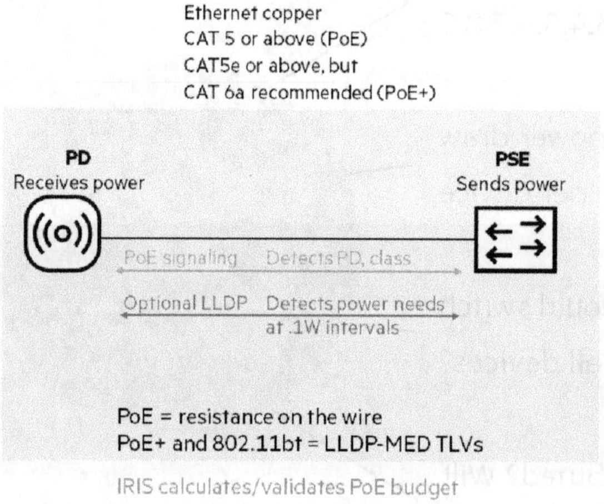

Standard	Class	Power available to PD (W)	Power required at PSE (PoE switch) (W)
802.3af	0	0.44-12.95	15.4
	1	0.44-3.84	4
	2	3.84-6.49	7
	3	6.49-12.95	15.4
802.3at	4	12.95-25.5	30
802.3bt	5	40	45
	6	51	60
	7	62	75
	8	71.3	90

PSEs classify PDs that do not indicate a class as 0, which can receive power up to the PoE maximum of 12.95W (perhaps slightly higher if the cable loss is low).

PSEs and PDs can negotiate the power that PDs require more precisely at 0.1W intervals using LLDP messages. PoE uses LLDP-MED TLVs for this purpose. PoE+ includes power management through LLDP as part of the standard.

The PSE supplies power to the PoE PD over a CAT 5 cable or above and to the PoE+ PD over a CAT 5e cable or above. PSEs use the resistance between the powered pairs of wires in these cables to detect whether the device is a PD, in which case the PSE knows it should power the device.

It is also by specific signature resistance that a PD indicates its class.

Some PDs might support Cisco Discovery Protocol (CDP) rather than LLDP for negotiating required PoE levels. CX switches also support this protocol.

Note

Although CAT5e cable can deliver PoE+, HPE Aruba Networking recommends CAT6a cable for less loss and better delivery.

Core and Aggregation Layers

With the access layer completed, let's look now at designing the core and aggregation layers.

Design the core for a 2-tier topology

To support enterprise resiliency requirements, you should generally deploy two core switches combined in a VSX cluster. In a 2-tier topology, the core connects every component to the data center and the Internet, including the access layer and HPE Aruba Networking gateways. The access layer itself could consist of individual switches and VSF fabrics.

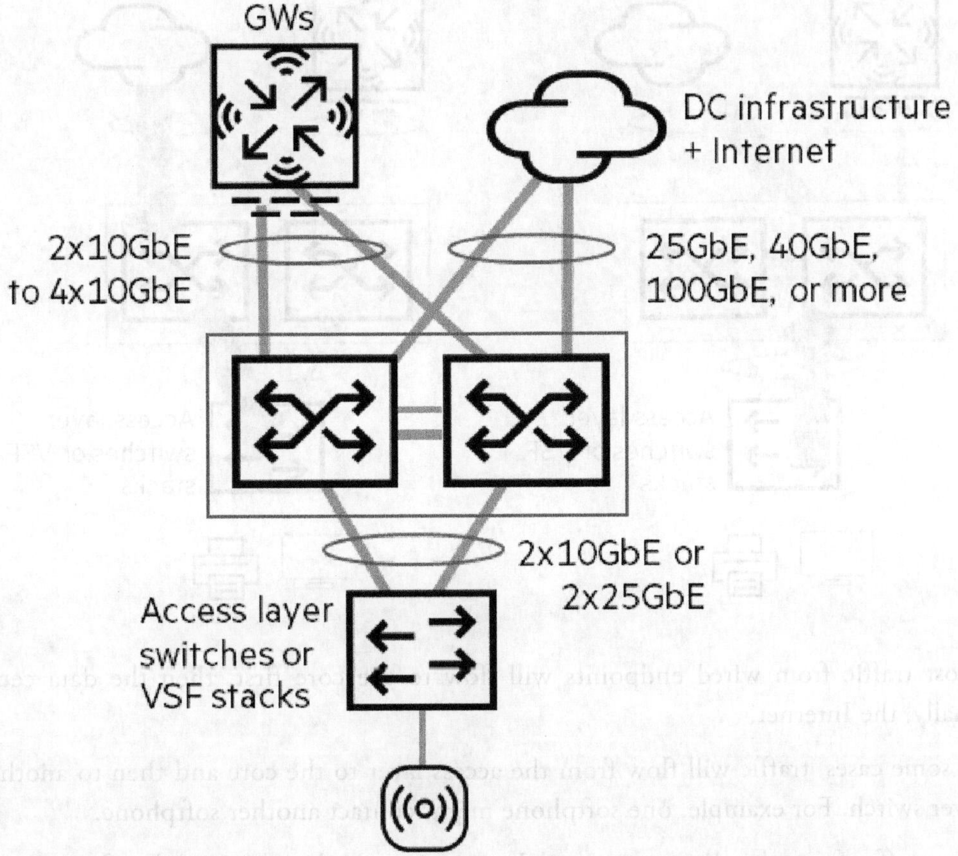

The typical bandwidth between the access layer and core can vary, depending on whether the access layer switches operate individually or combine in stacks or VSF fabrics. Usage requirements also affect this bandwidth. Links between the core and the GWs are about two to four 10GbE links. Depending on GW capabilities and environment bandwidth requirements, these links could scale from two 1GbE to eight 10GbE links. However, the ranges in the example are typical.

The example shows how the campus core connects to the data center. For larger networks and data centers, the campus core connects to a data center core, which supports the data center's 2-tier or 3-tier topology. Or the campus and data center cores might collapse with the same pair of core switches supporting the campus access layer and a few server access switches. The data center also typically provides a connection to the Internet.

Consider the wired traffic flow in a 2-tier topology

To understand the bandwidth requirements, you should understand how most traffic in your network will flow. The wired traffic flow is simple to understand.

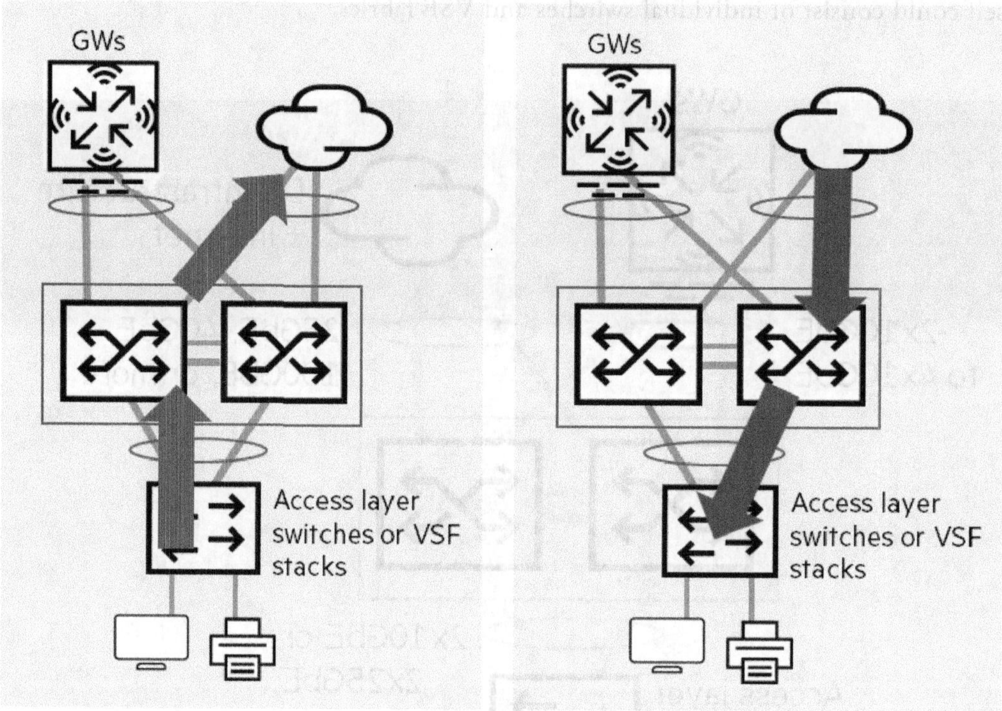

- Most traffic from wired endpoints will flow to the core first, then the data center, and finally, the Internet.
- In some cases, traffic will flow from the access layer to the core and then to another access layer switch. For example, one softphone might contact another softphone.
- The traffic from the data center and Internet to wired endpoints also flows through the core.

It is important to understand that all wireless traffic flows through the GWs.

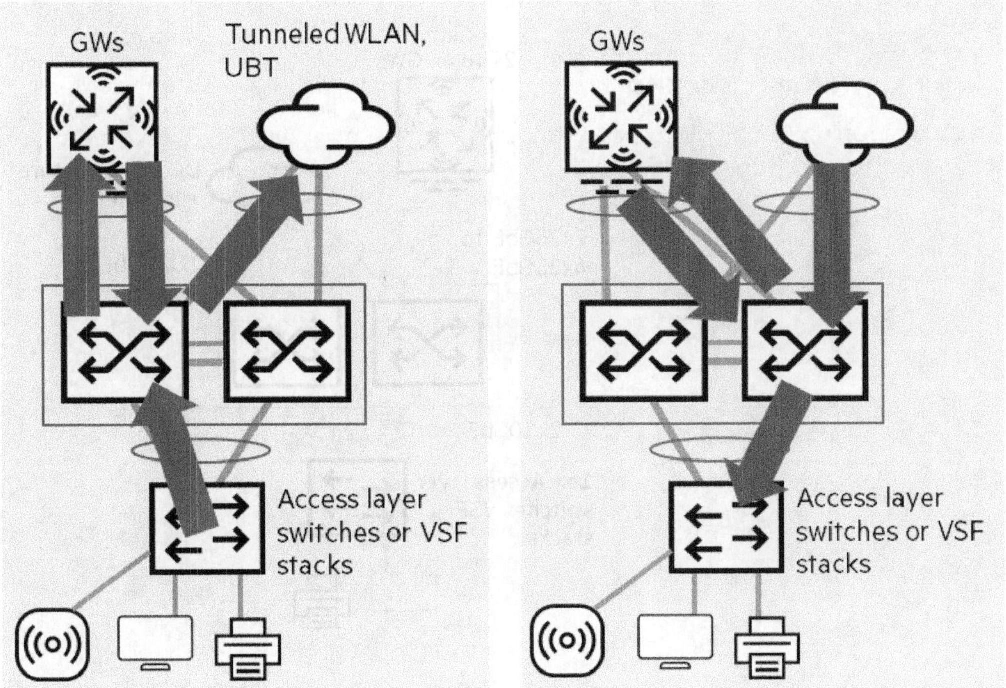

- Traffic from wireless endpoints flows through the access layer to the core and GWs, then back to the core and out to the data center or Internet.
- Traffic outbound to wireless clients from the data center or internet also flows both inbound and outbound on the GWs' links.

Plan appropriate oversubscription

Put together what you have learned.

Traffic to and from the access layer typically passes over the links between the core and the data center. Because the core is collapsed, the network can tolerate some oversubscription on links to the data center.

For example, you might have 16 access layer switches (or fabrics or stacks) with two 10GbE links each, for a total bandwidth of 320Gbps. However, you planned generous bandwidth at access layer uplinks, so you know that the links to the data center can tolerate some oversubscription. You might plan two 40GbE links on the core VSX cluster, which yields 4:1 oversubscription. This design works when you expect the access layer uplinks to operate well below their capacity most of the time.

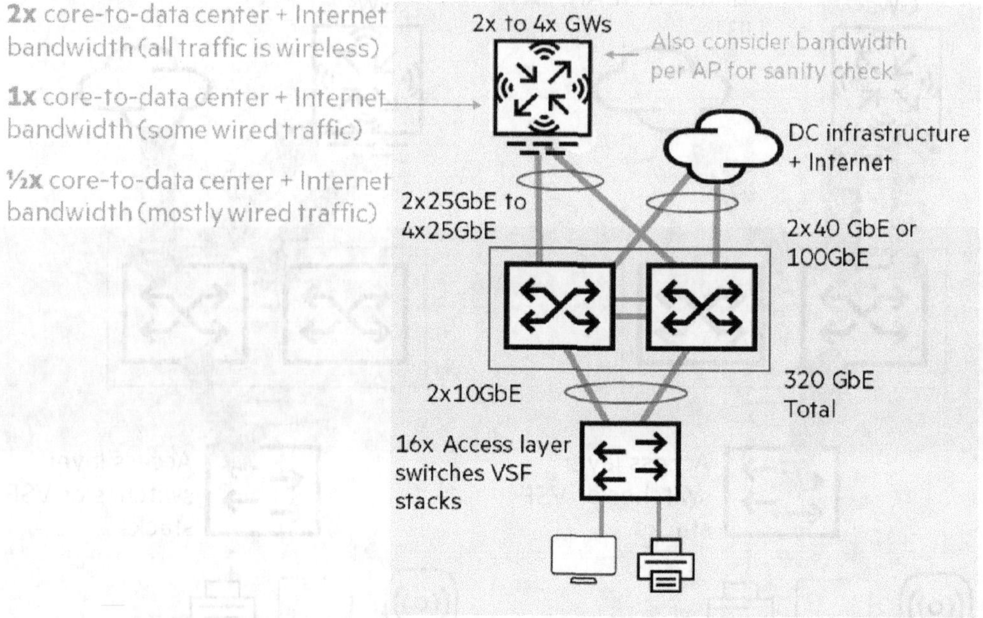

The GWs' bandwidth depends on wireless traffic and centralized overlay traffic. For example, if all traffic is wireless, plan about twice as much bandwidth to the GWs at the data center. All traffic to and from wireless endpoints must travel over the links in both directions. Here, you are counting the total bandwidth to all the GWs, not individual GWs.

You can also consider the bandwidth per AP for a sanity check. If you plan to scale the GWs' bandwidth to the maximum possible, but this maximum bandwidth does not seem enough for your environment, you might need to reconsider your GW plan. You could plan to deploy more GWs with fewer APs per GW.

This is the bandwidth per AP if you deploy the maximum number of APs on a GW and give the GWs their maximum bandwidth. The 7205 provides 48Mbps per AP. The 7210 and 7220 provide 40Mbps per AP. The 7240 provides 20Mbps per AP, and the 7280 provides 50 Mbps per AP.

In an example with four GWs, you might plan four 10GbE links for each, yielding 160Gbps bandwidth. (You would also select GWs with 40Gbps throughput for the firewall.) On the other hand, if you expect that some traffic will be wired-only, you can scale down the bandwidth requirements on the GWs. For example, you might plan only two 10GbE links on each of your four GWs.

If you expect light wireless traffic, you might plan to deploy just two GWs, and they can have two 10GbE links each.

If, on the other hand, you expect heavy wireless traffic, you might add a GW with 10 GbE links and distribute fewer APs to each GW.

These are just some examples, but they give you a sense of the issues affecting your plan.

Consider hardware requirements at the core

In a 2-tier campus topology, the VSX cluster or core switches typically route all traffic in the campus network. Therefore, the switch must support all of the VLANs, and its MAC forwarding tables and ARP tables must be large enough to store the MAC address of every device on the campus.

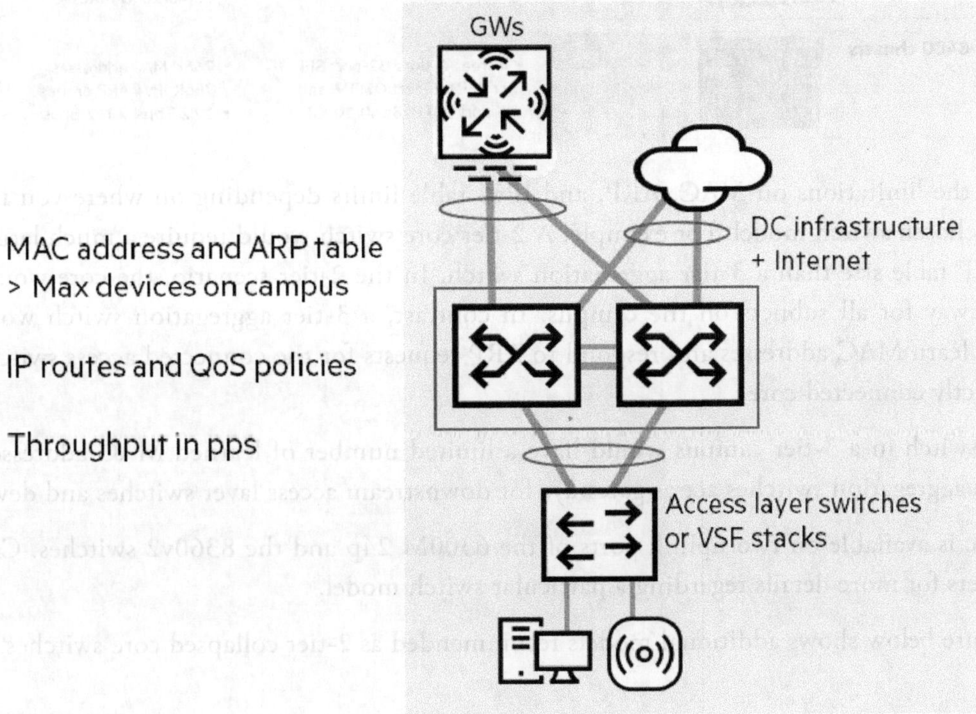

You should also consider factors such as the number of IP routes and QoS policies the switch can support. However, these capabilities are less likely to act as bottlenecks in most environments.

You should verify that your selected switches can support the throughput you expect the switches to handle. HPE Aruba Networking switch VSX supports the maximum bandwidth on the ports and far beyond. If the customer requires high performance, though, you might look beyond switching capacity to throughput in packets per second for small packets, which can give you a better sense of how the switch will perform.

Recommended 2-tier collapsed core switches

HPE Networking offers several switches for the core of a 2-tier network. The same switches that work at the core of a 2-tier network can provide the aggregation layer or core of a 3-tier network.

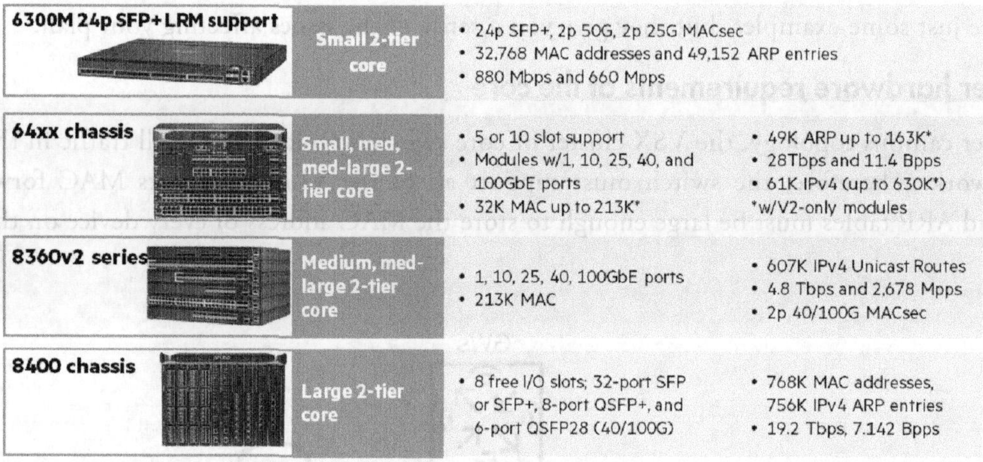

Review the limitations on MAC, ARP, and IPv4 table limits depending on where you intend to place a chosen switch model. For example, A 2-tier core switch would require a much larger MAC and ARP table size than a 3-tier aggregation switch. In the 2-tier scenario, the core would act as the gateway for all subnets on the campus. In contrast, a 3-tier aggregation switch would only need to learn MAC addresses and respond to ARP requests for the connected access switches (and the directly connected core).

A core switch in a 3-tier campus would have a limited number of learned MAC addresses, given that the aggregation switches act as gateways for downstream access layer switches and devices.

MACsec is available on two uplink ports of the 6300M 24p and the 8360v2 switches. Check the datasheets for more details regarding a particular switch model.

The figure below shows additional models recommended as 2-tier collapsed core switches.

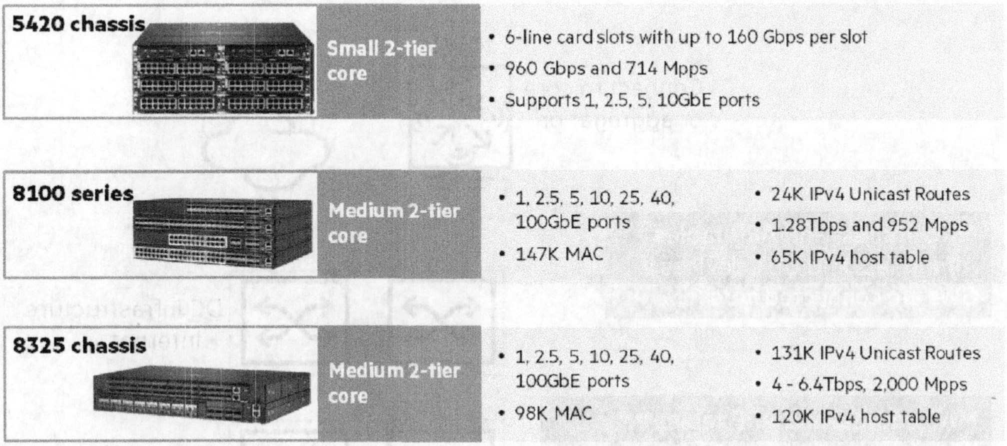

Design a 3-tier network

Many of the same considerations that apply to the core of 2-tier networks also apply to the core and aggregation layer of 3-tier networks. When a network has a separate aggregation layer, you should introduce the oversubscription at that layer. The core should not introduce any more oversubscription.

HPE Aruba Networking GWs might connect at the aggregation layer or the core, depending on the traffic flow, roaming needs, and the customer's desire for local resiliency.

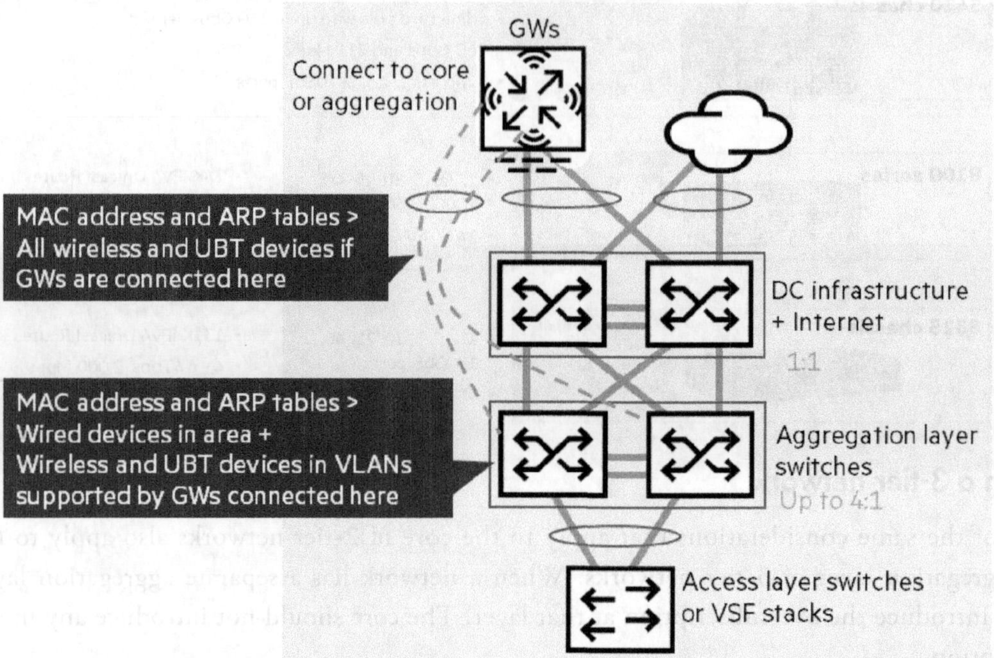

Remember that where you connect the GWs will affect the number of MAC addresses the aggregation and core switches must support. Typically, you should implement routing between the aggregation layer and the core, meaning VLANs for wired devices only extend to the aggregation layer. This limits the size of the MAC address tables for those VLANs.

When planning the GW bandwidth, you must remember what percentage of the traffic you expect from wireless devices and plan enough bandwidth for this traffic to enter and leave the GW in each direction.

Because APs tunnel traffic from wireless devices to GWs, all wireless VLANs exist in the core if the GWs connect to the core. The core must have a large enough MAC address table to support all devices. On the other hand, if the GWs connect to the aggregation layer, you can segment the wireless VLANs like the wired VLANs. However, this might affect the roaming or clustering plan. You could extend a wireless VLAN across the aggregation layer and core rather than routing between them. In this case, every core and aggregation layer switch would need a large enough MAC address to support all devices in the wireless VLANs.

Example medium-large 3-tier design

In this example, the aggregation layer consists of up to four pairs of CX 6300M series switches deployed as VSF stacks. Each aggregation VSX stack connects to the core with one 40GbE link on each member. The aggregation layer fabric connects with two 10GbE links to each access layer

VSF stack. The aggregation layer VSX stack can support between 8 and 16 access layer units, depending on the level of oversubscription you want. A pair of 6405 switches is deployed in a VSX cluster at the core.

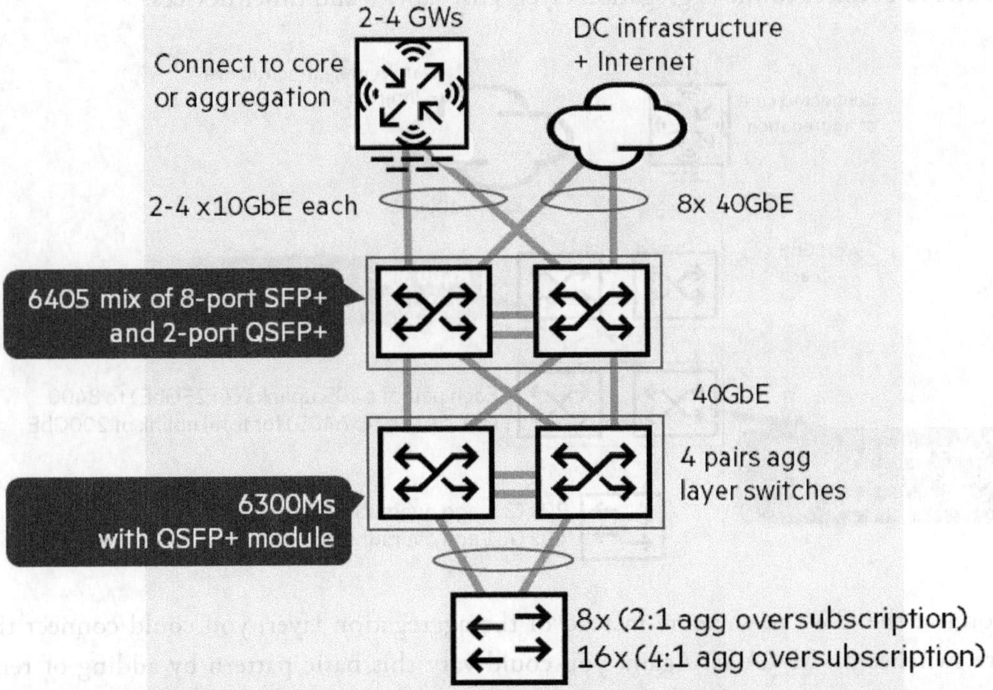

The fabric supports up to eight 40 GbE links in its link aggregation to the data center core, matching the amount it supports for the links to the aggregation layer switches. This allows the core to support up to four pairs of aggregation layer switches with two 40GbE ports each.

When the access layer connects on two generous 10GbE links, 4:1 oversubscription often works well. The access layer switch model is not shown but could be switched, such as CX 6200F series combined with VSF.

You could add another two 40GbE links between the core and the aggregation layer. In that case, the core could only support two pairs of aggregation layer switches to maintain non-blocking forwarding. Because the aggregation layer switches would then support twice as much bandwidth to the core, they would provide less oversubscription for the links to the access layer.

Example large 3-tier design

In this design, CX 6405 switches operate in VSX fabrics at the aggregation layer. They have 200 Gbps uplink bandwidth (and could scale higher). This VSX cluster can support up to 40 access layer switches (or fabrics or stacks), each with two 10GbE links, at a 4:1 oversubscription rate. However, in this example, the HPE Aruba Networking GWs connect at the aggregation layer, taking up some available access layer connections and leaving the VSX fabric capable of supporting 24 or 28 access layer switches. Two CX 8400 series switches at the core can connect to the data center core on high-speed 100GbE link aggregations. The CX 8400 series switches can use M-LAGs to connect to the aggregation layer VSX fabrics and other devices.

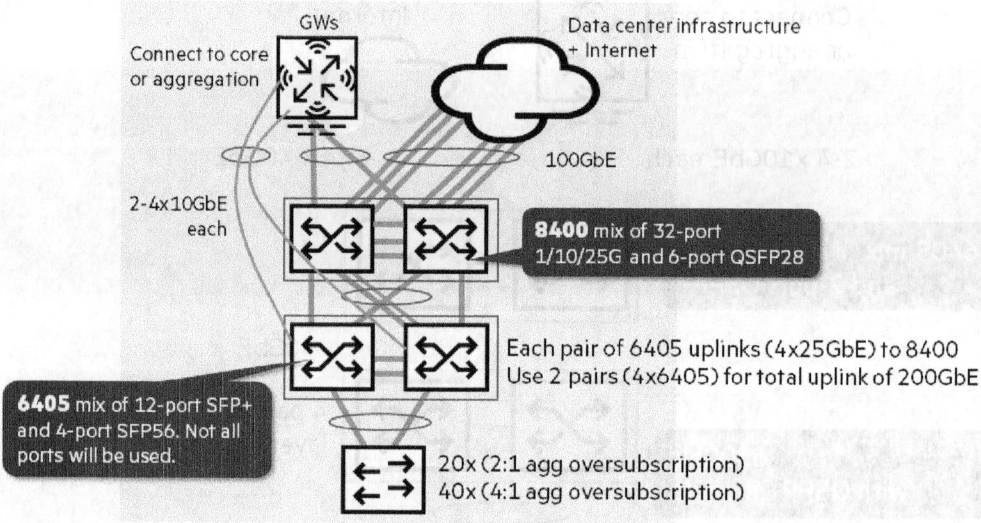

If you connect the GWs to the core instead of the aggregation layer, you could connect them to both core switches on M-LAGs. Again, you could vary this basic pattern by adding or removing links between the aggregation layer and the core to support the desired number of switches at the access layer and maintain the required oversubscription level.

Plan VSX links at the aggregation layer and core

You must plan adequate bandwidth for VSX and stacking links for these designs. Make sure to design the network such that links in each link aggregation balance out over both members. This helps you to minimize traffic on the VSX links. Because each member of the VSX cluster has a local link in every link aggregation to forward traffic, no unicast traffic crosses the VSX link during normal operation. The VSX link does need to handle multicast and broadcast traffic. In this example, one member receives a multicast and sends it over the VSX link. The other member is the designated multicast forwarder for the outgoing link aggregation, so it forwards the traffic.

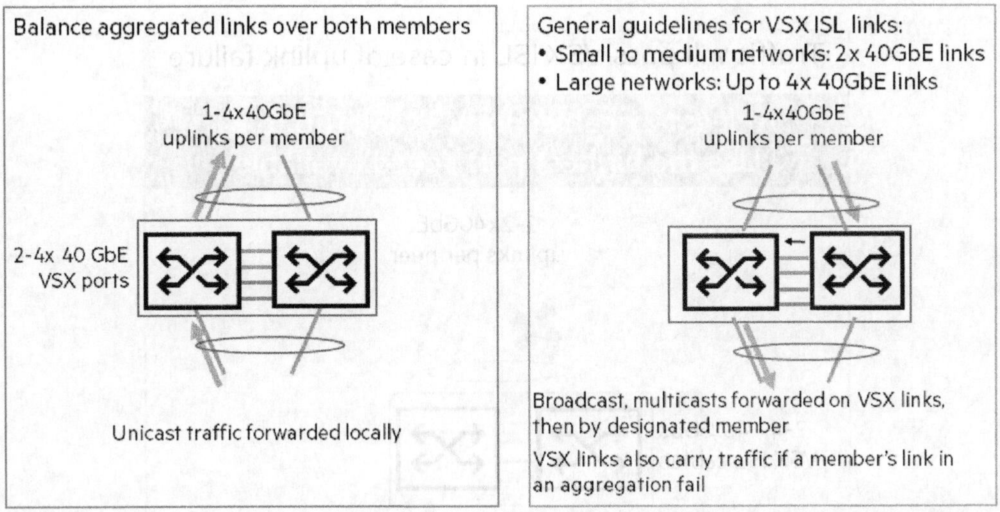

Remember also that if all of a member's local links in an aggregation fail, the switch must forward traffic destined out of that aggregation over the VSX link. As a general rule, plan two 40GbE links for the VSX link for CX 6405 switches at the distribution layer or the core of small to medium networks. Plan up to four 40GbE links for VSX links for CX 6405 switches at the core of larger networks.

Link aggregation on a VSX cluster has a single member that acts as the designated multicast forwarder. The cluster automatically picks which member will play this role for each link aggregation.

Plan aggregation & core bandwidth

You should remember that the pair of switches might exchange significant traffic.

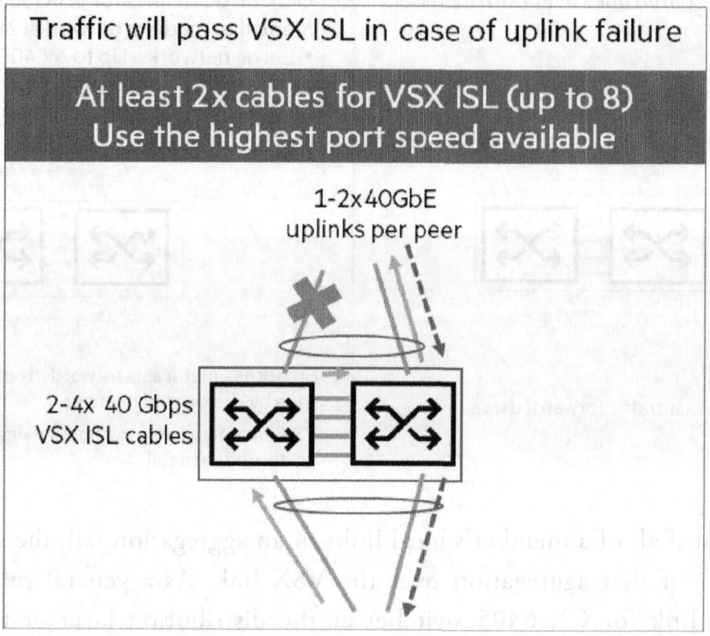

You will perform best if the VSX ISL links provide more bandwidth than the uplinks. Plan two VSX ISL cables for most deployments and up to four for higher bandwidth networks.

Learning check

Check your understanding of wired network design by answering the following questions.

Question #1

What is one reason to design a 2-tier campus LAN instead of using a 3-tier?

- O Cross-campus traffic has grown beyond the capacity of a single collapsed core.
- O Access aggregation points are not scaling to meet traffic demands.
- O The network has grown beyond a few building aggregation points.
- O The campus has small buildings with only a few wiring closets.

Question #2

What is the best practice for using VSX at the Core of a 3-tier design?

- O You should implement VSX at the core when you connect your GW cluster to the core.
- O You should never implement VSX at the core of a 3-tier redundancy core.
- O VSX-LAGs allow the collapsed core to connect directly to services such as gateways and services only so long as spanning-tree is used to prevent loops.

- You should implement VSX at the core when the aggregation layer is layer 3 only with OSPF routed traffic forwarding between the core and aggregation layers.

Question #3

A junior network designer is asking what is the simple difference between a main distribution framework (MDF) closet and an intermediate distribution framework closet (IDF).

- MDFs always have larger rooms than IDFs.
- MDF is a term used in Europe whereas the Americas refer to all network closets as IDFs.
- MDF is the point where traffic egresses the campus network, IDFs distribute that connectivity throughout the building.
- MDFs only connect to other MDFs, and IDFs only connect to other IDFs.

Question #4

Which access switch oversubscription option would be best suited for the mix of high-performance access points with Smart Rate ports and employees?

- 1:1
- 6:1
- 24:1
- 48:1

Supplemental: Switch management & control plane

Use this supplemental material to review and study on your own.

HPE Aruba Networking Central for AOS-CX management

HPE Aruba Networking Central features an even simpler, more powerful UI, to ease routine changes to interfaces, ports, VLANs, and more on the HPE Aruba Networking CX Switch Operating System switches.

Fast, error-free management via HPE Aruba Networking Central

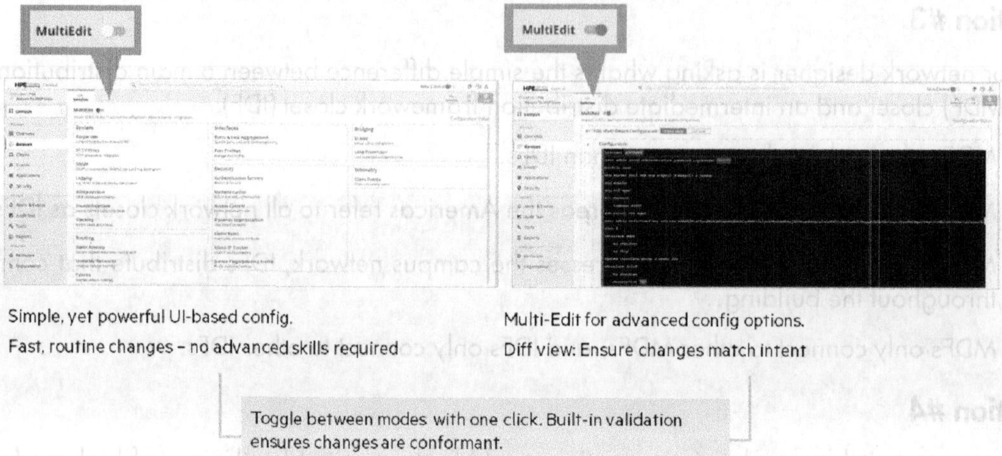

You can easily toggle to MultiEdit mode for advanced config settings, with assistances like command completion, syntax highlighting, and built-in validation of network policies. You have improved flexibility over implementing configuration changes, where you can flip back and forth using the UI or MultiEdit mode, depending on your needs and preferences.

Network automation with AOS-CX

The NAE engine is integrated with AOS-CX system configuration and time series databases. This enables you to examine historical trends and predict future problems due to scale, security, and performance bottlenecks. With that information, you can create software modules that automatically detect issues, take appropriate actions, and reduce the time spent on manual tasks and troubleshooting.

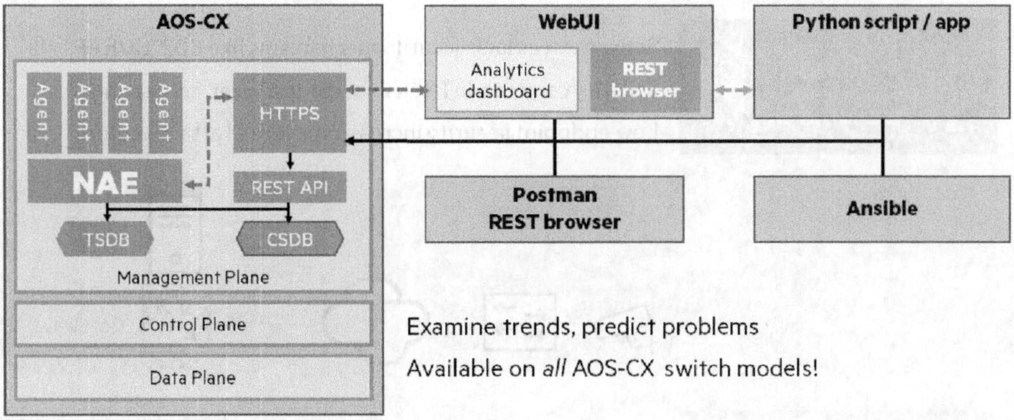

NAE uses Python-based agents developed by HPE Aruba Networking and the HPE Aruba Networking community to automate network monitoring and troubleshooting operations. Agents monitor the state, settings, and statistics for system components in the current state database (CSDB), for example, the number of bytes transmitted on an interface or the switch's operating temperature. The agent then writes ongoing results to the time series database (TSDB) for ongoing analysis.

You can interact with the NAE through the switch's web UI's analytics dashboard. You can also automate the NAE through the REST API using tools such as Postman, Ansible, or another choice. You can create or obtain Python scripts for new agents and load the agents on the switch using the web UI or the REST API.

Using the HPE Aruba Networking Central configuration UI, you can also deploy a pre-defined set of NAE scripts to CX switches. Custom NAE scripts can be pushed via template configurations as well.

Dynamic Segmentation with centralized overlay: problem solved

All modern desktops, laptops, tablets, and smartphones support robust security mechanisms like 802.1X/EAP (Extensible Authentication Protocol). However, many devices lack these security mechanisms. This includes specialty devices and devices that rely on PoE, such as security cameras, payment card readers, Internet of Things (IoT), and medical devices. This lack of endpoint security can pose a serious risk to your infrastructure.

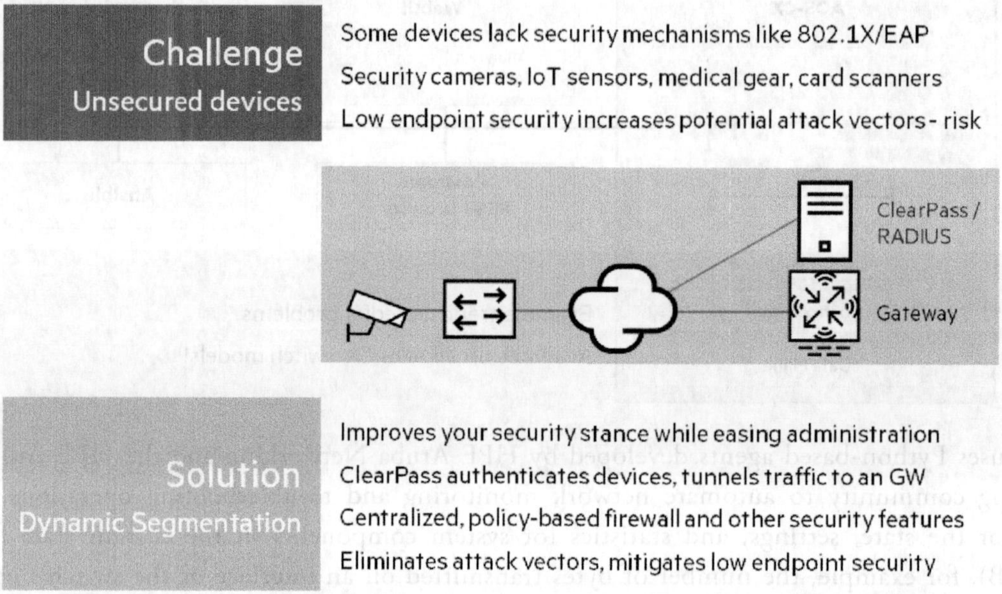

A switch with dynamic segmentation improves your security stance while easing administration. It authenticates these devices using HPE Aruba Networking ClearPass and tunnels the traffic to an HPE Aruba Networking gateway, centralizing formerly disparate access control lists (ACLs) and firewall rules into a single, intuitive policy system. You can continue to use these devices without compromising network security.

Dynamic Segmentation: operation overview

With dynamic segmentation, the switch authenticates the user by interacting with an authentication, authorization, and accounting (AAA) server, like ClearPass. ClearPass profiles each device deems appropriate endpoints acceptable and sends accept messages to the switch. The messages include tunnel information, the assigned VLAN, and the secondary role. The switch uses this information to establish the tunnel to the gateway.

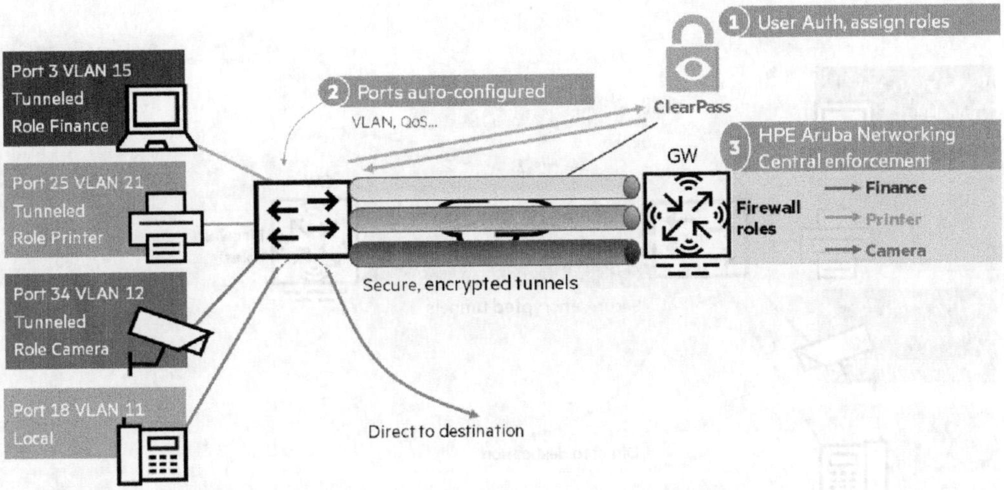

The user's traffic is tunneled to the gateways, where the gateway applies the appropriate policies, like firewall policies. In the figure, a user was profiled with the Finance role, so their port is configured for VLAN 15, and their traffic is securely tunneled to the gateway. The gateway applies centralized policy enforcement to limit finance traffic as appropriate. Printers and cameras are similarly treated, as shown in the figure.

ClearPass can also indicate if the switch should locally forward device traffic directly toward its ultimate destination without tunneling to the gateway. You sacrifice some of that centralized PEF enforcement but accommodate uber-sensitive devices to delay and jitter.

Network Segmentation

Network Segmentation is traditionally accomplished through VLANs and access control lists (ACLs) to enforce network security.

HPE Aruba Networking fortifies these standard practices through the following:

- The automated implementation of security policy through the overlay network
- Authenticated roles that identify and group access for each client and devices for policy consistency
- Segmenting campus traffic, both wired and wireless.

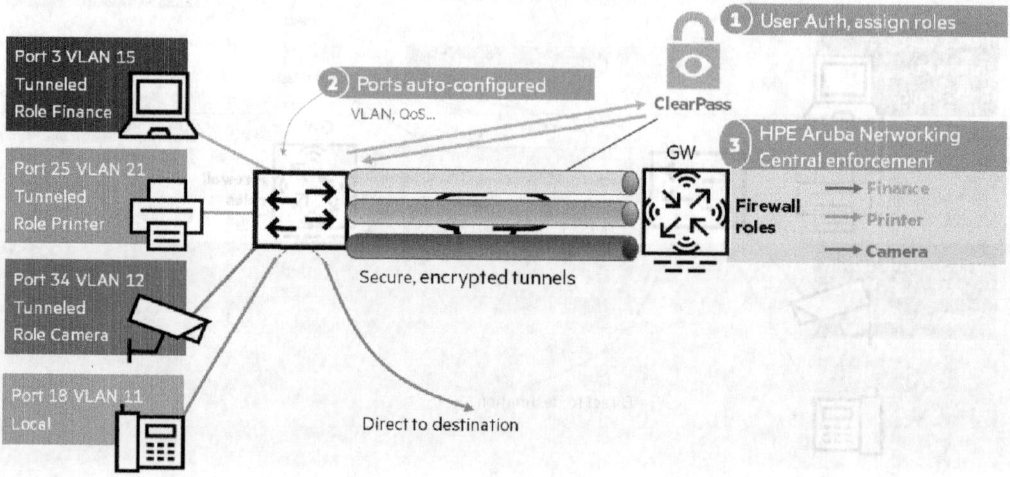

HPE Aruba Networking Dynamic Segmentation provides generic route encapsulation (GRE) tunneling of AP and switch traffic to a gateway cluster. The tunnel is applied dynamically through policies defined by user roles to achieve business requirements.

HPE Aruba Networking Central NetConductor also provides a distributed overlay architecture built using VXLAN-GBP to segment user traffic in the network. These tunnels are efficient and fault-tolerant, building an overlay fabric topology between any switch VTEPs and allowing policy enforcement through group-based policy (GBP).

VLANs, performance, and security

Virtual LAN (VLAN) is a fundamental tool that logically segments networks.

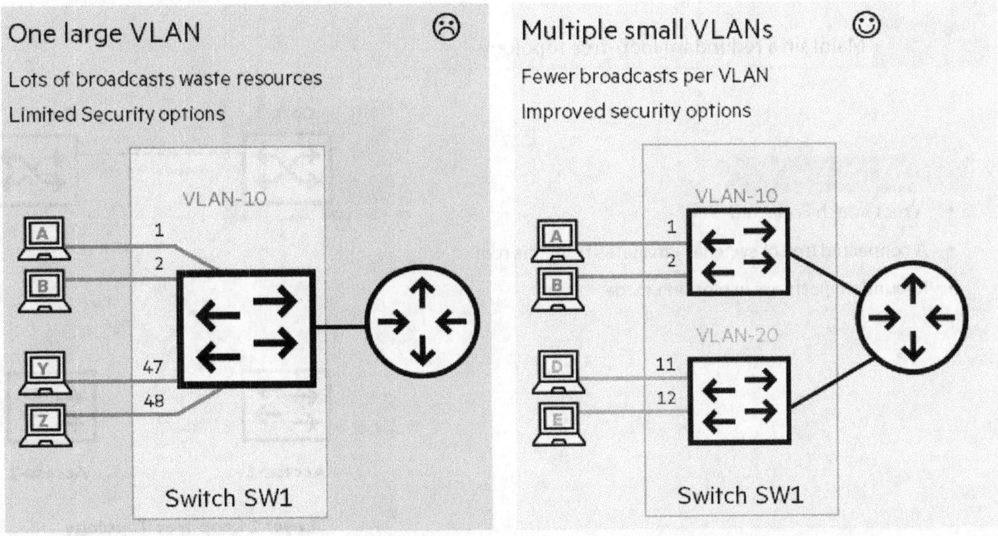

- Segments traffic at layer 2 by assigning a unique 12-bit VLAN id at the switch to restrict MAC reachability to groups with the same VLAN id. This creates a kind of macro-segmentation of users.
- Creates a separate IP broadcast domain and subnet per VLAN.
- Devices in the same VLAN are grouped by subnet and an IP ACL can then be applied to VLANs, layer 2, or layer 3 interfaces.

HPE Aruba Networking Central NetConductor can configure VLANs to create layer 2 segments as required.

Spanning-Tree

In the pursuit of high availability, the campus should provide layer 2 redundancy with multiple links between access to aggregation and from aggregation switches to the core. Adding these redundant uplinks assures that if a failure of one link occurs, there is at least one backup link that can be activated. There is a risk of introducing a layer-2 loop when adding another redundant uplink between a layer-2 access switch and an aggregation switch.

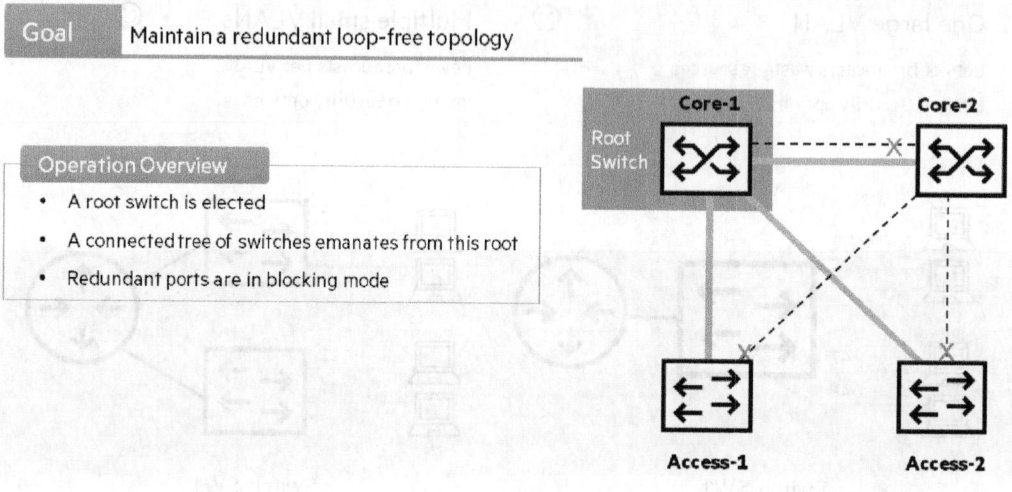

Layer 2 Loop-free Topology

Spanning-tree protocol (STP) can prevent such a loop regardless of how far the layer 2 domain extends. The version of STP enabled on HPE Aruba Networking 6xxx series switches by default is Multiple Spanning Tree Protocol (MSTP) IEEE 802.1s and can be enabled across any HPE Aruba Networking CX Switch Operating System switch.

Note that Gateways support Rapid-PVST+. While this is compatible with MSTP, Gateways should be placed on the campus edge across a layer-3 Core and in different switching domains from access switches running MSTP. Therefore, there should not be a layer-2 interaction between the two spanning-tree versions.

Spanning-Tree root bridge selection

Root Bridge Selection:

HPE Networking recommends enabling Spanning tree as a heavy-handed loop prevention mechanism and should be done regardless of network topology to prevent accidental loops.

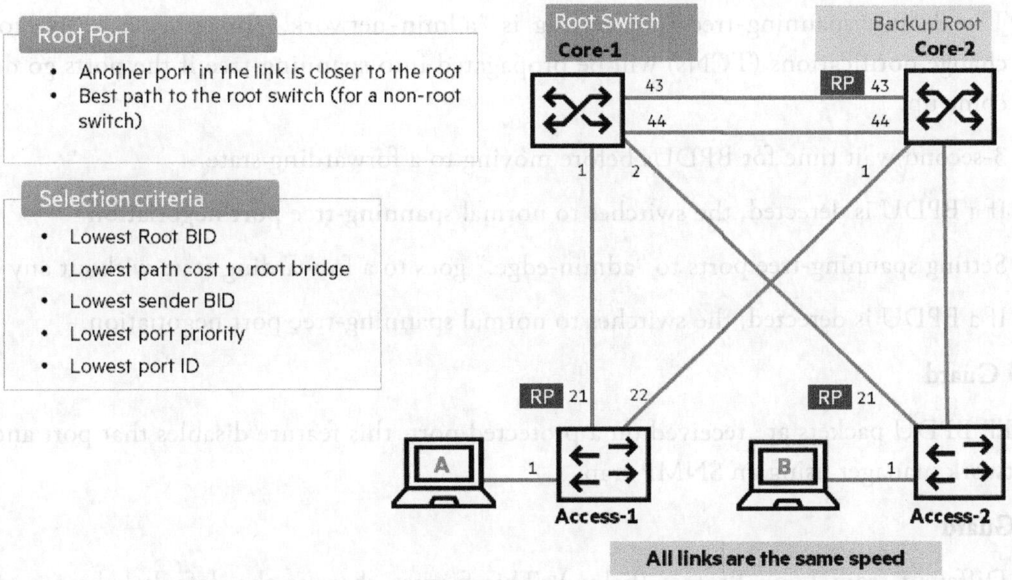

Access switches (and Gateways) should avoid becoming spanning-tree roots. Do this by assigning them high bridge IDs. Reserve lower bridge IDs for root bridges that are central to the network and act as VLAN default gateways. Root bridges should be assigned to aggregation switches in a 3-tier campus or the collapsed core in a 2-tier design.

Spanning tree root bridge placement:

Notice that if you have multiple aggregation switches across several buildings in a campus, each pair can be the root bridge for their building without causing contention. This is because spanning-tree does not extend over the layer 3 boundaries between each aggregation pair and the core.

Spanning-Tree features

Root Guard

- Once assigned to an interface, the port will be disabled when a superior BPDU is received.
- Assign root guard to aggregation switches' downlinks to prevent access switches from becoming root bridge.
- Optionally, assign root guard to access ports to prevent rogue devices from overtaking the root bridge role for added security.

Admin-Network versus Admin-Edge ports for access ports on a switch

- The default spanning-tree port setting is "admin-network." In either case, no topology change notifications (TCNs) will be propagated into spanning-tree if the ports go down or come up.
- 3-second wait time for BPDUs before moving to a forwarding state.
- If a BPDU is detected, the switches to normal spanning-tree port negotiation.
- Setting spanning-tree ports to "admin-edge." goes to a forwarding state without any delay.
- If a BPDU is detected, the switches to normal spanning-tree port negotiation.

BPDU Guard

If MSTP BPDU packets are received on a protected port, this feature disables that port and alerts the network manager using an SNMP trap.

Loop-Guard

- Different from Loop Protect (below). This feature changes the default behavior of spanning-tree when BPDUs fail to be received. Still, if the receiving port is not shut down, the port will fail to reach a forwarding state normally (such as when a unidirectional link error occurs). This can sometimes result in a loop in the network.
- Enabling the loop-guard will cause the port to go into an error-disabled state and avoid the loop.

Loop Protect

- The enabled ports send loop protection packets every 5 seconds by default. When a loop-protection enabled port receives a loop protocol packet, then there is a downstream loop, and the port takes one of three actions:
- Transmission disabled.
- The port that transmitted the loop detection packet is disabled. THIS IS THE DEFAULT
- Transmission and Reception disabled.
- Do not disable it; only trigger a notification and report using SNMP.

BPDU Filter

- Note that "BPDU-Filter" should only be used to disable spanning-tree on an interface when a port should never be blocked by STP. Be warned that this can easily cause loops.
- Fault monitor
- Enable the detection of excessive traffic or link errors automatically.
- Can log events, send SNMP traps, or temporarily disable a port.

- It should be enabled on all ports for data continuity; however, don't enable the "disable port" feature. Loop-protect and spanning-tree should be used instead to stop loops.

Access Control Lists (ACLs) - ingress

ACLs must be applied (using an apply access-list command) to take effect. ACLs can be applied to interfaces (including LAGs), VLANs, or the Control Plane. ACLs can be applied to an interface in two directions (inbound and outbond). The table in the slide shows the Different options that exist to apply an ACL.

ACL type	IPv4+6	IPv4+6	MAC	MAC
Direction	**In**	**Out**	**In**	**Out**
L2 interface (port)	Yes	Yes	Yes	Yes
L2 LAG	Yes	Yes	Yes	Yes
L3 interface (port)	Yes	Yes	Yes	Yes
L3 LAG	Yes	Yes	Yes	Yes
L3 interface (port) subinterface	Yes	Yes	Yes	Yes
L3 LAG subinterface	Yes	Yes	Yes	Yes
VLAN	Yes	Yes	Yes	Yes
Interface VLAN	Yes (routed)	Yes (routed)		
Management interface	Yes			
Control plane (per VRF)	Yes			

Notice that AOS-CX accepts multiple ACL types to be assigned to the same interface as follows:

- One MAC ACL inbound
- One MAC ACL outbound
- One IPv4 ACL inbound
- One IPv4 ACL outbound
- One IPv6 ACL inbound

One IPv6 ACL outbound

Note

If you apply an ACL of a type, in a direction that is already in use, the switch replaces the current ACL with the new ACL.

Ensuring resiliency in a traditional design

A traditional network requires rapid spanning tree protocol (RSTP) or multiple spanning tree protocol (MSTP) to block redundant links to eliminate loops and handle failover to a redundant link if the active link fails.

If the customer requires a degree of load-balancing across the links, you should deploy MSTP. In this example, you are planning a 2-tier network, so the MSTP region extends across the network. You should assign some VLANs to one instance and other VLANs to another instance. Then, set priorities on the redundant core switches such that one is the root for one instance and the other is the root for the other instance.

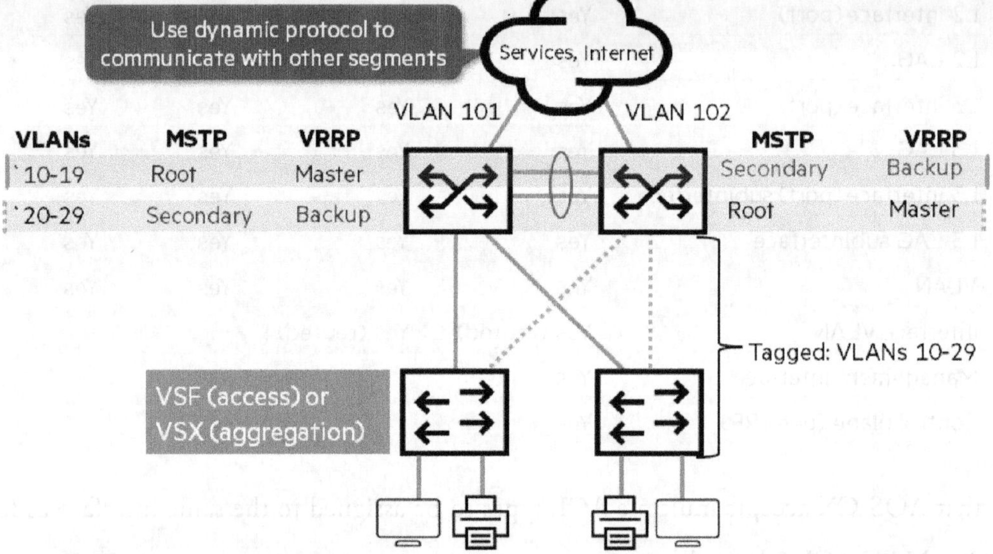

The core switches must implement VRRP to work together to provide default router services for VLANs. You should align the MSTP roles and the VRRP roles. Use VRRP priorities to ensure each core switch is the VRRP master in its VLANs. Finally, the core switches should use a dynamic routing protocol to communicate with devices in other segments, such as the data center. (Redundant static routes could provide an alternative.)

More complete recommendations for designing spanning tree and VRRP follow. These are provided for your reference in case you need to deploy a traditional network. In most cases, you will use the more modern VSF stacking-based design.

Spanning Tree

AOS switches support several different spanning tree protocols, including rapid per-VLAN spanning tree plus (RPVST+), RSTP, and MSTP. RPVST+ is a proprietary Cisco protocol, and HPE Aruba Networking recommends its use on AOS switches only when you need to integrate those switches into a heterogeneous network that already uses RPVST+ or PVST+. RPVST+ imposes a higher processing burden on switches than the standard spanning tree protocols because it uses a separate bridge protocol data unit (BPDU) and creates a separate tree for each VLAN.

To choose between RSTP and MSTP, consider whether the customer truly requires active use of multiple redundant links. If so, you should use MSTP. If not, RSTP can be simpler to design.

Then, use the following recommendations to avoid some of the most common problems:

Keep the topology simple. Ideally, limit the hops between each switch and the root to two. Also, minimize loops within loops. Flattening the architecture from three-tiers to two-tiers helps in this endeavor. If you are working with a 3-tier topology, implement routing at the aggregation layer. The pair of aggregation switches act as the MSTP roots. Avoid implementing STP across multiple segments in a WAN; route traffic between the segments instead.

Explicitly set the priority of the primary root to 0 and the priority of the secondary root to 1 (4096). You might want to set which switch has priority 2, but this is unnecessary.

Ensure all devices use the same standard for setting costs, ideally the 802.1t standard.

Make sure that the failure of a link never isolates a VLAN. Redundant switch-to-switch links should always carry the same VLANs. This prevents issues if one link fails. Also, avoid issues if the link fails between the MSTP root switches. If some VLANs are missing on the new path between the root switches, the root switches, which typically also implement VRRP, take over as VRRP master in those VLANs. The resulting IP address conflict can disrupt connectivity throughout the network. In the simplest and typically recommended setup, all switch-to-switch links in the MSTP region carry all VLANs. You should assess whether this setup will cause issues in your environment. If the network has many VLANs, it might congest the switch-to-switch links with excessive broadcasts. In that case, you can try another solution. (Ideally, you can implement VSF stacking and avoid the issue entirely, but these guidelines are for circumstances when you cannot.) You can explicitly define the backup path for each instance when a core-to-core link fails by configuring a low cost for the desired instances on the port or link aggregation that connects to the root for that instance. You may also set a favorable priority for this instance on the link if another switch offers an equal-cost root path. Then, ensure the backup path supports all VLANs in this instance. You could also establish a highly resilient link between the MSTP root switches.

If uneven link usage (one link cannot handle the full traffic load on its own) concerns you, follow these steps:

Implement MSTP.

Minimize the number of MSTP instances. There is no reason to create a new instance in an MSTP region unless you plan to have a new switch act as the root. Usually, two instances meet the need. Then, you have two switches, each acting as the primary root for one MSTP instance and the secondary root for the other MSTP instance.

Ensure all switches use the same MSTP mappings to establish the same MSTP region. If you add a VLAN to one switch, you must add it to all switches and map it to the same instance, even if some switches' uplinks do not carry the VLAN.

Assign VLANs to instances based on the expected traffic load.

Understand that instance 0, or the IST, is the instance that switches within an MSTP region used to connect to the Common Spanning Tree (CST). The CST establishes a tree between MSTP regions and between switches that implement different versions of STP (including Cisco's PVST+ or RPVST+). Even if you are not planning to carry traffic on instance 0, you should set the priorities for the switches that you want to be root and secondary root in this instance. This practice provides better behavior in case of misconfiguration of MSTP region settings.

VRRP

VRRP enables clients to use a single IP address for their default gateway address. This address is either a virtual IP address shared by two (or more) VRRP routers or an address owned by one of the routers but used by the standby router in case of the owner's failure. HPE Aruba Networking recommends using the former setup: each VRRP switch has its IP address separate from the virtual IP address. Although VRRP enables two (or more) routing switches to present a single virtual IP address to route a VLAN's traffic, the switches act as independent devices with their control plane and routing processes.

When you need to design VRRP, follow these best practices:

Align the VRRP master role with the MSTP primary root bridge and the VRRP back role with the MSTP secondary root bridge.

Use preempt mode, so after a failed default router is restored, that router takes over part of the routing burden again. However, set a preempt delay so the restored device can discover routes before assuming the master role. Typically, a preempt delay of 15 to 20 seconds is more than adequate, depending on the number of OSPF neighbors the device has and the complexity of the network.

Consider the best way to set up the routing protocol so that failover occurs seamlessly. You should connect each routing switch in the VRRP group to the upstream router in a different VLAN dedicated to that link. These VLANs should not run VRRP. If the network uses OSPF, each VRRP switch will establish an OSPF adjacency with the upstream router.

Allowing the upstream router to use OSPF ECMP to send traffic to both members of the VRRP group might cause issues for some applications (the traffic sent to the standby member would need to take a longer route, flowing back over the link between the VRRP members, assuming that MSTP roles are also aligned.) Test the solution with real applications and, if necessary, set costs on VLAN interfaces so that the upstream router prefers the VRRP master of each subnet as the next hop.

Rather than use OSPF on both VRRP members, some administrators prefer to avoid these complexities by configuring floating static routes on the upstream router or routers. One route to each VRRP subnet has a lower administrative cost for the master, and another route to that subnet has a higher administrative cost for the standby router.

You can also configure a track on the VRRP switches. The track monitors connectivity with a key upstream resource and lowers the VRRP priority if connectivity is lost. Then, if the VRRP master loses upstream connectivity but still has connectivity on the VLANs for which it is master, it concedes the master role to the standby router. The traffic then continues to reach its destination.

Static routes vs dynamic routing

Working with static routes is simple and easy for small networks that have a few subnets. However, when an organization has hundreds or even thousands of subnets, static routing is not an efficient method to administer subnets. There is no automatic route advertisement; you must manually configure everything. Other than the simple Floating Static routes you just learned, there's no dynamic failover mechanism.

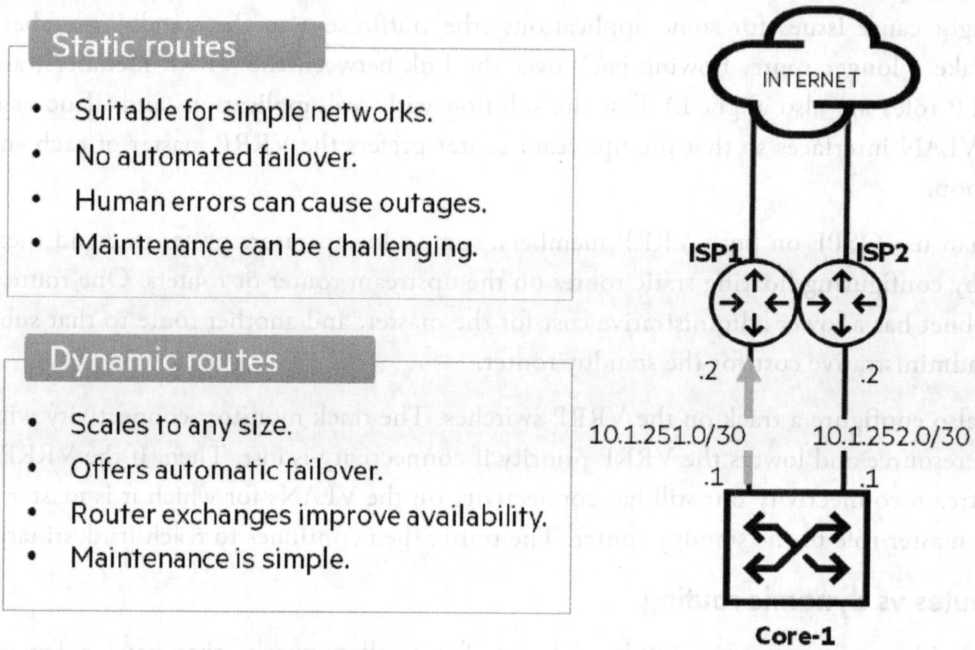

The human factor is also a big consideration. Administrators can cause serious problems if a route is placed in the incorrect device or if the route is not properly configured. This can create Layer 3 route loops, traffic blackholing, and lost connectivity.

Recovering from a routing device failure could be slow if Static routing is used. You must manually configure new alternative paths. This is inconvenient at best, especially when dynamic routing would automatically do this for you.

Dynamic routing protocols are far more scalable, handling thousands or millions of routes across multiple routing devices. They automatically failover alternate paths with little to no downtime. Routers constantly exchange messages that keep the network available and minimize your managerial workload.

OSPF routing

OSPF routing sample OSPF topology

Different departments across the campus will require scalable connectivity to share information. Campus networks extend typical single building 2-tier connections through a central Core. This redundancy between Core and Aggregation requires routing of Layer 3 connections. A routing protocol is needed to provide dynamic learning of subnets, resiliency when networks change, and loop prevention.

The most common routing protocols in enterprises today are Open Shortest Path First (OSPF) and Border Gateway Protocol (BGP). The latter, BGP, is the backbone of internet routing between large public networks and is used as the control plane for EVPN-VXLANs in the data center and campus overlays. OSPF is the standard choice for a campus infrastructure within an autonomous system or managed enterprise. It is a link-state dynamic routing protocol that scales.

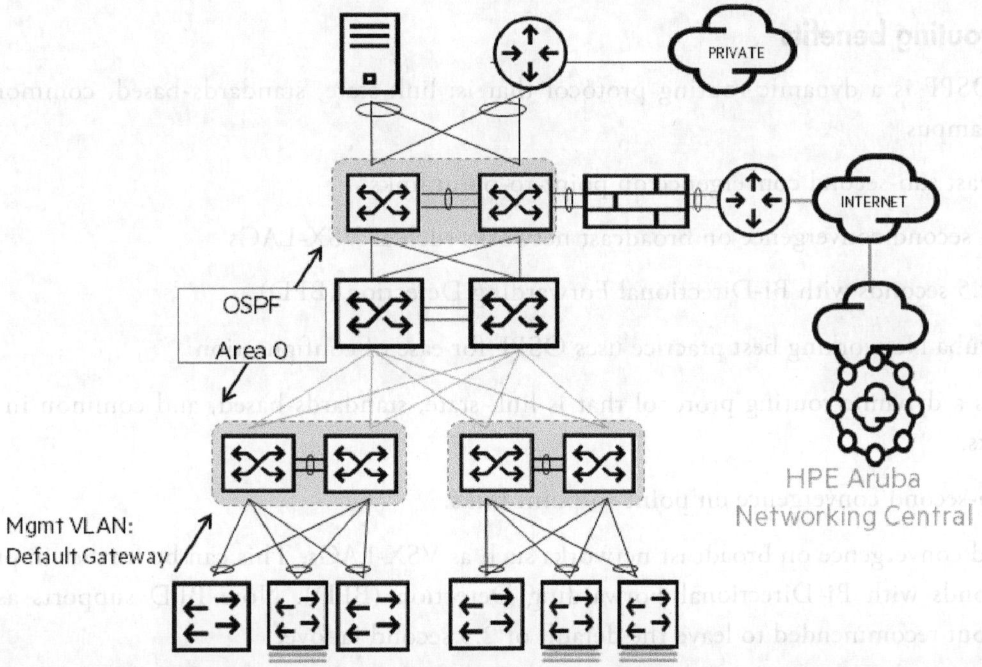

Using separate areas, OSPF routes small networks to larger WAN-connected campuses with branch deployments. Area assignment is normally done along the lines of one building to the next or to divide one location from another. A single backbone area can quickly get OSPF routing up and running in a single location or set of campus buildings. When the network scales from one campus to include remote or branch locations, multiple areas connected to a backbone are recommended.

OSPF enables remote site subnets and VLANs to be dynamically learned across the campus and to each other. Route summarization is possible If these branch locations are in their areas, separated by an Area Border Router. Consider using OSPF routing authentication to secure the network better. Public internet can be accessed using a simple static default route, or OSPF can propagate a default route to an edge router/gateway acting as the Autonomous System Boundary Router (ASBR).

Some larger enterprise customers may benefit from sharing external BGP routes between multi-homed ISPs. OSPF would still be designed and deployed to reach the edge BGP routers as ASBRs without redistributing public routes between BGP and OSPF, as OSPF will not scale to the full internet routing table. Most enterprise networks, however, would typically have a policy of a primary ISP and a backup assigned through static default routing instead of external BGP.

OSPF routing benefits

- OSPF is a dynamic routing protocol that is: link-state, standards-based, common in the campus
- Fast sub-second convergence on point-to-point links
- 4 second convergence on broadcast networks such as VSX-LAGs
- 2.5 seconds with Bi-Directional Forwarding Detection (BFD)

HPE Aruba Networking best practice uses OSPF for ease of configuration.

OSPF is a dynamic routing protocol that is link-state, standards-based, and common in campus networks.

Fast sub-second convergence on point-to-point links.

4 second convergence on broadcast networks such as VSX-LAGs. This can be further improved to 2.5 seconds with Bi-Directional Forwarding Detection (BFD). Note BFD supports as fast as 500ms but recommended to leave the default of 2.5 second failover.

HPE Aruba Networking underlay best practices use OSPF point-to-point links with ECMP between core and aggregation switches without the use of VSX on the core. Even with VSX on the aggregation switches, do not implement VSX-LAG facing the core as this would require a broadcast network and lose the benefit that point-to-point links would provide.

Single OSPF area: limited scalability

- Every router must maintain LSAs from every other router in the area in its LSDB
- LSDBs become large. SPF algorithm requires ever-more processing overhead
- No benefit: Segment 1 routers only have 2 paths to reach Segment 2 – via Core-1 and Core-2
- Stability: Flapping interface in Segment 1 affects every router in the entire area

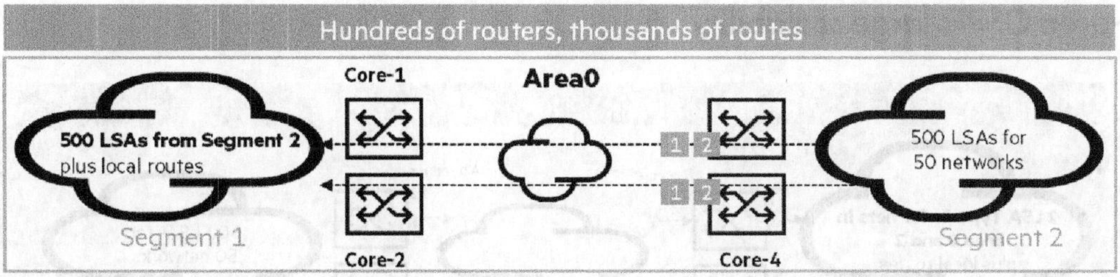

The figure shows a single area OSPF network with two segments. The segments might be a campus LAN and a data center, a campus LAN and a large branch office, buildings in a large 3-tier campus topology, or any other segments within a larger network. Segments 1 and 2 represent many routers that connect to two core routing switches. These core routers connect through a network backbone, which might also include more routers and routing switches. The figure only shows two segments for simplicity, but a large enterprise network might have several such segments.

Every OSPF router in an area must maintain LSAs from every other router in the area in its LSDB. The top example shows how a single area implementation forces each of these Type 1 and 2 LSAs to propagate across the system. Plus, the figure only mentions LSAs associated with segment 2. Each OSPF router must maintain LSAs for routers and networks in Segment 1 and the backbone.

As a domain grows, the LSDB becomes excessively large and difficult to process. Routers must dedicate valuable processing resources to running the SPF algorithm. Often, this resource consumption adds little value. As shown in the figure, networks often feature aggregation such that one segment connects to another through a few core devices. With only two exit points from Segment 1 towards Segment 2, the routing switches in Segment 1 have no reason to maintain so much granular information about networks in Segment 2.

In addition, suppose that two routing switches in segment 2 provide a connection to the subnet A. One of those switches has a flapping interface. It loses its connection to subnet A and sends an LSU, withdrawing that subnet from its list of links. Every routing switch in the network backbone and segment 1 receives this LSU, adjusts the LSDB, and re-runs the Dijkstra algorithm on the topology. However, these routers still reach subnet A through the same core routing switches. When the flapping interface returns, the switch sends another LSU, which propagates throughout the entire network, causing every OSPF router to recalculate shortest paths, again without any significant change.

What is the solution?

Multi-area OSPF: large scalability

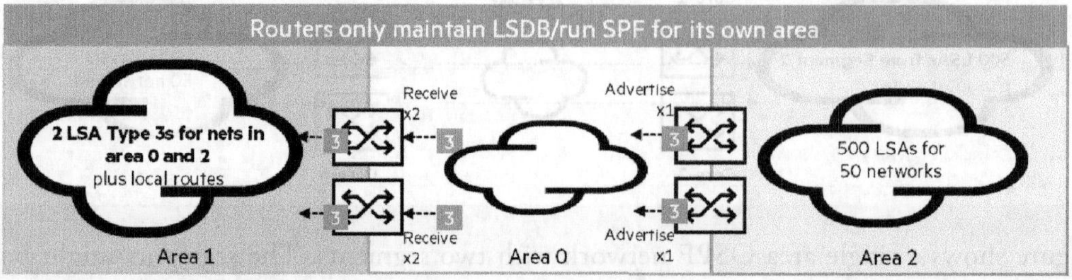

- Aggregate inter-area routes
- Much smaller LSDBs
- Reduce LSA/LSU flooding
- Minimize SPF recalculations
- Increase stability and control

OSPF areas segment the domain into more stable and manageable pieces. Each OSPF routing device maintains the link-state database (LSDB) and only runs the shortest path algorithm for its areas. OSPF area border routers (ABRs) forward routing information between areas. Type 1 and Type 2 link-state advertisements (LSAs) do not flow across area boundaries. ABRs summarize this huge number of Type 1 and Type 2 LSAs into a Type 3 summary LSA for each subnet.

Segmenting into multiple areas improves stability: routers do not have to re-run the SPF algorithm as often, and when they do, it runs on a smaller LSDB. ABRs do not automatically summarize the routing tables themselves. If you had 100,000 routes in your routing table with a single area design, you would still have 100,000 entries after using areas.

However, now that you have ABRs, you can configure route summarization, which can drastically reduce your LSDB and routing table sizes. Routers in Area 1 may only have a few dozen or a hundred routes in that area, plus a default route (0.0.0.0) to reach all other subnets via the ABRs. These benefits apply even to powerful routers that could support large areas.

This is especially true as it relates to network stability. With ABRs and route summarization, you have effectively segmented your network into smaller failure domains. A flapping route in Area 1 does not affect Area 0 or Area 2. This can be a big boost in network stability and ease of troubleshooting.

Special area types

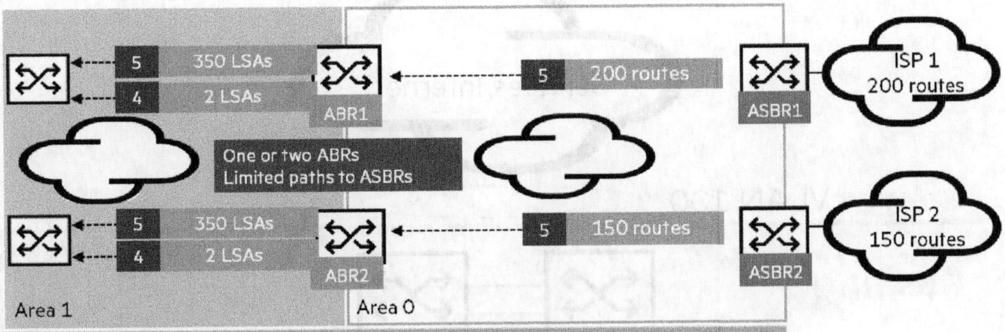

Area type	Intra-area routes	Type 3 LSAs inter-area routes	Type 5 LSAs external routes
Normal	☑	☑	☑
Stub	☑	☑ + default route	
Totally stub	☑	Default only	

Often, an area only has one or two ABR exit points: internal routers will choose the same next hop or hops for all external routes. This is when you might filter the Type 5 LSAs at the area boundary. You do so by configuring the area as a special type: stub or totally stubby. In a totally stub area ABRs only advertise a default route and do not inject any type 3, 4, and 5 LSAs.

Ensuring routing resiliency

- Dynamic routing with OSPF for networks with multiple paths
- OSPF graceful restart:
  - For switches with redundant mgmt modules, on VSX clusters
  - Adjust restart timer and redundancy switchover timers
- Dedicated VLAN per link or link aggregation between routers
- OSPF BFD if VLANs must extend over multiple devices

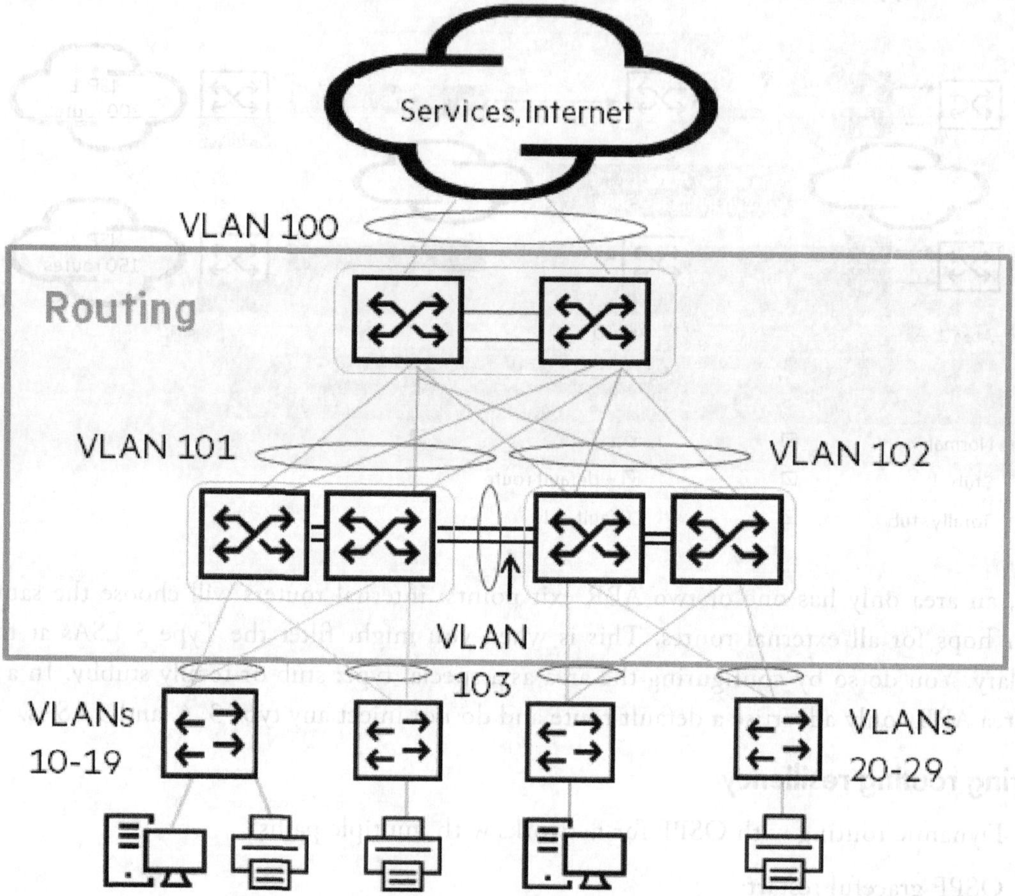

A redundant physical and layer 2 topology goes far toward supporting routing resiliency. Here, you see a network that uses VSX stacking at the aggregation layer and link aggregations, indicated by the links surrounded by circles. However, you must also ensure that the layer 3 solution is resilient.

- In a network with multiple paths, dynamic routing with OSPF provides the best convergence and failover capabilities.
- As one best practice, enable OSPF to restart any device that might continue functioning when its primary control plane fails.
- Such devices include modular switches with redundant management modules and VSX clusters.
- Also, make sure to set the OSPF restart timer longer than required for the switch to re-establish the routes but lower than the switch's redundancy switchover timer.

A link failure can cause a slight disruption even in a well-designed topology. You should minimize the time OSPF takes to detect a failed link to minimize convergence time. You can design direct, point-to-point wired connections between routing devices.

In other words, each link or link aggregation between routing devices has a VLAN and subnet dedicated to that link alone. If the link fails, the neighbor relationship fails, and routes reconverge. Segmenting each link between the layer 3 switches into its own VLAN also eliminates the need to run spanning tree over the links if you have redundant links, such as the link assigned to VLAN 103 in this example.

If you must extend a VLAN over multiple devices, consider using OSPF Bidirectional Forwarding Detection (BFD) in echo mode. Echo mode BFD enables OSPF neighbors to detect when they lose connectivity with a neighbor in under one second. The HPE Aruba Networking 6300, 6405, and 8400 switches, targeted for the aggregation and core layers, support this feature.

IP multicast

- Advantage of multicast over broadcast
- Rendezvous point
- Non-IGMP enabled switch behavior
- Dynamic multicast optimization for Wi-Fi
- Broadcast to unicast conversion
- ARP optimization

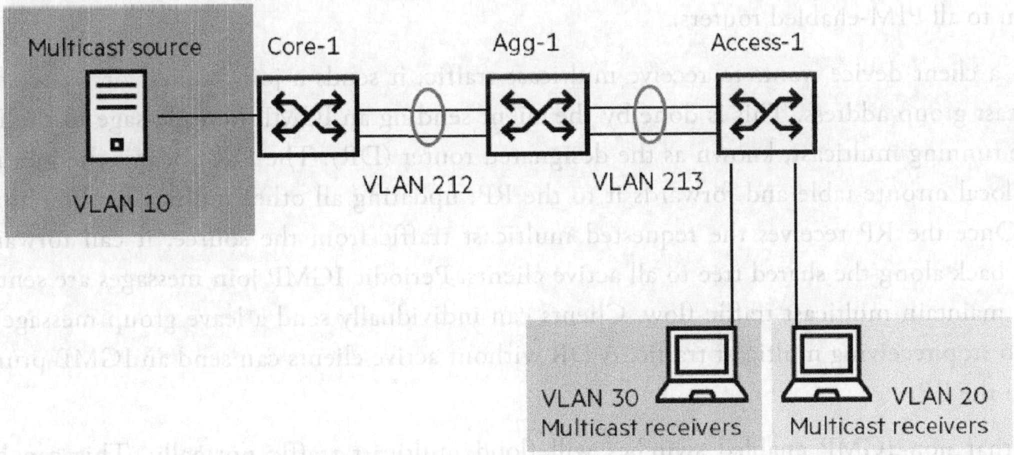

IP multicast has been a staple of modern campus networks with streaming media and network applications. Layer 3 switches and routers only need to enable internet group management protocol (IGMP) to forward multicast traffic and allow an efficient way for multiple receivers to request data from one or more sources serving that data. By contrast, a server could send the same data as often as needed using unicast through head-end replication. Or a server could broadcast the data to be replicated throughout the network. But both options would be inefficient and waste network resources.

One advantage of IP multicast over broadcasting traffic is that multicast can be routed between Layer 3 subnets using protocol-independent multicast-sparse mode (PIM-SM). This protocol enables an IP multicast-enabled router to act as the rendezvous point (RP) for the other multicast routers in the network. When a new source for multicast traffic becomes available, it is forwarded to the RP. The same is true for receiver reports asking for the group of traffic. The RP will immediately forward multicast traffic to the receivers, known as a shared tree path, until the shortest path tree (if one exists) can be established. The OSPF routing table is used for reverse path forwarding to confirm that the shortest path tree is the shortest path back to the source.

Multiple RPs can be configured for redundancy and load sharing across assigned multicast group addresses if Multicast Source Discovery Protocol (MSDP) is enabled. MSDP allows multicast routers to have inter and intra-domain active-active redundancy using an Anycast IP address as the RP. The anycast address allows switches to forward traffic to the closest device in the network based on the location of the user's initial request.

Use MSDP for intra-domain redundancy on the core switches of a campus network. MSDP and RP candidate announcements advertise the Anycast IP address to multicast neighbors.

RPs can be statically configured on all PIM routers or can be elected. When elected, neighboring multicast routers learn the RP address from the bootstrap router (BSR). The BSR is also elected from a list of candidate BSRs configured on the network, with only one BSR advertising RP information to all PIM-enabled routers.

When a client device wants to receive multicast traffic, it sends a join request or report for that multicast group address. This is done by the client sending an IGMP join message to the Layer 3 device running multicast, known as the designated router (DR). The DR updates the join request in its local mroute table and forwards it to the RP, updating all other multicast router along the way. Once the RP receives the requested multicast traffic from the source, it can forward that traffic back along the shared tree to all active clients. Periodic IGMP join messages are sent to the RP to maintain multicast traffic flow. Clients can individually send a leave group message at any time to stop receiving multicast traffic. A DR without active clients can send an IGMP prune message to the RP.

Note that non-IGMP enabled switches will flood multicast traffic normally. This can be prevented by enabling IGMP-snooping so that even Layer 2 switches will only forward multicast traffic out ports where an IGMP join message was detected. For Layer 3 switches, such as Core and Aggregation switches, IGMP should be enabled along with PIM-SM.

Dynamic Multicast Optimization for Wi-Fi

In 802.11, multicast and broadcast frames must be sent at the minimum mandatory data rate so all clients can receive the traffic. When enabled, dynamic Multicast Optimization (DMO) can overcome this limitation by converting multicast traffic to unicast for individual client devices. These frames can then be sent at the highest possible data rate supported by individual clients and are retransmitted should a frame be lost over the air.

Broadcast to Unicast conversion

This allows guaranteed delivery as unicast frames are acknowledged and resent if lost. In addition, this greatly reduces the channel duty cycle while allowing much faster bandwidth over the mandatory minimum broadcasts that must be used.

Note that DMO conversion of multicast to unicast is done in the AP/Gateway datapath. This can be CPU intensive if multiple multicast sources use the same datapath. To avoid this bottleneck, consider designing larger Layer 2 networks. If larger Layer 2 networks are unavailable, additional Gateways are recommended. Enabling DMO within a properly sized network should have no negative impact.

DMO will work with ARP optimizations set to "ARP ALL" and would otherwise drop all multicast traffic along with broadcasts.

IP multicast PIM-SM

HPE Aruba Networking recommends PIM-SM to route multicast traffic.

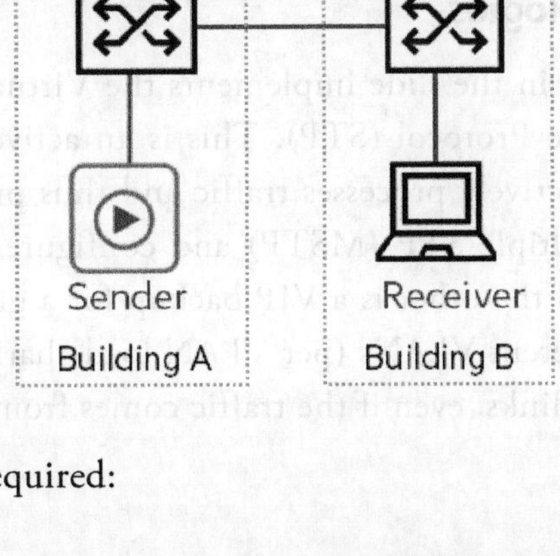

Additional multicast features required:

- Bootstrap router protocol (BSR)
- Rendezvous point (RP)
- Multicast Source Discovery Protocol (MSDP)
- Internet Group Management Protocol (IGMP)
- Configure IGMP on all client-facing links on layer 3 switches.
- Configure IGMP-snooping on all client-facing links for layer 2 switches.
- Configure PIM-SM on all links (routed or client-facing) on layer 3 switches.
- Enable the BSR protocol on all PIM multicast domain routers to share RP information dynamically.

Core switches should be configured as candidate RPs to establish the core as the root of the Rendezvous Point Tree (RPT). Configure core switches to announce an anycast loopback IP address as a candidate-RP to be selected by the BSR.

MSDP facilities active-active RP anycast redundancy by updating multicast source information between RPs. This enabled anycast redundancy so that multiple RPs could be active simultaneously for traffic flow optimization.

NOTE that all IGMP timers should match across all switches in the network, including any 3rd party vendor switches.

Supplemental: Stacking technologies

The previous sections outlined some of the important choices you can make when designing a wireless network. Virtual switching framework (VSF) stacking can affect your architectural decisions and play an important part in simplifying the architecture while enhancing its resiliency.

Switch virtualization technologies

The example on the left shown in the slide implements the Virtual Router Redundancy Protocol (VRRP) and the Spanning Tree Protocol (STP). This is an active/standby solution. This means that only one VRRP member actively processes traffic and thus produces a sub-optimal solution. Of course, you can enable Multiple STP (MSTP) and configure multiple VRRP groups. Thus, one switch is a VIP master, and the other is a VIP backup for a given VLAN. But this only load-shares between traffic from different VLANs (per VLAN load-sharing). What if you want switches to forward traffic and use all uplinks, even if the traffic comes from the same VLAN?

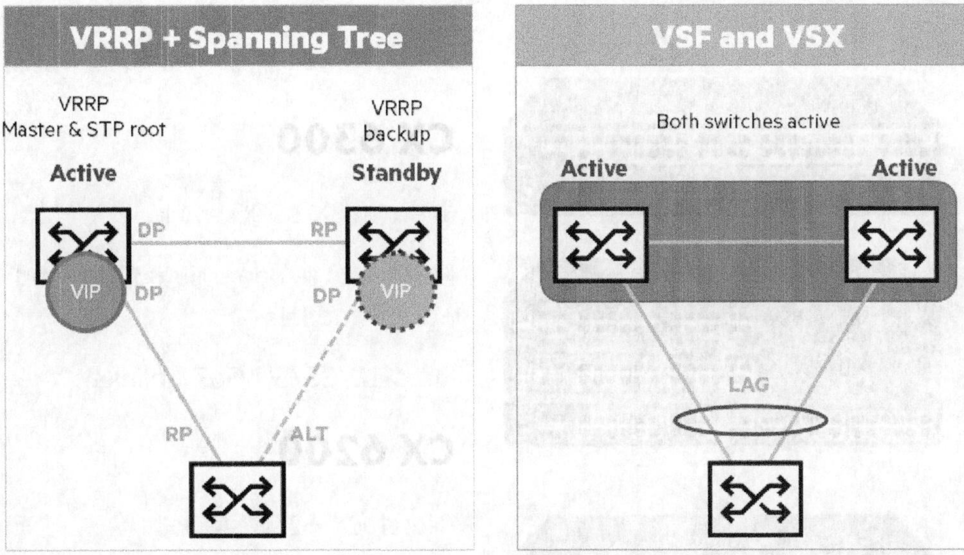

The solution is to use a switching virtualization technology and technologies that allow true active/active traffic processing, like HPE Aruba Networking's Virtual Switching Framework (VSF) and Virtual Switching eXtension (VSX). Depending on the technology, you can combine two or more switches to create one virtual switch. Now, you can use link aggregation instead of STP/VRRP between the access and core/aggregation layers! All uplinks can be used for load sharing—even traffic from the same VLAN. Both switches share the same IP address for IP routing, and both can do L3 forwarding. You get true active/active L2 and L3 traffic processing!

Virtual Switching Framework (VSF)

For resilient and scalable campus and branch access networks

- Simplified Management with virtual switch stack
- Expand VSF stack across campus and site using long range transceivers
- Fast stack with CX Mobile App using auto-detection
- Quick and simple stack configuration with HPE Aruba Networking Central UI

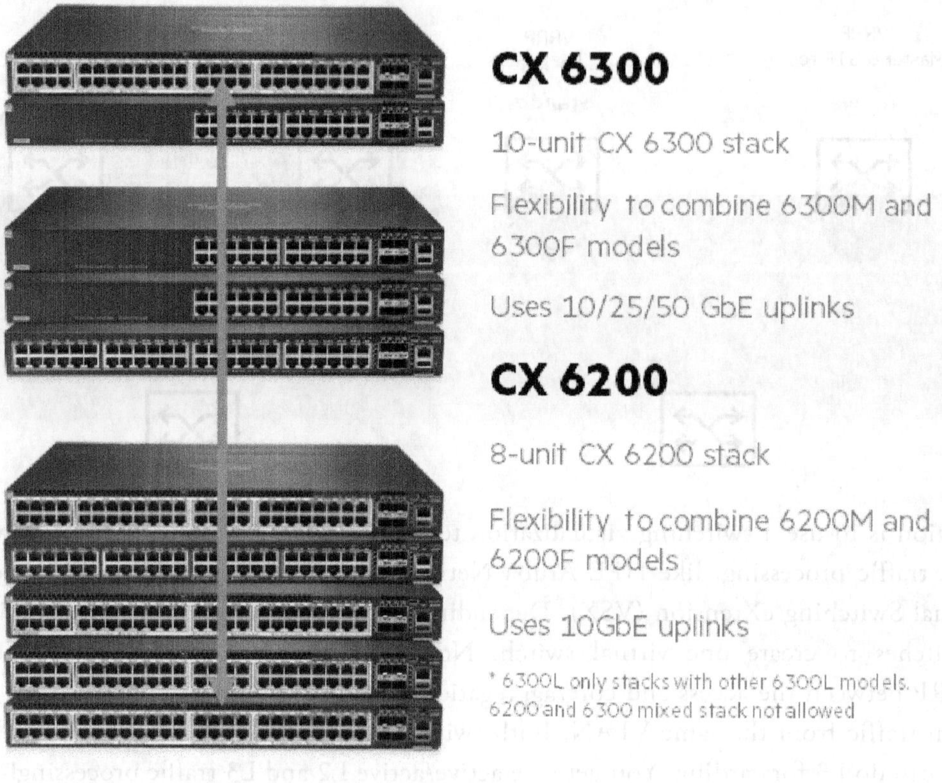

Key Takeaway – Simplicity and flexibility of VSF stacking is tremendous. Stacks can span across a campus with long range transceivers. The CX 6300 series allows flexibility of stacking both M and F models so customers can truly deploy what they need.

HPE Aruba Networking Virtual Switching Framework (VSF) - Virtualization is implemented so that two chassis' have a single control plane which simplifies configuration of the two devices but lowers network availability during chassis upgrades unless multiple management modules are in play. We feel that this approach best serves the needs of access layer deployments where simplicity outweighs the HA criteria.

Switch virtualization solutions

HPE Aruba Networking VSX is a virtualization technology running the AOS-CX operating system. This solution lets the switches present themselves as one virtualized switch in critical areas. Configuration synchronization is one aspect of this VSX solution where the primary switch configuration syncs to the secondary switch.

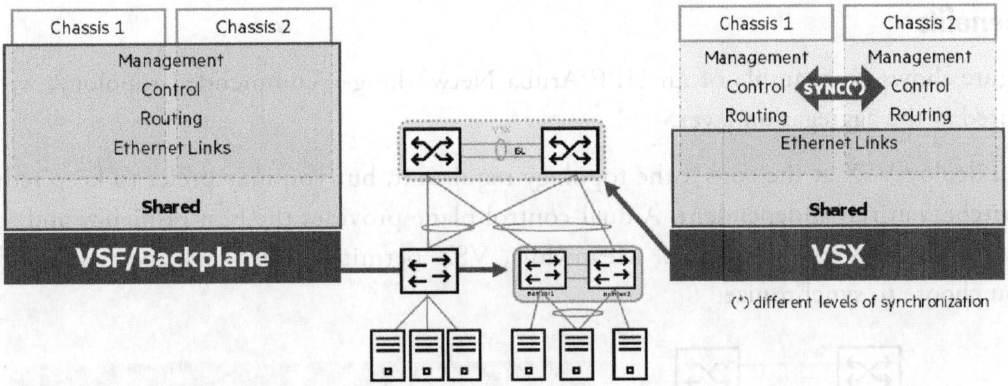

VSX virtualizes the control plane of two switches to function as one device at layer 2 and independent devices at layer 3. From a datapath perspective, each device does an independent forwarding lookup to decide how to handle the traffic. Some forwarding databases, such as the MAC and neighbor tables, synchronize between the two devices using a proprietary VSX control plane. Each switch independently builds some of the forwarding databases.

You will primarily use VSF at the access layer of a campus network.

AOS-CX switch virtualization: VSX vs VSF

Both VSX and VSF allow active/active traffic forwarding. A significant difference is that VSX delivers high availability during software upgrades, with near-zero downtime and continuous packet forwarding. This is because a pair of VSX switches can run different versions of the AOS-CX firmware, which is impossible with VSF.

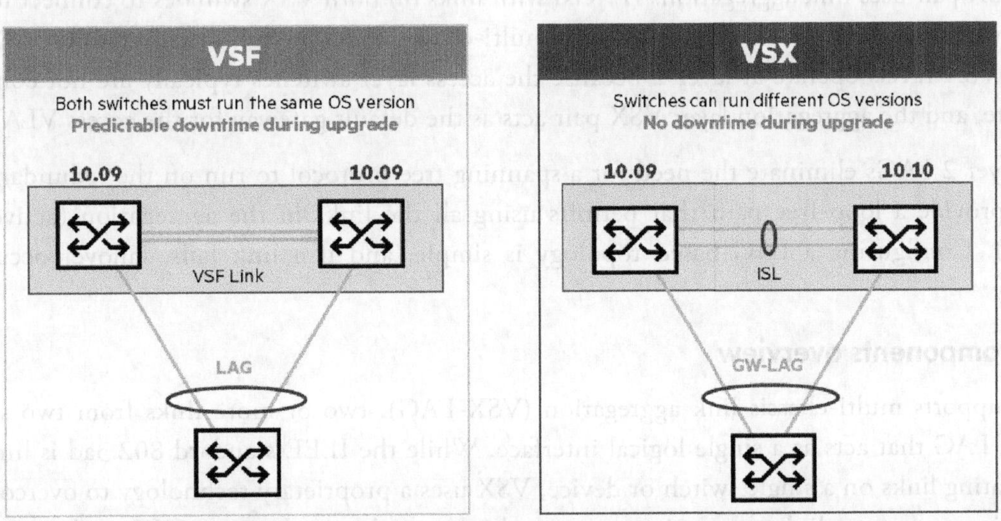

VSX is HPE Aruba Networking's core and aggregation layer technology, while VSF is an access layer switch stacking feature. Let's dig a bit deeper.

VSX benefits

The figure shows an example of an HPE Aruba Networking-recommended topology, with VSX configured at the aggregation layer.

You can deploy VSX at the core if the topology requires it, but you may prefer to keep redundant core switches entirely independent. A dual-control plane provides the best resiliency and software upgrades with near-zero downtime. Meanwhile, VSX permits unified management for features that you choose to synchronize.

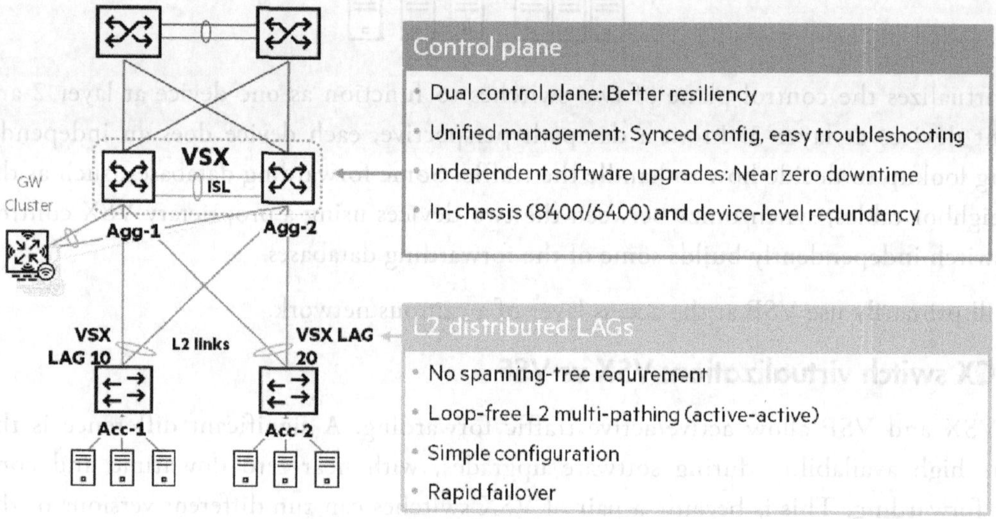

The VSX pair uses link aggregations (LAGs) with links on both VSX switches to connect to access layer switches. This type of LAG is called a multi-chassis (VSX-LAG), distributed, or VSX LAG. These VSX LAGs operate at layer 2 because the access layer switches typically are not configured to route, and the aggregation layer VSX pair acts as the default gateway for the access VLANs.

The layer 2 LAGs eliminate the need for a spanning tree protocol to run on the redundant links. They provide a loop-free path that permits using all the links in the aggregation (active-active design). Configuring a LAG-based topology is simple, and if a link fails, failover occurs very quickly.

VSX components overview

VSX supports multi-chassis link aggregation (VSX-LAG): two or more links from two switches form a LAG that acts as a single logical interface. While the IEEE standard 802.3ad is limited to aggregating links on a single switch or device, VSX uses a proprietary technology to overcome this limitation to support link aggregation across multiple switches in the same VSX stack.

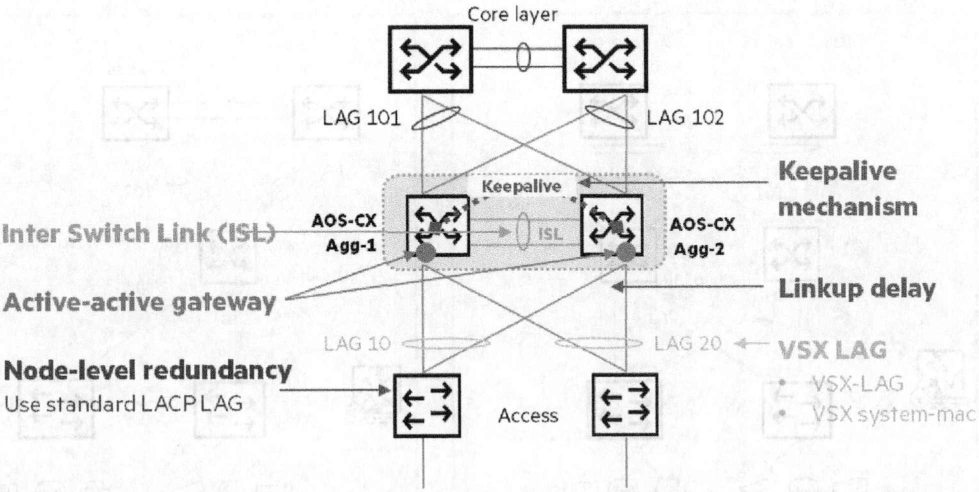

An Inter-Switch Link (ISL) connects the two switches, and a keep-alive mechanism is used for resiliency.

Also, VSX provides node-level network redundancy when one switch fails. You must configure the downstream device as a standard 802.3ad LAG interface: the access switches perceive the VSX pair as a single device with a single system MAC address. Downstream devices can be anything that supports 802.3ad Link Aggregation Control Protocol (LACP), including switches, APs, firewalls, gateways, and other products. In VSX, one device acts as primary and the other as secondary, but both devices forward traffic for an active-active gateway solution.

You will soon learn about the linkup delay feature, which enhances initial forwarding readiness.

VSX - data plane virtualization

Data plan virtualization supports VSX LAGs. Data plane virtualization makes the two switches in the VSX pair appear as a single device to the device on the other side of the link aggregation. In this example, those are both downstream access layer switches and upstream core switches.

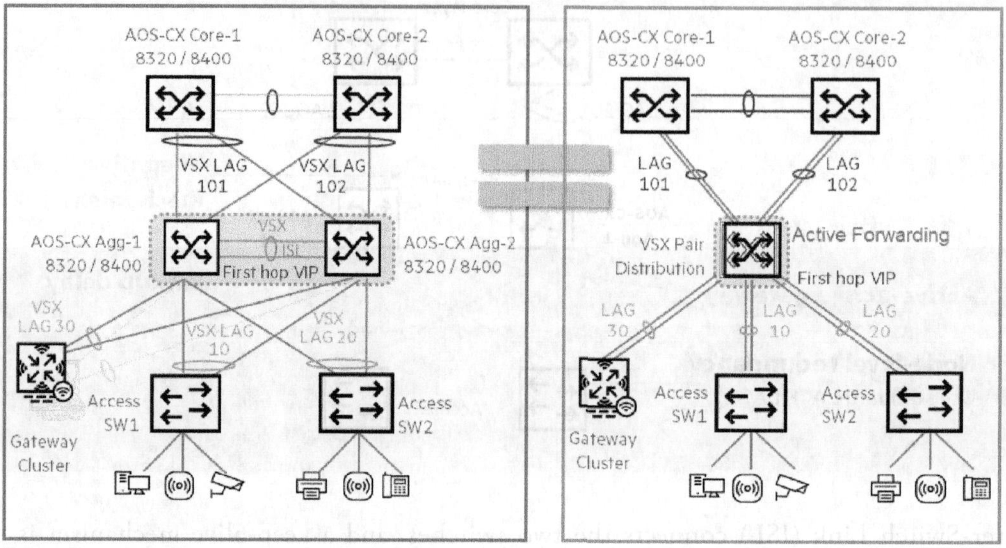

VSX supports a link-aggregation technique, in which two or more links across two switches aggregate together to form a LAG that will act as a single logical interface. While the IEEE standard 802.3ad is limited to aggregating links on a single switch or device, the VSX feature uses a proprietary technology to overcome this limitation. VSX supports link aggregation for links spanning multiple switches in the same VSX stack.

VSX redundancy

Control Plane	• Dual control plane • Dual Management Plane • In-chassis redundancy (10000-9xxx-8xxx-64xx)
L2 Distributed LAGs (Agg to Acc)	• No spanning-tree (MSTP/PRSTP) • Loop-free L2 multi-pathing
L3 Links (Core to Agg)	• Equal Cost Multi Path routing (ECMP) • L2 VSX LAG
Active Gateway	• Active-Active • Gateway (VIP)

A dual-control plane provides the best resiliency and software upgrades with near-zero downtime. At the same time, VSX permits unified management of the features that administrators choose to synchronize.

The VSX pair uses link aggregations (LAGs) with links on both VSX switches to connect to access layer switches. The name of this type of LAG is a multi-chassis, distributed, or VSX LAG. These VSX LAGs operate at layer 2 because the access layer switches do not route, and the aggregation layer VSX pair acts as the default gateway for the access VLANs. The layer 2 LAGs eliminate the need for multiple spanning tree protocol (MSTP) or rapid per-VLAN spanning tree (RPVST) on the redundant links. They provide a loop-free path that permits using all the links in the aggregation (active-active design). Configuring a LAG-based topology is simple, and if a link fails, failover occurs very quickly.

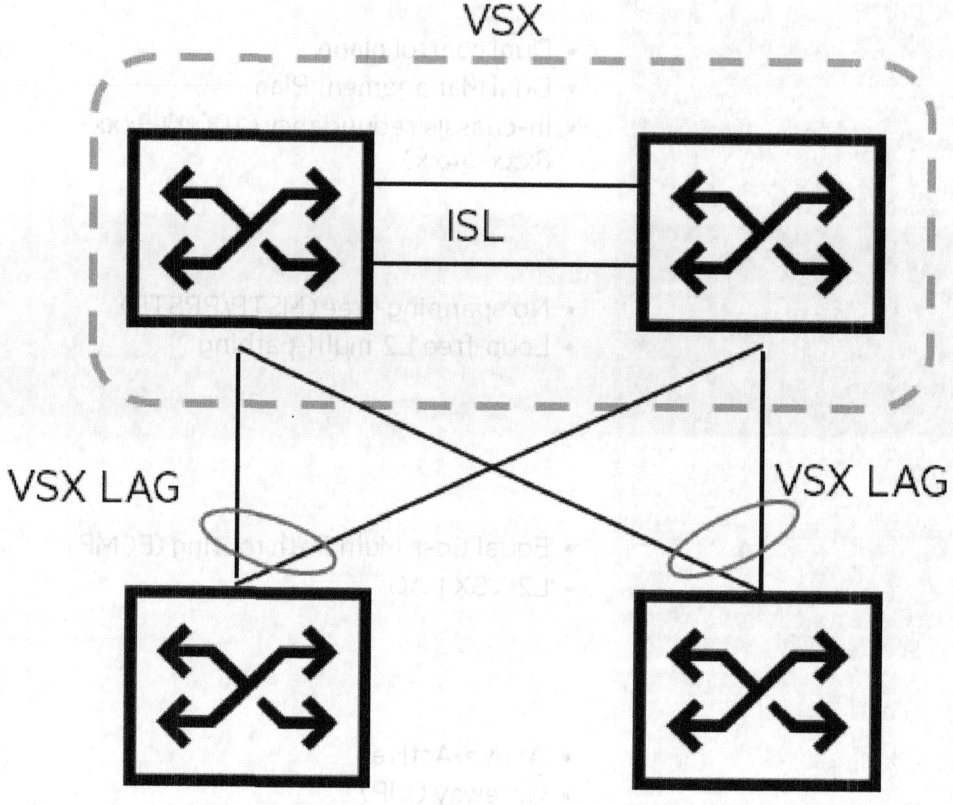

The VSX pair has layer 3 links to the core. Each VSX switch can connect independently to each core switch on a routed-only port (ROP) or a layer-2 port associated with layer 3 VLAN interfaces (switch virtual interfaces or SVIs) dedicated to that port. Alternatively, both VSX switches could connect to each core switch on a VSX LAG. This LAG would operate at layer 2 and associate with one or more VSIs. Whichever option you select, VSX can support active-active layer 3 routing through equal cost multipath (ECMP) and an efficient path for unicast traffic. With the final option, the combination of layer 3 ECMP and the VSX LAG creates a highly fault-tolerant system.

Remember that the VSX pair acts as the default gateway for the access VLANs. To do so, the pair uses the active gateway feature. This feature allows each switch in the pair to act as an active default gateway for the VLAN using a shared virtual IP address (VIP) and virtual MAC address. It eliminates the need for virtual router redundancy protocol (VRRP) or hot standby router protocol (HSRP). Simple to configure, the active gateway feature relies on VSX operations, so it does not add any protocol overhead. It also supports redundancy for DHCP relay functions.

VSX benefits: layer 3

The VSX pair has layer 3 links to the core. Each VSX switch can connect independently to each core switch on a routed-only port (ROP) or a layer-2 port associated with a layer-3 VLAN interface dedicated to that port, also known as a Switch Virtual Interface (SVI). Alternatively, both VSX switches could connect to each core switch on a VSX LAG. This LAG would operate at layer 2 and associate with one or more SVIs. Whichever option you select, VSX can support active-active layer 3 routing through Equal Cost Multipath (ECMP) and an efficient path for unicast traffic. The final option combines layer 3 ECMP and the VSX LAG, creating a highly fault-tolerant system.

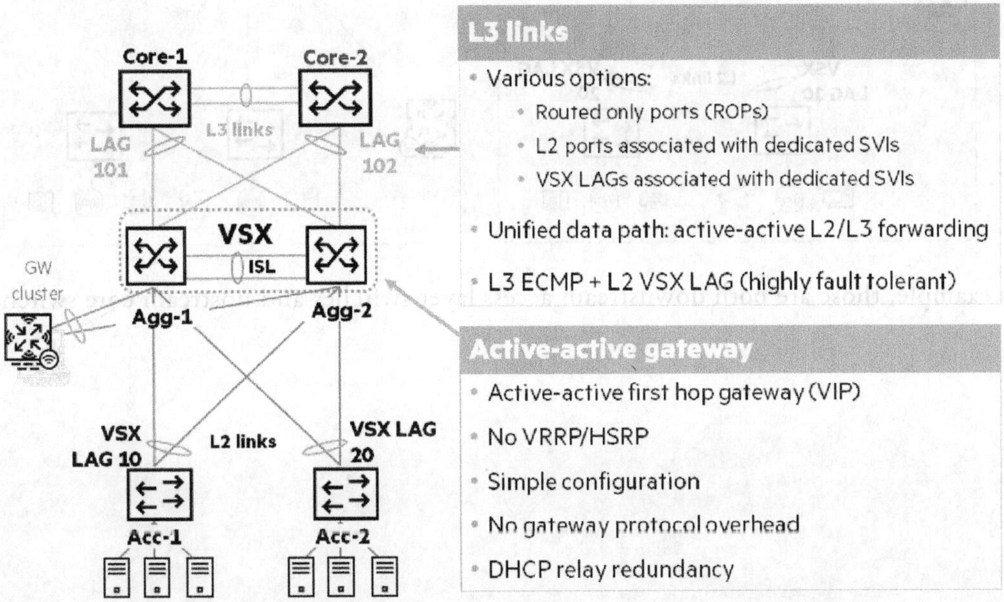

Remember that the VSX pair acts as the default gateway for the access layer VLANs. To do so, the VSX pair uses the active-active gateway feature. This feature allows each switch in the pair to act as an active default gateway for the VLAN using a shared virtual IP address (VIP) and virtual MAC address. It eliminates the need for Virtual Router Redundancy Protocol (VRRP) or Hot Standby Router Protocol (HSRP). Unlike VRRP, the active-active gateway feature allows both VSX switches to process traffic for the VLAN(s). Simple to configure, the active gateway feature relies on VSX operations. Thus, it adds no protocol overhead. It also supports redundancy for DHCP relay (IP helper) functions.

Data plane virtualization

Data plane virtualization supports VSX LAGs, so other LAG-connected devices perceive the VSX switch pair as a single device.

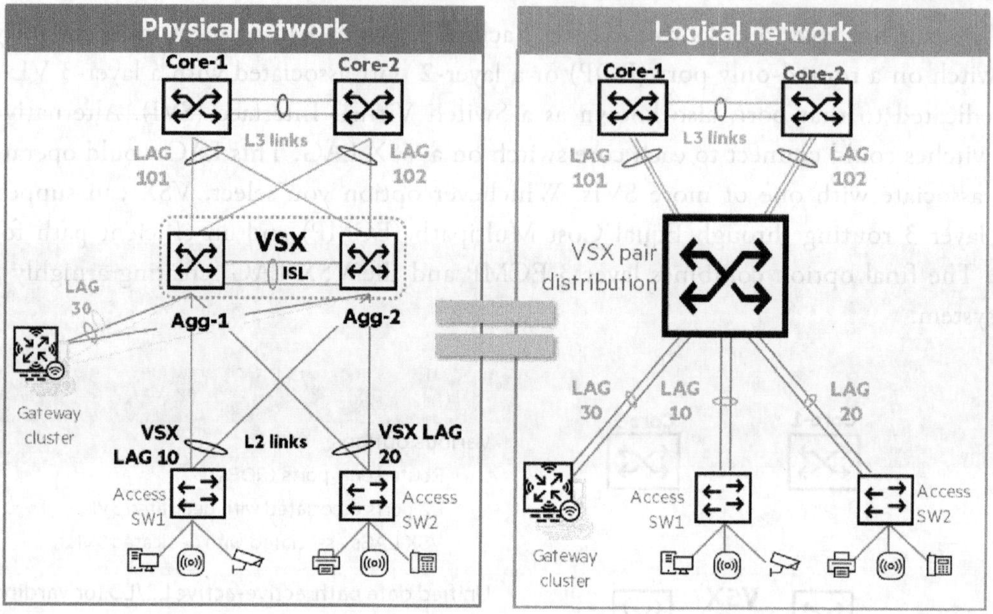

In this example, those are both downstream access layer switches and upstream core switches.

6 Network Segmentation

This module looks at IPv4 addressing for a campus deployment.

OBJECTIVES

After completing this chapter, you should be able to:

- ✓ Construct segmentation with VLANs & VRFs
- ✓ Complete an IP addressing design for wireless networks
- ✓ Develop IP routing protocol and VRF lite solutions

Segmentation with VLANs and VRFs

Learn to create network tenants to segment user traffic based on roles or functions within the organization. This can be imposed by assigning layer-2 VLANs and layer-3 VRFs. Roles for users, creation of overlays, and QoS prioritization will be identified in the upcoming modules.

Benefits of VLAN Segmentation for wired networks

While extending a VLAN as far as possible works well for wireless networks that need to support roaming, wired networks often benefit from more segmentation. First, create separate VLANs for wired users and devices and wireless networks. You can further divide users and their devices from there for several reasons. These include supporting access control by putting different types of users and devices in different VLANs. You may want to place specialized wired devices in their VLANs to ensure proper application of QoS and traffic management.

You can consider each user type or device type an access category. You should plan at least one VLAN per access category. However, sometimes you may plan multiple VLANs per access category because of the VLANs' other important function: limiting the size of the broadcast domain.

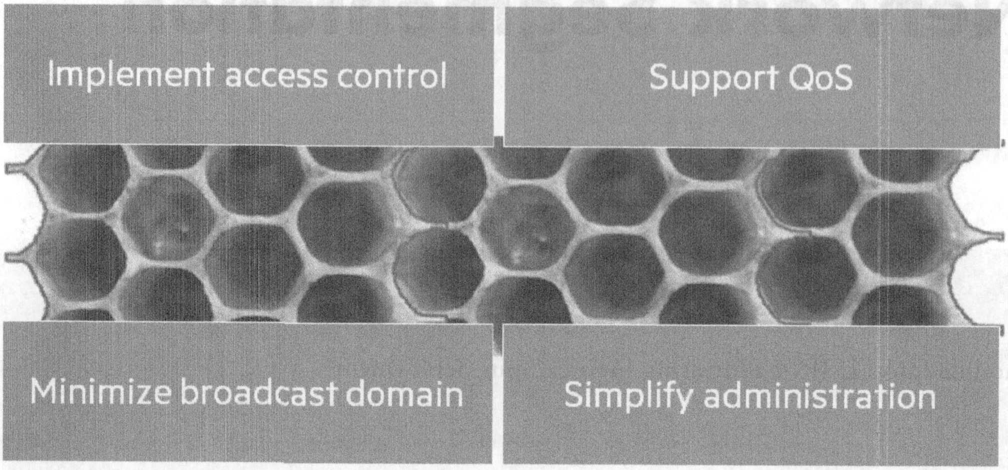

In IPv4, every device in the VLAN must process broadcast traffic, and broadcasts can impose a burden by congesting links and tying up endpoint processes. The more devices in a VLAN, the more broadcasts every device must handle. In addition, gratuitous ARP broadcasts can fill up devices' ARP cache. Similarly, multicasts extend through a broadcast domain. While you no longer have to make the broadcast domain as small as in the past, keeping broadcast domains a sensible size still makes sense.

Placing different types of users and devices in different VLANs can be very important for wired networks. The HPE Aruba Networking mobility gateway firewalls are role-based, but wired traffic might pass through a more traditional firewall, which uses the devices' source addresses to decide how to control the traffic. Therefore, placing different types of wired users and devices in different VLANs makes it easier for the firewall to control traffic differently.

For example, sometimes, by looking at the IPv4 address, the firewall can tell several things about a user, including the user's group, location, and whether the user is an employee, contractor, or guest.

A few organizations with large public IPv4 networks use public addressing within the campus. If the public IPv4 networks are non-contiguous, limited VLANs might need to correspond with each non-contiguous segment.

Out of band management network

One common example of network segmentation is creating an Out-of-band management (OOBM) network (right side of the figure). This network is a dedicated set of connections to network devices, separate from the end user data network. Often, a dedicated port is allocated just for this traffic. Furthermore, switching and routing can be segmented with a dedicated VLAN or VRF.

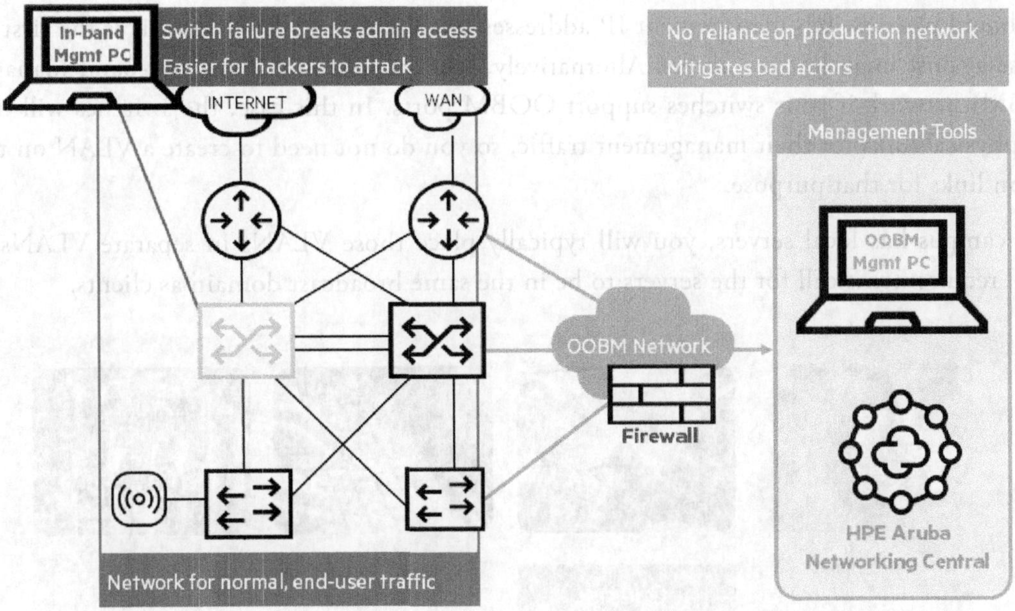

In-band management (left side of the figure) is where you remotely connect to and manage switches via a normal Ethernet port. While this may be convenient, in-band exposes management traffic to security and routing concerns. A switch or router misconfiguration could break your administrative access. Without tight security controls, hackers can access and control your devices.

You can monitor and configure network devices without relying on the production network. You establish a trust boundary, with only one entry point for the management interface, removing any other attack vector from bad actors. All AOS-CX switches support a dedicated management interface in a hard-coded "mgmt" VRF.

Isolate traffic for access control

Here is an example of grouping devices based on access control considerations. A basic design features a VLAN (or VLANs) for users. APs could be on their own VLAN but can also deploy on any VLAN available in the network. If APs implement IDS/WIPS rather than dedicated Air Monitors (AMs), APs can only detect rogues on the attached VLAN.

If the company offers wired access for guests, you should put employees and guests in different VLANs. Depending on the complexity of the guest solution, you might create VLANs for guest devices before the guests have registered and logged in. A company may also need a quarantine VLAN for devices it detects to have viruses.

You should place switch management IP addresses in an isolated VLAN, as this is the first line of defense against unauthorized access. Alternatively, you can establish an out-of-band-management (OOBM) network if your switches support OOBM ports. In this case, the switches will use separate physical links for their management traffic, so you do not need to create a VLAN on the production links for that purpose.

If the campus has local servers, you will typically place those VLANs in separate VLANs unless special requirements call for the servers to be in the same broadcast domain as clients.

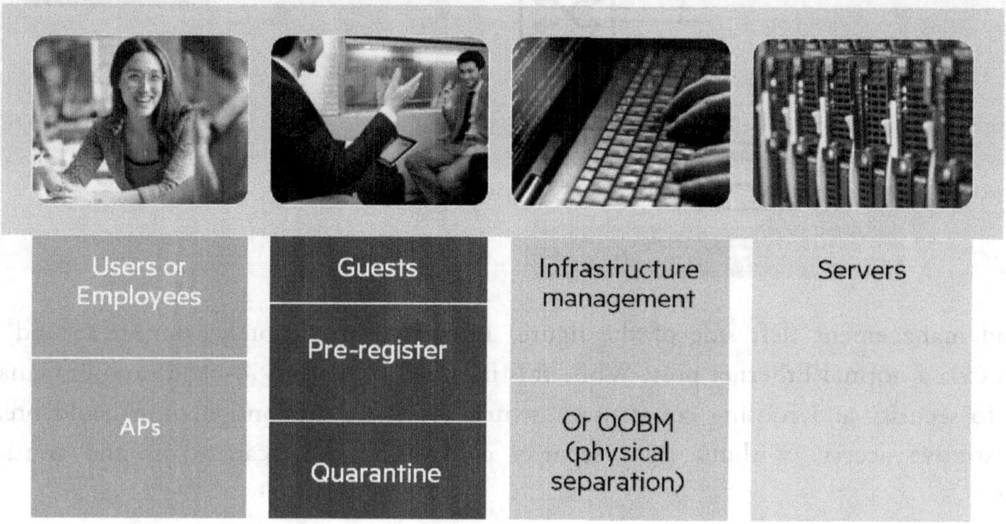

Users or Employees

This category includes the devices through which employees or other users connect to the network. Some companies might have more complicated requirements to place different types of employees in different VLANs, dividing the users by department. However, a simple user or employee category works for many customers. The more granular level of control might require authentication to identify the user and make the assignment. The authentication server could alternatively apply access control policies directly to the user session, allowing multiple types of users to share the same VLAN but have access to different resources.

APs/AMs

To support rogue AP detection, you should deploy AMs in the network and trunk/tag the ports with VLANs that have no APs connected. If you are only using APs, the APs will only be capable of classifying rogues in the same VLAN as the APs. Note that in AOS-10, there is no longer control/management traffic between APs and Controllers/Gateways. Instead, all control plane traffic is between the APs/GWs and HPE Aruba Networking Central over an HTTPS connection.

Guest

You must work with the customer to define the proper treatment for unknown or unauthorized users. The company might want a guest VLAN, which provides limited resources or Internet access. Some guest solutions also place devices not yet registered for access in a separate VLAN from those authorized for guest access. The customer might require a similar type of VLAN, called a quarantine VLAN. An endpoint integrity solution might place devices in this VLAN if the devices fail to comply with security policies, such as requirements for antivirus software. The VLAN typically provides access to limited resources that help the user to remediate the problem.

Management (Network infrastructure devices)

This category includes network infrastructure devices, such as switches, routers, and sometimes mobility gateways. It might also include servers that act as network infrastructure devices like traffic optimizers and load balancers. Routing switches and routers typically have IP addresses on other VLANs as well. However, the network infrastructure VLAN manages the devices. For this reason, this VLAN is often called the management VLAN.

Some customers prefer physically isolating the management traffic rather than separating it into a VLAN. You can create this physical isolation with an OOBM network. The OOBM ports are Ethernet ports that connect only to the switch management plane, not its data plane, ensuring complete, secure isolation. If you plan to use the OOBM ports, remember to plan for the OOBM network. You might add one or two switches to the proposal to support the connections. Those switches would then connect to any management stations or, if you want to provide management access from anywhere, to a core or aggregation level switch. The connection from the OOBM network to the rest of the network should be on a dedicated VLAN, and you should plan the appropriate ACLs to control access to this VLAN. For example, if you use role-based access control, ACLs for most roles could prevent access to this VLAN, but ACLs for the IT staff role could permit access. Also, remember that the switch configuration should set OOBM as the listening port for management protocols like Telnet and SSH.

Control Plane ACLs

Once an IP address is bound to an interface associated with a VRF, the switch may become exposed to management access from untrusted users or devices. This potential point of vulnerability can be mitigated by binding an Access Control List (ACL) to the control plane for that VRF. The control plane handles the device's management and routing functionality.

Servers

This category includes servers and devices like printers that provide services. A company with fewer servers might connect them directly to a campus LAN core switch or at least have this switch act as their router. In that case, the campus LAN must include one or more server VLANs. A true data center, on the other hand, might have a more complicated topology and connect to the campus LAN over routed connections.

Group traffic for specialized devices

You may want to create several types of access categories for specialized devices. A common example is Voice Over IP (VoIP). Placing voice traffic in a different VLAN from data traffic protects VoIP phones from other VLAN broadcasts. In addition, voice traffic requires a higher QoS than traditional data traffic. If you place all the traffic in its own VLAN (or VLANs), it is easier to ensure that QoS.

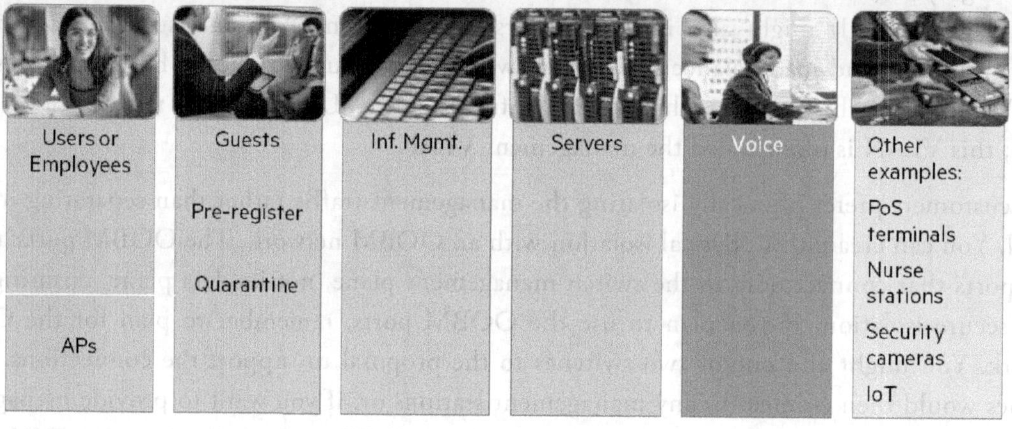

For similar reasons, you may want to group devices that handle much multimedia traffic in their own VLAN. Alternatively, you could group devices designed to run multicast applications to minimize multicast routing.

You can also group a company's specialized devices according to the application. For example, you could place the point of sales (PoS) devices in their access category. Such devices include nurse stations, security cameras, and Internet of Things (IoT) devices. These devices might or might not require high-priority handling. However, they form an access category of devices with similar needs and traffic patterns. You may want to segregate the devices for security reasons as well.

Minimize wired broadcast domain

Once you have a list of access categories, such as employees, guests, and voice, you should determine the estimated endpoints for each category. Then, you must decide whether to divide the access category into multiple VLANs based on broadcast domain limitations. The traditional guideline limits a wired VLAN to a /24 subnet, which provides enough IP addresses for 254 endpoints. For modern networks, you can often use /23 subnets, which support up to 510 endpoints, or /25 subnets, which support about 126 endpoints.

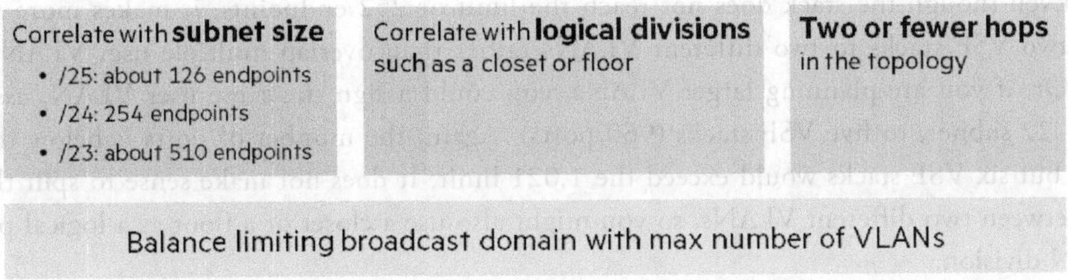

You should also correlate the VLAN with logical divisions in the access layer, such as a closet or floor. Limiting the number of hops a VLAN extends is also best practice. For example, it is usually best to terminate VLANs at the aggregation layer in a three-tier topology.

However, as you plan for larger campuses, keep in mind balancing the need to limit the broadcast domain with the need to stay under the maximum number of VLANs, theoretically 4,094. Of course, you can reuse VLAN IDs across routed boundaries. For example, you generally route all traffic between the campus and data center cores so the campus and data center can use overlapping VLAN IDs (but not subnets). More VLANs might come with a cost of their own, and you must balance the advantages of smaller VLANs with the disadvantages of more VLANs. For most enterprise networks using core routing, you should aim for a limit of about 200 to 250 VLANs per site. As long as you are not using MSTP, access layer switches generally only need to support the specific VLANs for their endpoints.

A few trends have allowed you to plan larger VLANs, such as ones associated with /22 subnets. Many endpoints now have more processing power, which reduces the impact of many broadcasts. Operating systems also tend to treat broadcasts more efficiently, and higher bandwidth for connections reduces the impact of broadcasts.

The need to avoid issues with spanning tree stood behind the guideline for limiting the extent of VLANs. Now, VSX clusters and VSF stacking can eliminate spanning tree design as a concern. VSF combines many ports into a single virtual switch so that you can extend a VLAN across those ports.

It is best practice to correlate the maximum number of endpoints that connect in a VLAN with a subnet size. You can calculate the endpoint count for each subnet from the number of addresses in the subnet minus three for the network address, broadcast address, and default router IP address. If you use VRRP for the default gateway, you must plan two additional IP addresses for the default gateway (for the actual IP addresses assigned to the redundant gateways).

As previously mentioned, you should use logical divisions as well. For example, you might have an access layer consisting of several VSF stacks with four members with 48 ports. Each stack has 192 ports for users. You could assign one user VLAN, associated with a /24 subnet, to each VSF stack. Even though the stack does not reach the limit of 252 endpoints, it makes more sense to assign two VSF stacks to two different VLANs rather than overlap multiple user VLANs across them. Or, if you are planning larger VLANs, you could assign the same user VLAN, associated with a /22 subnet, to five VSF stacks (960 ports). Again, the number of ports is below the maximum, but six VSF stacks would exceed the 1,021 limit. It does not make sense to split the sixth stack between two different VLANs, so you might also use a closet or a floor as a logical place for a VLAN division.

Finally, as you plan the VLAN size, keep some additional guidelines in mind. The user count does not always tell you how many ports you need for the users' access category. The company can hire more users. Offices might have multiple Ethernet jacks, and users might connect devices to all of them. Therefore, the maximum port count is about the number of edge ports at the site minus any devices you plan to put in their own VLANs, such as security cameras or VoIP phones that do not offer computer connections. If your count approaches the maximum size you selected for the VLAN, you should plan to divide the access category as a buffer for future growth. Generally, the same considerations apply to other categories. An inventory of devices might not tell you the eventual size of the VLAN.

A VLAN with fewer than 250 endpoints is fine if you have security reasons to create such a VLAN or if the division makes sense in the customer topology, such as a floor with 200 jacks. However, you do not need to divide access categories into VLANs smaller than this simply for the sake of having smaller VLANs.

Example: assign logical VLAN IDs

Although the VLAN ID does not matter beyond being unique within the layer 2 domain, choosing IDs in an organized fashion is often a good idea. Network administrators can easily remember the VLAN associated with each function.

Access category	VLAN Range	X=Location (Floor 1-5)
Switch Mgmt	10	
Voice	2x	21 - 25
Employees	3x	31 - 35
Guest	40	

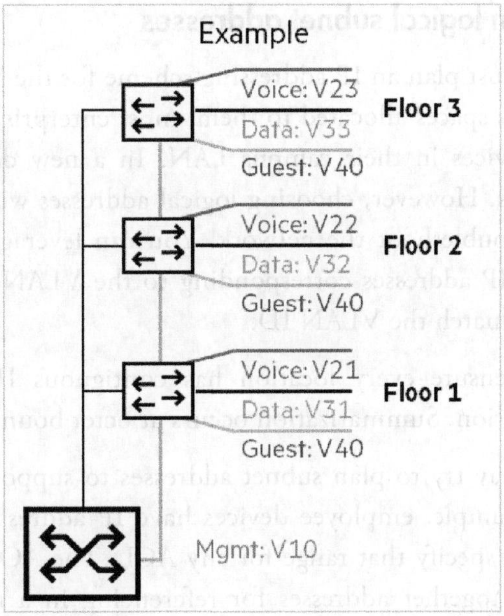

In this example, the architect uses the first digit in the VLAN ID to indicate the access category or VLAN type (e.g., switch management, voice, employees, and guests). You can use the second digit to indicate the location if you divide an access category into multiple VLANs in different network locations. In this example, each floor is associated with one VLAN, and the floor number is the second ID in the VLAN. The company does not support many wired guest devices, so those all go into the same VLAN.

This is just one example of how to assign VLAN IDs. You can adapt similar plans to the particular customer environment.

For example, in a multiple-building campus, you can use one digit to indicate the building and another to indicate the floor.

Switches support VLAN IDs from 1 to 4094. VLAN 1 is the default VLAN on AOS switches, meaning it is the VLAN to which all ports belong by default. Sometimes, administrators use VLAN 1 as the management VLAN. However, using a different VLAN ID provides better security. You may want to use VLAN 1 as a dead-end VLAN that uplinks do not carry. Then, if someone connects to a port whose configuration does not match the correct VLAN, the device cannot obtain network access.

Assign logical subnet addresses

You must plan an IP addressing scheme for the VLANs. Although some enterprises have public IP address spaces allocated to them, most enterprises use private IP addresses, defined by RFC 1918, for devices in their campus LAN. In a new deployment, you can choose any private network address. However, choosing logical addresses will help network administrators manage, maintain, and troubleshoot the network. You can leverage the meaning embedded into the VLAN ID by using IP addresses corresponding to the VLAN IDs. For example, the third octet in the address could match the VLAN ID.

Also, ensure every location has contiguous IP address blocks to enable simpler route summarization. Summarization occurs at octet boundaries and binary blocks within an octet.

You may try to plan subnet addresses to support summarized ranges within ACLs and firewalls. For example, employee devices have IP addresses between 10.1.16.1 and 10.1.19.254, enabling you to specify that range for any ACL. The AOS firewall supports aliases, which helps you easily group together addresses for referencing in a rule. Therefore, this consideration might be less important than route summarization.

Finally, remember that you cannot always balance all of the principles. Sometimes, you might choose not to correlate the subnet IP address and the VLAN ID to create contiguous IP addresses.

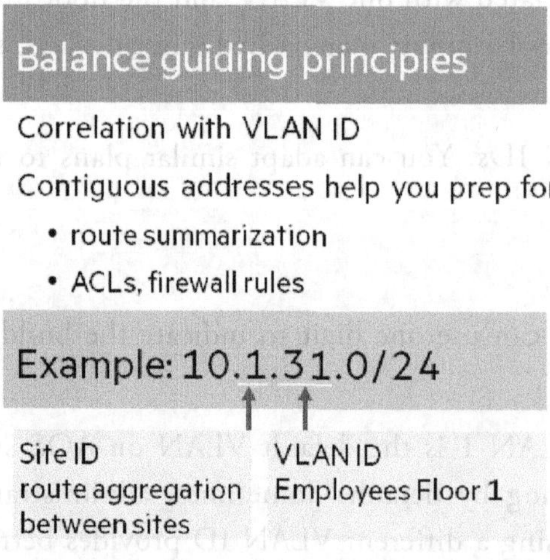

Private IP network	Number of subnets
10.0.0.0/8	65,536 /24 32,768 /23 16,384 /22
172.16.0.0/12	4096 /24 2048 /23 1024 /22
192.168.0.0-192.168.255.255	256 /24 128 /23 64 /22

Keep some additional tips in mind for each of these considerations.

Corresponding with VLAN ID

You could take different approaches for corresponding the IP subnet with the VLAN ID. For example, the architect used XY for the VLAN ID, in which X is the VLAN type, and Y is the closet or floor number. The network address also includes this information using one of these schemes:

- 10.X.Y.0/24
- 172.16.XY.0/24
- 192.168.XY.0/24

If the location is very large and will correspond with an OSPF area, you should put the location first in the VLAN ID to support route summarization. You may also choose to correspond with just part of the VLAN ID. For example, the first part of the VLAN ID might indicate the building number, while the second part indicates the third octet in the subnet address.

Preparing for Route Summarization

You should select a block large enough that, if the company needs to add networks to an area in the future, they can find addresses within that block. If you use the second octet in the 10.0.0.0/8 private address range to indicate the area, the area supports up to 256 /24 networks.

Do not only count the subnets currently in use. Also, think about the future. Planning for the future prevents a situation in which you would have to add a non-contiguous block to the network, thus introducing more route summaries per area. For example, if you need to add more subnets later, you might plan two /16 blocks for the area, like 10.0.0.0/16 and 10.1.0.0/16 or 10.2.0.0/16 and 10.3.0.0/16. In this case, the area should always start with an even number in the second octet so you can create one route summary (10.2.0.0/15) on the area border routers (ABRs).

If the area you need to summarize is smaller, you could divide it into smaller blocks, such as 10.1.0.0/17 (128 /24 subnets) or 10.1.0.0/18 (64 /24 subnets). This makes sense when you need to conserve IP addresses. On the other hand, you should also value logical numbering and ease of management. Choose a system that makes sense to you and the network administrators who will manage the system.

Summarization occurs at octet boundaries and binary blocks within an octet. For example, you can aggregate routes from 10.1.0.0/24 through 10.1.255.0/24 with 10.1.0.0/16. You can also aggregate routes from 10.1.0.0/24 through 10.1.15.0/24 to 10.1.0.0/20. However, routes from 10.1.0.0/24 through 10.1.19.0/24 require several aggregated routes. You generally want to aggregate routes between sites or different network segments, such as the campus and data center. When you use OSPF, area boundaries define where you can aggregate routes. One simple way to promote route aggregation is to assign 10.X to each OSPF area, with X corresponding to the area ID.

If you are making the VLAN ID correspond with the subnet address, you could use the same or different VLAN IDs between areas. For example, you could use VLANs 1000 to 1255 in area one (subnets 10.1.0.0/24 through 10.1.255.0/24) and VLANs 2000 to 2255in area two, or you could reuse VLAN IDs 1-255 in both areas because the areas have layer 3 boundaries between them.

Preparing for ACLs

Grouping devices in a VLAN can help you control a user or device's access rights. Sometimes, you can make it easier for administrators to configure the ACLs and firewall rules by using contiguous blocks of addresses for VLANs to which the same rules apply. In addition, simpler ACLs impose less of a burden on the device processing them. This could be important for a solution that filters a lot of traffic. However, this consideration might not be as important.

You can also combine both uses. The route summarization block should encompass the blocks you create for access rights. For example, you could assign blocks of 256 /24 subnets to each OSPF area and then divide those blocks into four blocks of 64 /24 subnets for various VLAN types.

Using Contiguous Blocks

As you choose contiguous blocks, remember to think in binary. In other words, blocks of addresses come in sizes corresponding to powers of two. You can have a block of two /24 subnets, a block of four /24 subnets, a block of eight /24 subnets, and so forth.

Guest Network

Also, some customers like to make the guest network as distinctive as possible. You might use an entirely different IP address block for this VLAN. For example, you could use 192.168.1.0/24 for the guest VLAN in a network using the Class A 10.0.0.0/8 block. A security appliance or firewall often routes the guests directly to the Internet, so their traffic does not need to fit within the internal routing scheme. A different classful network can help administrators distinguish and treat guest devices differently. It might make setting up the guest solution easier when the customer has separate teams. Finally, it can also help conceal your IP address scheme from a hacker.

Fit into an existing solution

Until now, you have been looking at best practices for planning wired VLANs and sub-netting for a new network. However, often, you are updating an existing network. Generally, you should follow the current design unless you have a good reason to change. For example, a site might use wired VLANs associated with /24 subnets. Although it is possible to use larger subnets, there is no pressing reason to change, and you do not want to interfere with the routing and DHCP scopes already in place. On the other hand, if the company has a very large VLAN and is experiencing issues, you could recommend dividing the VLAN.

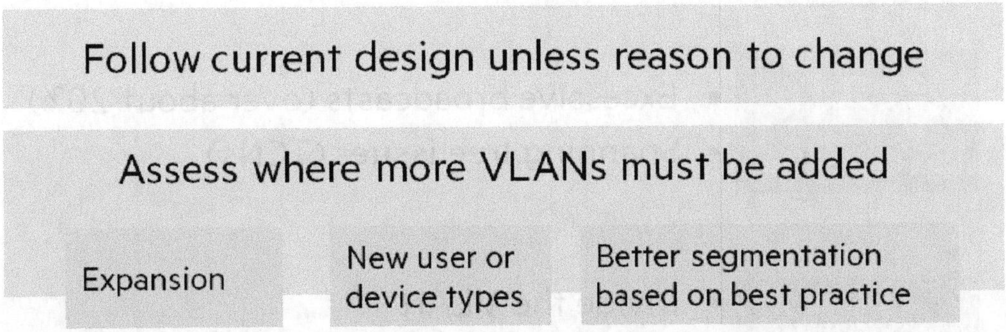

If you are expanding the number of ports, you may need to add more VLANs, fitting them with the current scheme, where possible. You should also look for new VLANs you might want to add. For example, if the company adds IP security cameras or IoT devices, you could add a VLAN for those specific device types. Or if the customer is not currently following best practices to isolate different types of users and devices – such as guests and employees, then you might recommend doing so.

Detect and address issues with large broadcast domains

You may investigate whether a large wired VLAN works well for the customer when the associated subnet uses a larger prefix length. Signs of issues could include slow connections and poor performance. However, many issues can cause such symptoms, from inadequate network capacity to poor quality of service (QoS) measures. One indication that the issues could stem from a large VLAN is an excessive number of broadcasts on links. Generally, you should expect broadcast traffic to be less than about 20 percent of the traffic.

If the network uses spanning tree, then issues with the spanning tree could also indicate that a VLAN extends too far across the topology. Once again, excessive broadcasts can indicate a problem. In addition, you can check switch logs for Topology Change Notices (TCNs).

In either case, it is important to take a systematic approach to searching for issues, checking port utilization across the network at many different times of the day. . A centralized solution like HPE Aruba Networking Central can help you to establish a pattern of behavior over time. You can also configure alerts and events.

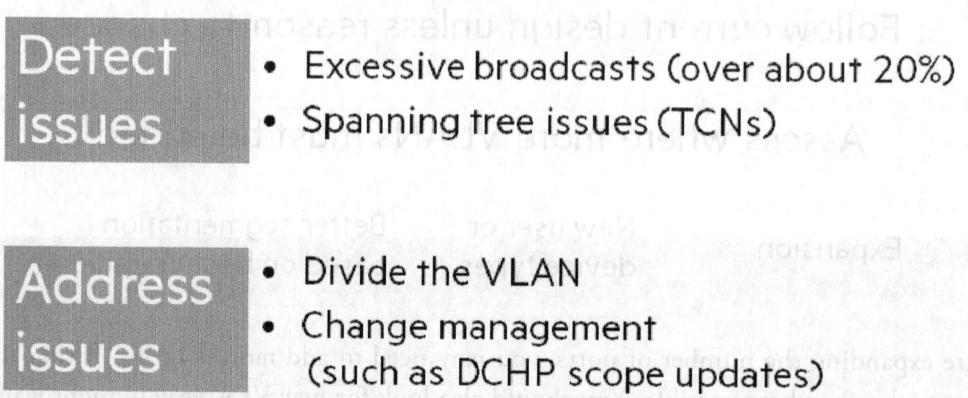

Once you have determined that a VLAN might be too large, you can address the issue by dividing the VLAN. As you do, take the proper change management steps with the customer to ensure the change has no unexpected effects. For example, VLAN 10 might associate with subnet 10.1.0.0/16 subnet, support 5000 devices, and extend across 100 access layer switches. You might divide this VLAN into 20 VLANs (101 through 120). If you want to re-use the former addressing space, you could each associate these VLANs with subnets 10.1.0.0/24 through 10.1.20.0/24. The customer IT staff would need to reprogram DHCP scopes for the new VLAN. You should also discuss with the customer whether any devices in the VLAN have static IP addresses, which might need updates. You would also need to check routing switches. In addition to adding the VLANs to switches that will act as default routers, you must ensure that other switches can reach the new VLANs. If you are dividing up the VLAN gradually rather than all at once, you should use subnets in a new addressing space to prevent overlap.

You must also consider how you will assign ports to the new, smaller VLANs. For the simplest approach, you could simply assign the new VLANs to switches based on their location. In this example, you might assign VLAN 101 to five switches in a similar location, VLAN 102 to the next five switches, and so on. However, you should discuss with the customer whether this approach meets their needs. Are there endpoints that need to be in the same VLAN so they can use broadcast-based protocols or exchange multicast more easily? If so, you must group those devices in the same VLAN.

Example 2-tier wired VLAN design

In this wired VLAN scheme for a two-tier topology, VoIP phones are in VLANs 21 and 22. Employees and APs are in VLANs 31 and 32. This topology shows only two access layer switches, but the network would have several access layer switches using VLANs 21 and 31 and several more using VLANs 22 and 32.

VLAN Name	VLAN ID
Switch Mgmt	10
Voice	2X
Users	3X

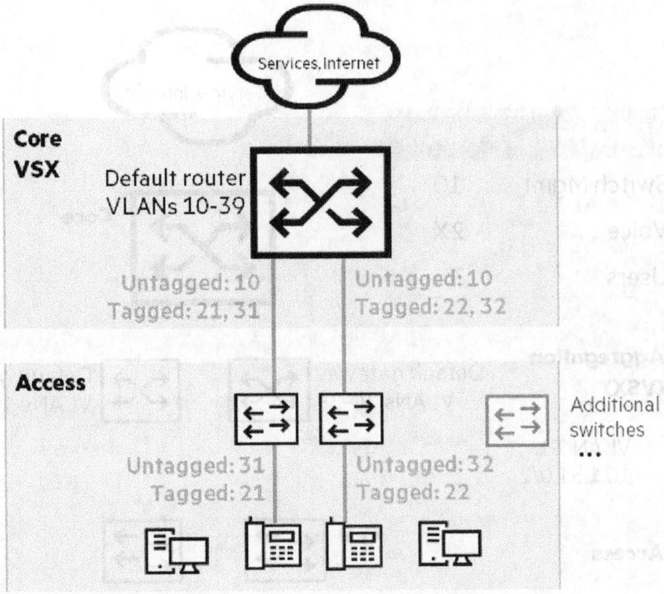

The two-tier design makes it simple to extend VLANs across the campus as desired. For example, you can extend VLAN 10 across the site and place the switch IP addresses. You can make the switch management VLAN the untagged VLAN on all switch-to-switch links rather than have VLAN 1 untagged on those links. This dead-ends the default VLAN, preventing users from connecting to it, or you could remove untagged traffic from the switch-to-switch links if you prefer.

Note

This network topology focuses on logical connections. The core switch is two physical switches in a VSX fabric, and the single links between the various components are link aggregations.

Example 3-tier wired VLAN design

You can create similar schemes in three-tier topologies, but you should most often route between the aggregation layer and core. Therefore, you cannot extend the same VLAN into segments served by different aggregation layer switches. However, you can re-use the VLAN IDs, which can be helpful for management purposes. Multiple access layer switches can share the same configuration file, for example. The VLAN IDs would then associate with different IP networks on aggregation layer switches.

For example, this network has an aggregation layer in each building on the campus. Both buildings are using VLANs 10-39. However, the VLANs terminate at the aggregation VSX stack, which provides routing for those VLANs, and in Building 1, VLAN 31 uses IP address 10.1.31.0/24, while in Building 2, VLAN 31 uses IP address 10.2.31.0/24.

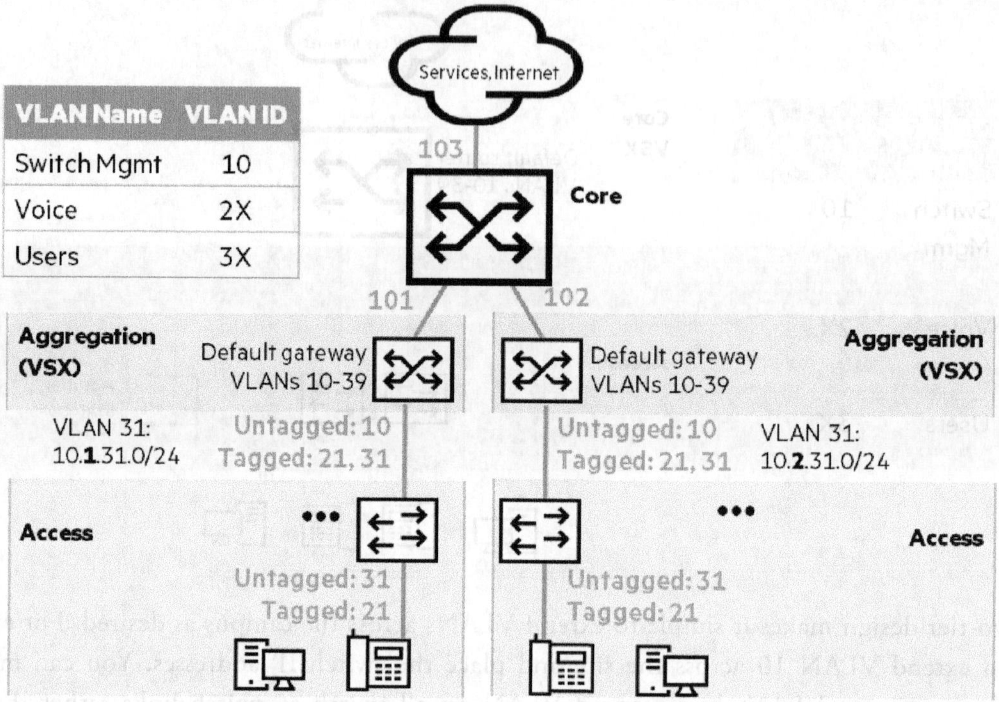

Finally, remember that routing switches must also have a VLAN for their connections to each other. It is best practice to terminate all the VLANs used by endpoints at the default router. Then, create a dedicated VLAN for each layer 3 switch-to-switch link. Using a dedicated link for each separate connection helps to make the routing solution converge more quickly.

For simplicity, the figure does not show all of the access layer devices in each building. However, in the real network, additional access layer switches would exist. If the building had enough switches, they would place devices in VLANs 22, 23, 32, 33, and so on.

Some administrators prefer to use different VLAN IDs across layer 3 boundaries, which is also a valid approach. For example, in this design, Building 1 could use VLANs 100-139, and Building 2 could use VLANs 200-239.

IP Addressing Design for Wireless Networks

Let's dive into IP addressing design for WLANs.

Advantages of single VLAN

There are many advantages of a single flat VLAN. The design is simple with one scope and easy to support, and there are no IPv6 issues. Roaming from AP to AP is simplified, and users never change their IP. Plus, you do not have to implement a complicated roaming scheme since the single VLAN efficiently distributes IP addresses to users.

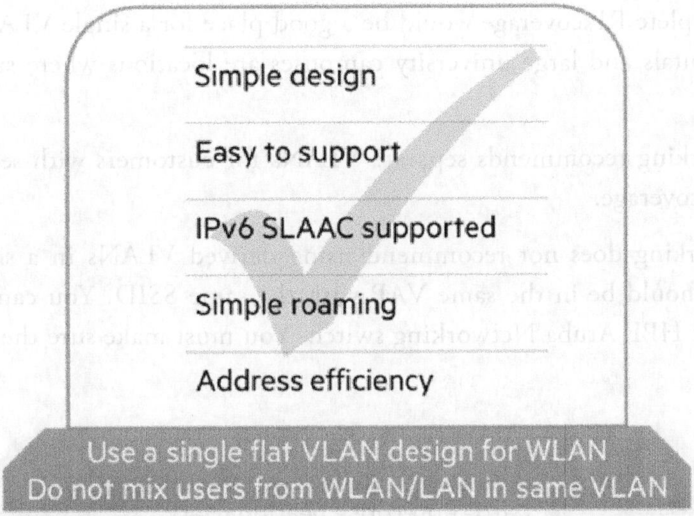

Single VLANs have a simple design and are easy to support with just one large subnet for all wireless clients connecting to an SSID. The network administrator does not need to configure multiple VLANs or DHCP scopes, extend VLANs across multiple devices, or configure redundancy for default gateways in each VLAN. On top of that, the network administrator does not need to configure VLAN or IP mobility for clients roaming across different Gateways. Since the client's VLAN remains the same, the client keeps using the same IP address.

The simplicity of a wireless LAN design makes it very easy for network administrators to support it. If issues arise, it is very easy to troubleshoot them. Network administrators do not need to determine which VLAN the client falls into or try to isolate an issue related to that VLAN or network-wide.

The single VLAN design solves IPv6 SLAAC challenges. Since multiple RAs do not come from IPv6 routers, there is no issue of the client getting the wrong v6 address. This creates a simplified roaming experience for users. No matter where the client roams, either on a different AP on the same or a different gateway, the client does not need to change its IP address. Also, the network administrator does not need to configure layer 2 or 3 roaming on the Gateway. This solves the roaming complexity for the network administrator.

Single VLAN gives the administrator address efficiency. With one VLAN, there is no issue of some VLANs being completely exhausted while other VLANs stay less utilized. In large campus environments with thousands of users, this helps prevent clients from not getting IP addresses due to some VLANs being drained.

Recommendation for single VLAN

A campus with complete RF coverage would be a good place for a single VLAN as it will simplify user roaming. Hospitals and large university campuses are locations where simplified roaming is an advantage.

HPE Aruba Networking recommends separate VLANs for customers with several buildings with noncontiguous RF coverage.

HPE Aruba Networking does not recommend using derived VLANs in a single VLAN design. The single VLAN should be in the same VAP with the same SSID. You cannot simply set up a single VLAN on the HPE Aruba Networking switch. You must make sure the core can handle the large VLAN as well.

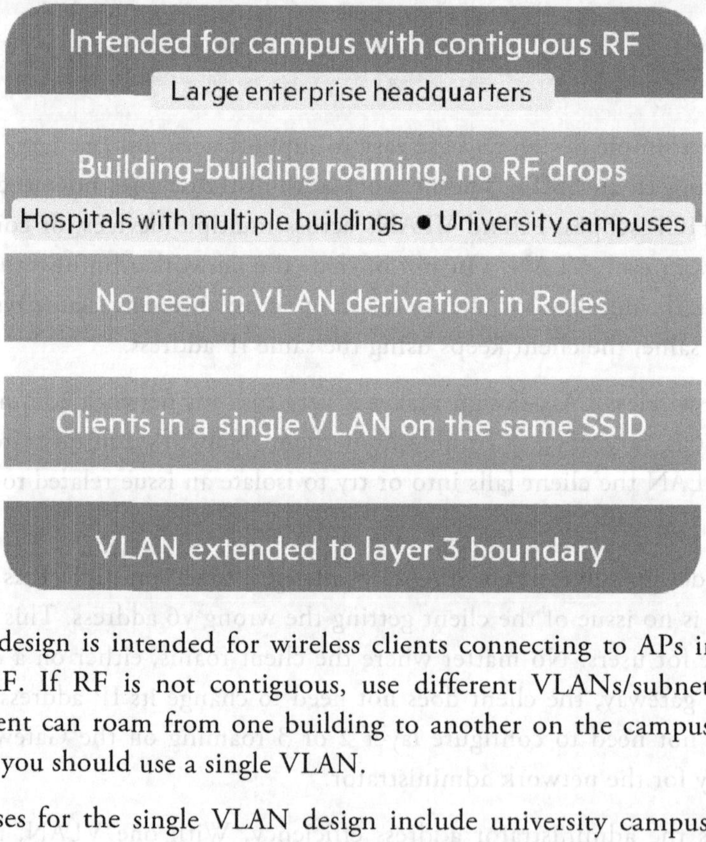

The single VLAN design is intended for wireless clients connecting to APs in the same campus with contiguous RF. If RF is not contiguous, use different VLANs/subnets for different RF domains. If the client can roam from one building to another on the campus without dropping off from the SSID, you should use a single VLAN.

The typical use cases for the single VLAN design include university campuses, large enterprise headquarters, and hospitals with multiple buildings. RF is not contiguous if the customer has multiple buildings and offices across different locations. Thus, they should use separate VLANs. The typical use case is branch offices or home offices.

Customers must not use VLAN derivation by user role or server rule. If clients connected to the same SSID are in different VLANs, the behavior is similar to VLAN pooling. Thus, all the concerns with VLAN pooling will apply to the design. The large subnet is for wireless clients connected to the same SSID only. Use different VLANs for different SSIDs (e.g., employee vs. guest SSID).

Suppose customers have multiple Gateways serving a campus network with contiguous RF. In that case, they should extend the client VLAN back to their core switch, router, or firewall where the default gateway exists. Customers using the single VLAN design must ensure that their uplink switches, routers, or firewalls can support the required number of ARP and MAC address table entries depending on the subnet size. You should ensure that the DHCP server can support a DHCP scope as large as the size of the client VLAN.

Key considerations

To set up a single VLAN, there are key considerations you need to verify. Is your DHCP server capable of creating a large scope and distributing these addresses? Many Linux-based servers are not capable of this process. Is your router capable of handling large ARP tables? Is your switch capable of large bridge and MAC tables?

Upstream firewalls will likely be configured based on VLAN rules. Also, there is the problem of broadcast and multicast, but HPE Aruba Networking has wireless solutions for these problems.

To support a large VLAN, other devices on the network also need to be capable of handling the workload. Most of the time, the wireless client VLAN configuration is layer 2 on the HPE Aruba Networking Gateway and uplink switch. The uplink router configures as a layer 3 interface working as a default gateway for that VLAN. While using large VLANs, the ARP table on the router can also be sizeable depending on how many clients connect to the WLAN. Thus, network administrators should find any router limitations and address them accordingly.

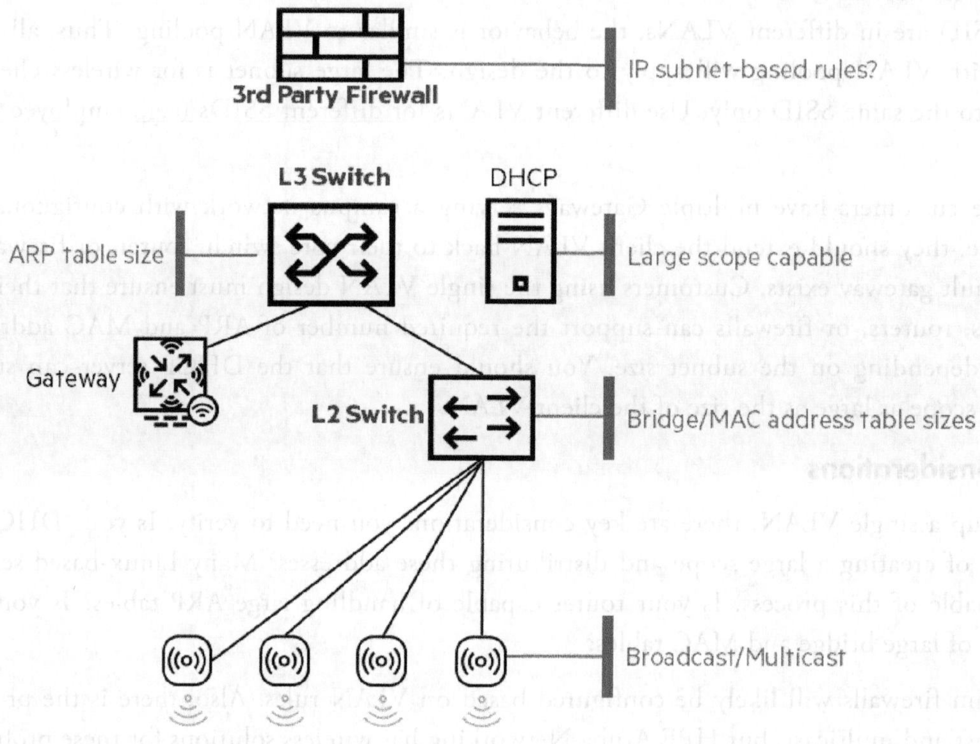

The same goes for switches. Although the Gateway uplink switch does not need to handle ARP entries, the bridge/MAC address table entries can be very large. The network administrator should consider this. In the case of a single VLAN design, the DHCP server needs to support large DHCP scopes, as big as /16, which many Linux-based DHCP servers cannot support. Thus, the network administrator needs to ensure that the DHCP server can handle a large DHCP scope. At the same time, the DHCP server needs to be powerful enough to lease out IP addresses quickly, which is particularly essential in a failover situation. For most of enterprise networks, IP subnets are the basis of existing firewall rules. Many network administrators still think this way. Administrators may configure routing, access control, and QoS inside the network based on the VLANs and IP subnet through routers and legacy firewalls.

Next-generation firewalls are user- and application-centric and deploy reasonably at large enterprise customers' Internet gateway and data centers. With the single VLAN design, all the traffic of wireless clients is in one subnet only. Hence, you need user- and application-centric firewalls to apply proper firewall rules. Another option is to use the HPE Aruba Networking Gateway's built-in stateful firewall to manage firewall policies for wireless clients.

HPE Aruba Networking wireless broadcast / multicast solutions

HPE Aruba Networking can limit broadcast and multicast by converting the necessary traffic to unicast and removing unnecessary broadcast and multicast traffic, which can also solve many IPv6 issues. Many of these features are the default configuration.

Convert Broadcast ARP to Unicast

Suppress ARP

Optimize Broadcast and Multicast traffic

Optimize Duplicate Address Detection

Drop Broadcast and Multicast Traffic

IGMP Snooping and Dynamic Multicast Optimization

IPv6 Traffic Optimization

Convert the broadcast ARP to unicast - when a router sends out an ARP broadcast, there is no need to send this frame to every AP. HPE Aruba Networking Central knows both the user and the AP associated with the user. The Gateway simply changes the ARP request to a unicast and sends it directly to the user. If you disable this feature, the Gateway does no conversion, and the ARP broadcast would transmit to all APs with a user in the subnet, causing degradation of the radios. This unicast feature is on by default.

Suppress ARP - APs will suppress any ARP that does not correspond to any user in the APs list, which saves the unnecessary broadcast and use of the radio. APs will broadcast on any of its wired ports or any wired AP port and on any VAP in split tunnel mode. This feature is on by default.

Optimize duplicate address detection - controls the flood of gratuitous ARP (GARP) and IPv6 DAD frames. If a user sends a GARP, APs will forward this frame to the router but not flood other APs. APs will forward GARPs or DADs from the router because this is important information for the clients. For example, if you have redundant gateways and lose the primary gateway, the secondary gateways notify the client that it is now the owner of the IP address. This feature is on by default.

Drop Broadcast and Multicast Traffic - Once ARP and DHCP packets convert to unicast, APs will block all unnecessary broadcasts and multicast. The default setting for this feature is disabled, but HPE Aruba Networking recommends enabling it.

IGMP Snooping and Dynamic Multicast Optimization - IGMP snooping keeps track of clients subscribed to multicast streams and converts multicast packets to unicast before sending them to wireless clients. Again, the default for this feature is disabled, but you should enable it since the network requires multicast streaming.

Multicast transmission optimization

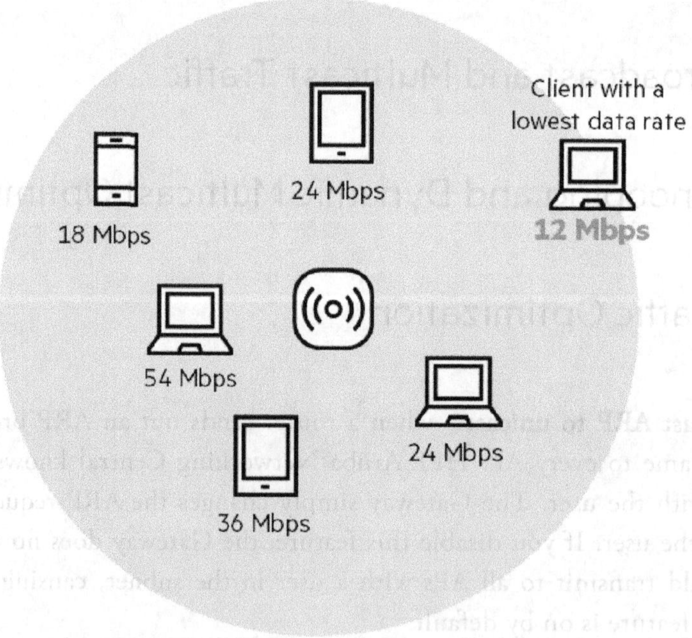

To turn on multicast transmission optimization, access the WLAN wizard. From the General tab, expand Advanced Settings, expand Broadcast/multicast. Turn on the toggle switch to allow the AP to select the optimal rate for sending broadcast and multicast frames based on the lowest of unicast rates across all associated clients.

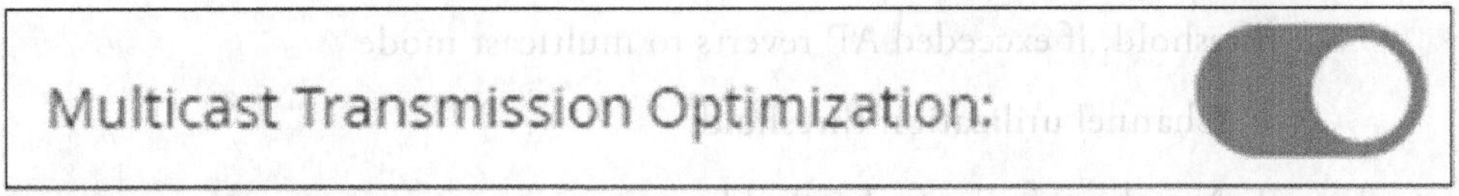

All broadcast/multicast traffic uses **12 Mbps** data rate

When this option is enabled, multicast traffic can be sent up to a rate of 24 Mbps. The default rate for sending frames for 2.4 GHz is 1 Mbps, and that for 5 GHz is 6 Mbps. This option is disabled, by default.

Dynamic multicast optimization

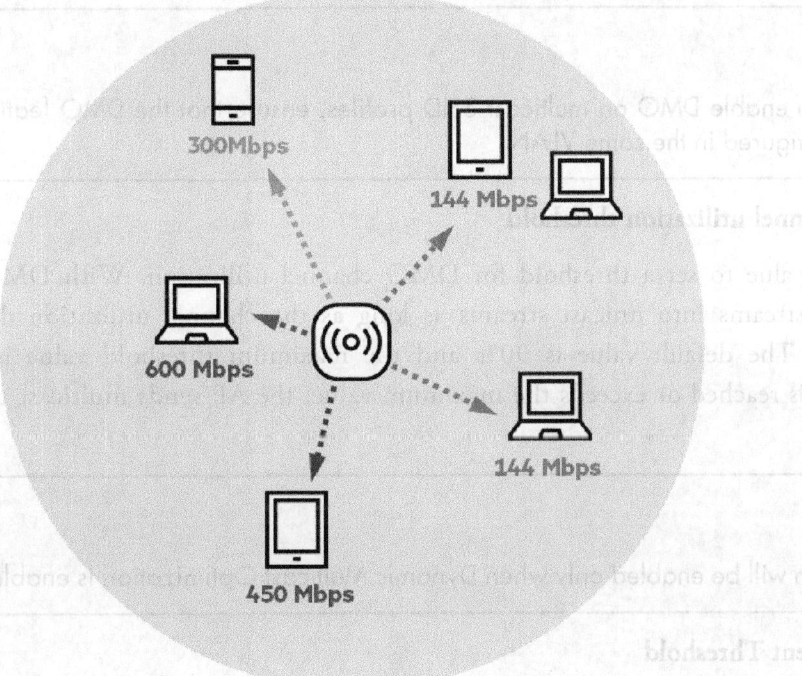

- Convert multicast to unicast
  - Clients receive traffic with a highest possible rate
  - Less airtime required to deliver the traffic
- Large number of multicast clients may cause a problem

- Set threshold, if exceeded AP reverts to multicast mode
    - Channel utilization threshold
    - Number of clients threshold

Dynamic multicast optimization (DMO) will convert the downstream multicast traffic to unicasts before forwarding it to the wireless clients. This allows the client to receive the traffic with the highest possible data rate and thus requires less airtime to deliver the traffic.

To enable DMO, access the WLAN wizard. From the General tab, expand Advanced Settings, expand Broadcast/multicast. Turn on the toggle switch to allow the AP to convert multicast streams into unicast streams over the wireless link. Enabling DMO enhances the quality and reliability of streaming video, while preserving the bandwidth available to the non-video clients.

Note

When you enable DMO on multicast SSID profiles, ensure that the DMO feature is enabled on all SSIDs configured in the same VLAN.

DMO channel utilization threshold

Specify a value to set a threshold for DMO channel utilization. With DMO, the AP converts multicast streams into unicast streams as long as the channel utilization does not exceed this threshold. The default value is 90% and the maximum threshold value is 100%. When the threshold is reached or exceeds the maximum value, the AP sends multicast traffic over the wireless link.

Note

This option will be enabled only when Dynamic Multicast Optimization is enabled.

DMO Client Threshold

Specify a value between 2 and 255 to set the DMO client threshold. The recommended initial DMO client threshold is 40, it should be then adjusted based on the client's experience

Single VLAN optimizes ipv6 traffic

Clients' IPv6 addresses can be configured automatically by listening to RA frames or sending RA frames to the router. All RA and RS messages are sent to specific multicast addresses. This could affect the wireless performance.

Periodic RAs are not an issue

Optimize duplicate address Detection Messages IPv6

- Stops forwarding DAD packets from clients
- Forwards DAD packets from router

A periodic RA sent in a flat VLAN is not a major issue. However, an issue could develop with VLAN pooling when users are on the same AP but in different VLANs because each VLAN would advertise RAs. The gateway will intercept the RAs and unicast to the clients to avoid this issue.

A client will forward DAD packets to inform others of its IP address. If all clients in a flat VLAN send out DAD frames, each must retain this information in its ARP table. Client devices have a limited ARP table, which could be an issue. HPE Aruba Networking gateways stop the forwarding of DAD frames to clients. However, DAD frames from the router are still forward.

Identity-based HPE Aruba Networking firewall

Role derivation will determine what role and rights the user will receive, regardless of the AP, SSID, or subnet the user falls into. For example, in this diagram, Pedro, Robert, Frank, and Susan are on the same SSID and are on the same subnet, but all have different roles. The IP address they have is irrelevant to what role they receive.

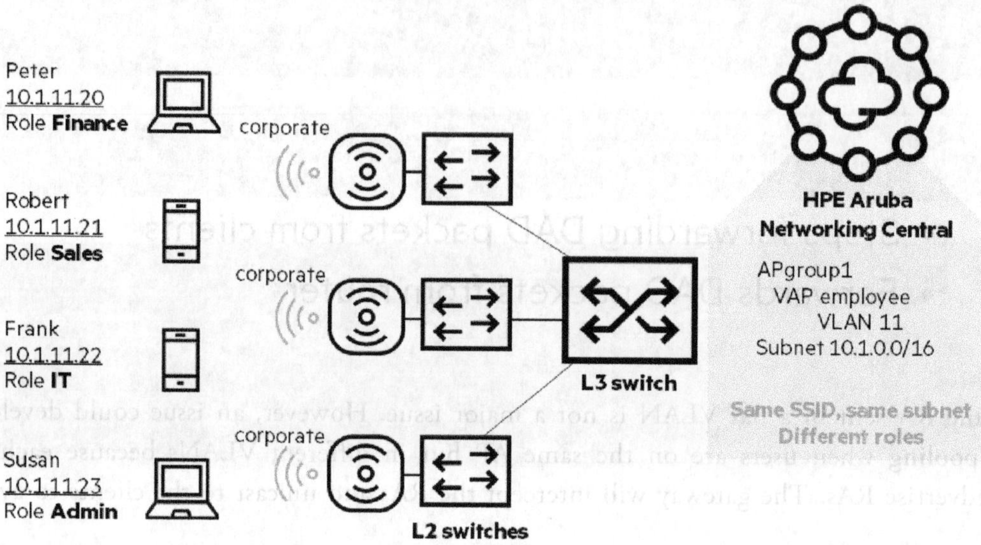

You can derive roles locally on the APs or by a RADIUS server or ClearPass. The user falls into the authentication default role if there is no locally derived or server-derived role.

Note

You can write user rules that say you receive a role assignment if you are on this AP or in this VLAN.

With HPE Aruba Networking networks, not only can we gain contextual insights about users, devices, and application states, but we can also influence the way the infrastructure behaves. For example, if we need to update the policy enforcement for a specific user's device, we can program the network to route traffic differently. This will improve data security and change how physical spaces interact with mobile devices based on location or identity.

IP Routing Protocols and VRF Lite

The section offers a brief overview of IP routing protocols.

OSPF overview

Many companies have multiple routers and routing switches that send traffic across the campus. Dynamic routing protocols enable routers to discover and update routes automatically.

Link state routing protocol

- Uses triggered updates
- Calculate lowest cost route based on topology
- Communicate updates and converge quickly
- Uses Dijkstra (SPF) algorithm

Hierarchical, area-based design

- Stable and scalable networks

OSPF is a link state protocol, where routers advertise information about their connected Layer 3 interfaces, or links, along with the cost (or metric) associated with that link. These link state advertisements (LSAs) propagate across participating routers, which use the advertisements to build a network topology, the routers connected to them, and associated costs. Using the topology, routers calculate their own lowest cost route to each destination subnet.

OSPF uses Dijkstras algorithm, also called the shortest-path first (SPF) algorithm to find the least cost, loop-free paths.

For scalability, OSPF supports a two layers hierarchical design.

OSPF is built upon a hierarchy of network areas. All areas for a given OSPF domain reside in the same AS. An AS is a collection of contiguous networks under common routing policy. An AS can be divided into multiple areas. Each area represents a collection of contiguous networks and hosts, and the topology of a given area is not known by the internal routers in any other area, this provides more stability and scalability in the network.

OSPF routing sample topology

OSPF Routing sample OSPF topology

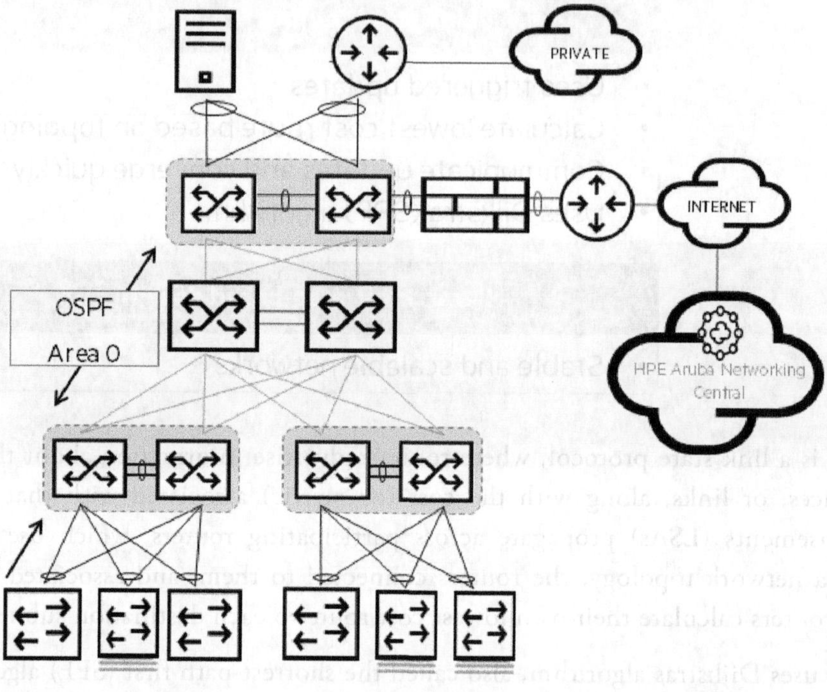

Different departments across the campus will require scalable connectivity to share information. Campus networks extend typical single building 2-tier connections through a central Core. This redundancy between Core and Aggregation requires routing of Layer 3 connections. A routing protocol is needed to provide dynamic learning of subnets, resiliency when networks change, and loop prevention.

The most common routing protocols in enterprises today are Open Shortest Path First (OSPF) and Border Gateway Protocol (BGP). The latter, BGP, is the backbone of internet routing between large public networks and is used as the control plane for EVPN-VXLANs in the data center and campus overlays. OSPF is the standard choice for a campus infrastructure within an autonomous system or managed enterprise. It is a link-state dynamic routing protocol that scales.

Using separate areas, OSPF routes small networks to larger WAN-connected campuses with branch deployments. Area assignment is normally done along the lines of one building to the next or to divide one location from another. A single backbone area can quickly get OSPF routing up and running in a single location or set of campus buildings. When the network scales from one campus to include remote or branch locations, multiple areas connected to a backbone are recommended.

OSPF enables remote site subnets and VLANs to be dynamically learned across the campus and to each other. Route summarization is possible If these branch locations are in their areas, separated by an Area Border Router. Consider using OSPF routing authentication to secure the network better. Public internet can be accessed using a simple static default route, or OSPF can propagate a default route to an edge router/gateway acting as the Autonomous System Boundary Router (ASBR).

Some larger enterprise customers may benefit from sharing external BGP routes between multi-homed ISPs. OSPF would still be designed and deployed to reach the edge BGP routers as ASBRs without redistributing public routes between BGP and OSPF, as OSPF will not scale to the full internet routing table. Most enterprise networks, however, would typically have a policy of a primary ISP and a backup assigned through static default routing instead of external BGP.

BGP overview

For autonomous systems to share routing information and route traffic between networks, BGP is a crucial protocol for interdomain routing on the Internet. It offers the policy-based routing capabilities, scalability, and flexibility required to manage the intricate Internet routing infrastructure.

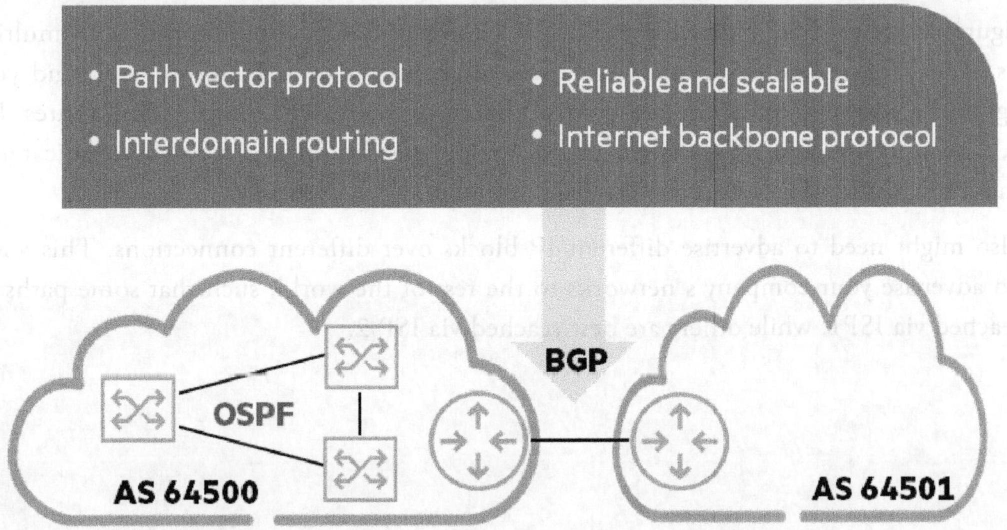

Some of it's features as follows,

Path Vector Protocol: BGP is a path vector protocol that uses a vector of AS numbers to represent the path to reach a destination network.It takes into account multiple factors, such as policy decisions and route attributes, to determine the best path for routing traffic between ASes.

Interdomain Routing: BGP is primarily used for interdomain routing, facilitating the exchange of routing information and traffic between different autonomous systems on the Internet. It enables ISPs and organizations to advertise their network prefixes (IP address ranges) to other ASes and receive routing updates from them.

Policy-Based Routing: BGP allows network administrators to implement policy-based routing decisions based on factors such as route preferences, traffic engineering, and business requirements. Policies can be defined to influence route selection, control traffic flow, and enforce security measures.

Reliable and Scalable: BGP is designed to be highly reliable and scalable, capable of handling the vast and complex Internet routing infrastructure. It uses TCP as its transport protocol, providing reliability and error detection mechanisms for exchanging routing information.

Internet Backbone Protocol: BGP is the protocol that forms the backbone of the Internet routing infrastructure, enabling connectivity between thousands of networks and ASes worldwide. It allows ISPs to establish peering relationships and exchange routing information with each other to ensure global connectivity.

Multi-homed ISP connections

The figure illustrates BGP's primary use case - a company must communicate with multiple ISP routers. Another use case is when a large company connects to a single provider, and you need more granular path manipulation and control than is available with simple static routes. In other words, You only need to do so when the company specifically needs to communicate routing information with the ISP.

You also might need to advertise different IP blocks over different connections. This allows the ISPs to advertise your company's networks to the rest of the world, such that some paths may be best reached via ISP1, while others are best reached via ISP 2.

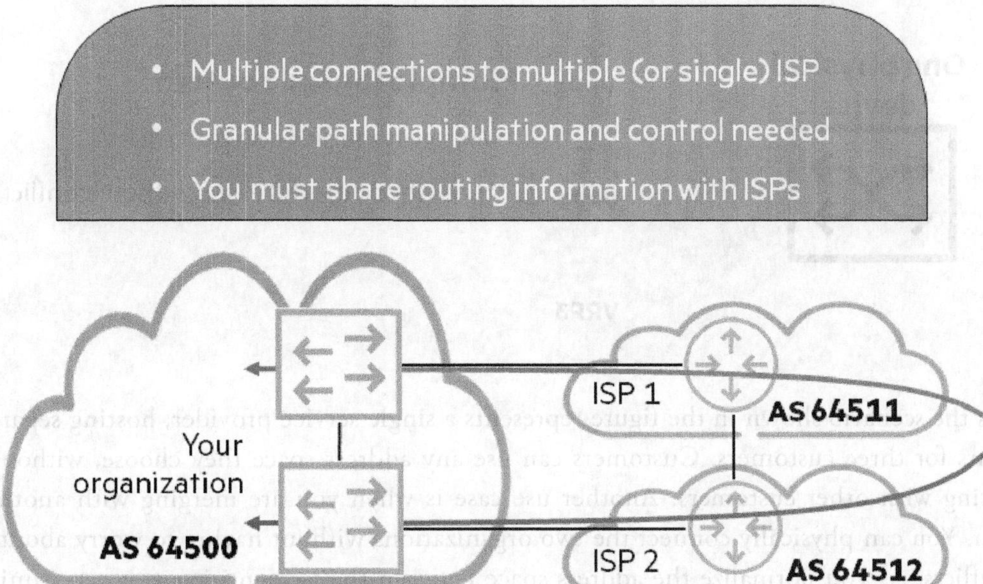

The figure's dual-ISP example shows a different router connected to each ISP. This improves resiliency over a single connection but can be slightly more complex because it requires two BGP routers, which communicate with each other as well as the ISP routers.

Note that In many cases, BGP is not required. If you only connect to a single ISP, and no granular path control or manipulation is needed, you can simply use static routes, reaching all internet sites via your ISP routers. The ISP handles announcing the block of public IP addresses assigned to the company.

VRF lite overview

Virtual Routing and Forwarding (VRF) allows a single physical device to host multiple logical routers, with complete L3 isolation. Because the device maintains separate route processes and tables for each VRF, they can have overlapping address spaces. Network functionality is improved because network paths can be segmented without requiring multiple routers.

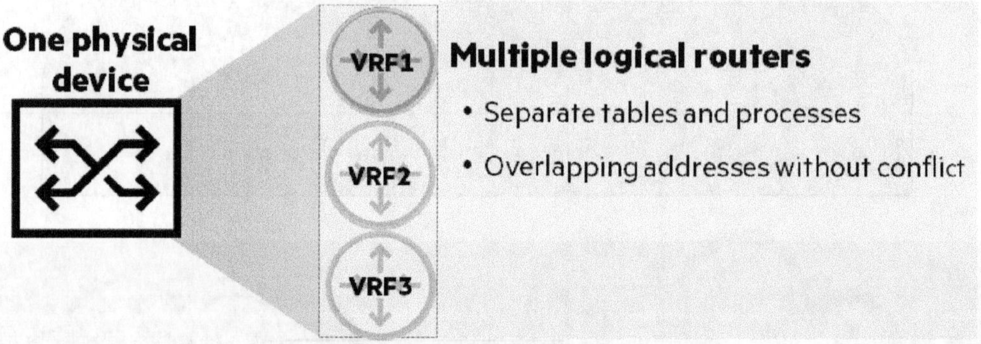

Perhaps the scenario shown in the figure represents a single service provider, hosting separate environments for three customers. Customers can use any address space they choose, without fear of conflicting with other customers. Another use case is when you are merging with another organization. You can physically connect the two organizations without having to worry about addressing conflicts. As you normalize the address space between the two entities, you can eliminate the VRF configuration as appropriate. Another common use case is to maintain logical separation of business units. For example, a banking company might maintain separation between their retail, lending/mortgages, investments, and internal applications.

Use cases

- Provider that supports multiple customers
- Integrate two merging companies
- Logical separation of business units
- Separate business functions for security

A common use case is to separate business functions. Place network management functions in one VRF, with the rest of the network in another VRF. Management network routes are unknown to the rest of the network, providing a logical separation between the two—you get enhanced security. Or in a data center place public resources in one VRF, with private or internal services in a separate VRF. Even though the both sets of services are in the same data center, they are logically separated since their routes are in different routing tables, thus enhancing security. Let's analyze some scenarios in more detail.

Learning check

Check your understanding of network segmentation by answering the following questions.

Question #1

What is a common reason for placing voice-over IP (VoIP) devices in a separate VLAN?

- ○ VoIP traffic compression requires a unique voice VLAN assigned by the switch.
- ○ Mixing VoIP with normal data traffic will give both traffic types higher quality of service (QOS) even without tagging the class of service (COS) in the VLAN header.
- ○ Placing VoIP traffic in the same VLAN with data traffic protects the VoIP phones from that data's broadcast frames.
- ○ VoIP traffic requires special handling through class of service (COS) quality of service assignment tagged in a VLAN header.

Question #2

What is the best practice when planning wired VLANs and sub-netting for network design?

- ○ You should follow the current design unless you have a good reason to change.
- ○ Always change subnets to use a 192.168.0.0/16 range if a small business, 172.16.0.0/12 range if a medium business, and 10.0.0.0/8 for a large business.
- ○ Minimize the number of VLANs regardless of segmentation requirements.
- ○ The third octet of a subnet must correspond with the VLAN assignment. For example, VLAN 25 must utilize 10.0.25.0/24.

Question #3

Which three of the following are all key considerations when leveraging a single VLAN? (select three)

- ☐ Is your subnet range a class A address range?
- ☐ Is your DHCP server capable of maintaining a larger scope?
- ☐ Is your router capable of handling large ARP tables?
- ☐ Is your switch capable of large bridge and MAC tables?
- ☐ Is your firewall capable of handling large number of NAT entries?

Supplemental: Multi-area OSPF

Use this supplemental section to explore multi-area OSPF.

Multi-area OSPF: large scalability

- Aggregate inter-area routes
- Much smaller LSDBs

- Reduce LSA/LSU flooding
- Minimize SPF recalculations
- Increase stability and control

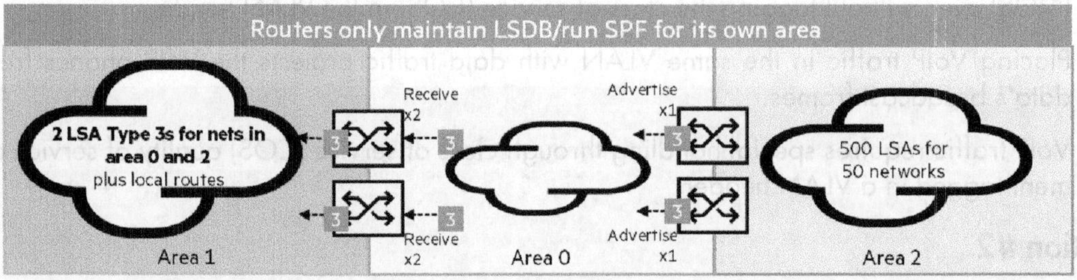

OSPF areas segment the domain into more stable and manageable pieces. Each OSPF routing device maintains the LSDB and runs the shortest path algorithm only for its own areas. OSPF ABRs forward routing information between areas. Type 1 and Type 2 LSAs do not flow across area boundaries. ABRs summarize this huge number of Type 1 and Type 2 LSAs into one Type 3 summary LSAs for each subnet learned from the Type 1 and 2 LSAs.

If you had 100,000 routes in your routing table with a single area design, you would still have 100,000 entries after using areas.

However, now that you have ABRs, you can configure route summarization, which can drastically reduce your routing table sizes. Routers in Area 1 may only have a few dozen or a hundred routes in that area, plus a default route (0.0.0.0) to reach all other subnets via the ABRs. These benefits apply even to powerful routers that could support large areas.

This is especially true as it relates to network stability. With ABRs and route summarization, you have effectively segmented your network into smaller failure domains. A flapping route in Area 1 does not affect Area 0 or Area 2. This can be a big boost in network stability and ease of troubleshooting.

Need for route and LSA aggregation between areas

Benefits of multi-area

- Decrease the size of OSPF routers' LSDBs
- Reduce SPF calculations
- Allows aggregate routes

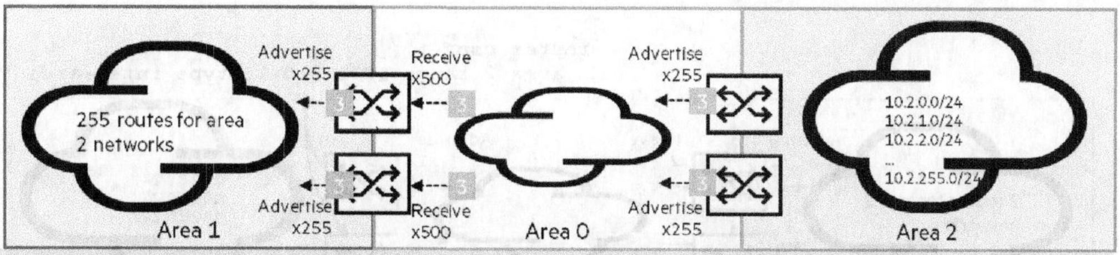

How can you aggregate all these type 3 LSAs?

Simply dividing an AS into multiple areas decreases the size of OSPF routers' LSDBs and reduces the complexity of SPF calculations. However, to achieve the true benefits of multiple areas, you should further reduce the number of LSAs advertised between areas.

In this example network, both area 1 and area 2 have only two ABR exit points toward the backbone and other areas. Internal area 1 routers and backbone routers might benefit from some detail in the information they receive about area 2. For example, each ABR's Type 3 LSAs reflect the metrics of reaching the network in question, enabling routers in other areas to select the best ABR for reaching each network in Area 2.

However, with only two entry points into Area 2, routers (particularly internal Area 1 routers) probably receive too much granularity about each subnet. This information clutters the routers' LSDB and routing tables.

In addition, small changes in the area 2 topology could flood LSUs through the backbone and area 1, causing routers to re-run the SPF algorithm. For example, network 10.2.40.0/24 in area 2 becomes unavailable. The two Area 2 ABRs withdraw their Type 3 LSA for that specific network in Area 0, and the two Area 1 ABRs do the same in Area 1. However, the Area 1 routers have no other path to this network. It might be best not to inform routers beyond area 2 of the small change. If they continue to forward traffic for this network, the area 2 ABRs simply drop it.

You should aggregate these Type 3 LSAs to hide the unnecessary information and to smooth over small changes, thereby gaining the true benefits of a multiple-area OSPF configuration. Although this example focuses on aggregating advertisements for Area 2 networks, you would aggregate summaries of Area 1 or Area 0 networks for similar reasons.

Problem solved: ABR route aggregation

Solution: ABR area summary

- Generate a single Type 3 LSA for the specified range
- Suppress any Type 3 LSAs within that range

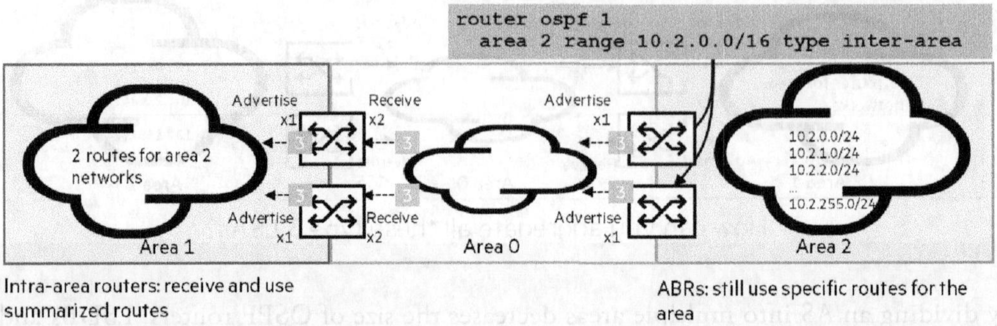

You can configure an ABR to aggregate its summary routes. Rather than generate a Type 3 LSA for every network in an area, it generates a few Type 3 LSAs, each including many networks in its range. In other words, the Type 3 LSA advertises a network with a shorter prefix length, such as a /20 or a /16 subnet that includes many /24 subnets.

You configure area summaries on the ABR that connect to the area to be aggregated. In the figure, you configure area summaries for Area 2 on the two ABRs connected to Area 2. These ABRs then advertise the summaries to other backbone routers and ABRs in area 0. The other ABRs can then copy them into their own areas. In this example, the aggregated summary 10.2.0.0/16 shows up as a route on ABR A and internal routing switches in Area 1.

Similarly, you could configure the ABRs connected to Area 1 to summarize Area 1 routes and inject them into Area 0. All ABRs connect to Area 0, so you can configure them all to summarize Area 0 routes and inject them into another area.

You see the drastic reduction in routes. Before summarization, Area 1 had 255 entries for Area 2. Now, it has two entries.

The figure shows the syntax to configure this aggregation.

When you enter the area x range command, the ABR automatically suppresses the generation of individual Type 3 LSAs for subnets within the specified network. However, it continues advertising individual Type 3 LSAs for networks that do not fall into the range. Thus, it is best to assign contiguous IP addresses within an area – it ensures easy route aggregation. The ABRs for Area 1 now receive one Type 3 LSA with the aggregated summary from each of the two ABRs that connect to Area 2. Each area 1 ABR selects the best LSA for the summary range and creates an inter-area route to 10.2.0.0/16 through the best next-hop. Each area 1 ABR then generates an area 1 Type 3 LSA to advertise this inter-area route to routers in area 1. The internal area 1 routers now have two LSAs for all area 2 networks, and they learn routes to only the one aggregated network (up to two routes if the ABR advertises equal metrics and the internal routers support Equal-Cost Multi-Path [ECMP]).

Note

After configuring area summaries on an area's ABR, you can see the summaries in the LSDB and routing tables for routers not connected to that area.

You typically want to create as few area summaries as possible, but this is not always possible. Areas might have grown organically and without planning, creating non-contiguous ranges or ranges that do not fit neatly within a CIDR subnet. For example, area 2 might include networks in the 10.2.0.0-10.2.255.255 range and 10.3.0.0-10.3.23.255. Area 3 includes networks in the 10.3.0.0-10.3.19.255 range and the 10.3.30.0-10.3.39.255 range. Due to the non-contiguous and overlapping ranges, you cannot simply use 10.2.0.0/16 for the Area 2 summary and 10.3.0.0/16 for the Area 3 summary. Instead, you must use several area summaries for Area 2 and several summaries for Area 3.

It is important to note that examples throughout this module use the /16 prefix length for summary ranges to illustrate the concept in a simple way. When you create a range for your area, choose an appropriate range. For example, your area might include 16 /24 subnets, 10.1.0.0/24 to 10.1.15.0/24. This equals the 10.1.0.0/20 network; specify that in the area <ID> range command.

Special area types

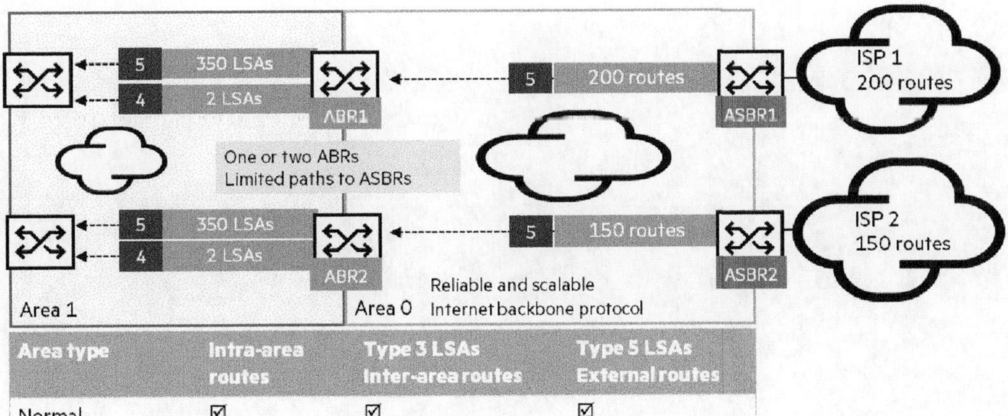

Area type	Intra-area routes	Type 3 LSAs Inter-area routes	Type 5 LSAs External routes
Normal	☑	☑	☑
Stub	☑	☑ + default route	
Totally stub	☑	Default only	

Often, an area only has one or two ABR exit points - internal routers will choose the same next hop or hops for all external routes. This is when you might filter the Type 5 LSAs at the area boundary. You do so by configuring the area as a special type: stub or totally stubby. In a totally stub area ABRs only advertise a default route and do not inject any type 3, 4, and 5 LSAs.

7 Campus Security

This module looks at campus security design elements for an HPE Aruba Networking security-first, AI-powered networking deployment.

Note

HPE Aruba Networking security-first, AI-powered networking was formerly known as the Edge Services Platform (ESP).

OBJECTIVES

After completing this chapter, you should be able to:

- ✓ Compose security policy
- ✓ Use HPE Aruba Networking ClearPass Policy Manager Platform & Cloud Auth
- ✓ Provision Onboarding and PKI
- ✓ Design for client traffic security
- ✓ Implement profiling & NAC
- ✓ Describe HPE Aruba Networking SASE

HPE Aruba Networking Security

HPE Aruba Networking security-first, AI-powered networking architecture is the industry's first cloud-native platform designed to automate, unify, and secure the EDGE.

HPE Aruba Networking design tenets

HPE Aruba Networking security-first, AI-powered networking uses:

- AIOps to reveal and resolve issues before they can impact users and critical business applications.
- ZTS (Zero Trust Security) to modern protection methods and address the complexity of IoT attack vectors.
- Unified Infrastructure to bring HPE Aruba Networking Central cloud management and troubleshooting onto a single platform.

We've already covered HPE Aruba Networking Central cloud management. Let's dive into AIOps and ZTS.

HPE Aruba Networking AIOps: Overview

Consider how HPE Aruba Networking includes AIOps and the essential building blocks needed to deploy a self-optimizing, AI-powered network and how these building blocks work together to solve today's evolving networking challenges reliably.

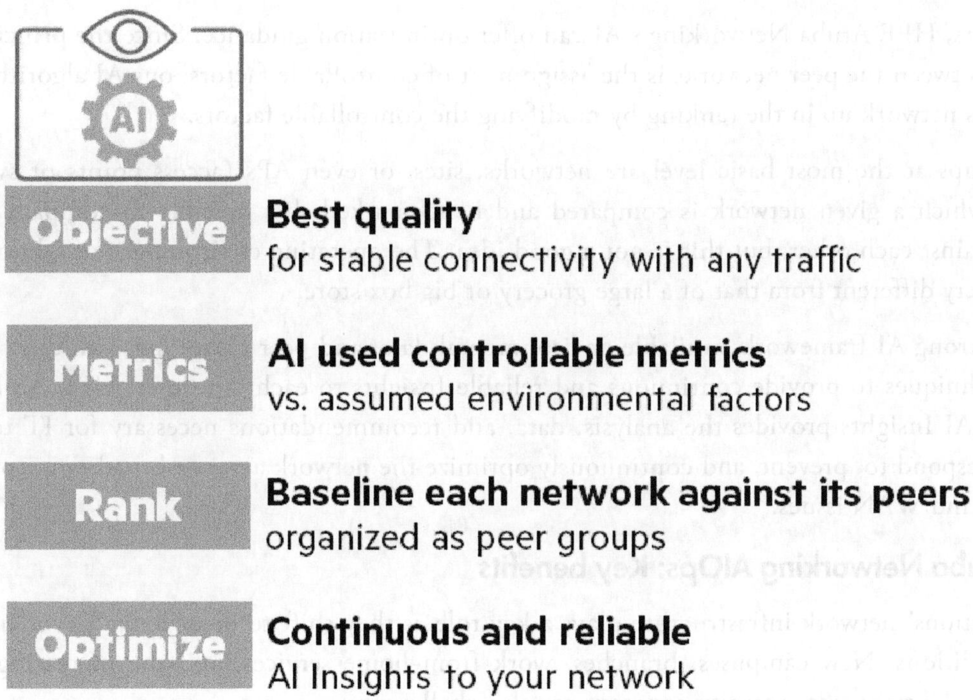

A self-optimizing network aims to provide its users with stable connectivity and the ability to satisfy traffic demands to experience the highest quality of service. In practice, this perceived user experience is very difficult to measure as no two sites are identical – from an environmental and user behavior perspective. HPE Aruba Networking uses measurable metrics to build, train, and evaluate our AIOps, such as throughput, latency, or resource efficiency.

A metric or numeric representation of how well the network is doing and how it compares to other networks is chosen per use case. HPE Aruba Networking also identifies the factors that influence the chosen metric and separates them into two types: those that the AI used is allowed to control (controllable factors) and those that are assumed as given (environmental factors). To optimize radio transmission parameters for the access points in a Wi-Fi network, the controllable factors include RF channel bandwidth and RF transmission power levels. Environmental factors include the spacing between adjacent access points in the deployment. Equipped with the definitions of a measurable objective function and its controllable and environmental factors, HPE Aruba Networking instructs the network to continuously collect this data and store it in the cloud.

To evaluate performance fairly, AIOps baseline each network against its peers, a set of other networks with the same environmental factors belonging to the same customer or others. For example, if a network ranks in the 10th percentile of its peers, that is, it performs worse than 90%

of its peers, HPE Aruba Networking's AI can offer optimization guidance. Since the principal difference between the peer networks is the assignment of controllable factors, our AI algorithms can move this network up in the ranking by modifying the controllable factors.

Peer groups at the most basic level are networks, sites, or even APs (access points or switches) against which a given network is compared and benchmarked. It's possible to compare all networks against each other, but this is not a good idea. The operating environment of a local coffee shop is very different from that of a large grocery or big box store.

With a strong AI framework, available environmental data, and peer clustering, you can combine these techniques to provide continuous and reliable Insights to each customer. HPE Aruba Networking AI Insights provides the analysis, data, and recommendations necessary for IT teams to rapidly respond to, prevent, and continuously optimize the network across a broad range of wired, wireless, and WAN issues.

HPE Aruba Networking AIOps: Key benefits

Organizations' network infrastructure plays a key role as they shift to meet today's evolving business conditions. New campuses, branches, work-from-home services, and the increasing use of IoT strain IT resources. Security concerns are also challenging.

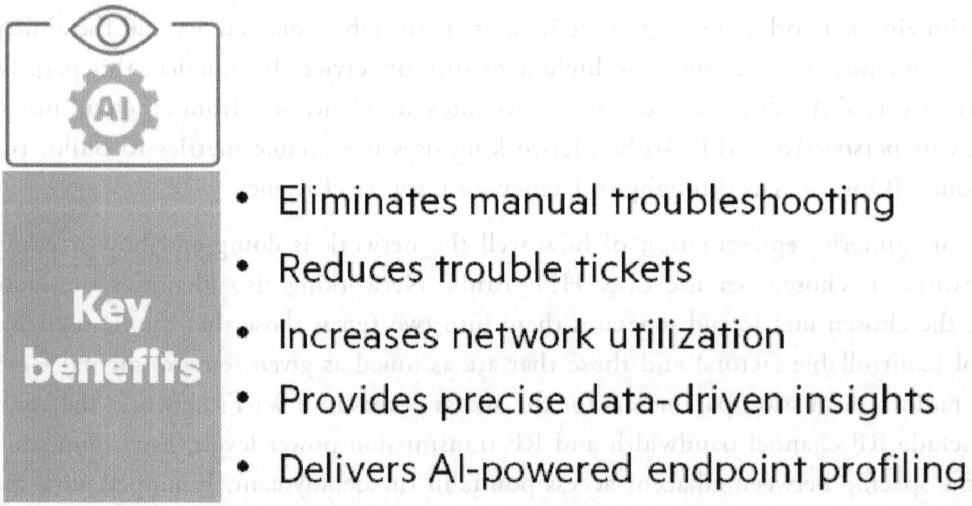

When you consider everything that IT is tasked with, it's clear that a new network operations approach is needed. One that makes life easier for network admins by reducing trouble tickets and ensuring SLAs deliver world-class user experiences. A way to quickly resolve connectivity problems using automated root cause analysis, precise recommendations, and remediation is needed so that IT can focus on driving business value instead of dealing with mundane tasks. One where AI predicts and preempts issues before they happen.

Key benefits

- Eliminates manual troubleshooting and reduces mean time to resolution by up to 90%.
- Reduces trouble tickets by identifying issues before they impact the business.
- Increases network utilization by 25% per site and peer-based configuration optimization.
- Provides precise data- driven insights and recommendations with greater than 95% accuracy.
- **Delivers AI-powered endpoint profiling (IoT and compute) with up to 99% accuracy.**

HPE Aruba Networking Zero Trust Security components

HPE Aruba Networking takes a multi-faceted approach to security. HPE Aruba Networking ZTS solutions monitor and control every user and device that connects to the network, not just once but continuously.

A broad range of authentication options, including Authentication, Authorization, and Accounting (AAA) and non-AAA options, ensure that only authorized devices receive network access. HPE Aruba Networking defines a user's network experience not by location or access type but by role. HPE Aruba Networking's "one role, one network" approach means that users receive consistent and appropriate access rights based on their role, whether connected from a wired device at their desks, a wireless device in a conference room, or a remote device from home.

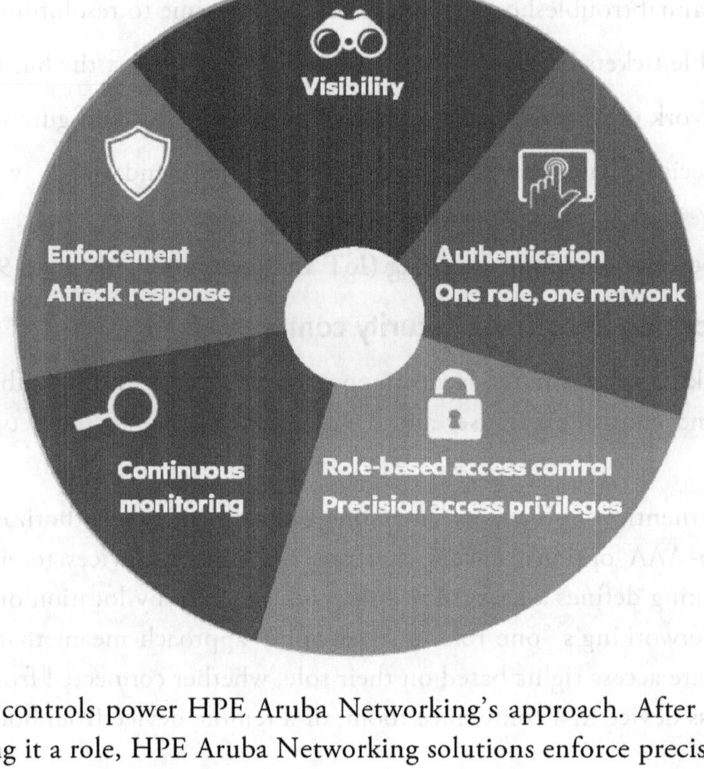

Role-based access controls power HPE Aruba Networking's approach. After authenticating every client and assigning it a role, HPE Aruba Networking solutions enforce precise access privileges to micro-segment every client. While based on identity, these privileges also take context into account. For example, HPE Aruba Networking solutions can enforce different privileges for a user on a laptop and the same user on a smartphone in keeping with the different security profiles of those devices.

Because HPE Aruba Networking understands that a ZTS solution can never authorize a client and then forget about it, connected devices are continuously monitored. HPE Aruba Networking solutions collect real-time telemetry information to detect intrusions and other threats. With HPE Aruba Networking ZTS, companies can monitor a broad range of clients, traffic, and behaviors because not only do HPE Aruba Networking solutions have embedded capabilities, but they also integrate with more than 150 third-party products.

Every threat detected through monitoring requires a response. HPE Aruba Networking solutions make those responses faster and more powerful. When a triggering event occurs, such as detecting a malware-infected device, HPE Aruba Networking Solutions can take various actions, including locking an infected device out of the network or placing it in a quarantine segment for remediation.

Finally, because you can only secure something if you know it's there and understand what it is, HPE Aruba Networking grants customers a high level of insight into the devices in their environment. HPE Aruba Networking solutions discover and profile all the devices connected to a network. Advanced, custom fingerprinting technologies identify each client's precise type and operating system.

HPE Aruba Networking solutions that support ZTS

Here, you can see which HPE Aruba Networking solutions support which pieces of ZTS. ClearPass Policy Manager (CPPM) provides some visibility benefits, while ClearPass Device Insight (CPDI) provides complete visibility for a company's heterogeneous network.

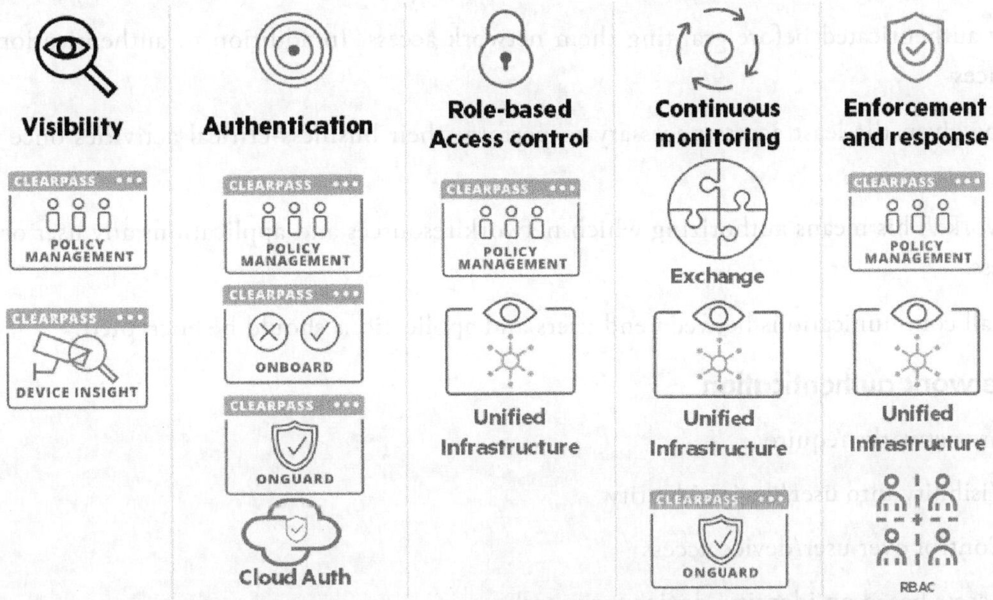

CPPM provides a broad range of sophisticated authentication methods, and HPE Aruba Networking Client Onboarding Software and Client Onboarding Software add further capabilities. Building on its authentication capabilities, CPPM delivers role- and context-based access control based on highly flexible policies. The HPE Aruba Networking Unified Infrastructure receives the instructions from CPPM and enforces them with a powerful firewall.

The Unified Infrastructure also helps to deliver continuous monitoring. Companies can extend monitoring with various third-party solutions that are part of ClearPass Exchange. Based on monitoring results, CPPM can make new decisions to respond to threats, and the Unified Infrastructure can enforce those decisions and mitigate threats.

Additional details about 360 Security Exchange: https://www.hpe.com/us/en/networking/technology- partner- programs.html/program/360-security-exchange

HPE Aruba Networking campus security policy concepts

An HPE Aruba Networking campus implementation focuses on the following:

- Zero-trust
- User role-based authentication
- Campus-wide micro-segmentation
- Consistent policy enforcement at every point in the network

Basic principles of good networking dictate that, when possible, all devices and users should be identified and

properly authenticated before granting them network access. In addition to authentication, users and devices

should be given the least access necessary to perform their business-critical activities once they're on

the network. This means authorizing which network resources and applications any user or device can access.

Finally, all communications between end users and applications should be encrypted.

Why network authentication

Zero trust networks require

- Visibility into user/device identity
- Control over user/device access
- Access based on identity, device type, and more

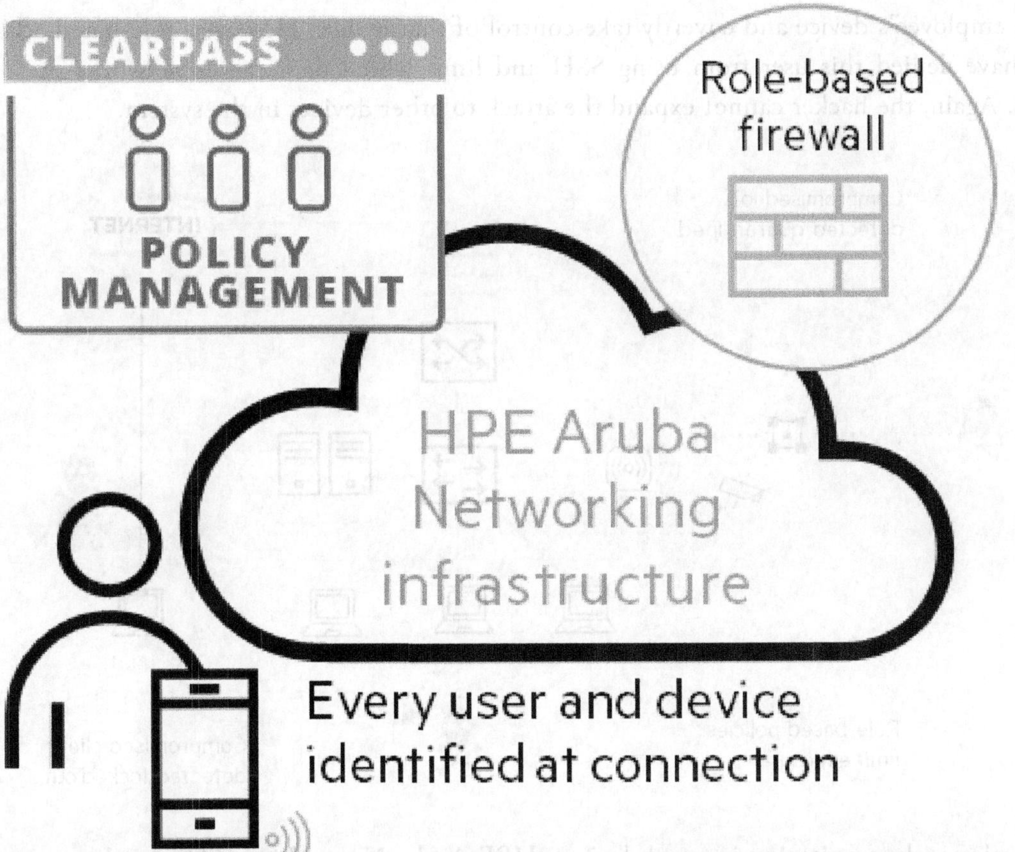

You have learned about HPE Aruba Networking security-first, AI-powered networking platform. Zero-trust networking is one of the elements for delivering security in zero-trust environments. A key component of this security strategy involves network authentication. A company gains visibility into its environment by identifying every user and device at the internal perimeter. Such visibility lays the foundation for control. First, the company can take action to control access to the internal network, ensuring that only authorized users and devices can connect. Next, the company can enhance this control by assigning the appropriate access controls to each client based on knowledge about who is using the client and other factors such as device type and location. This is called network access control or NAC.

HPE Aruba Networking ZTS in action

Now, look at more examples of how HPE Aruba Networkingg ZTS can secure a network against internal threat vectors. Perhaps hackers succeed in compromising an IoT device. Because HPE Aruba Networking ZTS enabled the company to tightly control the types of traffic that IoT devices can send, the hackers will find it much more difficult to use the devices to reach sensitive resources. Similarly, perhaps a bad actor uses phishing and social engineering to install malware on an employee's device and covertly take control of that device. However, role-based access controls have denied this user from using SSH and limit which data center networks the user can access. Again, the hacker cannot expand the attack to other devices in the system.

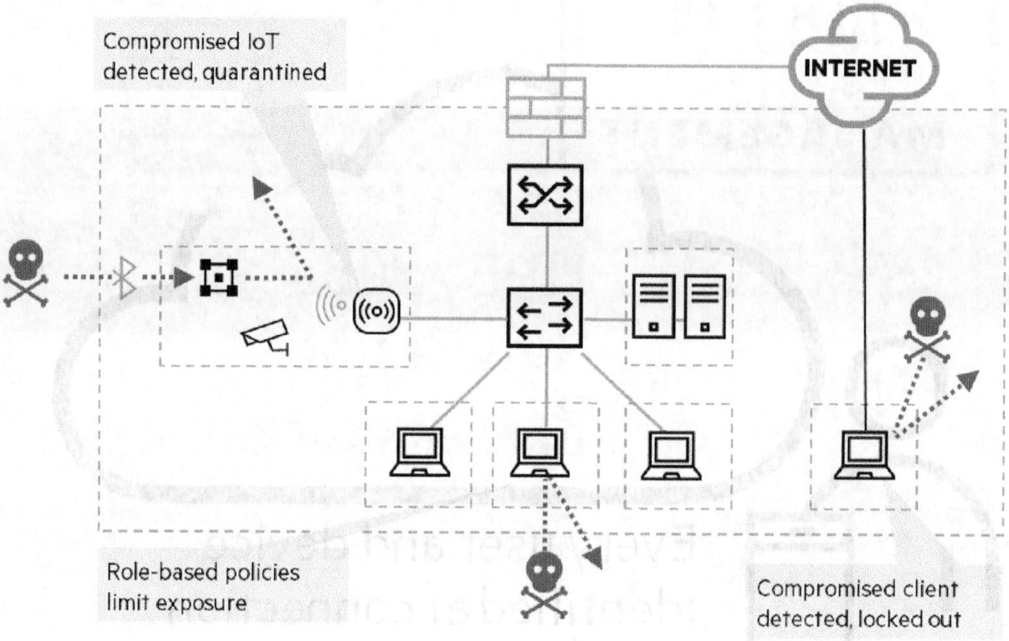

By continuously monitoring internal devices, HPE Aruba Networking ZTS can detect suspicious behavior, such as an IoT device that suddenly begins trying to establish SSH sessions with a broad range of IP addresses. The network can then take action to lock out the device behaving suspiciously.

This is the substance of Zero Trust Security. The network never lets a device in and then forgets about it. By restricting each device to the lowest level of appropriate resources and continuously monitoring whether devices continue to merit that access, ZTS minimizes real damage to the network.

ClearPass Policy Manager & Cloud Auth

This section covers using ClearPass and Cloud Authentication with HPE Aruba Networking Central to support security.

ClearPass Policy Manager overview

ClearPass Policy Manager, part of the HPE Aruba Networking 360 Secure Fabric, provides role- and device-based secure network access control for IoT, BYOD, corporate devices, employees, contractors, and guests across any multivendor wired, wireless, and VPN infrastructure.

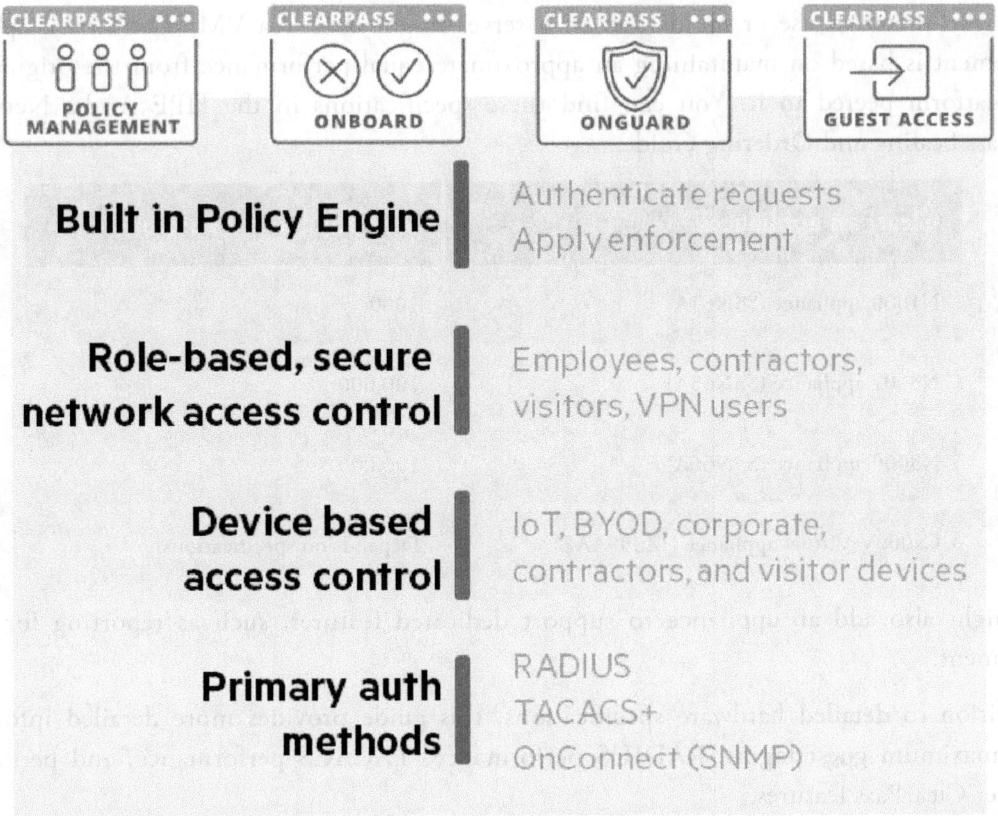

Policy Manager provides:

- Role-based secure network access control for employees, contractors, visitors, and VPN users.
- Device-based access control for IoT, BYOD, corporate, contractors, and visitor devices.
- Built-in Policy Engine that matches an authentication request before applying an Enforcement Policy

The primary authentication methods used by CPPM are:

- RADIUS
- TACACS+
- OnConnect (SNMP)

ClearPass Policy Manager appliance options

You can get ClearPass as an appliance or a VM. The choice of the appliance depends primarily on the performance and load required. The N1000 supports a maximum of 1000 concurrent sessions, the N3000 and N3001 support a maximum of 100,000 concurrent sessions.

Those who prefer to use or build their own server might choose a VM. The virtual appliances requirement is based on maintaining an approximate equal performance from the original hardware platform peered to it. You can find these specifications in the HPE Aruba Networking ClearPass Scaling and Ordering Guide.

ClearPass appliances	Max concurrent sessions
N1000 appliance (S3N67A)	1000
N3001 appliance (S3N65A)	100,000
N3000 appliance (S3N66A)	100,000
Cx000V Virtual appliance (JZ399AAE)	Depends on specifications

You might also add an appliance to support dedicated features, such as reporting for a large deployment.

In addition to detailed hardware specifications, this guide provides more detailed information about maximum guest logins, RADIUS performance, TACACS performance, and performance for other ClearPass features.

The VM works with several hypervisors, including VMware ESXi and Microsoft Hyper-V. Refer to the current HPE Aruba Networking ClearPass Scaling and Ordering Guide for exact versions.

ClearPass Policy Manager 6.12 licensing

With ClearPass 6.11, licenses and appliances are uncoupled. You can propose appliances based on the performance requirements and then choose the licenses the customer needs more granularly. Each hardware or virtual appliance requires a perpetual platform activation key (PAK), which does not expire. (The PAK for a hardware appliance is registered with no change licenses and is generated based on the serial number.)

You add other licenses to the ClearPass appliance to support the desired functions.

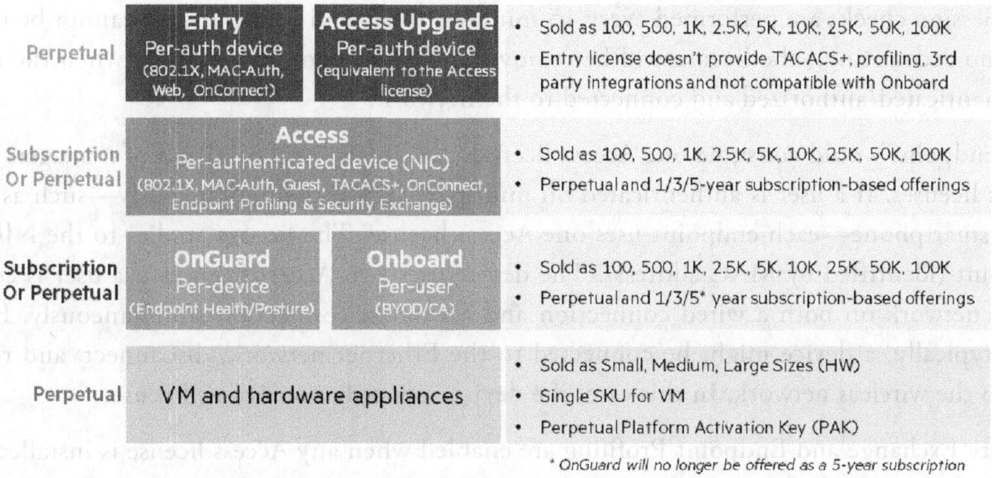

Access licenses support ClearPass's Policy Manager and Guest functions. ClearPass requires one license for each endpoint concurrently authenticated or authorized, no matter how the device authenticates. If the customer uses ClearPass OnGuard Software, you should propose one license per endpoint to which ClearPass deploys an agent. For example, if the ClearPass OnGuard Software persistent agent is to be installed (persistent agent) or used (dissolvable agent) on five endpoints within 24 hours, five ClearPass OnGuard Software licenses are required.

ClearPass Onboard Software licenses are calculated per user. A user can onboard several devices but still only requires one license.

Each license is sold in increments ranging from the granular 100 to 10,000. You can also offer them as perpetual licenses or one-year, three-year, or five-year subscriptions. One exception applies: ClearPass OnGuard Software licenses do not support a five-year subscription.

Access licenses:

Every time a user connects a device to the network and is authenticated and authorized by ClearPass, the device uses one Access license. An employee laptop could authenticate with 802.1X, an IoT device could authenticate with MAC-Auth, and a guest could log in through the captive portal. Access license consumption is based upon a concurrent session per endpoint model. The 360 Security Exchange, standard endpoint visibility (also known as device fingerprinting), and TACACS+ are enabled when at least 100 Access licenses are installed but do not consume any Access licenses when used.

A session is considered active when an endpoint is authenticated/authorized and actively connected to the network. When a new endpoint establishes a session, an Access license is removed from the pool. An Access license is returned to the pool when the endpoint discontinues the session. Session checks are performed every 15 minutes. If the end of the session cannot be identified (e.g., no accounting), the license will be removed from the pool for 24 hours when the endpoint is authenticated/authorized and connected to the network.

Each endpoint would consume one Access license. TACACS+ authorization of managers also uses Access licenses. If a user is authenticated on multiple endpoints simultaneously—such as a laptop and a smartphone—each endpoint uses one Access license. The license applies to the NIC on the endpoint (identified by MAC address). The device uses two Access licenses if a user is connected to the network on both a wired connection and a wireless connection simultaneously. However, more typically, a device might be connected to the Ethernet network, disconnect, and then connect to the wireless network. In this case, the device uses only one Access license.

Security Exchange and Endpoint Profiling are enabled when any Access license is installed but not restricted to licensed capacity limits.

ClearPass OnGuard Software Licenses:
ClearPass OnGuard Software license consumption is based on a per-endpoint model and applies to persistent and dissolvable agents. For example, within 24 hours, ClearPass installs persistent agents on two endpoints and uses dissolvable agents on three endpoints; five ClearPass OnGuard Software licenses are required. Plan one license for each endpoint the customer wants to apply posture assessments.

ClearPass Onboard Software Licenses:

ClearPass Onboard Software license consumption beginning with ClearPass 6.11 is based upon an active certificate per-user model. For example, if a user has four devices with an active certificate each, only one ClearPass Onboard Software license is required. If, over time, three out of the four devices are retired and their associated certificates revoked, the fourth active device certificate will still keep the ClearPass Onboard Software license associated with the user. The intentional onboarding of large numbers of devices by a single user to avoid purchasing ClearPass Onboard Software licenses violates the End-User Software License Agreement.

License Aggregation and Subscriptions:
Customers can install multiple licenses of the same type on a standalone appliance or cluster for increased licensed capacity—for example, 100+100+100 = 300. Subscription licenses (including support) are tracked on licensed capacity and term. If a customer installs two subscription licenses six months apart, the total term will be one and a half years; however, in the last six months, the licensed capacity will drop to the remaining valid subscription.

Entry Licenses:

Entry licenses (introduced in Policy Manager 6.8) include features designed for Basic NAC use cases.

- 802.1X
- MAC-Authentication
- Web Based User Registration and Authentication (captive portal authentication)
- Multi-Factor Authentication (MFA)
- OnConnect
- System APIs

Entry licenses are perpetual licenses from 100 to 100K concurrent endpoints. Multiple Entry licenses can be

installed on a standalone appliance or cluster for increased licensed capacity, for example, 100+100+100 = 300.

Entry licenses can be upgraded to Access using the Access Upgrade licenses.

Upgrading to Access would add the following features:

- TACACS+ for Device Administration (e.g., Router, Switch, Controller, Firewall, etc.)
- 360 Security Exchange (previously ClearPass Exchange)
- Standard endpoint visibility (also known as device fingerprinting)

ClearPass Policy Manager sizing & scaling

Features required when creating a BOM:

- Access, Onboard, and/or OnGuard licenses
- Concurrent # of endpoints/users
- # of users that can onboard devices
- #of concurrent users with OnGuard installed, connecting to the network daily

ClearPass Policy Manager licensed features fall into three categories: Entry or Access, ClearPass Onboard Software, and ClearPass OnGuard Software. Depending on the features required, you may use one or all of them in a given environment.

Entry Licenses

Entry licenses (introduced in Policy Manager 6.8) include features for basic NAC use cases, including 802.1X, MAC-Authentication, Web Based User Registration and Authentication (captive portal authentication), and Multi-Factor.

Authentication (MFA), OnConnect, and System APIs. Entry license consumption is based on concurrent sessions for authenticated/authorized endpoints. For example, ten users authenticating would consume ten licenses.

Access Licenses

Access licenses (introduced in Policy Manager 6.11) include features for Secure NAC use cases, including 802.1X, MAC-Authentication, Web Based User Registration and Authentication (captive portal authentication), Multi-Factor Authentication (MFA), TACACS+, OnConnect, System APIs, 360 Security Exchange (previously ClearPass Exchange), and standard endpoint visibility. Access license consumption is based on concurrent sessions for authenticated/authorized endpoints. For example, ten authenticated users would consume ten licenses. The 360 Security Exchange, standard endpoint visibility, and TACACS+ are enabled with just a minimum of 100 Access licenses for unlimited use.

ClearPass Onboard Software Licenses

ClearPass Onboard Software licenses enable automated provisioning, including creating unique device identity certificates. Onboard license consumption is based on the number of users with at least one active certificate each.

For example, five users with two device certificates each would consume five licenses. ClearPass Onboard Software can be installed on top of Entry or Access licenses. ClearPass OnGuard Software licenses enable health-check capabilities to ensure endpoint posture compliance and network safeguards before devices connect.

ClearPass OnGuard Software

ClearPass OnGuard Software license consumption is based on the number of endpoints it is installed in and consumed for Twenty-four hours upon first health check. For example, five persistent and five dissolvable agents perform a health check, and ten licenses are consumed for 24 hours. OnGuard can only be installed on top of Access licenses.

Step 2 – Calculate the Number of Licenses Needed

Entry and Access licenses

To better understand how Entry or Access licenses are consumed, consider the following use case:

- 6,000 endpoints using a mix of username/password and certificate (Corp) based authentication
- 2,000 IoT endpoints that use MAC address authentication

- 1,000 guest endpoints that use self-registration or social logins

Given that all authentication methods are equal, we have 9,000 endpoints to consider. However, we are only

concerned with the maximum number of users concurrently authenticated/authorized.

If we believe that ALL the endpoints will be concurrently connected to the network, we will need to license for 9,000.

However, given the network data available (e.g., DHCP max pool size and lease times, max firewall session usage, etc.), we can determine that only 6,000 endpoints are ever concurrently connected to the network, and therefore we only need 6,000 Entry or Access licenses.

ClearPass Onboard Software licenses

To better understand how ClearPass Onboard Software licenses are consumed, consider the following use case:

Five hundred users can onboard their devices as per the BYOD policy. It is estimated that these 500 users have a total of One thousand five hundred devices based upon network usage.

We just need 500 Onboard licenses since the user count is all we care about. The 1,500 devices do not matter. Additionally, ClearPass Onboard Software licenses are consumed regardless of whether the device is connected to the network. The license is consumed as long as at least one active certificate is associated with a given user.

ClearPass OnGuard Software licenses

To better understand how ClearPass OnGuard Software licenses are consumed, consider the following use case:

2,500 endpoints that have ClearPass OnGuard Software installed and connect daily

ClearPass OnGuard Software is going to be installed on 2,500 endpoints, and these endpoints will be connecting at least once daily; we will need 2,500 OnGuard licenses.

License summary

Given the above examples, we will need the following licenses assuming perpetual licenses:

- 6,000 Access Licenses: JZ404AAE (Includes 5,000) + JZ402AAE (Includes 1,000)
- 500 ClearPass Onboard Software Licenses: JZ437AAE (Includes 500)
- 2,500 ClearPass OnGuard Software Licenses: JZ475AAE (Includes 2,500)

ClearPass Policy Manager sizing & scaling: specifications

Load/performance features required when creating a BOM: burst rate for any given application that will be used and max concurrency			
Appliance specifications	**N3000 appliance (S3N66A)**	**N3001 Appliance (S3N65A)**	**N1000 Appliance (S3N67A**
Hardware model	HPE ProLiant DL325 Gen11 Server	HPE ProLiant DL325 Gen11 Server	Advantech FWA-2012
CPU	AMD EPYC 9124 with 16 cores (32 threads)	AMD EPYC 9124 with 16 cores (32 threads)	Intel Atom® C3758R with 8 cores (8 threads)
Memory	128 GB	128 GB	16 GB
Hard drive storage	(3) SSD (960 GB SATA) RAID-5 controller	(3) SSD (960 GB SATA) RAID-5 controller	(1) SSD (1 TB) M.2
Out-of-band management	HPE iLO Advanced	HPE iLO Advanced	n/a
Network interfaces	4 x 1GbE	4 x 10/25GbE SPF28	4 x 1GbE

When sizing the appliances, use two types of metrics; burst rate for any given application that will be used and max concurrency. These values differ depending on the appliance. ClearPass hardware appliances have the specifications shown in the table above.

ClearPass Policy Manager sizing & scaling: performance

Load/performance features required when creating a BOM: burst rate for any given application that will be used and max concurrency			
	N3000 Appliance (S3N66A)	**N3001 Appliance (S3N65A)**	**N1000 Appliance (S3N67A**
Performance summary			

Load/performance features required when creating a BOM: burst rate for any given application that will be used and max concurrency			
	N3000 Appliance (S3N66A)	N3001 Appliance (S3N65A)	N1000 Appliance (S3N67A
ClearPass version	6.12.0		
Max concurrent sessions	100,000	100,000	1,000
Guest web login	287 auths/second	287 auths/second	62 auths/second
Guest self-registration	74 requests/second	74 requests/second	7 requests/second
OnBoard	8 requests/second	8 requests/second	1 request/second
OnGuard	68 client/second	68 client/second	12 client/second
EAP-TLS with AD	608 auths/second	608 auths/second	83 auths/second
TACACS with AD	404 requests/second	404 requests/second	81 requests/second

The ClearPass hardware appliances have the following performance characteristics shown in the table. Virtual appliance performance for the same models (e.g., C3000V) may differ based on hypervisor load (e.g., amount of vCPU oversubscription), CPU architectures, and storage types (e.g., SAN versus Internal).

Note that if customers size their virtual deployments for ClearPass, they can expect comparative results with an equivalent hardware deployment of Policy Manager.

ClearPass Policy Manager sizing & scaling: virtual appliances

Load/performance features required when creating a BOM:			
	C1000V Virtual Appliance (JZ399AAE)	C2000V Virtual Appliance (JZ399AAE)	C3000V Virtual Appliance (JZ399AAE)
General virtual appliance specifications			
CPU	8 reserved virtual CPUs (Recommended to have Passmark® score of 3000 or higher)	8 reserved virtual CPUs (Recommended to have Passmark score of 9600 or higher)	24 reserved virtual CPUs (Recommended to have Passmark score of 16,000 or higher)
Memory	8 GB	16 GB	64 GB
Min. network interfaces	2x 1GbE virtual switched ports	2x 1GbE virtual switched ports	2x 1GbE virtual switched ports
Hard drive storage	1000 GB disk space required	1000 GB disk space required	1800 GB disk space required
Minimum functional IOP rating (40 to 60 read/write profile for 4K random read/write	75	105	350

Identify the load/performance Required:

When sizing the appliances, we are concerned with two types of values, burst rate for any given application that will be used and max concurrency. These values differ depending on the appliance. ClearPass hardware appliances have the following specifications.

Virtual appliances require similar resource specifications to ensure a consistent ClearPass experience regardless of hardware or virtual appliance. Beginning with ClearPass 6.7, the virtual appliance SKU is a single SKU that can be used for all virtual variants of virtual appliances. The virtual variant merely adds the letter 'V' to the end of the model number.

ClearPass Policy Manager is currently supported (as of version 6.11) on the following hypervisors and virtual private clouds:

- VMware vSphere Hypervisor (ESXi) 6.7 and 7.0
- Microsoft Hyper-V Server 2016, Microsoft Hyper-V Server 2019, or Windows Server 2016 with Hyper-V
- KVM on CentOS 8, Ubuntu 18.04 LTS
- Amazon Web Services (please refer to the Deploying Policy Manager in AWS for supported instance types)
- Microsoft Azure

ClearPass Policy Manager hardware and virtual appliances

The HPE Aruba Networking ClearPass solution uses different models for different-sized deployments based on the number of authentication requests. This can include both hardware and virtual appliances.

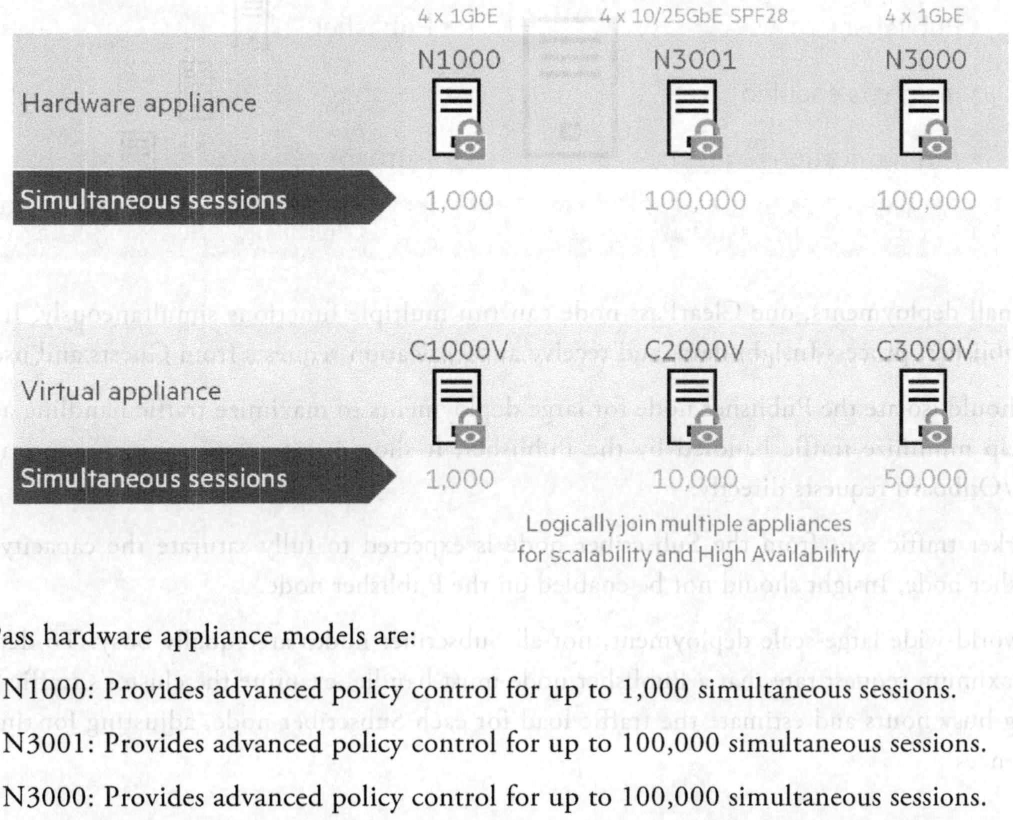

ClearPass hardware appliance models are:

- N1000: Provides advanced policy control for up to 1,000 simultaneous sessions.
- N3001: Provides advanced policy control for up to 100,000 simultaneous sessions.
- N3000: Provides advanced policy control for up to 100,000 simultaneous sessions.

The virtual appliances are distributed as a single image, which becomes CLABV (for testing purposes), C1000V, C2000V, or C3000V, depending on the allocated virtual hardware.

When demand exceeds the capacity of a single instance, or you need a High Availability deployment, you can logically join multiple instances to handle this workload, logically joining them into a cluster. Let's check it out.

CPPM Publisher node sizing

The Publisher node must be sized appropriately for three key reasons:

- It handles database write operations from all subscribers simultaneously.
- It must be able to process remote work directed to it during guest account creation and onboarding.

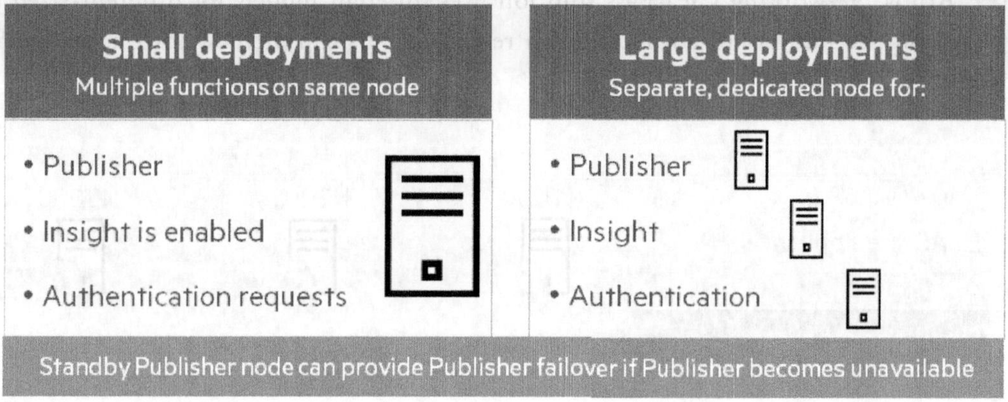

For small deployments, one ClearPass node can run multiple functions simultaneously. It can be the Publisher, process Insight data, and receive authentication requests from Guests and users.

You should isolate the Publisher node for large deployments to maximize traffic handling abilities. To help minimize traffic handled by the Publisher, it should not receive any authentication or Guest/Onboard requests directly.

If worker traffic sent from the Subscriber node is expected to fully saturate the capacity of the Publisher node, Insight should not be enabled on the Publisher node.

In a world-wide large-scale deployment, not all Subscriber nodes are equally busy. To determine the maximum request rate that a Publisher node must handle, examine the cluster's traffic pattern during busy hours and estimate the traffic load for each Subscriber node, adjusting for time zone differences.

CPPM Subscriber node guidelines

- Guest and onboard clients should use nearest Subscriber
- NAD/NAS-Subscriber RTT less than 600ms
- Subscribers should be workers for:
  Auth requests: RADIUS, TACACS+, Web-Auth
  OCSP requests
  Static content delivery: Images, CSS, JavaScript
- Ensure EAP-TLS performs localhost OSCP checks
- Validate sufficient available bandwidth to Publisher
  Avoid sending "worker traffic" to the Publisher

Guidelines for Subscriber node deployment are as follows:

Guests and ClearPass Onboard Software clients should be directed to the nearest Subscriber node. From the client's point of view, the internal API call to the Publisher is handled transparently. The best response time for static resources is obtained if the server is nearby. The recommended delay between a NAD/NAS and a Subscriber node should be less than 600ms.

Subscriber nodes should be used as workers that process the following:

- Authentication requests (for example, RADIUS, TACACS+, Web-Auth).
- Online Certificate Status Protocol (OCSP) requests.
- Static content delivery (for example, images, CSS, JavaScript).

Avoid sending "worker traffic" to the Publisher, as the Publisher services API requests from Subscribers, handles the resulting database writes, and generates replication changes to send back to the Subscribers.

If ClearPass Onboard Software is used, ensure that the EAP-TLS authentication method in Policy Manager is configured to perform localhost OCSP (Online Certificate Status Protocol) checks.

For reliable operation of each Subscriber, ensure sufficient bandwidth is available for communications with the Publisher. For basic authentication operations, there is no specific requirement for high bandwidth. However, the number of round- trips to complete an EAP authentication could cause delays for the end user.

ClearPass Policy Manager Insight: report configuration

Enabling a cluster node as Insight Primary allows other Insight-enabled nodes to subscribe to this node's Insight report configuration. You then define reports and alerts on the Insight Primary node.

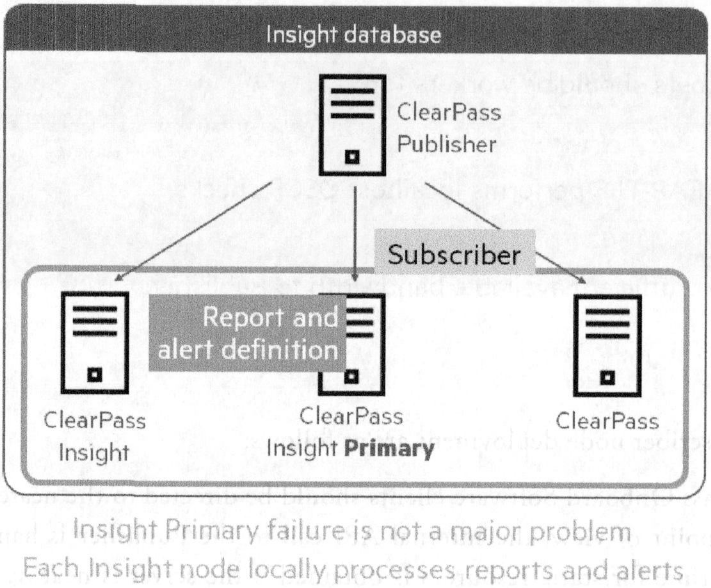

If the Insight Primary node fails, reports are still produced. This is because all nodes in the cluster send a copy of their NetEvents data to all ClearPass Insight-enabled nodes.

Note

Insight Primary was previously known as Insight Master.

ClearPass Policy Manager Insight: placement considerations

You get great resilience by enabling ClearPass Insight on multiple nodes. But be careful! For every node where ClearPass Insight is enabled, all the other cluster nodes subscribe through NetEvents to send data to the ClearPass Insight database. Remember, the data sent to this database can be extremely high. If you use Insight to process authentication requests in a cluster, be especially careful where you enable ClearPass Insight.

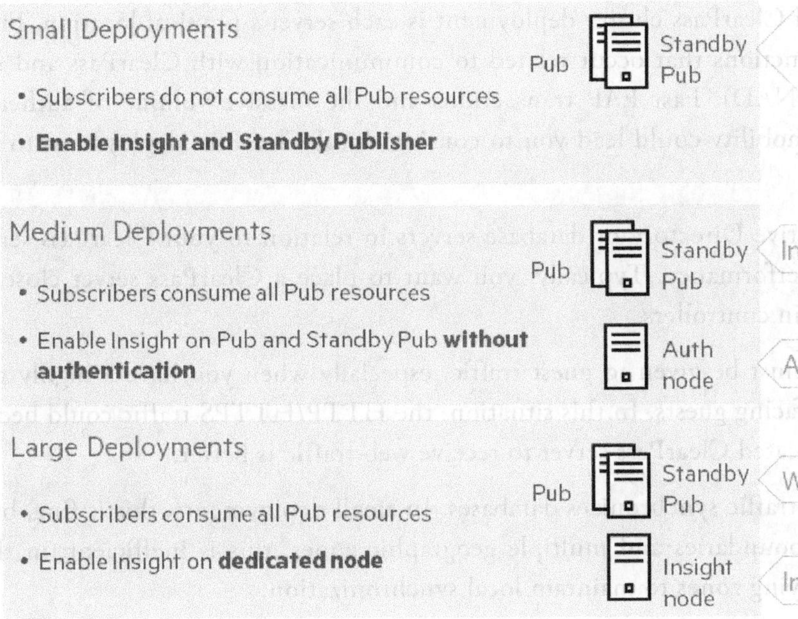

Monitor resource consumption when enabling Insight, especially when deploying Insight on a Publisher node! The resulting impact will dictate which deployment type you should use:

For small deployments, where Subscriber traffic does not consume all Publisher resources, enable ClearPass Insight on the Publisher and Standby Publisher for high availability.

For medium deployments, if Subscriber traffic consumes all Publisher resources, you can enable ClearPass Insight on the Publisher and Standby Publisher for high availability. Just be sure to move the authentication process to other nodes.

For Large deployments, place ClearPass Insight on its dedicated node. This removes a lot of processing and I/O from the Publisher, allowing it to handle the maximum amount of worker traffic.

CPPM server placement: centralized or distributed

ClearPass server placement is also an important aspect to consider. You can centralize them at a Data Center or distribute them at each site. To decide properly, you should answer the following questions.

- How mobile are the clients?
- What is the EAP latency between NAD and the ClearPass servers?
- How close are the Authentication sources from your ClearPass servers?
- What is the Guest traffic density?
- How intense is the Inter-cluster communication?

An important aspect of ClearPass cluster deployment is each server's physical location. First, consider all the various functions that occur related to communication with ClearPass and the Network Access Devices (NAD). Fast EAP transactions and the massive volume of authentication generated by wireless mobility could lead you to consider local ClearPass deployment to improve overall performance.

The location of the Active Directory or database servers in relation to your ClearPass servers can affect authentication performance. Typically, you want to place a ClearPass server close to your Active Directory domain controllers.

Special consideration must be given to guest traffic, especially when you have a highly transient site with many public-facing guests. In this situation, the HTTP/HTTPS traffic could become significant. Placing a dedicated ClearPass server to receive web traffic is better.

ClearPass Inter-cluster traffic synchronizes databases. In small deployments, this is fine, but when a cluster spans WAN boundaries and multiple geographic zones, this is inefficient; in this case, you can consider deploying zones to maintain local synchronization.

CPPM features

ClearPass Policy Manager (CPPM) is the central policy engine for controlling all access forms. It supports multiple authentication methods, including integration with Microsoft Active Directory, a portal for guests, certificate authentication, and multi-factor authentication. CPPM allows network device admins to define custom policies based on a user's identity or device type. It also enables them to define policies based on context, such as a user's location or the time of day. And CPPM integrates with third-party solutions to enhance security orchestration.

Authentication with CPPM can be enhanced with ClearPass Onboard Software and ClearPass OnGuard Software.

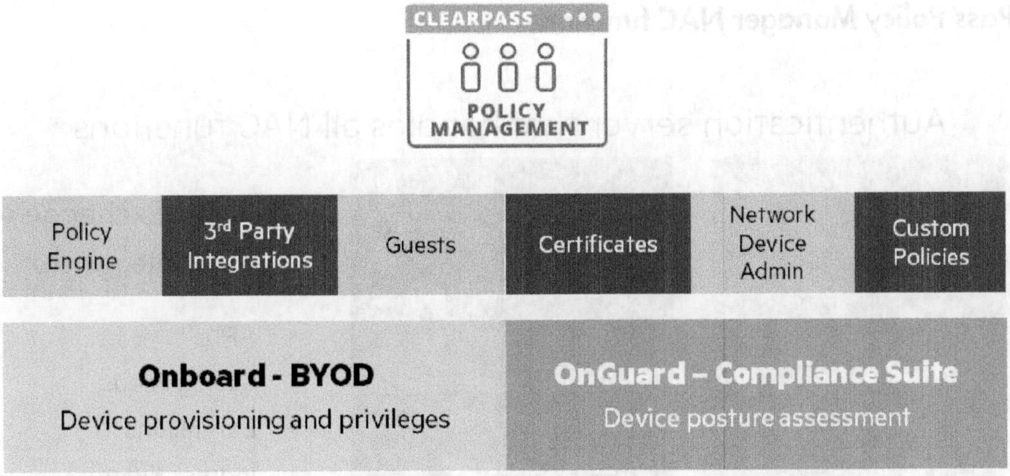

ClearPass Onboard Software simplifies the process of provisioning devices with the settings and certificates required for secure authentication. Onboard provides a self-service app that enables users to provision their own devices rather than making this another task for IT.

ClearPass OnGuard Software scans endpoints to determine their security posture. For example, it can verify firewall settings and ensure a device's patches are up-to-date. ClearPass OnGuard Software communicates information about an endpoint's security posture to CPPM. CPPM can then use that information to affect its access control decisions, such as quarantining the device.

ClearPass Policy Manager NAC functions

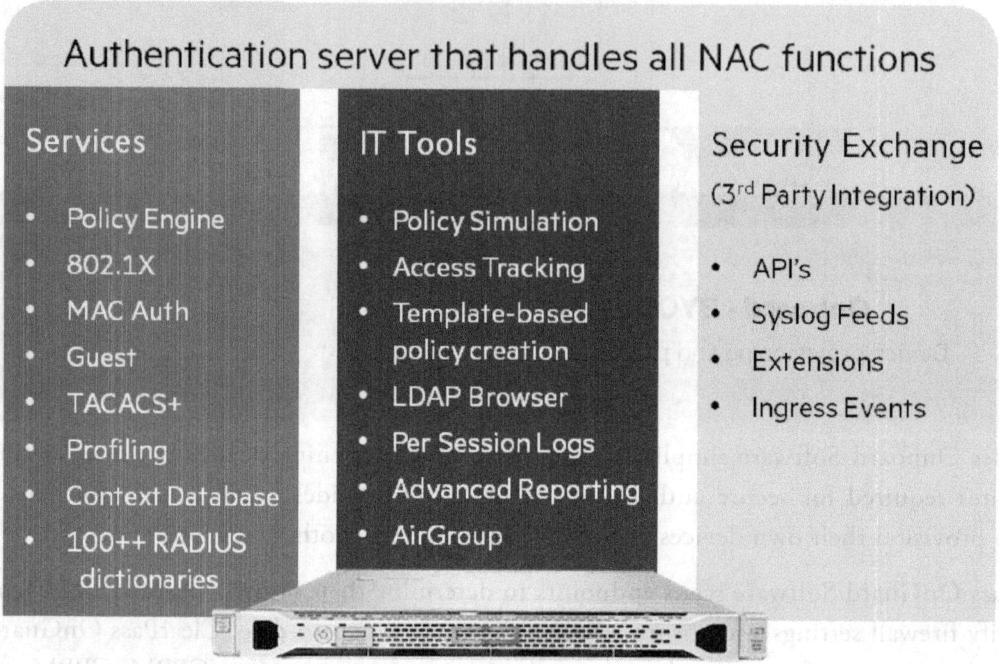

HPE Aruba Networking ClearPass is an authentication server that can easily handle all network access control (NAC) functions.

- It is a decision-making point when clients connect to the network and go trough authentication and authorization.
- One of the great things about ClearPass is that we include many features within the base software set.
- HPE Aruba Networking has had TACACS and RADIUS servers for over ten years, so they're tried and tested.
- The policy engine, basic guest, and device registration tools work without requiring dedicated appliances or extra licenses.

The visibility dashboard, policy creation templates profiling, and troubleshooting services have been there since we've been shipping the product.

- The exchange interoperability components (API and syslog messaging) are included.
- Customers are free to use built-in integration templates or create their workflows.

- And, if they want, customers can try any of the three expandable applications because we include 25 enterprise licenses with the base platform.
- ClearPass is available as an appliance or a virtual machine installation.

Introduction to CPPM services

When CPPM processes a request, it follows a set pattern. Starting with service selection, CPPM matches the request to the proper service. Then, CPPM gathers information about the request, such as the users' credentials, account attributes, endpoint profile context, the client's health status, and the source and type of request. The service also gathers profile information about the device making the request.

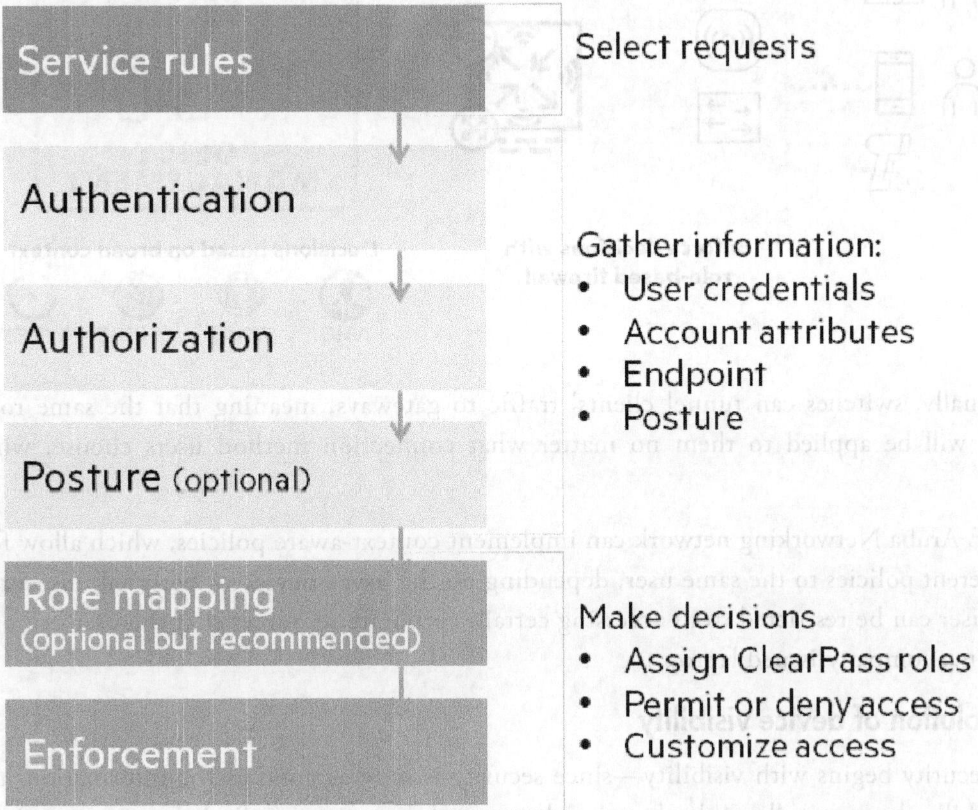

Finally, with all the data collected, the service roles mapping and enforcement processes can make decisions and reply to requests. The service implements the organization's desired access policies by evaluating the data collected and matching it to rules built into the enforcement policies. Services are custom-built to support various scenarios that the organization wants to support. In this way, CPPM can process a large number of different requests from different networks.

Role-Based Access Control

To implement role-based access control, customers rely on HPE Aruba Networking infrastructure to take action based on the roles that HPE Aruba Networking ClearPass Policy Managerassigns. HPE Aruba Networking APs and gateways provide a role-based Policy Enforcement Firewall (PEF), which filters all devices' traffic based on the CPPM-assigned role. For example, the PEF might allow a user with the HR role to access a subnet of servers that hold academic records while preventing other users on the same network from such access. The PEF might further restrict IoT devices from sending and receiving traffic from a very limited set of IP addresses.

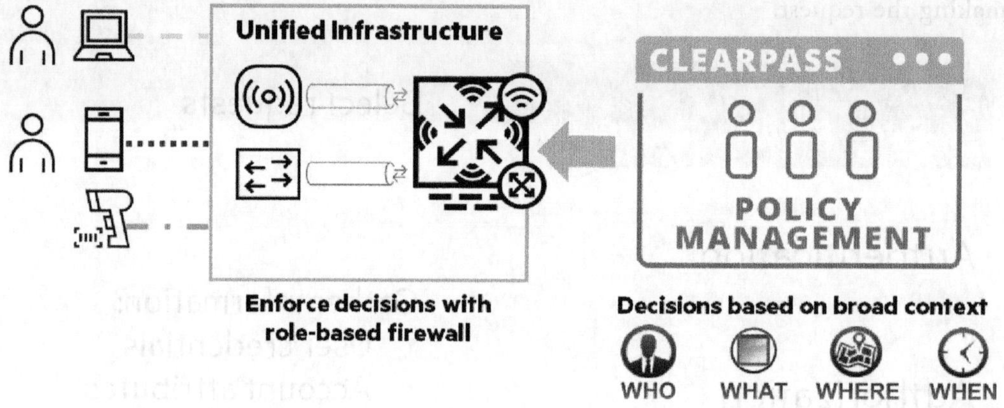

Additionally, switches can tunnel clients' traffic to gateways, meaning that the same role-based policies will be applied to them no matter what connection method users choose, wireless or wired.

An HPE Aruba Networking network can implement context-aware policies, which allow for applying different policies to the same user, depending on the user's device. A personal, insecure smartphone user can be restricted from accessing certain corporate resources that the user can otherwise access on a company-issued laptop.

The evolution of device visibility

Good security begins with visibility—since security is only as good as the information it's based on. Visibility has naturally evolved as time has passed, but you may find that some customers are still relying on antiquated visibility.

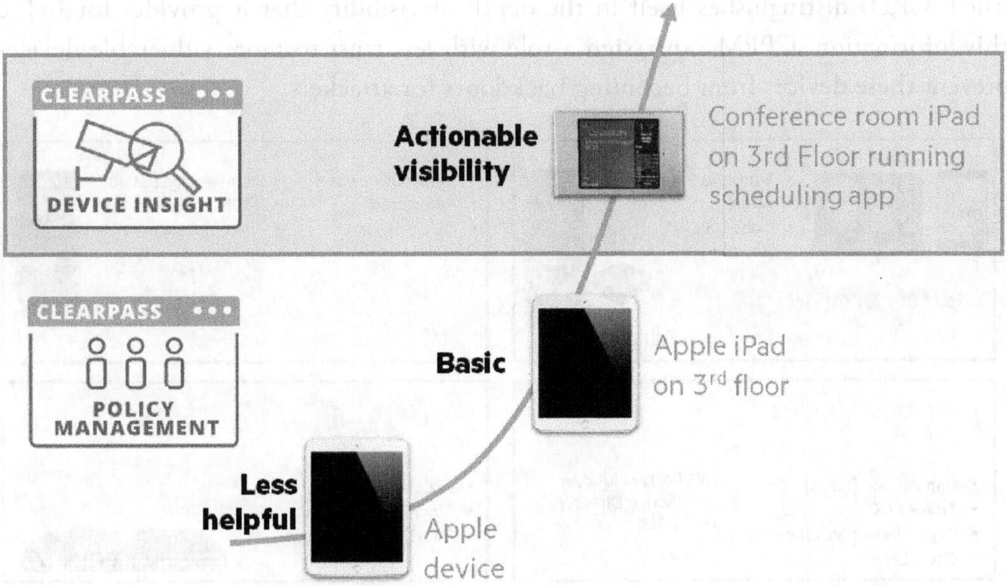

This level of visibility is often based on MAC address and DHCP. It might, for example, only inform you that an Apple iPad has joined the network. With the rising sophistication of exploits, this level of visibility is simply not enough. Going one step up, many customers are still utilizing only basic visibility. This class of visibility relies on technologies such as SNMP and is based on network scans. It offers more specific information about a device. For example, you might see that an Apple iPad has joined the network on the 3rd floor. CPPM provides this level of basic device visibility.

With ClearPass Device Insight, HPE Aruba Networking leads customers to actionable visibility. This level of visibility is based on a client's behavior and provides granular insights about each device on the network. For example, you would see that an Apple iPad has joined the network in the conference room on the 3rd floor and is now running a scheduling app. Actionable visibility provides the kind of detailed information that organizations need to protect themselves from today's attacks.

Sources of usable context

CPPM has several sources of information on which to draw for "context" for its policies. First, CPPM gains information from the devices themselves. CPPM can implement device profiling or integrate with CPDI for more advanced profiling. Device profiling yields information such as the device OS and category. For example, HPE Aruba Networking Solutions can distinguish between a laptop and a smartphone, and they can determine the hardware vendor for the smartphone. As you learned, CPDI distinguishes itself in the depth of visibility that it provides for IoT devices. Using this information, CPPM can assign a role with less trust to more vulnerable devices, helping to prevent these devices from becoming backdoors for attackers.

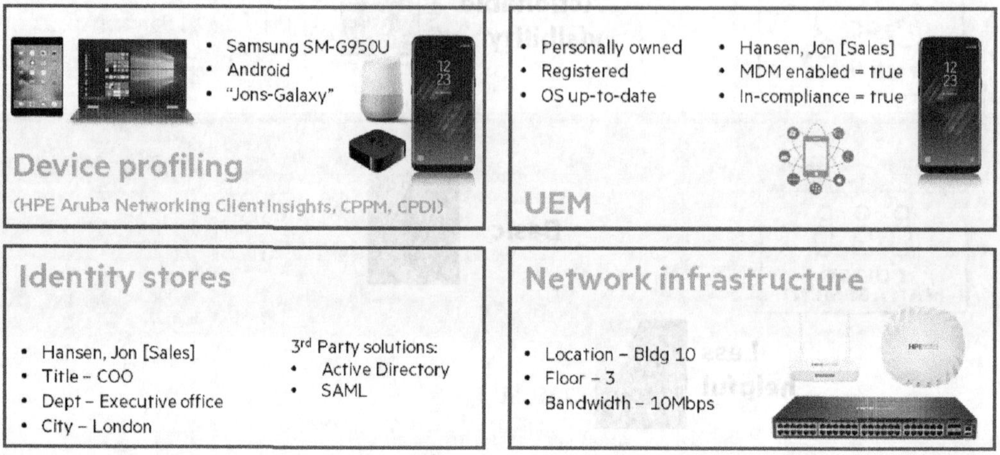

CPPM can also integrate with unified endpoint management (UEM) tools to gather context. With this data, CPPM can determine if a device's OS is up-to-date or if the device complies with policies. If the OS is not up-to-date or the device is not compliant, the HPE Aruba Networking network can react accordingly to protect the network from this vulnerability.

CPPM also uses identity stores to obtain a more user-centric context. For example, an employee's name, title, department, and location will be available through sources like Active Directory, which allows CPPM to limit users' access precisely based on their actual needs.

Finally, CPPM can also obtain contextual information from network infrastructure devices. Based on the information the network access device (NAD) includes in its RADIUS request, CPPM can determine the client's location and apply location-based policies. It can also use accounting information from NADs to track clients' bandwidth consumption. By integrating ClearPass with HPE Aruba Networking Central, you can leverage HPE Aruba Networking Central Client Insights for device profiling information.

Introducing the HPE Aruba Networking 360 Security Exchange

Here, you see just some of the many third-party solutions that can integrate with CPPM. These solutions span many categories, including security and authentication. Companies can even let guests log in with credentials from popular social media platforms.

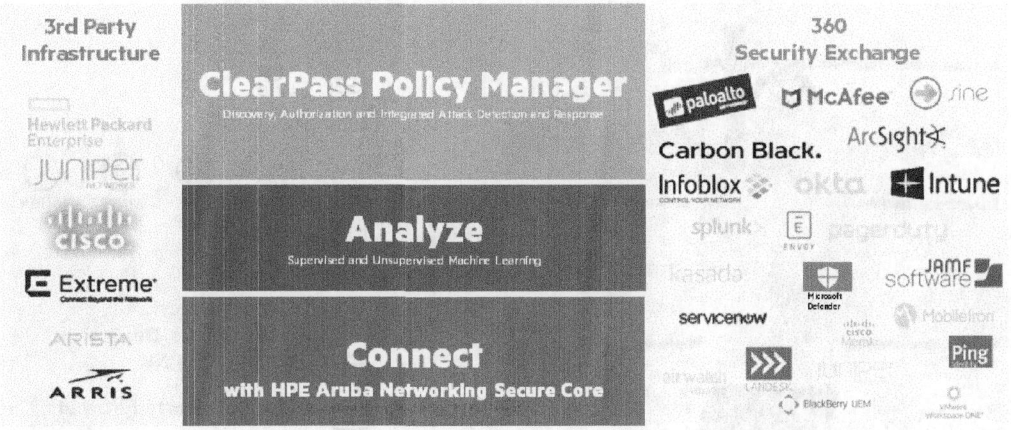

By integrating with logging solutions, CPPM helps to create audit trails and support companies' organization-wide security efforts. Integration with Enterprise Mobility Management (EMM) solutions can help to provide important context, as you learned earlier, and CPPM integrates with many commonly used ones.

Check these links for more info:

https://www.hpe.com/us/en/networking/technology-partner-programs.html

https://arubanetworking.hpe.com/techdocs/NAC/clearpass/integrations

The benefits of integration

By integrating with third-party solutions, CPPM can assign policies with much greater accuracy based on who the user is, what device is in use, where the client is located, and when the device is trying to access network resources. With third-party integrations, CPPM can derive users' identity primarily from apps that users regularly interact with, like Active Directory. CPPM relies on other third-party solutions, especially mobility management solutions, and its integrations to further derive user identity and all the other relevant information about that client.

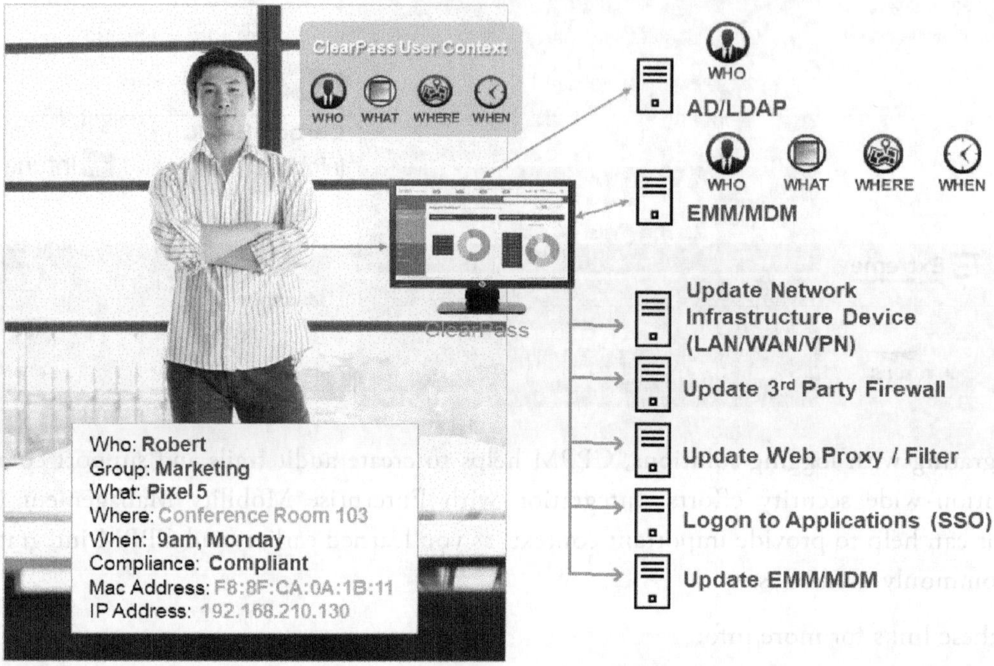

Consider a company using AD for its directory service and MobileIron MDM for security and management. CPPM can utilize integrations with these solutions to build a profile of the user, who, for example, tries to access files stored on a corporate server using a mobile device. Rather than being an anonymous user on a mobile device, CPPM now identifies this user as Robert from Marketing. He uses his Pixel 3 in Conference Room 103 at 9 AM Monday to access corporate resources. The device is determined to be compliant with security policies. Its MAC and IP address are naturally known as well. Now, CPPM can update enforcement devices, the firewall, and the web proxy to apply appropriate policies for Robert. CPPM can adjust Robert's ability to log on to applications and even update MobileIron to ensure Robert and his device are still secured effectively.

ClearPass Policy Manager guest access

ClearPass supports visitors and employee personal devices, allowing customized access in the form of:

- captive port pages
- branding and logos
- sponsor-based approvals
- social login
- self-registration

Onboarding has integration with MDM/UEM solutions and comprehensive integration with 3rd party vendors using the HPE Aruba Networking 360 Security Exchange Program.

Review of cloud authentication & policy for HPE Aruba Networking ClearPass

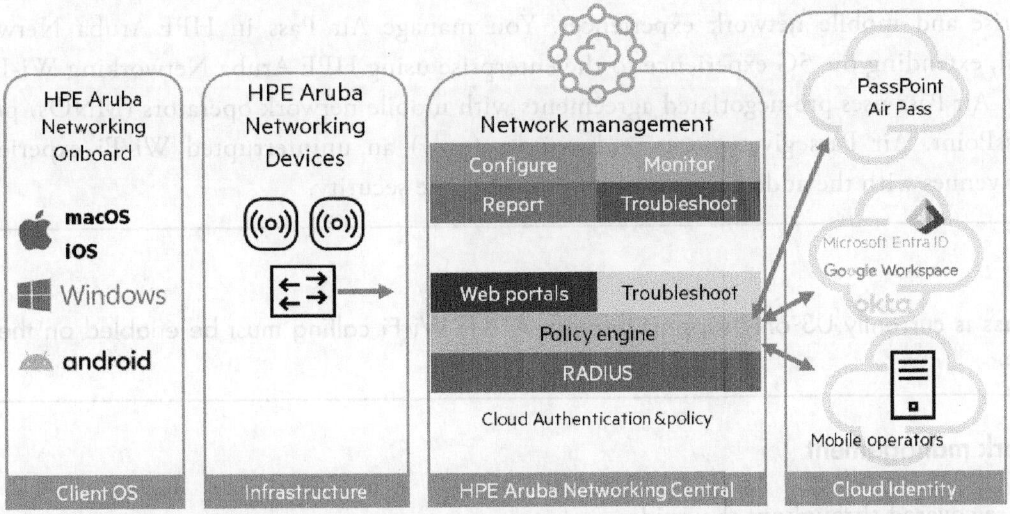

Cloud Authentication and policy creation in HPE Aruba Networking ClearPass involve four main components:

Cloud Identity

HPE Aruba Networking ClearPass enables end users to connect to Wi-Fi networks securely and automatically. The application integrates with a company's existing cloud identity store, such as Google Workspace or Azure Active Directory, to authenticate user information and assign appropriate access. You get frictionless Wi-Fi connectivity for known and new users, visibility into traffic patterns, and fine-grained policies to control access.

End-User device onboarding

Easily configure employee devices for seamless wired and wireless connections. Managed through HPE Aruba Networking ClearPass, end users are authenticated through the company's cloud identity store with an enrollment link provided on HPE Aruba Networking ClearPass.

Using company credentials to log in, the user is redirected to the identity store for authentication. Configure client devices using HPE Aruba Networking ClearPass Onboard Software, a client app supported on macOS, Windows, iOS, and Android operating systems. With the Enterprise PassPoint Profile installed on the device, the user automatically connects with appropriate access based on your HPE Aruba Networking Central configuration when the user walks into the network range. The client app provides automatic renewals: no additional onboarding steps and upkeep from the end user. You can change and update policies at any time.

Automated access for guests and visitors

HPE Aruba Networking Air Pass is the industry's first seamless cellular roaming solution to unify enterprise and mobile network experiences. You manage Air Pass in HPE Aruba Networking Central, extending the 5G experience to the enterprise using HPE Aruba Networking Wi-Fi technology. Air Pass uses pre-negotiated agreements with mobile network operators (MNOs) powered by PassPoint. Air Pass gives guests and visitors (only) an uninterrupted Wi-Fi experience in indoor venues with the added benefit of enterprise-grade security.

Note

Air Pass is currently US-only support through AT&T. Wi-Fi calling must be enabled on the user's device.

Network management

They are covered throughout the guide.

Cloud Authentication key benefits and features

It is critical to apply consistent security controls and ensure users have seamless access to apps and data in the office, at home, and on the go. HPE Aruba Networking Central simplifies this IT process with cloud-based authentication, extending its ability to deliver a single point of visibility and control over all network infrastructure and related security services. Featuring an easy-to-use interface and dashboard, HPE Aruba Networking Central makes it easy to onboard new clients and monitor and troubleshoot user connection issues.

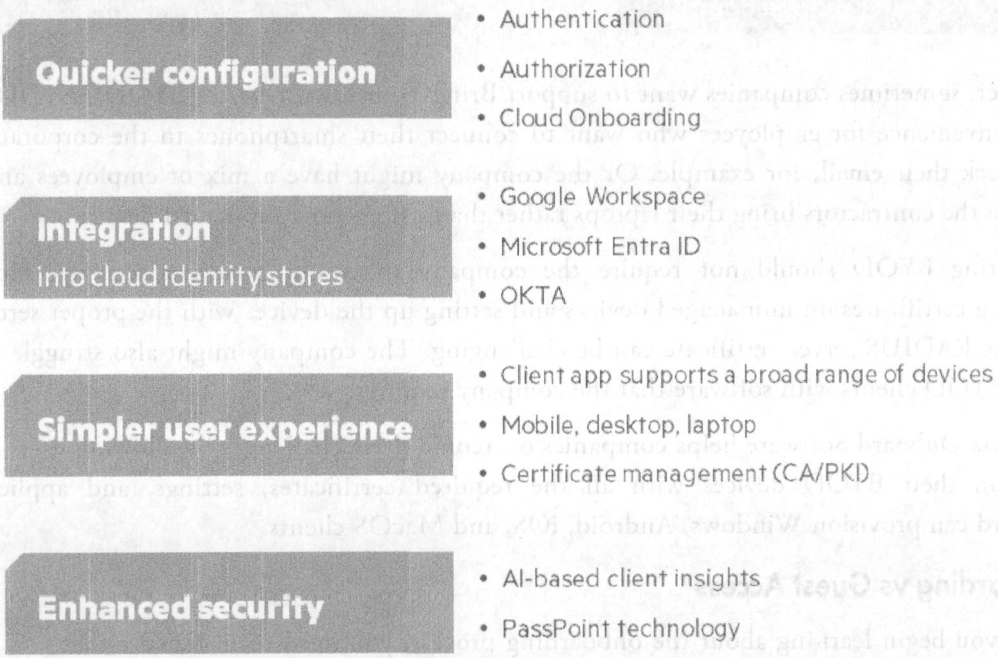

You configure fine-grained policies in HPE Aruba Networking Central. Thus, when end users authenticate, they are authorized for appropriate network access. To help enhance security, you get MAC-based authentications and AI-based client insights, which capture and profile all connected devices. This gives you continuous, in-depth visibility into the connected client behavior, enabling a stronger security posture than using MAC addresses alone.

Onboarding and PKI

Let's take a quick look at how Onboarding and PKI affect design.

Why use ClearPass Onboard Software?

You have learned how to support certificate-based authentication for users on devices joined to a Windows domain. Those are managed devices, and it is generally easier to implement certificate-based authentication for managed devices. Admins have the tools to provide the devices with the proper certificates.

However, sometimes companies want to support Bring Your Own Device (BYOD). BYOD might be a convenience for employees who want to connect their smartphones to the corporate LAN and check their email, for example. Or the company might have a mix of employees and contractors; the contractors bring their laptops rather than using company-issued devices.

Supporting BYOD should not require the company to compromise on security. However, installing certificates on unmanaged devices and setting up the devices with the proper settings to trust the RADIUS server certificate can be challenging. The company might also struggle to provision BYOD clients with software that the company requires.

ClearPass Onboard Software helps companies overcome these challenges by allowing users to self-provision their BYOD devices with all the required certificates, settings, and applications. Onboard can provision Windows, Android, iOS, and MacOS clients.

Onboarding vs Guest Access

Before you begin learning about the onboarding process, you need to make sure that you understand the difference between users who require onboarding and guests. In both cases, you might have users bringing smart devices and trying to get access to the network.

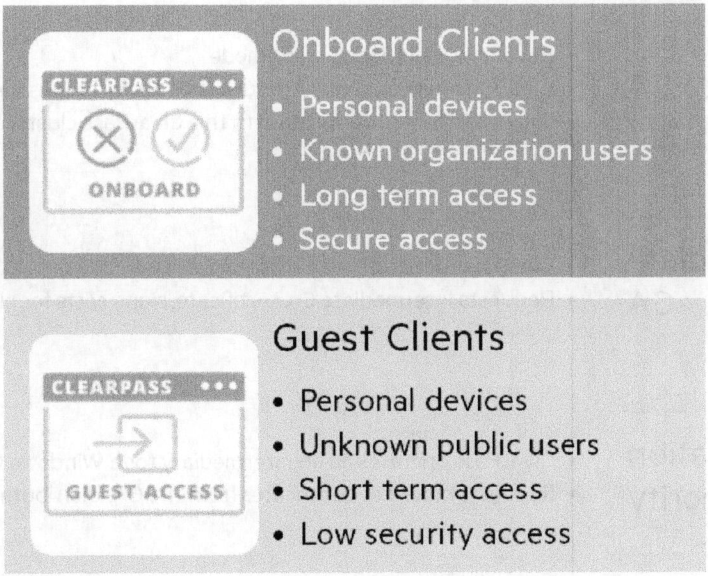

However, in onboarding, the typical users are people known to the organization, like company employees, university staff, and faculty members. These individuals are bringing their personal devices to use on the secure network. Guests are typically users who are unknown to the organization and do not exist in a corporate identity store like Active Directory. For example, guest users include guests who check in and out of hotels, customers visiting a company's office, or visitors at a hospital. These users may bring their own personal devices. Another difference is that in onboarding, you are trying to get employee mobile devices connected to a secure network like an 802.1X network; you want to control what devices they can use on this network. A guest network typically uses an open SSID, and anyone can connect.

Options for onboard certificate authorities

You can configure ClearPass Onboard Software as one of three types of certificate authorities. In every case, ClearPass Onboard Software is the authority that issues the certificate to the client. The difference is the location of the certificate's root.

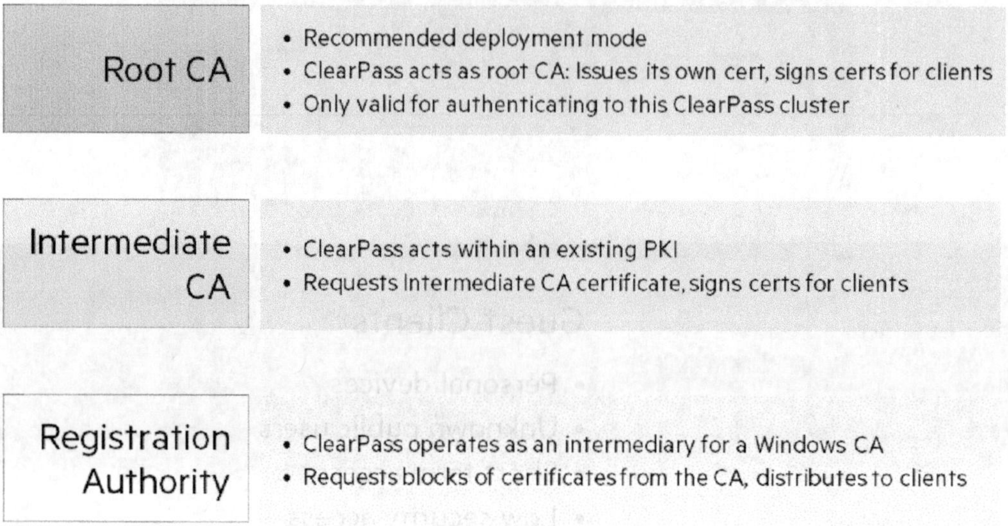

Root CA

You can set up ClearPass Onboard Software as its own root CA. It issues its own certificate and also issues certificates to clients. It is generally recommended that you use this straightforward option. Many companies want to limit how users can use the certificates deployed to them through the onboarding process. They should be able to use the certificates to connect to the network but not gain other privileges. When ClearPass Onboard Software issues the certificates from its own PKI, they are only valid for authenticating to that ClearPass cluster.

Intermediate CA

However, if a company wants to use its existing PKI to issue certificates, you can set up ClearPass Onboard Software as an intermediate CA within that PKI. ClearPass Onboard Software requests an Intermediate CA certificate, which the root CA must sign. Onboard can then sign and issue certificates to clients. These certificates will be trusted throughout the existing domain.

Registration Authority

ClearPass Onboard Software can also act as a registration server for a Windows CA. ClearPass Onboard Software requests a block of certificates from the Windows CA. That CA already signs these certificates. Onboard then distributes the certificates to clients.

One question is whether you should put a public certificate on ClearPass Onboard Software. The answer is no because of the contained nature of ClearPass Onboard Software and the fact that Onboard loads all the certificate trust client stores; the client will be able to validate the self-issued certificates from ClearPass.

Securing Client Traffic

Turn your attention now to how a proper design should consider securing client traffic.

Need for secure wireless communications

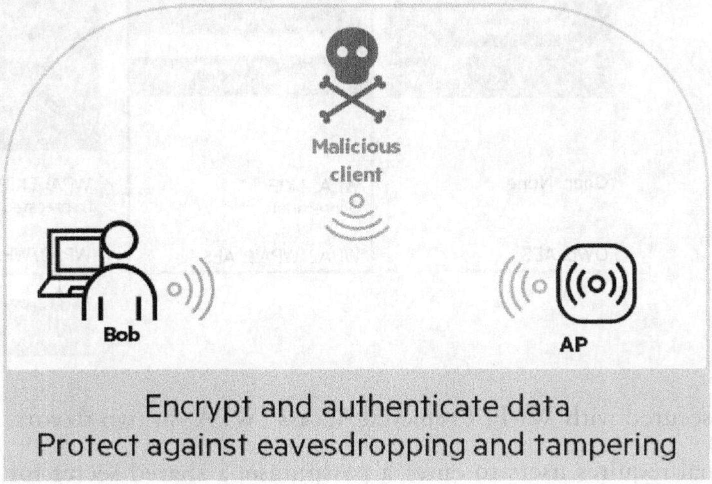

Because it is trivial for anyone in the area to collect wireless communications, you must take extra measures to secure all communications with encryption and authentication. Otherwise, you are susceptible to hacks and attacks.

Review WLAN security options

Every WLAN operates in one of three broad categories, regardless of the additional authentication types you layer.

802.11 WLANs can be open: users need not submit authentication credentials to connect. The client requests a connection, and the AP permits that connection, which is very simple to implement. Traditionally, open networks offer no encryption. But today, Opportunistic Wireless Encryption (OWE) can now provide Advanced Encryption Standard (AES) encryption for open networks, as you soon learn.

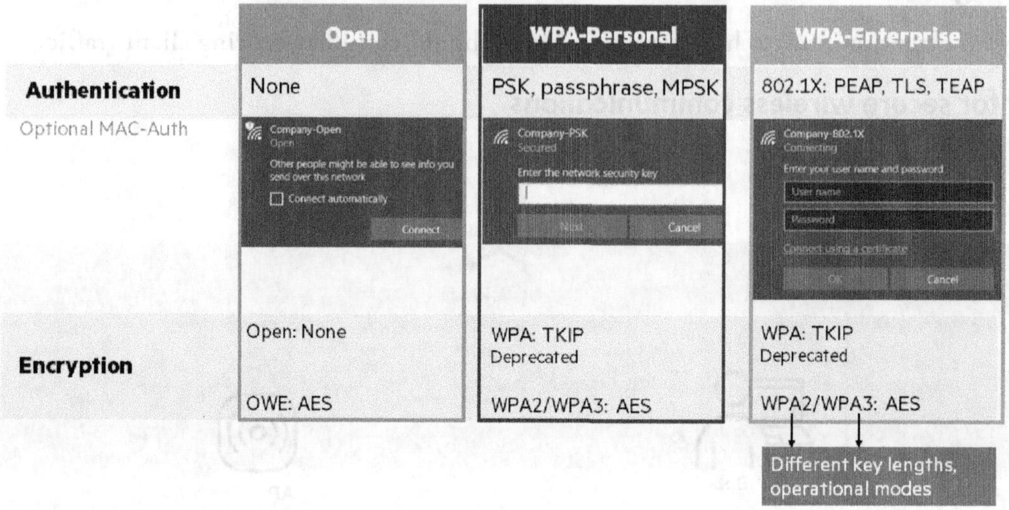

Some WLANs are secured with Wi-Fi Protected Access (WPA) in two flavors.

- WPA-Personal requires users to enter a passphrase, a shared secret for all authorized users to connect. The passphrase is sometimes called a pre-shared key (PSK), used as the foundation to encrypt and authenticate traffic.
- WPA-Enterprise is recommended due to its superior 802.1X-based authentication. With 802.1X, a client authenticates to a backend authentication server using individual credentials: a username, password, and a digital certificate.

Both the Personal and Enterprise flavors do authentication and encryption. WPA comes in three versions, and the version determines the supported encryption algorithm. WPA originally used Temporal Key Integrity Protocol (TKIP), based on RC4 encryption. WPA and TKIP are now deprecated.

WPA2 and WPA3 both support AES, but WPA3 can impose different requirements for the AES key length and mode of operation. Note that later versions can also have backward compatibility with earlier versions. For example, a WPA2 WLAN could permit clients to use TKIP or AES; however, it should only allow AES for security.

The WPA versions have additional differences, as you will soon learn. For now, just understand that the user experience for all three versions is similar. The user enters a passphrase to gain access to a WPA-Personal WLAN but submits individual credentials to gain access to a WPA-Enterprise WLAN. Finally, note that you can implement MAC-auth with any of these options. However, the main focus of this module is the base security options.

Note

Wired Equivalent Privacy (WEP) was a pre-WPA security standard. Static WEP uses a passphrase to authenticate users and derive a key for encrypting data. Dynamic WEP used 802.1X for authentication and a key established during the 802.1X exchange to encrypt data. However, both forms of WEP are proven to be insecure and should not be used in today's networks.

HPE Aruba Networking MPSK ClearPass

Some companies have headless devices that do not support 802.1X authentication, so these devices must connect to WLANs in Personal mode. However, the devices might also lack support for WPA3, so WPA2-Personal, with all its flaws, is the only option for them. HPE Aruba Networking provides multi Preshared keys (mPSK) to enhance security for these devices.

A detailed explanation of mPSK configuration is beyond the scope of this guide. The figure shows at a high level how the solution typically works.

Note

MPSK Local and HPE Aruba Networking Central NAC MPSK are available as another option

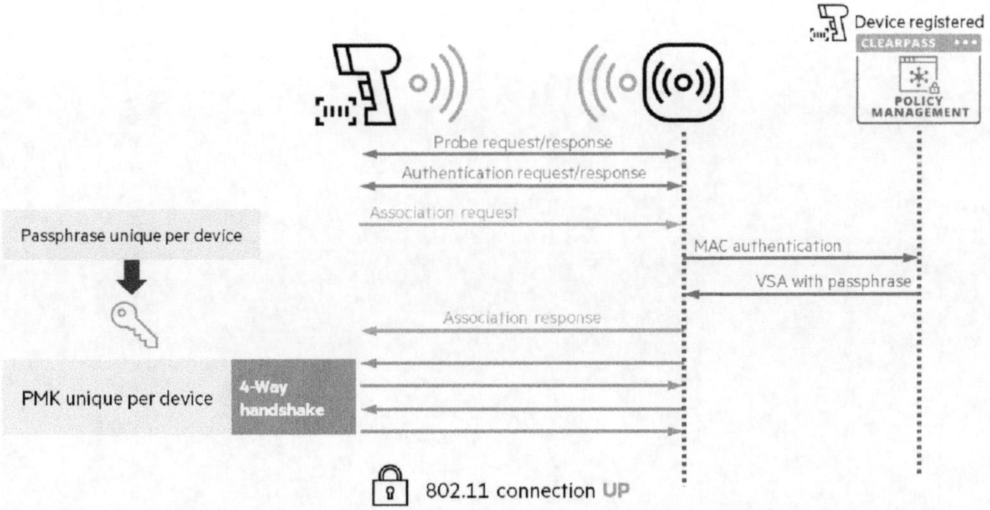

Step 1: A user connects to the ClearPass Self-Service Registration page on a different connection before attempting to connect the headless device. The user authenticates and specifies the device's MAC address on that page. ClearPass assigns the device a passphrase, or PSK, and displays that passphrase to the user.

Step 2: The user configures the client device to use the passphrase.

Step 3: The client can now connect to the WPA2-Personal WLAN. From the client's point of view, the connection proceeds like a typical WPA2-Personal connection.

Step 4: However, when the client sends its 802.11 association request, the HPE Aruba Networking Instant Access Point or Mobility Controller performs MAC-auth to ClearPass. ClearPass responds with the passphrase assigned to the client in an Aruba-MPSK-Passphrase VSA. The AOS device sends the 802.11 association response and performs the 4-way handshake using the PMK derived from the passphrase individual to the client device.

Visitor wireless features

- Same visitor/employee infrastructure
- Visitor traffic tunneled from AP to gateway
- Captive portal page for self-registration
- Preconfigured username/pass or social login
- Lobby Ambassador used to provision guests

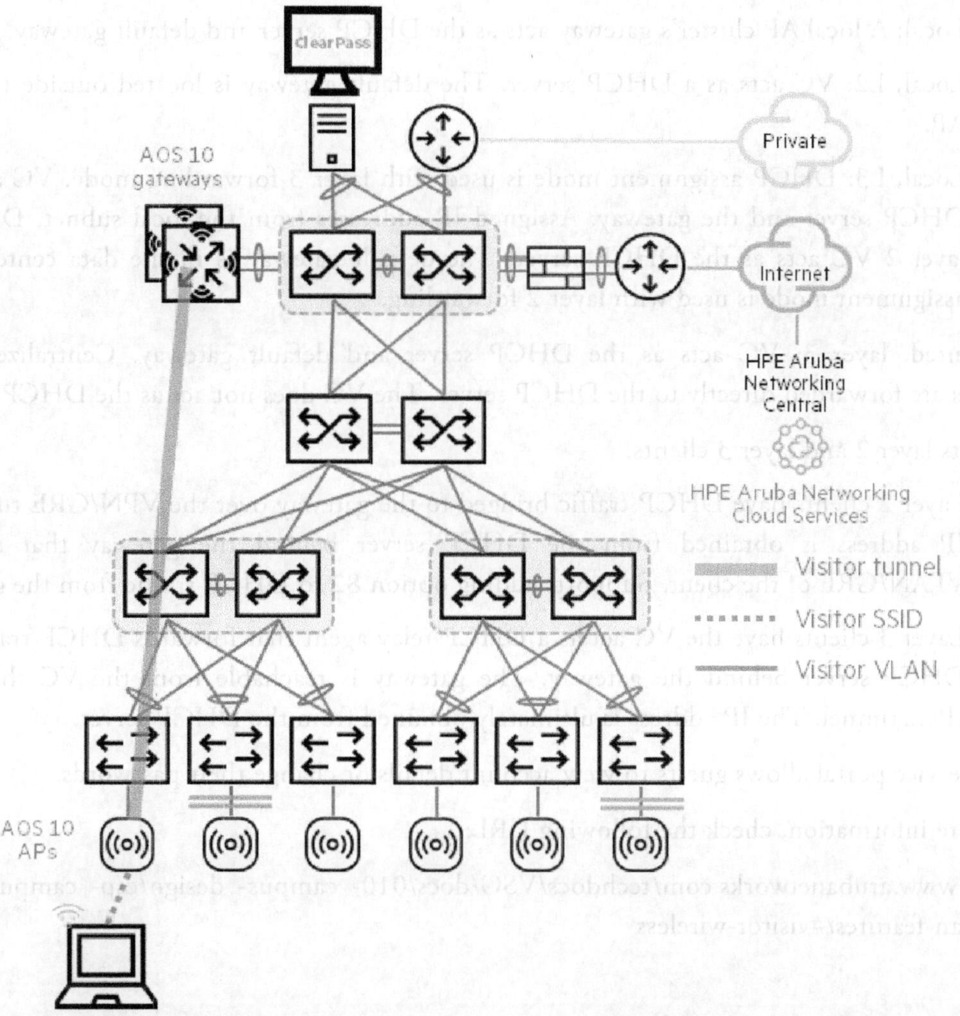

HPE Aruba Networking security-first, AI-powered networking supports visitor and employee access over the same wireless infrastructure without visitor access compromising employee network security. Visitor traffic can be tunneled security from the AP to the gateway in a separate VLAN with Internet-only access. This provides internet-only access to visitors through an open wireless SSID and supports a captive portal page for Self-registration, preconfigured guest usernames and passwords, or social login. Note that self-registration is also called self-provisioned access. If a customer desires more control over visitor account creation, a Lobby Ambassador can be a trusted employee with rights to provision and manage guest accounts.

Below are DHCP options where addressing can be assigned by the local AP, or an external DHCP server can be used.

- Local: A local AP cluster's gateway acts as the DHCP server and default gateway.
- Local, L2: VC acts as a DHCP server. The default gateway is located outside the Instant AP.
- Local, L3: DHCP assignment mode is used with layer 3 forwarding mode. VC acts as the DHCP server and the gateway. Assigned IP addresses from the local subnet. Distributed, layer 2 VC acts as the DHCP server. The default gateway is in the data center. DHCP assignment mode is used with layer 2 forwarding.

Distributed, layer 3: VC acts as the DHCP server and default gateway. Centralized DHCP requests are forwarded directly to the DHCP server. The VC does not act as the DHCP server.

Supports layer 2 and layer 3 clients.

- Layer 2 clients have DHCP traffic bridged to the gateway over the VPN/GRE tunnel. The IP address is obtained from the DHCP server behind the gateway that serves the VLAN/GRE of the client. Supports adding option 82 to DHCP traffic from the gateway.
- Layer 3 clients have the VC act as a DHCP relay agent that forwards DHCP traffic to the DHCP server behind the gateway. The gateway is reachable from the VC through the IPsec tunnel. The IP address is ultimately obtained from the DHCP server.

A self-service portal allows guests to view account details or change their passwords.

For more information, check the following URL:

https://www.arubanetworks.com/techdocs/VSG/docs/010- campus- design/esp- campus- design-048-wlan-features/#visitor-wireless

User roles overview

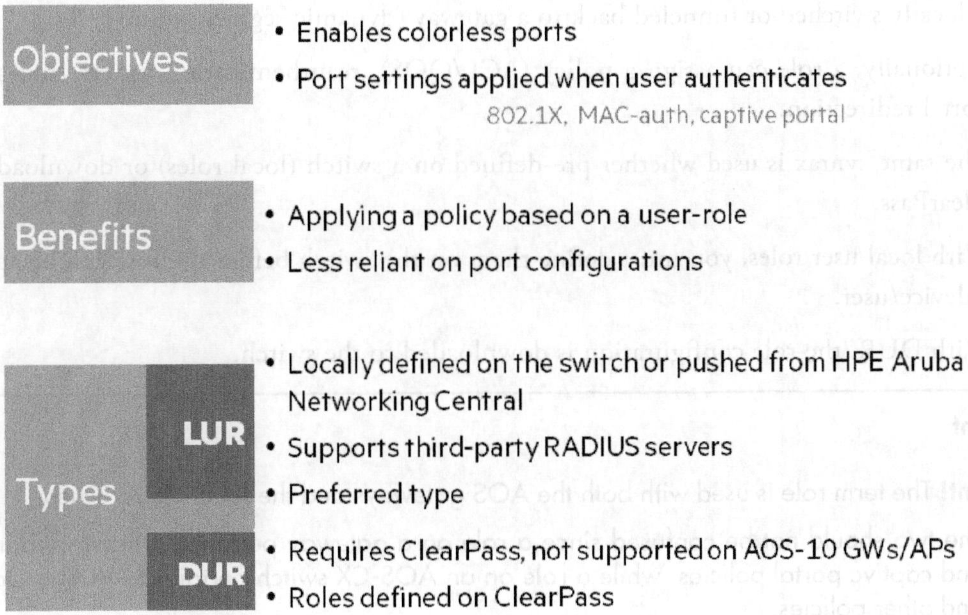

The benefits of assigning roles to users and devices are well-known within the HPE Aruba Networking wireless world.

- HPE Aruba Networking has adapted this feature to control port access on the AOS-CX switches.
- As with HPE Aruba Networking Wireless, you can simplify your configuration by grouping policies into a role many devices or user types can reference. Additionally, when ClearPass is incorporated, that who, what, how, where, and when context can be used to control user access rights.

You can configure roles locally on the switch using a local user role (LUR).

- When a device connects and authenticates successfully, an Aruba-User-Role name (VSA) is passed to the switch, and then the switch will apply the Aruba-User-Role to the port to which the device is connecting.
- Or configure downloadable user roles (DUR) on a ClearPass server.
- The ClearPass HPE-CPPM-Role VSA is used to instruct the switch to download a new or updated role using HTTPS.

A role is a role.

- At a minimum, a role dictates the tagged or untagged VLAN assignment and if the traffic is locally switched or tunneled back to a gateway (dynamic segmentation).
- Optionally, a role can assign a policy (ACL/QOS), reauthentication timers, and a captive portal redirection.
- The same syntax is used whether pre-defined on a switch (local roles) or downloaded from ClearPass.
- With local user roles, you must define them on the switch before the role can be applied to a device/user.
- With DUR, the role configuration is downloaded to the switch.

Important

Important! The term role is used with both the AOS gateways and the AOS-CX switches.

- The two should not be confused since a role on a gateway pertains to firewall, bandwidth, and captive portal policies, while a role on an AOS-CX switch pertains to VLAN, QoS, ACL, and other policies.
- In other words, roles on both devices apply policies, but the policy types differ.

Note

In the current release (10.4), HPE Aruba Networking doesn't support role download for a gateway or a secondary role to the gateway when implementing dynamic segmentation.

- The gateway would have to download itself.
- Also, since HPE Aruba Networking Central is used for centralizing configurations and policies, it is more common today to define the role information in HPE Aruba Networking Central and have it push that configuration to the switches.
- The switches would then use LUR when interacting with an AAA server.

Role-based AOS GW/AP firewall

The role-based AOS firewall can enforce different rules based on user identity and assigned role rather than simply on features in packet headers. For example, the firewall can restrict contractors' web traffic to a limited range of IP addresses while allowing engineers' traffic to flow more or less unrestricted.

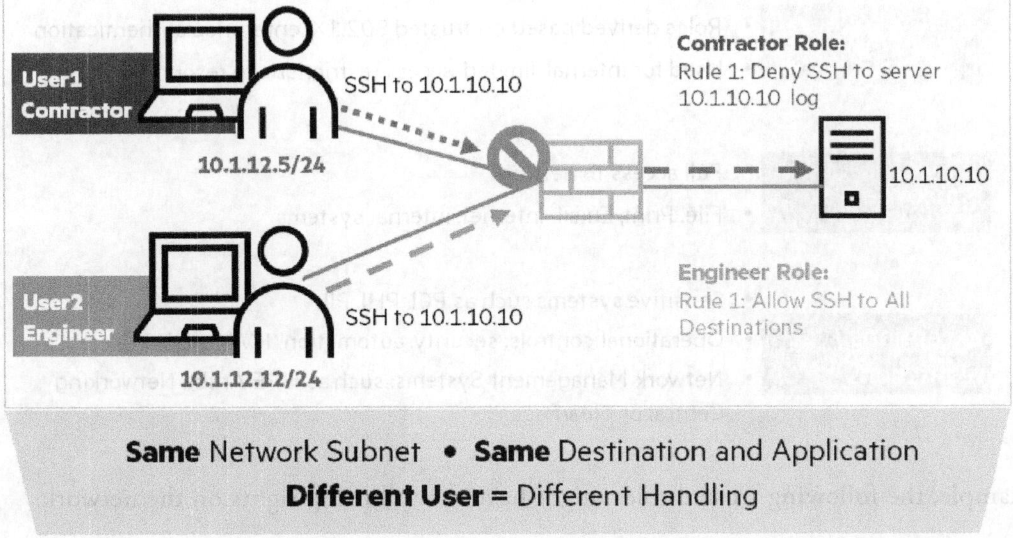

With less sophisticated firewalls, you would try to create such behavior by ensuring contractors and managers receive IP addresses in different ranges. You would then create different rules based on the source IP address. However, such an approach makes planning user connections more complex: you need more VLANs/subnets and a way to get users in the right VLAN every time. With the AOS firewall, you can place users in the same subnets but easily create different rules for different users.

The AOS firewall determines the user role associated with a device as the device authenticates and accesses the network. It might assign this role independently or have help from ClearPass, the HPE Aruba Networking authentication server. By sending all internal user traffic through the role-based AOS firewall, you can change from a model in which you protect a single perimeter to one in which you place every user within their unique and properly customized firewall.

Designing campus role-based policy

Plan policy enforcement to minimize interference with the functionality of the business. This relates to implementing trusted roles with the least privilege in mind. Roles that fall under the "Trusted" category are roles derived based on trusted 802.1X enterprise authentication. They should be used for limited internal restricted access to resources.

Role categories: Trusted

- Roles derived based on trusted 802.1X, enterprise authentication
- Used for internal, limited-access restrictions to resources

Employee

- Full access to network
- File, Print, Email, Internet, internal systems

Exceptions / restrictions

- Sensitive systems such as PCI, PHI, PII.
- Operational controls: security, automation, HVAC, lighting
- Network Management Systems: such as HPE Aruba Networking Central or ClearPass

For example, the following trusted roles might have the following rights on the network:

Employee

- Full access to network resources including File, Print, Email, Internet, and internal systems.

Exceptions/restrictions

- Sensitive systems such as Payment Card Industry (PCI), Protected Health Information (PHI), and Personally Identifiable Information (PII).
- Operational controls such as security, automation systems, HVAC, and lighting.
- Network Management Systems – such as HPE Aruba Networking Central or ClearPass.

IT

- Standard employee access.
- Network Management systems
- Exceptions/restrictions
- PCI, PHI, and PII data.

Critical

- Assigned to a single account.
- Used only when access to authentication services is disrupted.
- Role necessary for specific systems that require rights to perform repair, recover, and restoration.

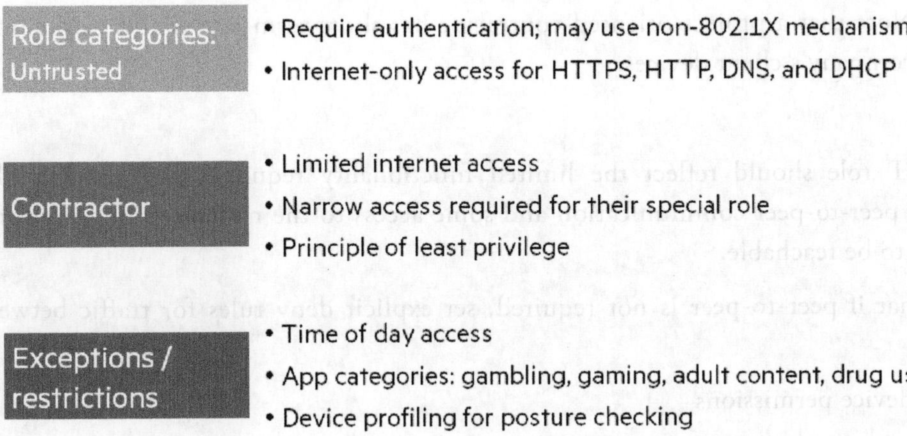

Untrusted

- Require authentication but may use non-802.1X mechanisms such as PSK, MAC, or guest captive portal.
- Visitor
- Internet-only access for HTTPS, HTTP, DNS, and DHCP.
- Exceptions/restrictions:
    - Bandwidth limits
    - Network Management Systems (NMS)
    - Application categories include gambling, gaming, adult content, drug use, and more
    - Time of day limits

Contractor

- Limited internet access.
- Narrow access is required for their special role as a contractor. Principle of least privilege.
- Exceptions/restrictions:
    - Network Management Systems (NMS).
    - Time of day access
    - Application categories include gambling, gaming, adult content, drug use, and more.
    - Device profiling for posture checking.

- Note that a project-oriented contractor has short-term access and works from their company's client devices.

IoT Devices

- The IoT role should reflect the limited functionality required to function. This may include peer-to-peer communication and some access to the management place for all IoT devices to be reachable.
- Note that if peer-to-peer is not required, set explicit deny rules for traffic between endpoints.
- Office device permissions
- Limited access to NMS
- Print servers, DHCP, and DNS.
- Exceptions/restrictions:
  - No access to the internet or other internal networks.
  - Building management systems (BMS) hosts.
  - Access to DHCP and DNS.
  - Access to other BMS hosts.
  - Exceptions/restrictions:
  - No access to the internet or other systems.

Review application visibility

One of the challenges of controlling user access with a firewall is that so many modern applications run over HTTP/HTTPS.

- For example, a user might be accessing social media, working in a shared Wiki, sending chats over a company app, and attending video conferences—all over the Web.
- With traditional ACLs, you must allow none or all applications that use HTTP/HTTPS.
- However, the AOS application visibility feature provides deep packet inspection (DPI), which gives visibility down to the application layer.

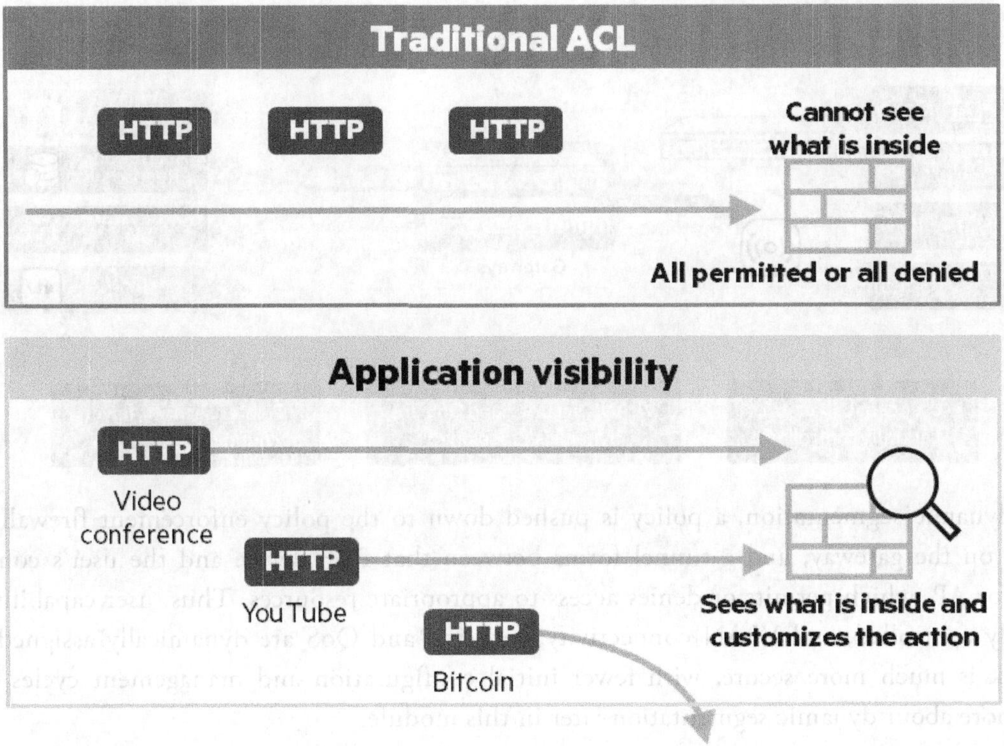

The first step to using application visibility is enabling Deep Packet Inspection (DPI) globally, which requires a reboot.

- You must also ensure that DPI is enabled on the role (it is by default).
- To filter by Web category and application, you must also enable WebCC globally and on the role (it is enabled by default).
- Also, make sure that you have the proper licenses for these features.

Dynamic Segmentation with centralized overlay

Dynamic segmentation can simplify policy enforcement in a wireless and wired network. Dynamic segmentation allows a switch to act like an AP: just like an AP, it will tunnel traffic to a gateway, where the gateway can use its advanced firewall capabilities to control traffic. Likewise, you can use the same or different policies for users who need to make connections for both wired and wireless access.

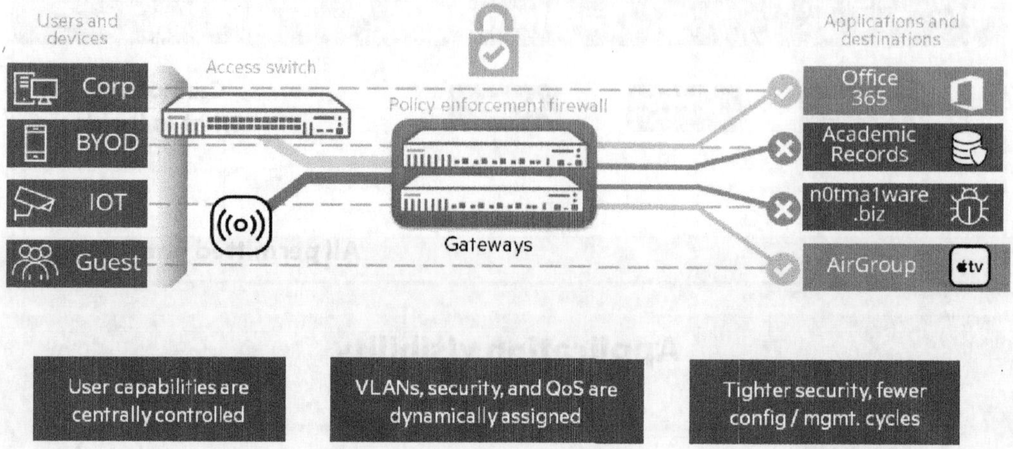

With dynamic segmentation, a policy is pushed down to the policy enforcement firewall (PEF) feature on the gateway, and a tunnel forms between that PEF device and the user's connected switch or AP, which permits or denies access to appropriate resources. Thus, user capabilities are centrally controlled, and VLAN connectivity, security, and QoS are dynamically assigned. Your network is much more secure, with fewer initial configuration and management cycles. You'll learn more about dynamic segmentation later in this module.

Customizing authenticated user access

Beyond the security benefits, port-based authentication helps you to provision the network edge more quickly. You learned about the old days when you had to configure each port with the appropriate settings for the intended user: tedious and error-prone. This can be done with 802.1X, MAC-auth, and a captive portal.

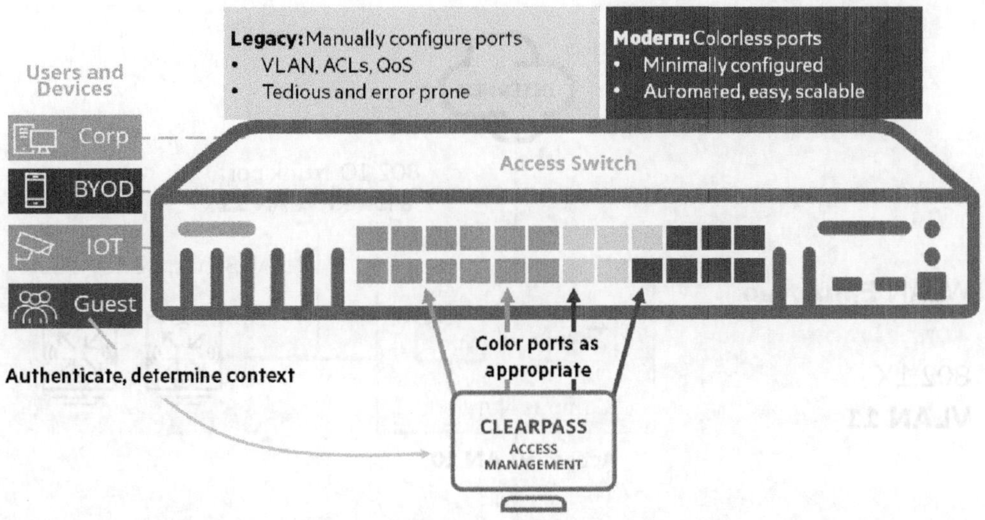

With 802.1X, the RADIUS server and the switch (authenticator) automatically determine the user identity and other context characteristics: who, what, how, where, and when. A RADIUS server policy can use this information to configure the appropriate authorization policy settings automatically: VLAN, ACLs, QoS/CoS, rate limits, etc. The network edge becomes more flexible and adaptable.

Because the switch automatically applies these settings to a user session, as dictated by the RADIUS server, these settings are sometimes called RADIUS settings or dynamic settings. For example, a dynamic VLAN assignment is the VLAN assigned to a user session by the RADIUS server when the user authenticates to the network. A dynamic ACL is the (port-based) ACL applied to a user session by the RADIUS server when the user authenticates, and so on. If the interface is disabled or the user disconnects, the port reverts to its original configuration and is moved to an unauthenticated state. Therefore, if a new device connects to the port, it must authenticate, and a new policy would be dynamically assigned to the port.

Access point tunnel forwarding mode

The figure in the slide shows that the AP receives a client's traffic and then is redirected (tunneled) to the gateway cluster. Gateways analyze and process clients' traffic. From the rest of the network's perspective, the client's traffic is generated on the gateway and not on the AP. As we discussed, this solution uses a GRE tunnel between the AP and the gateway; therefore, it is important to permit IP protocol number 47 between these two devices on intermediate filtering or firewall devices.

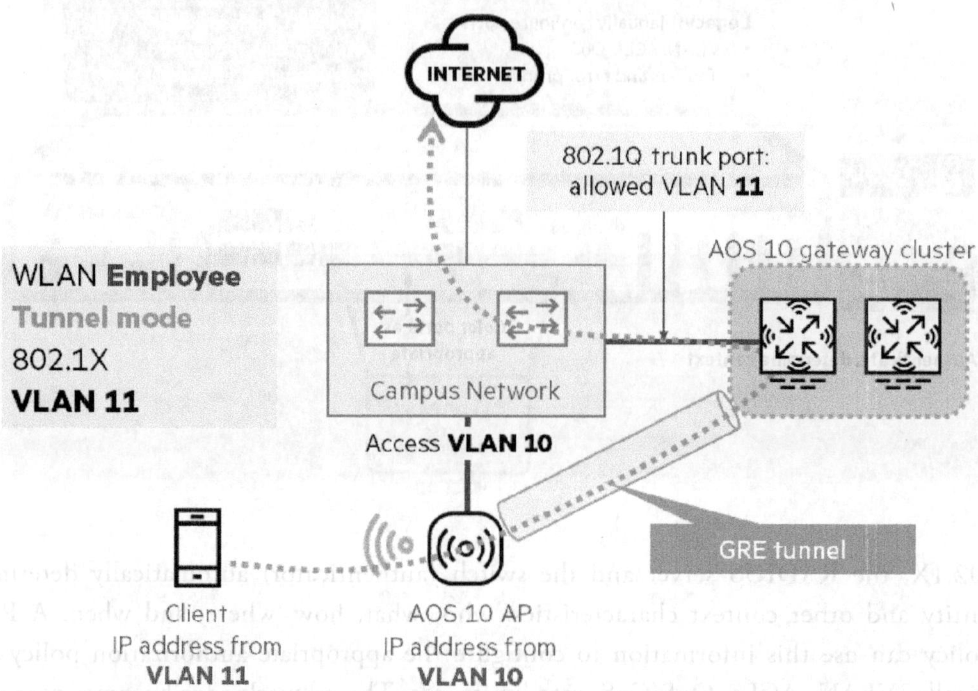

APs SSIDs and user traffic from switch ports are tunneled to a centralized gateway cluster. HPE Aruba Networking Central configures and pushes the policy to the AP, switch, and gateway. The gateway is the enforcement point, allowing higher throughput, firewalling, and DPI engine capabilities. Generally, HPE Networking recommends a gateway cluster if a campus is deploying more than 500 APs and 5,000 clients.

Notice that the client receives an IP address in VLAN 11, and this VLAN must be defined and configured on the gateway side and should be permitted on the link from the gateway to the campus switches. VLAN 11 is not required on the link between the AP and the switch since the AP is tunneling the VLAN traffic to the gateway. GRE acts as an overlay and provides a logical connection.

Access point bridge forwarding mode

- Example of bridge forwarding for a 2-tier campus network
- HPE Aruba Networking Central pushes policy to edge
- AP is enforcement point

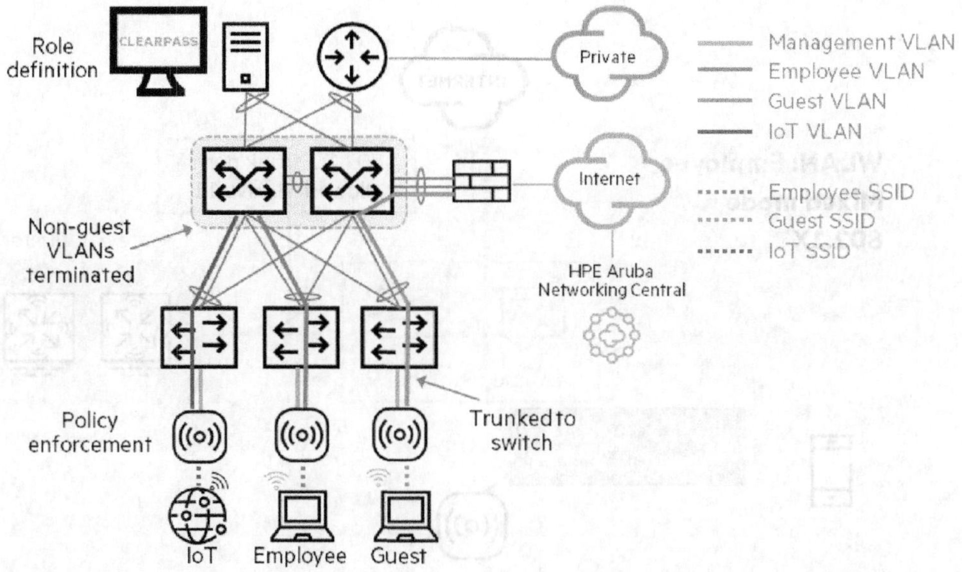

This diagram shows an example of bridge mode in a 2-tier wired campus network. If a client is assigned an AP-only VLAN or a role for local forwarding on a switch, then that user traffic is not tunneled to a gateway. APs using bridge mode require all VLAN trunked to a connected switch port. HPE Aruba Networking Central is where the policy is configured and pushed to APs. Instead of the gateway, the AP is where the firewall and DPI are enforced.

Note that VLAN assignment rules are created to control the use of the forwarding mode. By default, an implicit rule forwards all traffic to the Gateway; therefore, all traffic is tunneled. To change this default behavior, you must define at least one rule for bridging traffic.

Bridge mode SSIDs should be considered if less than 500 APs with fewer than 5,000 clients are deployed. Otherwise, utilize tunneled/mixed forwarding to gateways.

Access point mixed forwarding mode

In mixed mode, the AOS-10 AP decides which forwarding role is applied to each device that connects to the SSID. This decision is based on the authentication result (particularly the RADIUS attributes received from the AAA authentication server) or the client's MAC address.

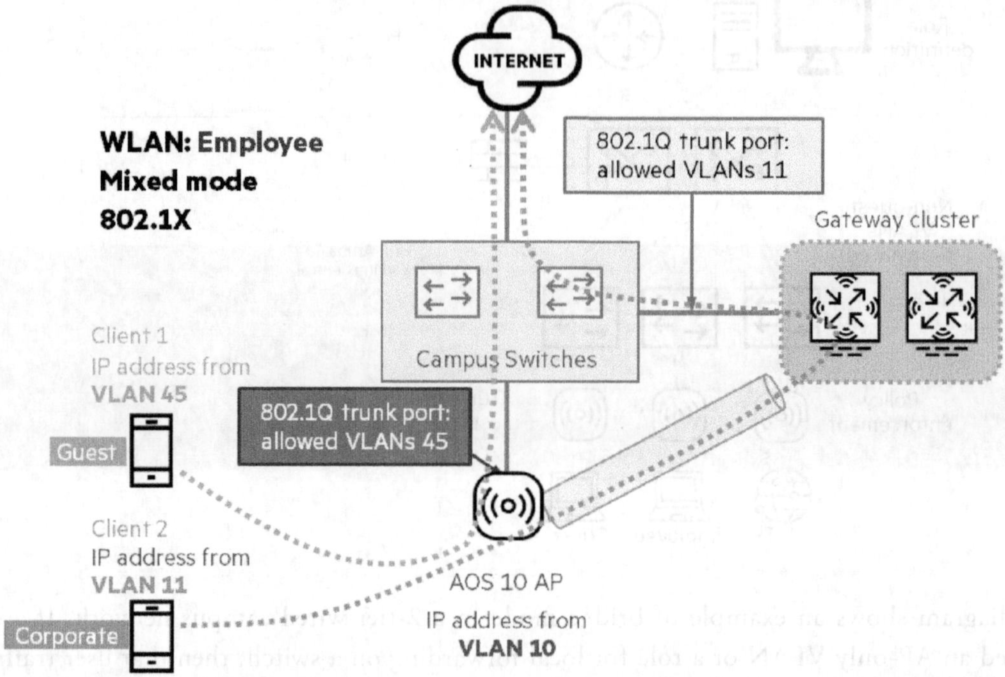

One advantage of mixed mode is that you are not required to broadcast two SSIDs to use two forwarding modes, representing an RF resource saving for the APs. Mixed mode thus allows the WLAN to perform more efficiently, resulting in a more positive user experience. A standard attribute (such as filter-id) or an HPE Aruba Networking VSA (such as user role) can indicate which role assignment. The key in configuring which VLANs tunnel versus which are bridged depends on which VLANs on the AP match those on the gateway. Those VLANs that match are tunneled. Those VLANs that exist only on the AP are bridged.

Note that HPE Aruba Networking ClearPass is recommended but not mandatory.

Profiling and NAC

This section covers device profiling and network access control (NAC).

ClearPass Policy Manager endpoint profiling

The Endpoint Profiler is another module built into the ClearPass Policy Manager. The Profiler's job is to gather contextual information about any device on your network. You can use this information for reporting or build it into your access control logic.

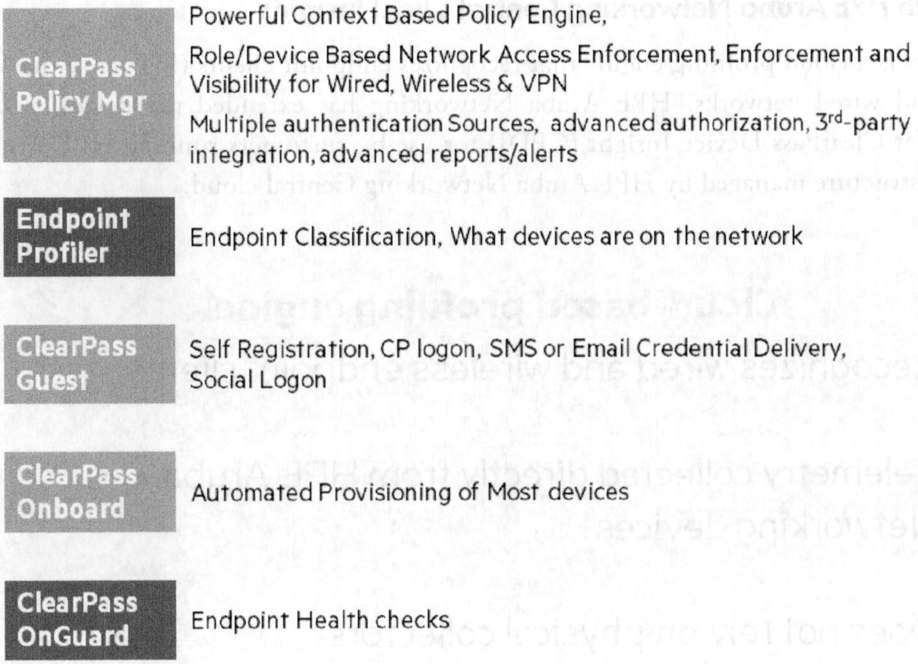

The Profiler can profile endpoints passively, meaning that ClearPass just monitors client data as it flows on the network and gathers endpoint context from the profiler. ClearPass has multiple active profiling tools that allow you to scan static networks or inactive networks where other forms of profiling don't work.

Client Onboarding Software is the easiest way to support a BYOD environment. Based on its internal certificate authority, Onboard provides an intuitive and easy-to-use portal for users to configure their devices on your network securely. To enforce the concept that every device should be granted access on its own unique merits, Client Onboarding Software assigns a unique identity to each device that goes through the Onboard process. For auditing capabilities, this identity includes information about the client device and the user that put the device on the network. Client Onboarding Software is an extra licensed module in ClearPass.

Posture checks and client system health validation have become necessary for most networks. Using an agent, either deployed through a web browser to the client or permanently as a small software applet, ClearPass OnGuard Software can provide security status validation on your wired, wireless, and VPN networks. ClearPass OnGuard Software makes it easy to enforce the organization's device compliance policies. ClearPass OnGuard Software is an extra licensed module in ClearPass.

Profiling with HPE Aruba Networking Central Client Insights

Client Insights is a cloud profiling engine that recognizes endpoint clients (IoT, etc.) that connect to wireless and wired networks. HPE Aruba Networking has extended the device recognition engine built for ClearPass Device Insight (CPDI) for use by customers running HPE Aruba Networking infrastructure managed by HPE Aruba Networking Central cloud.

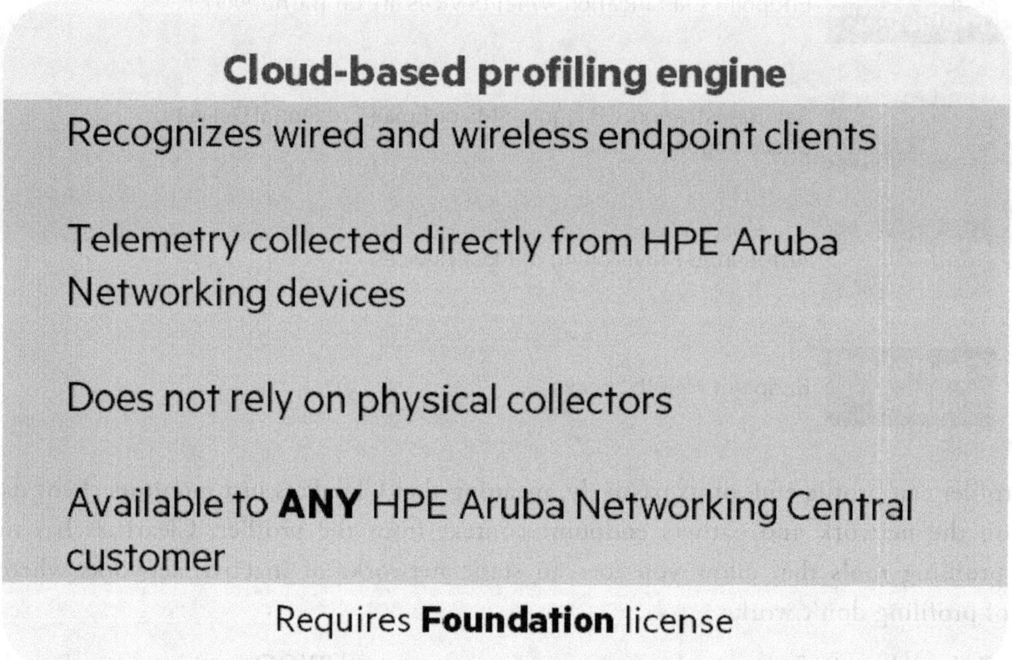

Note: Included under NetConductor Client Profiler

Instead of relying on physical collectors used by CPDI, HPE Aruba Networking now utilizes the telemetry data collected directly from HPE Aruba Networking APs, gateways, and switches. HPE Aruba Networking Central customers can now see profiling information displayed in the GUI's Profile tab within the Clients page. There's also a separate profile tab for each client.

Foundation licensing is needed for AI-powered profiling and visibility into static/flow-based attributes. Advanced licenses will offer the ability to utilize tags for granular security features when available.

The AOS-CX software preinstalled on all HPE Aruba Networking CX Switch Operating System switches includes an active, perpetual set of AOS-CX native enterprise features at no additional cost. The AOS-CX native enterprise features include everything needed to deploy, connect, and troubleshoot an enterprise network. For those networks that require enhanced visibility and assurance, the HPE Aruba Networking CX Advanced feature pack enables HPE Aruba Networking CX Edge Insights for AOS-CX CX 6300 series and CX 6400 series, offering deep visibility with application recognition and application-based policies, from OSI Layer 2 to Layer 7. The HPE Aruba Networking CX premium feature pack is supported on CX 10000 series only and enables scalable Network Address Translation (NAT) and IPSec VPN services.

Note

For networks managed by HPE Aruba Networking Central, the HPE Aruba Networking CentralAdvanced license enables all features supported by the HPE Aruba Networking CX Advanced feature pack, so no additional purchase of an HPE Aruba Networking CX Advanced feature pack is necessary.

HPE Aruba Networking AI-powered Client Insights

- Leverages existing infrastructure
- No installation of collectors/agents
- ML-based classification
- DPI used for hard-to-detect IoT devices
- Controlled crowd-sourcing for validation
- Behavior analytics

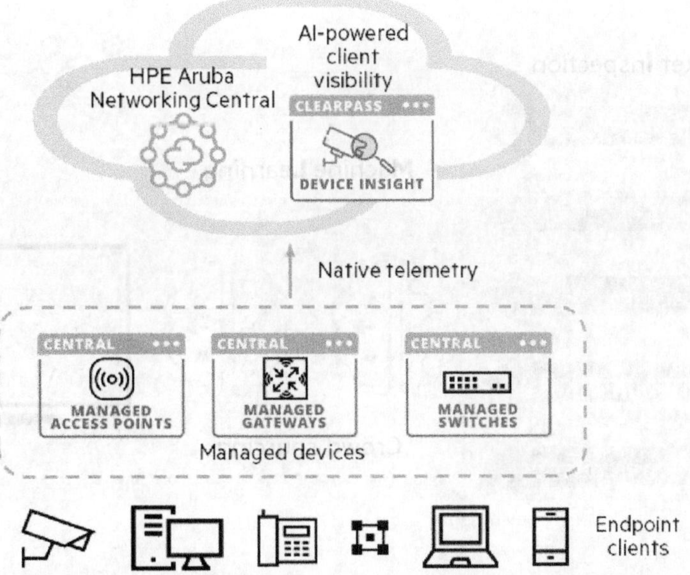

HPE Aruba Networking's network management solution, HPE Aruba Networking Central Cloud, now includes AI-powered Client Insights, which offers the industry the most granular profiling and visibility. Client Insights leverages native infrastructure telemetry from access points, switches, and gateways, as well as clients, without requiring the installation of physical collectors or agents. ML-based classification models fingerprint, identity, and accurately profile clients across the wired and wireless infrastructure.

These capabilities are further enhanced with Deep Packet Inspection (DPI), which provides additional context and behavioral information that helps to identify those hard-to-detect IoT devices accurately. By leveraging DPI, Client Insights can utilize a broader set of device attributes for more accurate identification. Client attributes that include communication and behavioral patterns are analyzed to build clusters of similar devices dynamically. Machine learning models constantly learn and update these attributes to update fingerprints and provide dynamic classification recommendations. Controlled crowd-sourcing technology is used to validate fingerprints at multiple customer sites before adding them to the HPE Aruba Networking classification database. This increases the precision and comprehensiveness of the classification engine.

HPE Aruba Networking Client Insights

HPE Aruba Networking Central cloud now includes AI-powered Client Insights, which offers the industry the most granular profiling and visibility. Client Insights leverages native infrastructure telemetry from access points, switches, and gateways, as well as clients, without requiring the installation of physical collectors or agents. ML-based classification models fingerprint, identity, and accurately profile clients across the wired and wireless infrastructure.

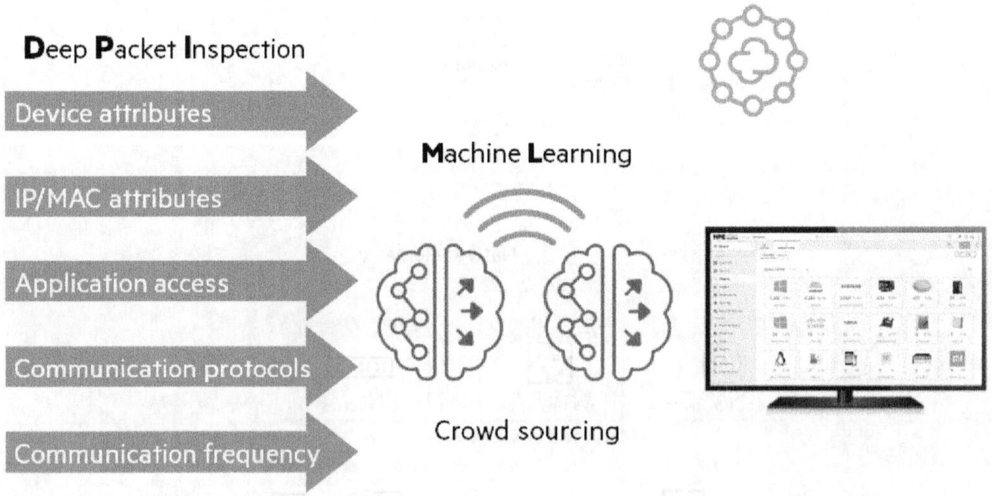

Feature include:

- Granular profiling and visibility
- Utilizes infrastructure telemetry from clients, APs, switches, and gateways without requiring physical collectors or agents.

- ML-based classification modes can fingerprint, identify, and profile various clients (wired or wireless).
- Addresses the rising number of IoT devices that otherwise would be difficult to detect and manage.
- Detects device attributes such as device type, vendor, hardware version, and behaviors.
- Integrates with HPE Aruba Networking Central and HPE Aruba Networking ClearPass Policy Manager Platform to enable comprehensive policy control and real time enforcement.

Microsoft Endpoint Manager

Microsoft Endpoint Manager is an MDM and MAM system that runs in the Microsoft Azure cloud. Microsoft manages various operating systems, including Apple iOS, iPadOS, Apple MAC, Google Android, and Windows. For a complete list, refer to Microsoft documentation.

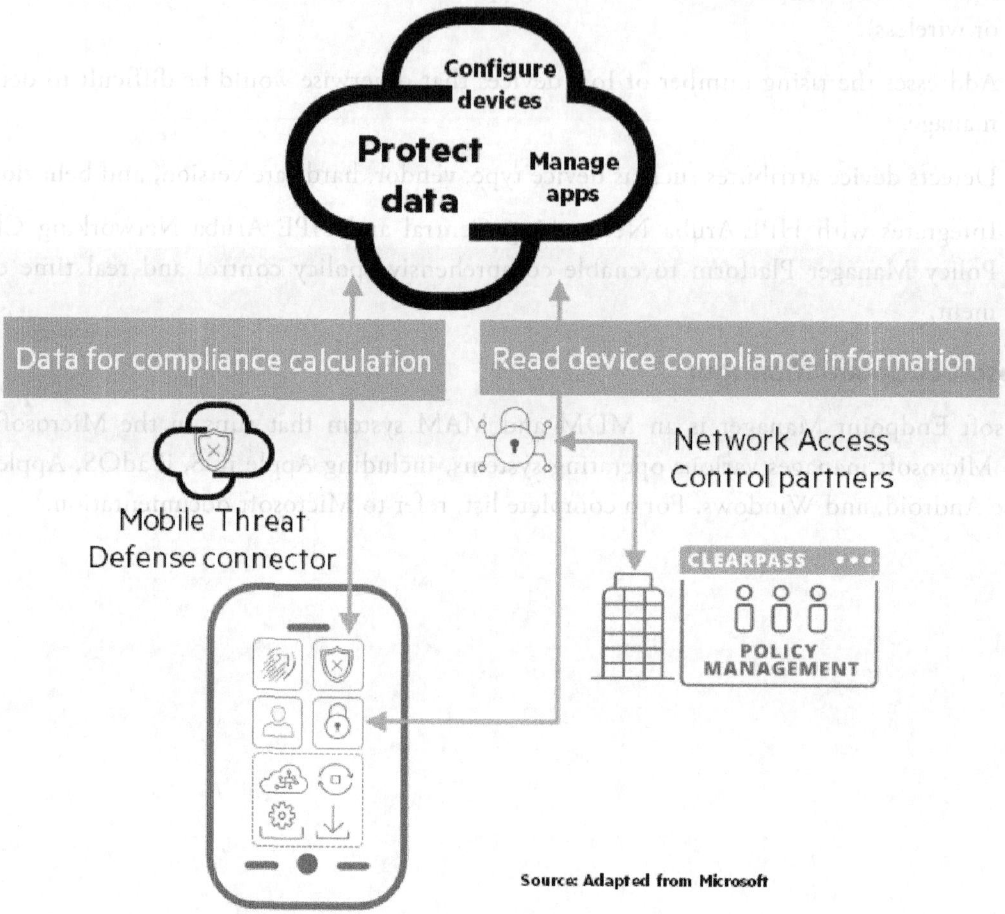

Endpoint Manager integrates with Azure Active Directory to regulate users' access to the corporation's Microsoft Office 365 and other SaaS applications. Admins can create application protection policies to control which applications users in various groups can access and what users can do within those apps. The company can further create different policies for devices enrolled in Endpoint Manager and unmanaged devices. Endpoint Manager also provides profiles and policies to configure mobile devices.

Endpoint Manager's ability to assess a mobile device's compliance status is of most interest to you for this guide. Endpoint Manager integrates with Mobile Threat Defense, which feeds information about the devices' security compliance and posture. The Endpoint Manager can then provide that information to a Network Access Control partner such as Policy Manager. (Policy Manager must run version 6.7.2 or above to integrate with Endpoint Manager.)

HPE Aruba Networking SASE

This section covers HPE Aruba Networking Secure Service Edge (SSE)

Unified SASE with HPE Aruba Networking

With HPE Aruba Networking security-first, AI-powered networking solutions, deliver innovative business results with comprehensive, security-first, AI-powered networking that connects and protects your business from edge to cloud without sacrificing high performance. Designed with Zero Trust principles, organizations benefit from advanced visibility, centralized policy management, data protection, threat defense, access control, and intelligent automation.

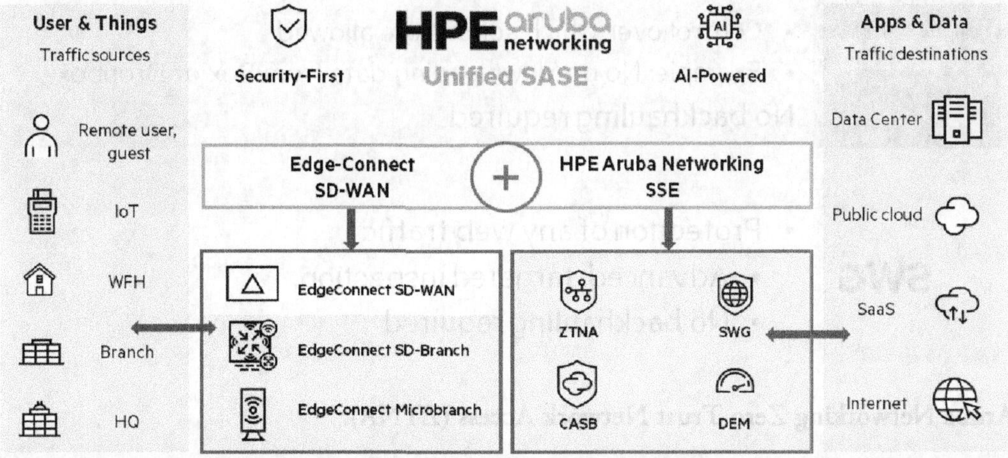

With HPE Aruba Networking EdgeConnect SD-WAN, IT teams can deliver WAN and cloud security controls directly to the application at the network edge—rather than routing data through the data center—while SSE ensures that Zero Trust security controls can be applied to all people and devices, no matter where they connect—on campus, in the branch, at home, or on the road. AI-powered network and security insights can help you manage and protect at scale, freeing network and security teams to focus on creating business advantage.

How HPE Aruba Networking SSE is modernizing remote access

HPE Aruba Networking SSE takes a more modern approach to security remote access. This solution consists of three main cloud services (as well as other services).

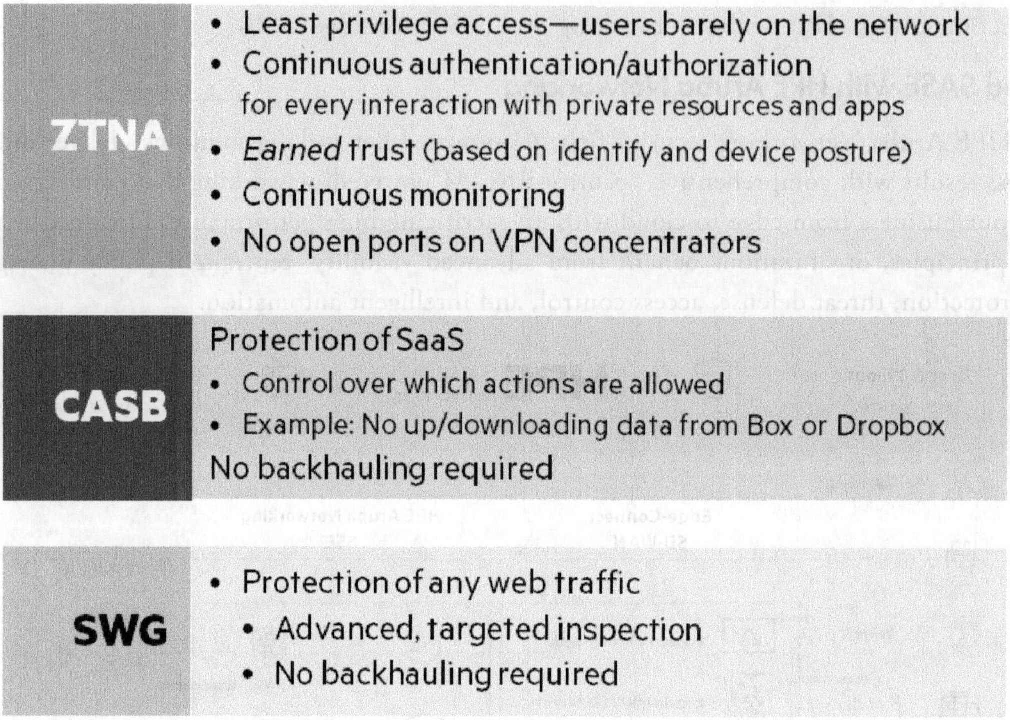

HPE Aruba Networking Zero Trust Network Access (ZTNA)

ZTNA closes security gaps by taking a new approach to granting users' access to private resources and applications.

- Rather than placing remote clients on the internal network—and then granting those clients intrinsic trust—ZTNA provides least privilege access.
- Users are barely on the network.
- Instead, they receive brokered access to specific authorized applications.

To determine whether to grant that access, HPE Aruba Networking ZTNA enforces continuous authentication and authorization for every interaction with an asset.

- For example, if a user wants to access an internal Sharepoint folder, the user must authenticate to receive that access.
- Then when the user also wants to run the company's private accounting app, the user must authenticate and be authorized for that app as well.

- (Note, that this authentication and authorization can sometimes occur in the background from the user perspective.) In short, users continually earn trust based on factors such as their identity and the posture (or health) of their device.

HPE Aruba Networking ZTNA also provides continuous monitoring, helping admins achieve better visibility into who is accessing which applications when.

This solution replaces client-to-site VPNs, but it is not a VPN.

- HPE Aruba Networking ZTNA does not require VPN concentrators at the organization's main office or data center.
- Therefore, the organization avoids the potential vulnerability of open ports.

HPE Aruba Networking Cloud Access Security Broker (CASB)

HPE Aruba Networking CASB protects users' access to software as a service (SaaS) applications.

- This solution can inspect users' traffic and enforce controls to protect the organization.
- For example, the company might choose to prevent users from uploading or downloading data from Box, Dropbox, and other file sharing SaaS applications.
- As a cloud service, HPE Aruba Networking CASB enforces this security efficiently without forcing organizations to backhaul traffic to their own site for inspection.

HPE Aruba Networking Secure Web Gateway (SWG)

HPE Aruba Networking SWG protects any web traffic.

- Organizations can establish rules for permitting or blocking users' access to various web site categories based on their internal policies.
- They can also implement advanced inspection on traffic destined to particular categories or particular sites, detecting malware. Again,
- HPE Aruba Networking SWG provides all these benefits without backhauling traffic to a main office or data center.

Secure Service Edge (SSE)

Security Service Edge (SSE) is a comprehensive cloud-based solution that ensures secure access, visibility, and governance across all business applications. It represents a unified security model that integrates core security components such as: Zero Trust Network Access (ZTNA) for safeguarding private applications, Secure Web Gateway (SWG) for managing web access, Cloud Access Security Broker (CASB) for securing SaaS applications, and Digital Experience Monitoring (DEM) for ensuring an optimized digital experience across all access needs. This convergence of services epitomizes the future of secure connectivity — a singular, cloud-centric service that equips IT departments with the authority to regulate access to users and applications, irrespective of their physical location, the devices they use, or the networks they connect to.

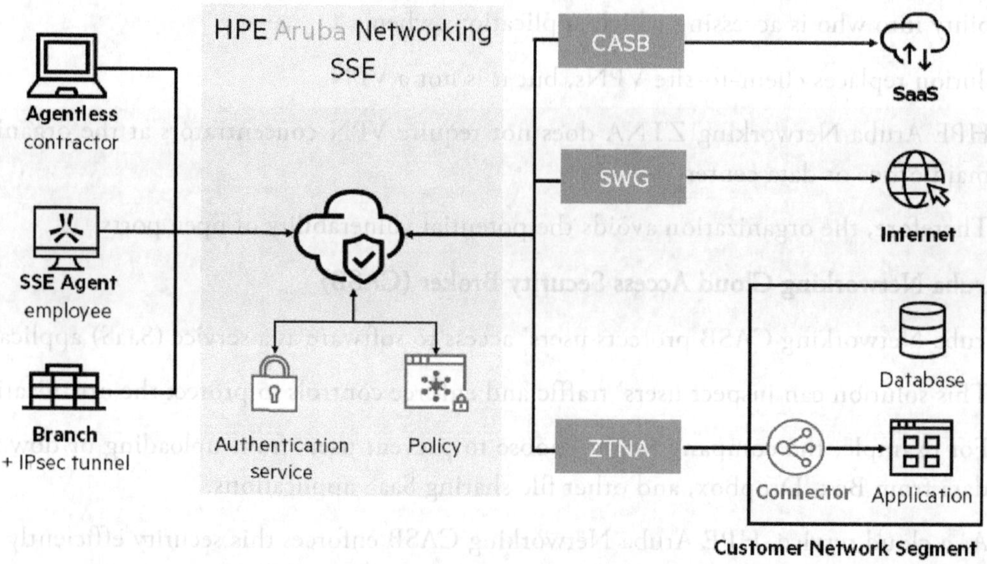

While HPE Aruba Networking SSE might appear to be just one more acronym in the expansive IT lexicon, it is significant for any contemporary workplace, making up half of SSE, with the other half composed of WAN Edge Services.

HPE Aruba Networking SSE management console

Each company receives a "workspace" dedicated to them within HPE Aruba Networking SSE cloud.

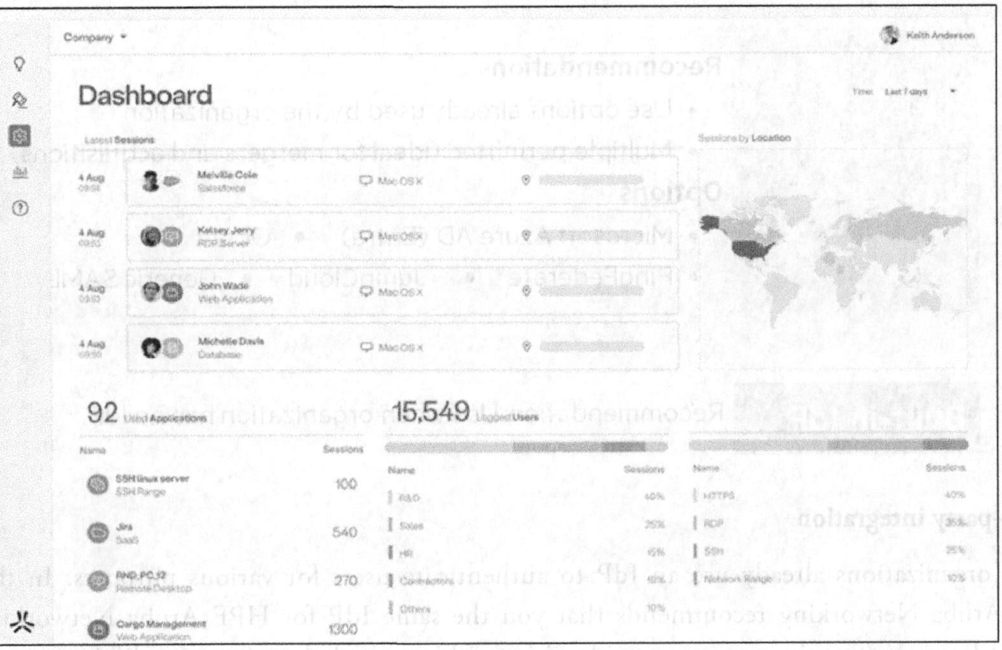

You access your workspace and set up HPE Aruba Networkingg ZTNA, CASB, SWG, and other features at the HPE Aruba Networking management console. This console provides role-based access control (RBAC), allowing admins to assign different administrative rights to different users. You will practice accessing and navigating this console in the lab.

HPE Aruba Networking SSE and Identity Providers (IdPs)

HPE Aruba Networking SSE uses Identity Providers (IdPs) to authenticate and authorize users for various forms of access. It integrates with several common third-party IdPs, as well as provides its own built-in IdP.

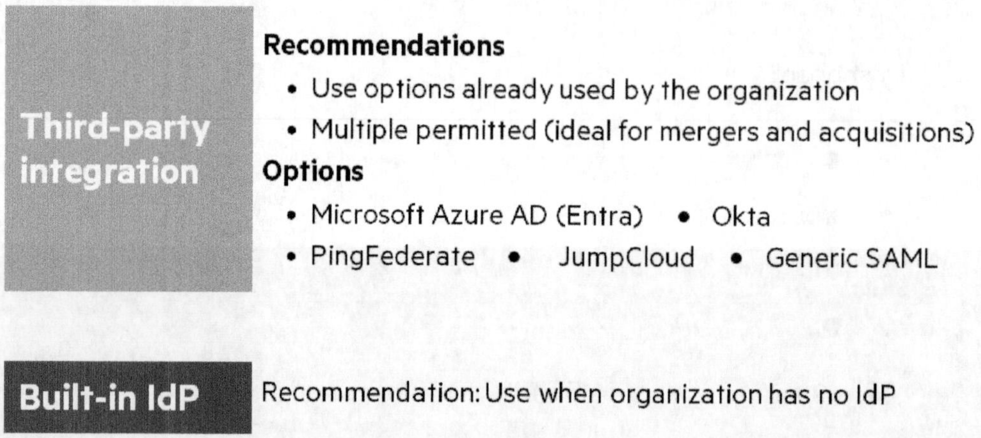

Third-party integration

Many organizations already use an IdP to authenticate users for various purposes. In this case, HPE Aruba Networking recommends that you the same IdP for HPE Aruba Networking SSE. Then admins can continue to manage identities within a single location, simplifying processes for the customer and minimizing potential issues, such as forgetting to revoke an account when a user leaves the organization. Users also benefit; they can log in with familiar credentials.

HPE Aruba Networking SSE supports integration with multiple IdPs in the same workspace, which might be required if mergers and acquisitions have resulted in the company using more than IdP. You simply attach the desired IdP to each application.

As of the release of this guide, HPE Aruba Networking SSE supports these IdPs:

- Microsoft Azure AD (renamed Entra)
- Okta
- PingFederate
- JumpCloud

If the organization uses a different IdP, which supports Security Assertion Markup Language (SAML), you can set up a generic SAML integration to accommodate that IdP.

Built-in IdP

For organizations without their own IdP, you can leverage the IdP built into the solution. This option also works well for labs and demos.

HPE Aruba Networking SSE service structure

The service can be purchased using any of the following user-based or site-based subscription tiers and feature add-ons. The End Customer purchases any of the user-based subscription based on total number of users and customer success package requirements. For site-based subscription, the End Customer purchases the subscription based on the total anticipated bandwidth requirements per site.

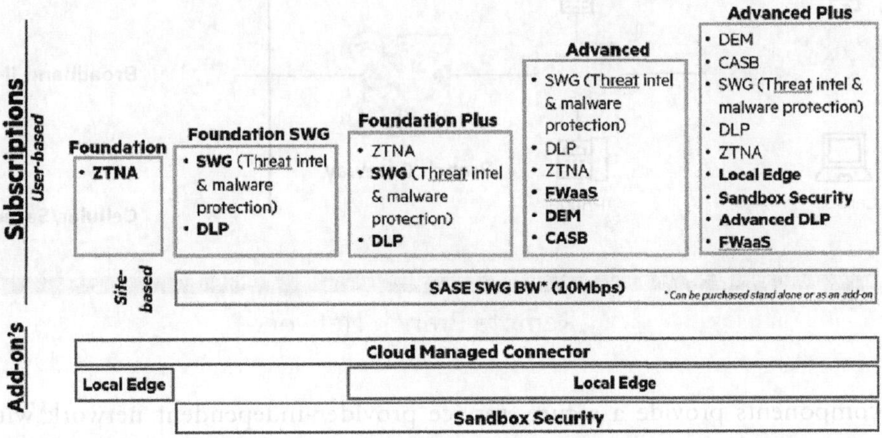

This slide contains a quick overview the service structure across user/site-based subscriptions and feature add-ons.

Read the solution descriptions for more details.

https://www.hpe.com/us/en/collaterals/collateral.a50011048enw.html

https://www.hpe.com/psnow/doc/PSN1014795496BOEN

EdgeConnect SD-Branch

HPE Aruba Networking EdgeConnect SD-Branch is a comprehensive solution that enhances branch office networking through a software-defined approach. This solution has two major components: SD-LAN and SD-WAN.

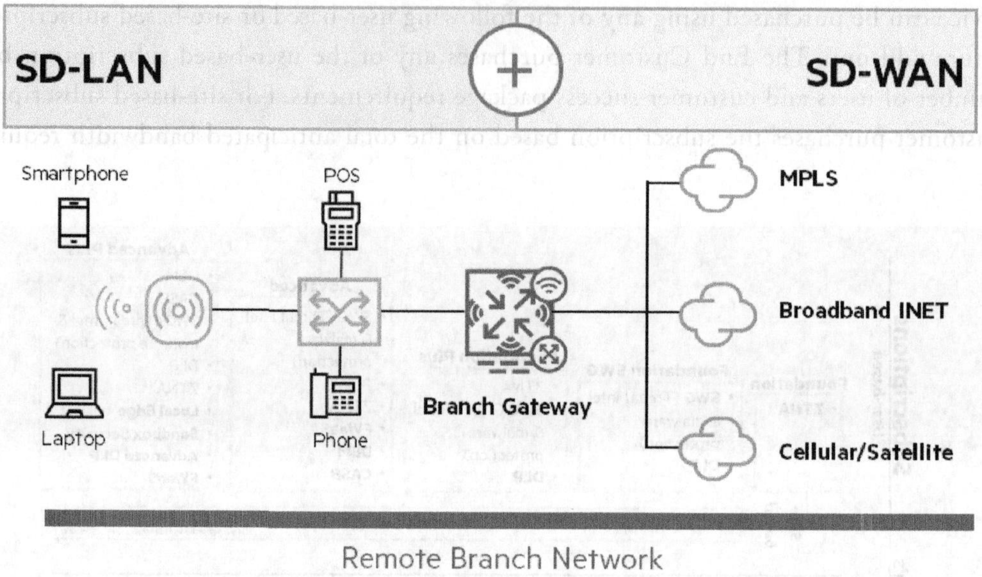

SD-WAN components provide a secure, service provider-independent network with enterprise-level performance across diverse wide-area network (WAN) technologies.

SD-LAN provides secure wire and wireless networks that enforce policies based on the connecting user or device.

EdgeConnect Software-defined branch (SD-branch) represents a technological shift toward agile, open, and cloud-integrated solutions.

EdgeConnect SD-Branch secure connection

Remote communication must be secure, so the EdgeConnect SD-Branch solution relies on IPsec tunneling. Unlike traditional networking, the solution orchestrates (automates) tunnel creation using the Tunnel Orchestrator service (formerly known as Overlay Tunnel Orchestrator – OTO) hosted in HPE Aruba Networking Central. This solution speeds up connectivity and scales deployment.

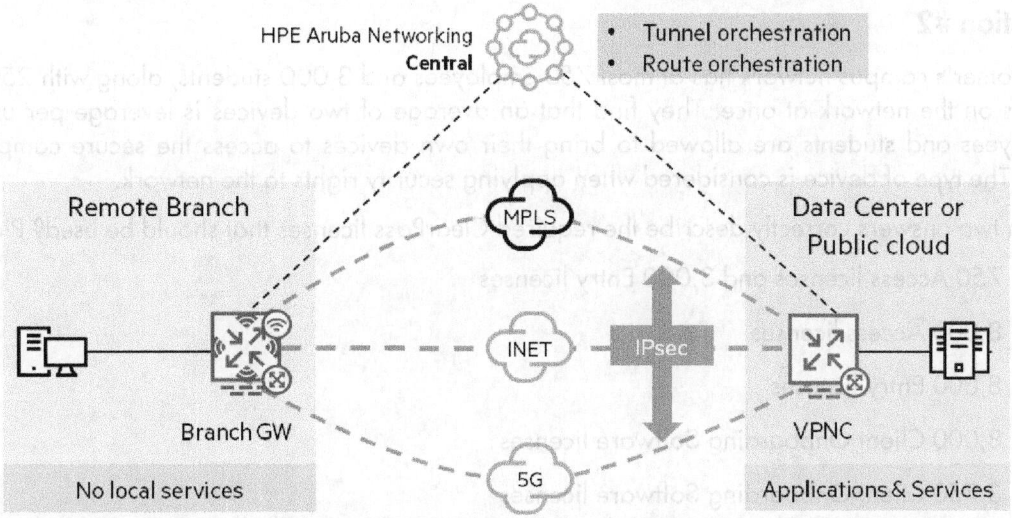

Routing is another critical consideration for connectivity; after all, border devices (branch gateways and VPN concentrators) must know the next hop to reach a destination. The Routing Orchestrator (formerly known as Overlay Router Orchestrator – ORO) is the routing service that simplifies routing. Similar to the Tunnel Orchestrator service, the Routing Orchestrator runs on HPE Aruba Networking Central.

The image displays three WAN services, on each, an independent IPsec tunnel is automatically created.

Learning check

Check your understanding of campus security by answering the following questions.

Question #1

Which one is not a part of HPE Aruba Networking security-first, AI-powered networking security?

- ○ Zero Trust Security (ZTS)
- ○ Artificial Intelligence Operations (AIOps)
- ○ Unified infrastructure with HPE Aruba Networking Central cloud management
- ○ Unified infrastructure with AirWave on-premises management

Question #2

A customer's campus network has at most 750 employees and 3,000 students, along with 250 guest visitors on the network at once. They find that an average of two devices is leverage per user. All employees and students are allowed to bring their own devices to access the secure campus network. The type of device is considered when applying security rights to the network.

Which two answers correctly describe the required ClearPass licenses that should be used? Pick two.

- ☐ 750 Access licenses and 3,000 Entry licenses
- ☐ 8,000 Access licenses
- ☐ 8,000 Entry licenses
- ☐ 8,000 Client Onboarding Software licenses
- ☐ 3,750 Client Onboarding Software licenses

Question #3

What is the recommended ClearPass Insight server placement for medium deployments?

- ○ Enable Insight on the publisher and the standby publisher as long as Subscriber traffic does NOT consume all Publisher resources.
- ○ Place ClearPass Insight on its own dedicated node.
- ○ Enable ClearPass Insight on the Publisher and Standby Publisher for high availability. Just be sure to move the authentication process to other nodes.
- ○ Enable ClearPass Insight on a second Publisher in its own cluster outside of the campus cluster.

Supplemental: AAA Overview

Use these notes to review AAA services when it is convenient for you.

AAA services

A common NAC solution is based on running AAA services on the authentication Server. So, the three A's of AAA services.

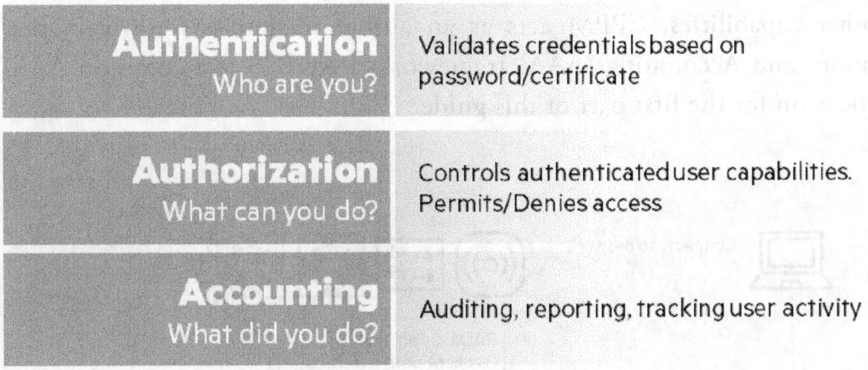

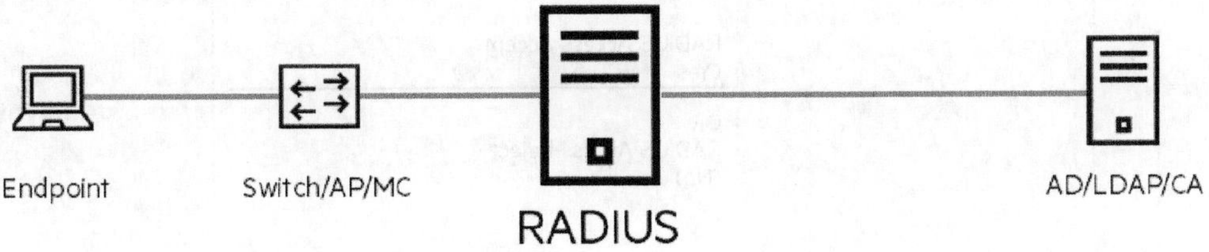

First up is authentication: who are you? The RADIUS or TACACS+ server permits or denies access based on your credentials. Perhaps your username is Maria, with password Secret123: access is granted if those are valid credentials. The user might alternatively submit a digital certificate with a valid username signed by a trusted certificate authority (CA).

Now that you're on the network, what can you do? That's authorization. User Maria is a member of the marketing group so that she can access all standard business applications, plus the marketing application and database. Other resources are off-limits for Maria. You might define a policy to authorize Maria to be assigned to a certain VLAN with a certain firewall and quality of service (QoS) settings, which the switch port dynamically applies.

You control who can access the network and what they can do. Now we need accounting: what did they do? With accounting, you have auditing, reporting, and tracking of user activity: when and from where they logged in when they logged out, and what resources they accessed during that time. This established non-repudiation; the user cannot deny their actions on the network.

Review Authentication, Authorization, Accounting (AAA)

CPPM provides the foundation for HPE Aruba Networking Zero Trust Security, so you will now learn some fundamental concepts that will help you use this powerful tool to secure the company's networks.

Among other capabilities, CPPM acts as an authentication server within the Authentication, Authorization, and Accounting (AAA) framework. RADIUS is a common AAA protocol, which you will focus on for the first part of this guide.

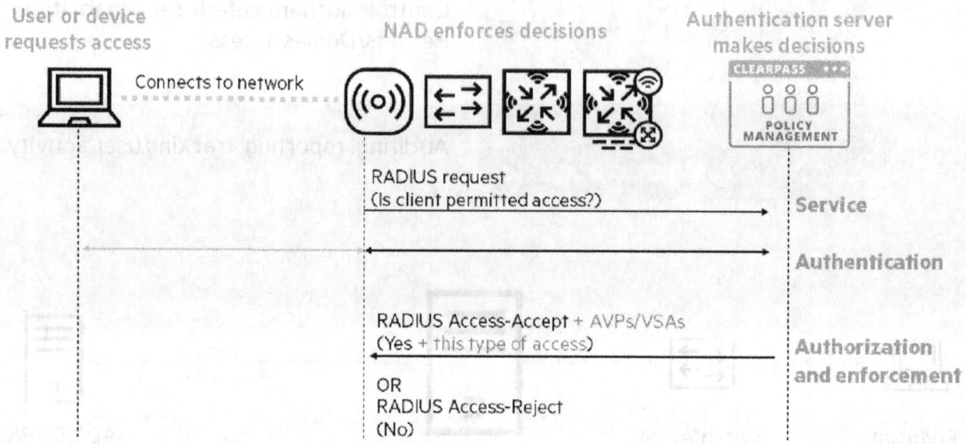

In AAA, a user or device—often called a client—requests access to a service. The client can request any service, but most often in this guide, the client requests network access by connecting to the network. In that case, the AP, switch, and gateway to which the client connects are the Network Access Device (NAD). (Sometimes, the NAD is called the Network Access Server (NAS). The NAD sends a RADIUS request to its authentication server to ask whether the client is permitted access.

As an authentication server, CPPM must first authenticate that the client is who and what it claims to be and, second, authorize the client for the correct level of access. It matches the RADIUS request to a service and uses that service to process the request. First, it implements the authentication method listed in the service. Often, that method calls for CPPM to exchange several more RADIUS messages with the NAD. If the NAD is enforcing 802.1X, the NAD acts as a go-between for the client and CPPM, which exchange Extensible Authentication Protocol (EAP) messages.

As a result of these communications, the client has passed or failed authentication. The CPPM service further processes the request and determines whether an authenticated client is allowed access and at what level. If CPPM decides the client is allowed, it sends the NAD a RADIUS Access-Accept. This message often includes attribute value pairs (AVPs) or vendor-specific attributes (VSAs), further defining the access level. If CPPM determines that the client is unauthorized, it sends a RADIUS Access-Reject.

Controlling network access

The following describes the job of a AAA server such as HPE Aruba Networking ClearPass Policy Manager Platform.

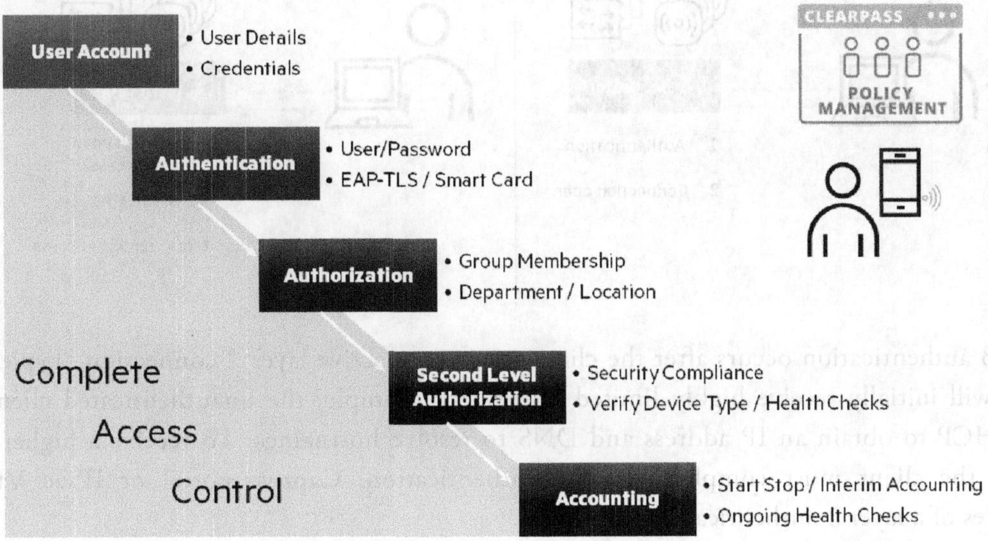

When a user joins the organization, they are given an account. This account details their relationship to the organization and their authentication credentials. These credentials are often stored in a Microsoft AD or LDAP database.

During authentication, the user provides credentials, which the network validates against the data store. This is typically done via a AAA server like ClearPass. Then, the network authorizes the user's account details and client device. The network may grant or deny access based on a complete picture of the client. If access is granted, the system continues to monitor client activity using start, stop, and interim accounting messages, with ongoing compliance checks.

ClearPass AAA-based deployment provides a complete access control system with three common authentication methods by leveraging all available information and implementing sound access rules.

Layer 2 vs layer 3 authentication

You can divide network authentication methods into two broad categories: layer 2 and layer 3. Layer 2 authentication methods authenticate users and devices before they receive an active layer 3 connection. If the client fails authentication, the client is disconnected at layer 2, preventing the client from sending or receiving any traffic. MAC authentication (MAC-Auth) and 802.1X are examples of layer 2 authentication methods.

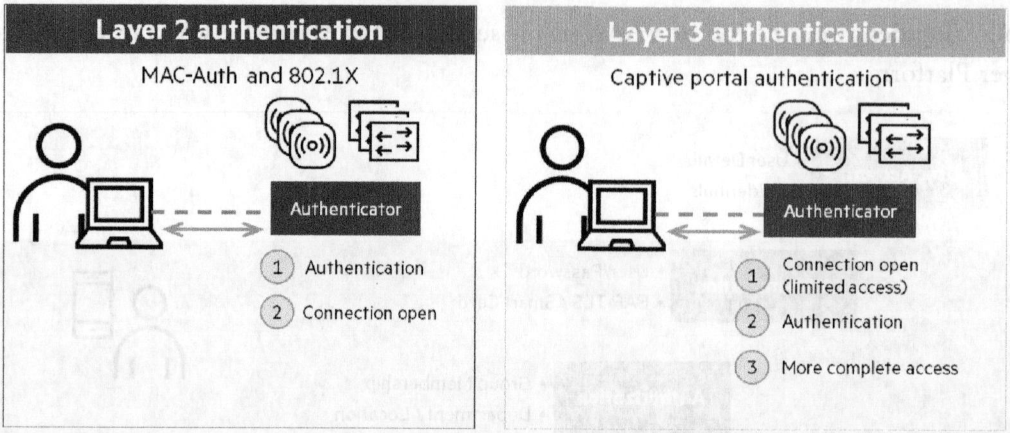

Layer 3 authentication occurs after the client receives an active layer 2 connection. Typically, the client will initially receive highly limited access. For example, the unauthenticated client might use DHCP to obtain an IP address and DNS to resolve hostnames. To receive a higher level of access, the client must complete layer 3 authentication. Captive portal or IPsec VPNs are examples of a layer 3 authentication method.

Port-based authentication use cases

One important use case for port-based authentication is to secure the edge: you must prevent unauthorized users from connecting to open switch ports. This may be required for compliance with corporate security policies and government regulations.

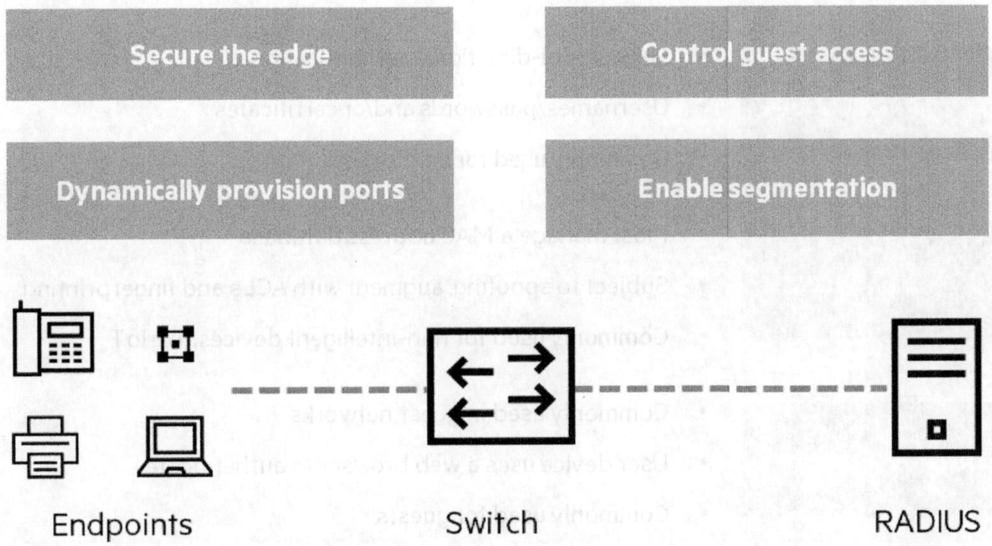

A second use case is to control guest access. You must ensure that guests are placed in the proper guest VLAN and must log in for legal and tracking purposes. A third use case that can help reduce administrative overhead while improving security is to dynamically provision access ports based on a user's identity or device.

Dynamic segmentation establishes least privilege access to applications and data by segmenting traffic based on identity and associated access permissions. This is a fundamental concept of both zero trust and SASE frameworks.

Note

Originally, HPE Aruba Networking used the term tunneled node to describe the dynamic segmentation feature. Today, HPE Aruba Networking uses the term dynamic segmentation when referring to the original tunneled node feature.

Common authentication methods

The authentication process validates a user or device identity to permit or deny network access. Typically, you only wish to permit authorized users with the proper network credentials. A MAC address, passphrase, username, password, or certificate can authenticate users with credentials preconfigured in an external source, like an LDAP or Active Directory (AD) data store. This is like checking a passport to verify a traveler's identity.

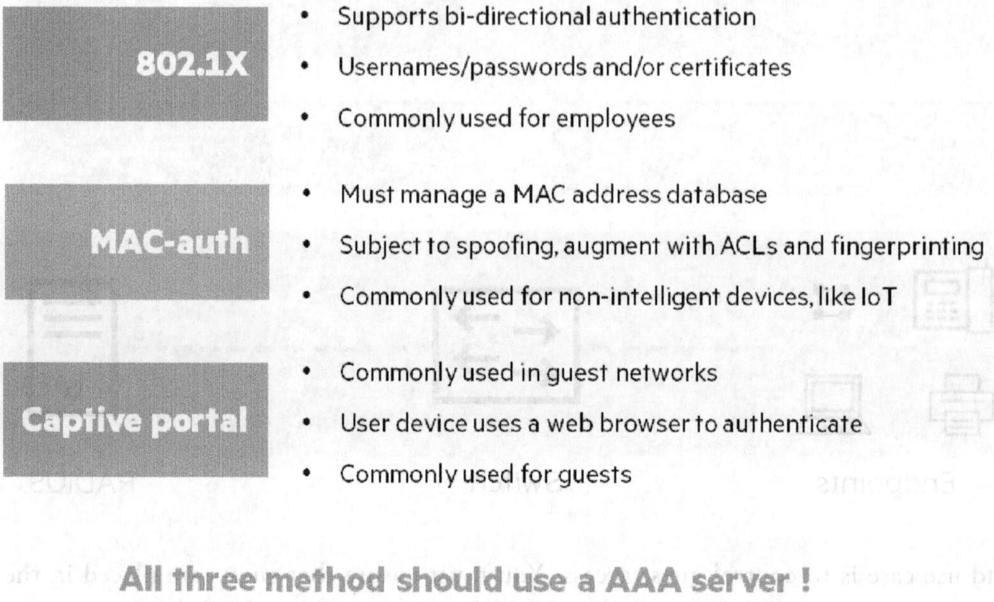

Let's briefly review the three common authentication methods:

- 802.1X
- MAC-auth
- Captive portal (commonly referred to as a web portal)

802.1X

802.1X is the most common method used to validate clients on enterprise-class networks, used in conjunction with an Extensible Authentication Protocol (EAP). There are many flavors of EAP, including PEAP, EAP-TLS, EAP-TTLS, and others.

You can think of 802.1X and EAP as "sister protocols" that work as a team. EAP uses 802.1X to transport the user credentials between the user device and the NAS. 802.1X happens at Layer 2, where users are authenticated before an IP address is assigned. This is independent of the layer 3 protocol. Most NAS solutions will then forward the EAP information to a centralized AAA server using the RADIUS protocol.

Note

A Network Access Service (NAS) is a common name used in AAA deployments and refers to the authenticator in 802.1X—the networking device that interacts directly with the user device.

802.1X was covered earlier.

MAC-auth

With MAC authentication, the network access device (NAS) validates an endpoint MAC address against a repository of valid MAC addresses. Endpoint MAC addresses are burned in ROM at the factory but are copied to RAM during bootup, and this is the MAC address an endpoint adds to Ethernet or 802.11 Wi-Fi frames. This means it is relatively easy for bad actors to change or "spoof" their MAC address to gain unauthorized access.

Since the MAC addresses in Ethernet and Wi-Fi frames are not encrypted, hackers can use packet capture software to find valid MAC addresses and change their MAC addresses accordingly. Due to these security weaknesses, MAC Authentication is only used for endpoints that cannot support better authentication methods like 802.1X. This includes specialty devices like older medical equipment, IoT sensors, and scanners.

To improve overall security with MAC-auth, use ACLs and firewall policies to strictly limit what devices can access. You can also use device fingerprinting to examine device protocol information, like DHCP and HTTP payload information, and then use this information to identify additional information about the device, like the product, operating system, and other information. This information can be used to determine which policy should be applied for the device's access.

Captive Portal

Some systems use layer 3 authentication, such as captive portal web authentication or VPN software. Captive portal is commonly used in guest networks to register and restrict their access, as well as onboard employee devices, like a new desktop, laptop, or smartphone. Captive portal is web-based and thus only requires a web browser on the user's device. Given that the web is based on the HTTP protocol, only minimal layer 3 access is needed to perform the captive portal: DHCP, DNS, and restricted HTTP perform the captive portal process.

MAC-auth overview

Objective: Control access for devices with limited capabilities

- Simple provisioning for printers, IoT devices (VLAN)
- Based on client source MAC address
- Minimal security—easily spoofed

- Second form of authentication, after 802.1X
- Captive portal with MAC-auth

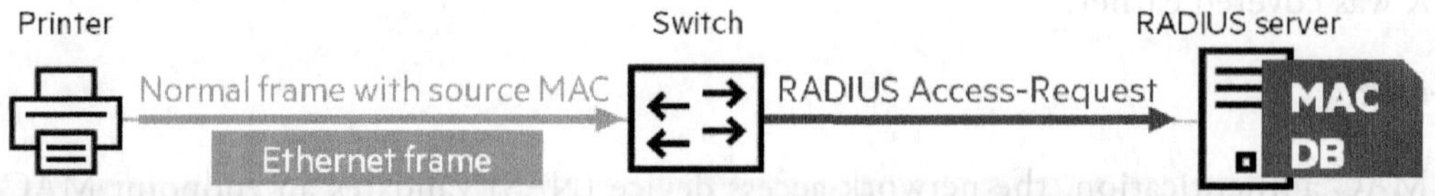

MAC authentication (MAC-auth) enables access control for devices with limited capabilities, with simple provisioning for printers and IoT devices that may not have an extensive user interface or lack support for 802.1X. In addition to permitting a device network access based on its group, the authentication server can assign dynamic settings, such as a VLAN ID and QoS priority, to the device's session.

These devices simply connect to a normal Ethernet frame, which includes their own MAC address as the source in those frames. In most cases, the device is probably generating a DHCP request. The switch harvests this source MAC address and sends a RADIUS Access-Request message to the RADIUS server. No client configuration is required on the client side.

When you need rigorous security, you should use 802.1X, whenever possible, since it supports a username and password and digital certificate authentication at layer-2. Often, it is trivial for a hacker to discover MAC addresses used in your network, whether by eavesdropping or accessing a legitimate device. The hacker can then easily spoof an authorized MAC address. MAC-auth might also be more difficult to manage than other forms of authentication, requiring directory or domain administrators to create accounts for each MAC address for each device. Meanwhile, 802.1X can use the user accounts and certificates already in the directory or domain.

MAC-auth relies on a RADIUS server. There is no need to configure MAC addresses on each switch: a RADIUS server or external server contains the MAC database (DB) for centralized control. You can define up to three RADIUS servers for redundancy if access to the primary server fails. It also means the same credentials are used for authentication, regardless of which switch or switch port is the current access connection into the LAN.

The switch acts as a port-access authenticator using a RADIUS server for authentication and uses the CHAP and PAP protocols to share the information with the RADIUS server. Inbound traffic is processed by the switch alone until authentication occurs. Some traffic from the switch to an unauthorized client is supported (for example, broadcast packets) before authentication occurs. And just like with 802.1X, you can make exceptions to LLDP and CDP messages if required.

Captive portal authentication

Look briefly at captive portal authentication as the primary example of Layer 3 authentication.

You are probably quite familiar with captive portal authentication, often implemented for guest wireless networks in airports, hotels, and coffee shops. Captive portal authentication provides an easy way for users to log in to a network without special requirements on the part of their devices or expertise on the part of the users. A user simply connects to an open network. The user then opens a Web browser, which is directed to a login page, in which the user can enter credentials or sometimes simply accept an agreement to log in. The user's client then receives more complete access to the network.

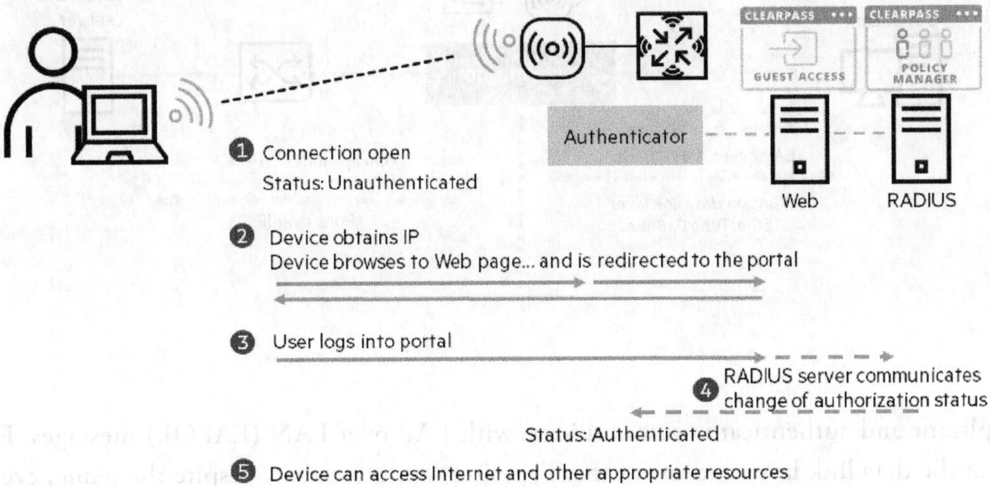

The process by which the network implements a captive portal can have many nuances based on how the company wants to implement the feature. The figure shows a high-level process that gives you a basic idea. In this example, the authenticator is an HPE Aruba Networking Mobility Controller (MC), which controls the access point (AP) to which a client connects. (HPE Aruba Networking Instant APs and switches also support captive portal). As shown here, MCs support an internal captive portal but can also integrate with ClearPass.

When a new client connects to the captive portal WLAN, the MC grants the client limited access. The client obtains an IP address and attempts to browse to a web page. The communications flow through the AOS firewall, redirecting the client to a login page on ClearPass Guest. The user logs in, and ClearPass Guest authenticates the user against ClearPass Policy Manager. ClearPass Policy Manager informs the MC of the user's new authentication status, and the MC assigns the user a new role based on that status. Now, the authenticated user can browse the Internet or use other network resources, as dictated by the user role.

You can attend other HPE Aruba Networking training if you want to learn more about captive portal authentication, as supported by HPE Aruba Networking networks.

802.1X authentication

In a moment, you will see how this process translates to 802.1X. To understand the process, you must understand the two communication protocols used in 802.1X.

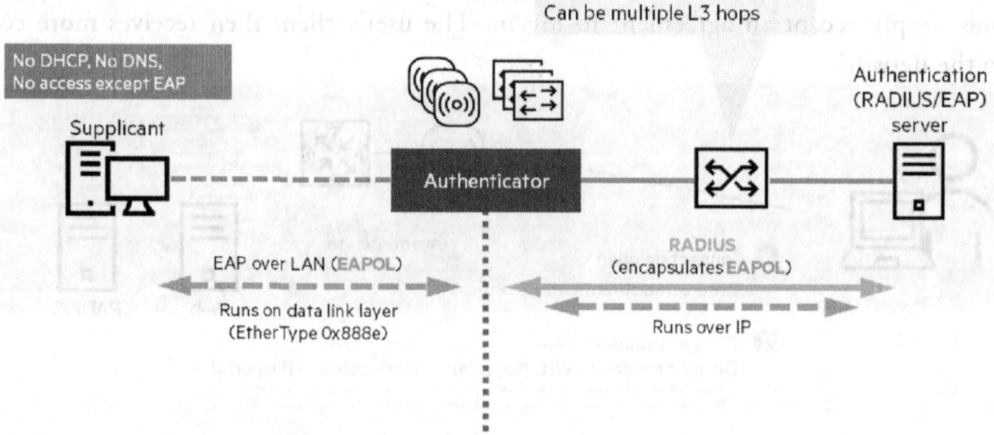

EAPOL

The supplicant and authenticator communicate with EAP over LAN (EAPOL) messages. EAPOL operates at the data link layer and uses EtherType 0x888e. Note that, despite the name, even wireless connections use EAPOL messages. In practice, the messages are often called simply EAP messages.

802.1X almost entirely shuts down the layer 2 link for unauthenticated clients. The client cannot even send DHCP or DNS messages into the network. The authenticator accepts nothing except EAP. This means that usually, the client has no IP address. It can only communicate on the local link. The supplicant is either directly connected to the authenticator, or if the authenticator is an HPE Aruba Networking gateway, an AP tunnels the supplicant's messages to the authenticator (gateway).

RADIUS

The authenticator reaches the AAA server over an IP network, and multiple layer 3 hops can exist between the devices. Like the security guard in the previous example, the authenticator acts as a relay between the supplicant and the authentication server. The authenticator encapsulates EAP messages from the client with RADIUS access requests, which it sends to the AAA server. Similarly, the AAA server encapsulates EAP messages for the client in RADIUS-challenge messages, which it sends to the authenticator. The authenticator decapsulates the EAP messages and passes them on to the client.

In this way, the supplicant and authentication server act as the two endpoints in an EAP exchange, with the authenticator transporting messages between them.

In RADIUS terms, the authenticator is the network access server (NAS), network access device (NAD), or RADIUS client. However, the rest of this discussion will use "authenticator" to refer to the device for consistency.

Examples of EAP methods

Many specific EAP authentication methods exist within the flexible framework. Some common ones are listed in the table. One important point is that each method provides mutual authentication—both the server and supplicant authenticate. The supplicant must authenticate the server to avoid man-in-the-middle (MITM) attacks in which the supplicant connects to a hacker device rather than the legitimate network. EAP-TLS is the preferred method and the one that you should implement for enterprise networks.

EAP method	Credentials	Comments
EAP-TLS	Server = Certificate Supplicant = Certificate	• Recommended option • Widely supported - Windows native • TLS handshake authenticates server and supplicant
PEAP	Server = Certificate Supplicant = Username and password (typical)	• TLS handshake creates secure tunnel (outer method) • A less secure authentication (MSCHAPv2) handles inner method
EAP-TTLS	Server = Certificate Supplicant = Username and password (typical)	• Similar to PEAP in concept
EAP-TEAP	Server = Certificate Supplicant = Username and password or certificate	• Emerging method (Windows 10, ClearPass 6.9 support) • TLS handshake creates a secure tunnel (inner method) • Enables "chaining" of machine and user auth with tokens

EAP-Tunneled Layer Security (TLS)

One of the most secure EAP methods, EAP-TLS, uses a TLS 1.2 handshake to authenticate the server and supplicant with digital certificates. The Windows native client supports EAP-TLS, as do many other supplicants. ClearPass also supports EAP-TLS. While deploying certificates to clients can be challenging, methods exist for easing and automating this deployment. You will look at some of these methods later.

Protected EAP (PEAP) and EAP-Tunneled TLS (TTLS)

PEAP was developed to provide some of the security of EAP-TLS without forcing stations to have digital certificates. It is the option that the Windows native client uses when the user logs in with a username and password.

Tunneled Transport Layer Security (TTLS) is a similar but less commonly used method. (TTLS was developed by Funk Software and Certicom; PEAP was developed by Microsoft, Cisco Systems, and RSA Security.) ClearPass supports both PEAP and EAP-TTLS.

TTLS and PEAP function in similar ways. Both methods involve a two-phase authentication process. In the first phase, the outer method creates a secure tunnel using TLS. The authentication server then initiates a second authentication method, called the inner method, inside the secure TLS tunnel. PEAP and EAP-TTLS can use several different options for the inner method. (In fact, they can even use EAP-TLS as the inner method; however, doing so does not typically make sense because the goal of PEAP and TTLS is to protect a less secure inner method.) EAP-Microsoft CHAPv2 (MSCHAPv2) is the most common option. EAP-MSCHAPv2 is not secure enough to run over an untrusted network but is acceptable when protected by the TLS tunnel.

EAP-Tunneled EAP (EAP-TEAP)

Like PEAP and EAP-TTLS, EAP-TEAP uses TLS as an outer method to create a secure tunnel. It then uses an inner method to authenticate the client. The client can authenticate with a username/password or a certificate, depending on the inner method implemented. EAP-TEAP also supports "chaining" of multiple authentications, which can be useful when a company wants to ensure users connect on authorized machines. The client can authenticate with machine credentials, proving the device is authorized. The client obtains a token during this authentication. The client can then authenticate with user credentials, including the token from the machine authentication, proving that both the user and device are authorized. EAP-TEAP is supported in Windows 10 and ClearPass 6.9.

EAP-PEAP considerations

EAP-PEAP Advantages

Like EAP-TLS, EAP-TTLS and PEAP provide strong, mutual authentication. Because TTLS and PEAP use encrypted tunnels to secure usernames and passwords (rather than require digital certificates on stations), you can implement these methods more easily than you can TLS.

TTLS has one unique benefit: it always protects the username. Depending on how PEAP is implemented, the username might be transmitted in plain text in the initial EAP/response identity packet, allowing a hacker to detect the user's identity and possibly lock the user out of his or her account.

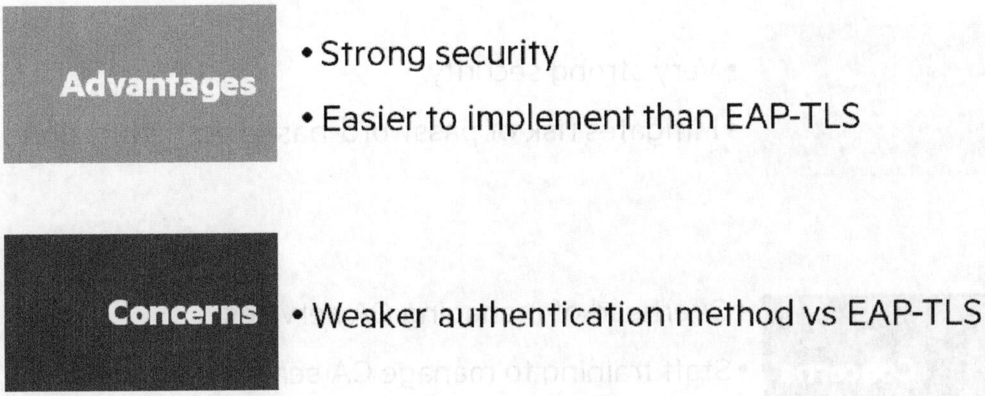

Concerns

As compared to EAP-TLS, EAP-TTLS and PEAP have the disadvantage of a weaker authentication method.

EAP-TLS considerations

The figure summarizes the advantages and concerns related to EAP-TLS. Please click on each box in the figure to explore.

Advantages

EAP-TLS is one of the most secure EAP methods because it provides mutual authentication with Public Key Infrastructure (PKI) digital certificates. Digital certificates rely on extremely strong asymmetric keys and are validated by trusted Certificate Authorities (CA). The digital certificate is typically stored on a laptop or a smart card, providing stronger security. Although you can set up requirements to force users to create stronger, and therefore more secure, passwords, you cannot prevent users from telling people their password or writing it on a paper displayed in plain sight. Generally, it is harder for users to leak certificates inappropriately; systems administrators might, for example, implement security measures to prevent users from copying certificates.

Hackers often run dictionary attacks on passwords, phish for passwords, or use other social engineering techniques to get users to leak their passwords. Although no credentials can be totally secure—malware might be able to steal certificates; for example—digital certificates are more resistant to many more common attacks.

Someone who steals a laptop or smart card might be able to access the certificates within. Still, they are usually protected by an additional layer of security, such as a password, which the user must enter before the supplicant can use the certificate. In addition, the user can immediately report the theft, allowing you to revoke the compromised certificate.

Advantages

- Very strong security
- Mitigates risk of password-based authentication

Concerns

- Overhead of managing CA services
- Staff training to manage CA services
- Certificate distribution and management

Concerns

EAP-TLS is not vulnerable to eavesdropping and replay attacks that affect less secure EAP methods such as EAP-MD5. Still, security comes at the cost of purchasing and managing digital certificates—substantially more expensive than managing passwords. Maintaining a large number of certificates requires specialized software and trained IT staff.

In addition, installing certificates on all clients can be difficult in environments with multiple operating systems. Bring Your Own Device (BYOD) environments can make managing certificates even more difficult. However, HPE Aruba Networking ClearPass BYOD onboarding capabilities can help companies surmount these challenges. ClearPass Onboard Software can provide new user devices with limited access and help users.

Validating domain users are on domain computers with EAP-TLS

However, with EAP-TLS (or PEAP), computer and user authentication occur separately. CPPM has to cache the [Machine Authenticated] role for the client across sessions.

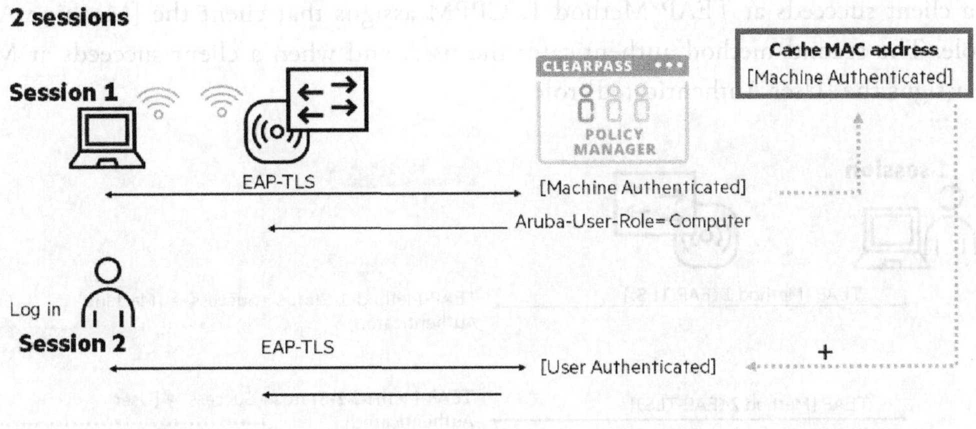

Challenge: Windows client does not always trigger sessions correctly

When a client authenticates as a computer, CPPM marks the client as Machine Authenticated and caches the client's MAC address with this status. When a user later logs in, CPPM marks the client as User Authenticated and also matches the client with the cached Machine Authentication. Role mapping and enforcement policies can dictate that the [User Authentication] status plus [Machine Authentication] delivers full privileges. In this example, CPPM tells NADs to place the client in the "employee-full access" role.

If a user logs in on a client that does not have a cached Machine, Authentication, role mapping, and enforcement policies can assign that client a different Aruba-User-Role, which provides limited privileges.

This EAP-TLS solution can work but can also introduce complexity and sometimes fail. The Windows client must figure out when to trigger machine or user authentication as separate sessions. Consider a situation in which a user logs into a domain computer for the first time. If the company is using certificate auto-enrollment, it can take time for the computer to obtain the new certificate. The user authentication might time out, leaving the user without full access.

Validating users are on domain computers with TEAP

Windows 10 2004 and CPPM 6.9 support EAP-TEAP, or TEAP, which provides a more graceful solution to validating whether domain users connect on domain computers. Whenever the Windows client authenticates, it does so as both a machine and a user.

Much as with PEAP, TEAP creates a TLS session and carries another EAP method within the session. However, TEAP supports the chaining of multiple authentication processes. The supplicant "chains" a token from the earlier authentication to the subsequent authentication. This way, the machine and user authentication can occur within the same session. The Windows OS using TEAP runs two methods whenever a user logs in. The first method authenticates the computer.

When a client succeeds at TEAP Method 1, CPPM assigns that client the [Machine Authenticated] role. The second method authenticates the user, and when a client succeeds at Method 2, CPPM assigns the [User Authenticated] role.

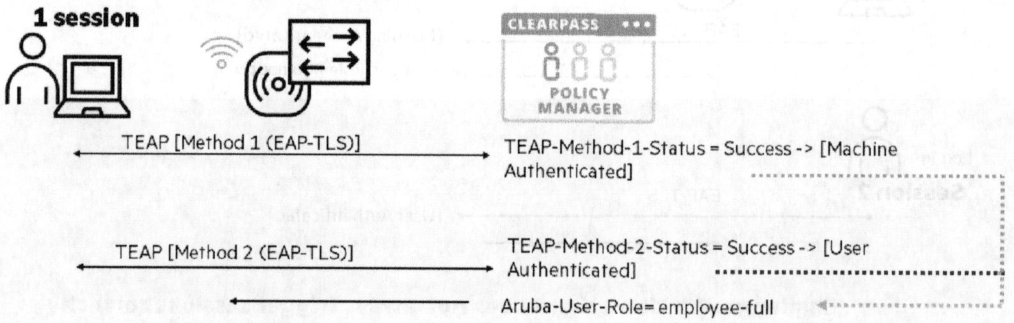

Windows client always authenticates as Machine and User

TEAP supports multiple types of inner methods. However, HPE Aruba Networking recommends EAP-TLS for both Method 1 and Method 2. The computer authenticates with its computer certificate, and the user authenticates with the user certificate.

TEAP makes it easier for the Windows domain client to authenticate consistently as both a computer and a user. The client is less likely to fail to authenticate as a user and get stuck in a computer role. On CPPM's side, CPPM no longer needs to cache the [Machine Authenticated] role.

As of the release of this guide, Windows AD does not support configuring TEAP on domain computers through GPOs. However, Microsoft Endpoint Manager does support pushing TEAP to clients.

WPA/WPA2-personal issues

WPA2-PSK has two main, related flaws. First, the PMK and all other keys ultimately derive from a human-selected passphrase. Humans rarely select good passphrases because they cannot remember good passphrases. So, it is important that any key that derives from a passphrase cannot be subject to a brute-force dictionary attack. However, WPA2-PSK is subject to brute-force dictionary attacks!

PMK derived directly from admin-selected passphrase

- Subject to offline dictionary attacks from captured handshakes

PMK shared by all users

- Passphrase often shared publicly
- Easy to derive another client's PTK

An attacker can capture a valid handshake between a client that knows the password and the AP. Two of the handshake messages include a MIC, which is a hash of the plaintext message with the KCK (one of the keys derived from the PMK). An attacker takes this data offline and uses the massive parallel processing power now available with GPUs and hacking farms to try every possible passphrase until one produces keys that create the right MIC.

Even if the passphrase is not a common word and includes special characters if it is eight characters or under, processing power is high enough now that brute force attacks can quickly crack the PSK. While companies could attempt to use very long and random strings, doing so is not feasible in practice.

Another issue with WPA2-Personal is that the password is often freely shared. While every client technically has its own PTK, anyone can easily derive another client's PTK. You just need the client's and AP's MAC addresses and the nonce values transmitted in plaintext during the handshake. Anyone who knows the passphrase can decrypt other devices' traffic. While you could hope that no one trusted with the passphrase would do this, that is often not the case.

Wireless security: WPA3

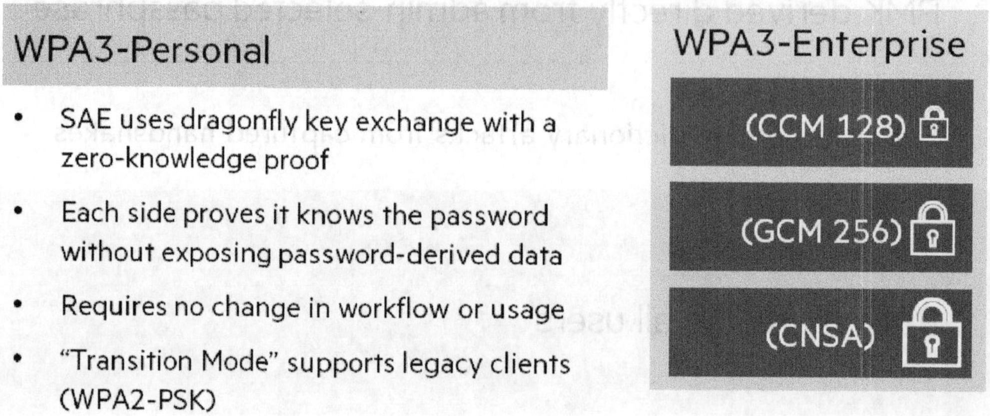

WPA3-Personal

- A replacement for WPA2-PSK
- SAE uses the Dragonfly key exchange to perform password-authenticated key exchange (RFC 7664) with a zero-knowledge proof. This exchange is resistant to active, passive, and dictionary attacks.
- Enabling "Transition Mode" supports backward compatibility with legacy clients connecting with WPA2-PSK.
- Each side proves it knows the password without exposing password-derived data. User experience is identical to providing the password in WPA2-PSK.

WPA3-Enterprise (CCM 128)

- Dynamic keys using 802.1X
- Backwards compatible with WPA2
- Enables optional support for 802.11w Protect Management Frame (PMF).
- Supports bridge, tunnel, and mixed-mode SSIDs.
- Popular choice for transitioning customers that still need to support client devices that only support WPA2.

WPA3-Enterprise (GCM 256)

- Requires new key management with SHA-256, new ciphers, and PMF.
- Legacy clients (WPA2) are not supported.
- Recommended for sites requiring stronger key management and encryption.

WPA3-Enterprise (CNSA)

- It uses the Commercial National Security Algorithm (CNSA) 192-bit.
- Key management with SHA-384
- Mandatory PMF endpoint support.
- Requires compatible EAP servers such as HPE Aruba Networking CPPM 6.8 or later versions and EAP-TLS.
- Supports bridge, tunnel, and mixed-mode SSIDs.
- Primarily used by military and government agencies.

WPA3-Personal SAE

Instead of using a passphrase as the source of a PMK, WPA3-Personal uses Simultaneous Authentication of Equals (SAE) to generate a different PMK each time a station connects.

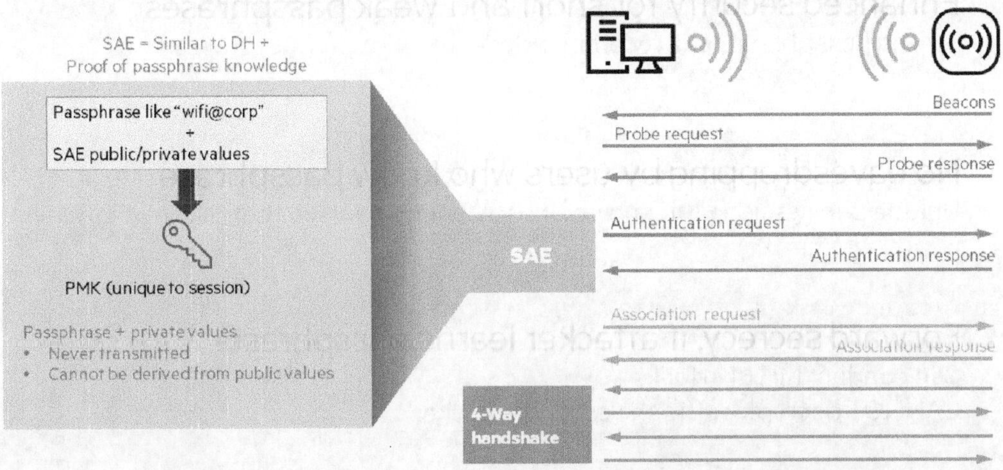

The AP advertises SAE as the authentication method for a WPA3-Personal SSID in its beacon and probe response frames. This is a new 802.11 authentication method (value 3); WPA3-Personal is the only security option that uses an 802.11 authentication method other than open-system. (Shared-key is a deprecated, insecure option that should never be used.)

The 802.11 SAE process involves four frames. You do not need to know the precise exchange but should understand the high-level concepts. SAE is based on the Dragonfly Key Exchange defined in IETF RFC 7664. This key exchange process uses some of the same cryptographic concepts as Diffie-Hellman and can use either Finite Field Cryptography (FFC) or Elliptic Curve Cryptography (ECC). AOS uses group 19, a 256-bit Elliptic Curve group. The added twist distinguishing SAE from Diffie-Hellman is that the peers can only derive the same shared key—the PMK—if they know the same passphrase.

After going through SAE, the client and AP have authenticated each other and derived a PMK unique to the session—the client associates. The familiar four-way handshake then occurs.

WPA3-Personal benefits

WPA3-Personal offers many improvements over WPA/WPA2-Personal.

Passphrases not subject to offline dictionary attacks
New SAE process for every guess at passphrase

Enhanced security for short and weak passphrases
Strong passphrases still recommended

No eavesdropping by users who know passphrase
Unique per-session PMK known only by participants in that SAE exchange

Forward secrecy, if attacker learns passphrase:
Can conduct MITM attacks
CANNOT decrypt traffic she collected earlier

The SAE exchange is not subject to offline dictionary attacks because it is computationally infeasible to derive the passphrase from the values transmitted in the SAE exchange. While an attacker can still take the four-way handshake offline, the PMK is no longer directly derived from the passphrase, making it more resistant to a dictionary attack. Cracking the PMK will not be useful because it is different for each session. If an attacker wants to launch a dictionary attack, she must go through the SAE process and handshake each attack infeasible. Therefore, administrators can choose weaker passphrases while still maintaining good security. SAE is designed to work with a large pool of passphrases known to hackers. However, it is still best practice to choose good passphrases.

WPA3-Personal also protects users' communications from eavesdropping by other authorized users. Every session has a unique PMK, known only to the client and AP that established the session.

WPA3-Personal further provides forward secrecy, meaning that a future passphrase compromise does not compromise data transmitted earlier. If an attacker learns the passphrase somehow—perhaps through a social engineering attack—they can conduct honeypot AP and MITM attacks. However, they cannot return to the encrypted data they collected earlier and decrypt it. This is because the attacker was not part of the SAE exchange that derived the PMK for those exchanges.

WPA/WPA2/WPA3-enterprise

You will now move on to WPA/WPA2-Enterprise.

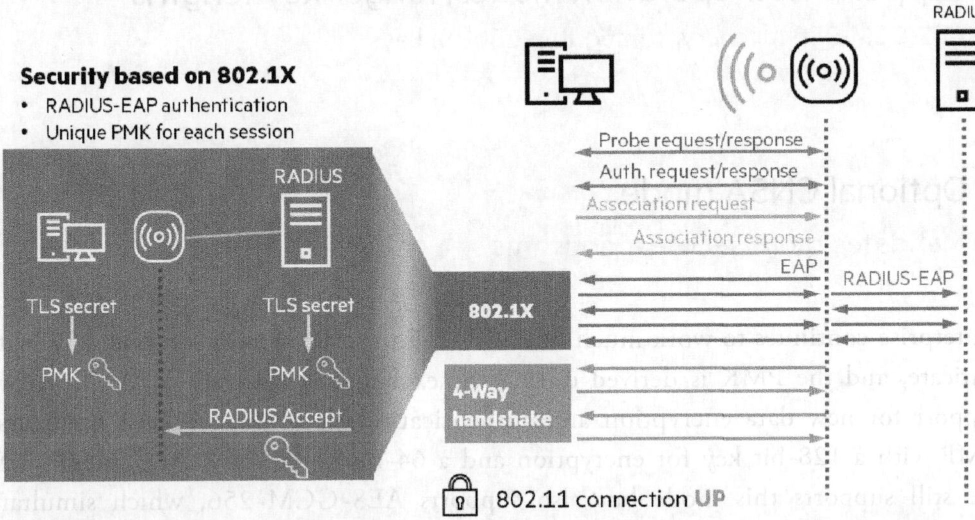

A connection to a WLAN that enforces this form of security begins with open-system 802.11 authentication and an 802.11 association. 802.1X then shuts down the connection to all but EAP traffic. EAP authentication proceeds, as you examined previously, enabling the client and RADIUS/EAP server to authenticate each other. In addition to authentication, the EAP process generates a TLS master secret (whether EAP-TLS, PEAP, or another EAP method is used). The client and RADIUS/EAP server use this secret to derive the PMK.

Because the AP or gateway is the authenticator that needs the PMK for encrypting and decrypting wireless traffic, the RADIUS server sends the PMK to the authenticator in the RADIUS Access-Accept message. It encrypts the PMK with the RADIUS shared secret. Both the client and the authenticator know the PMK for that session. The client and AP completed the four-way handshake, and the connection was up.

WPA3-enterprise: overview

Same basic framework as WPA2-Enterprise

Supports WPA-2 AES CCMP: 128-bit key encryption, 64-bit key integrity check

Supports new operation modes, longer key lengths

AES-GCM-256 for encrypt/integrity – 256-bit keys

Optional CNSA mode

Mandates more secure algorithms

WPA3-Enterprise continues to work much like WPA2-Enterprise. Stations use 802.1X with EAP to authenticate, and the PMK is derived during the exchange. However, WPA3-Enterprise has added support for new data encryption and authentication algorithms. WPA2-Enterprise used AES CCMP with a 128-bit key for encryption and a 64-bit key for ensuring integrity. WPA3-Enterprise still supports this mode but also supports AES-GCM-256, which simultaneously encrypts and ensures integrity for data with 256-bit keys. The key new feature is (CNSA) mode.

How the RADIUS server sends dynamic settings

You define dynamic policy settings on a RADIUS server, like ClearPass. These settings apply to certain users based on their identity and context.

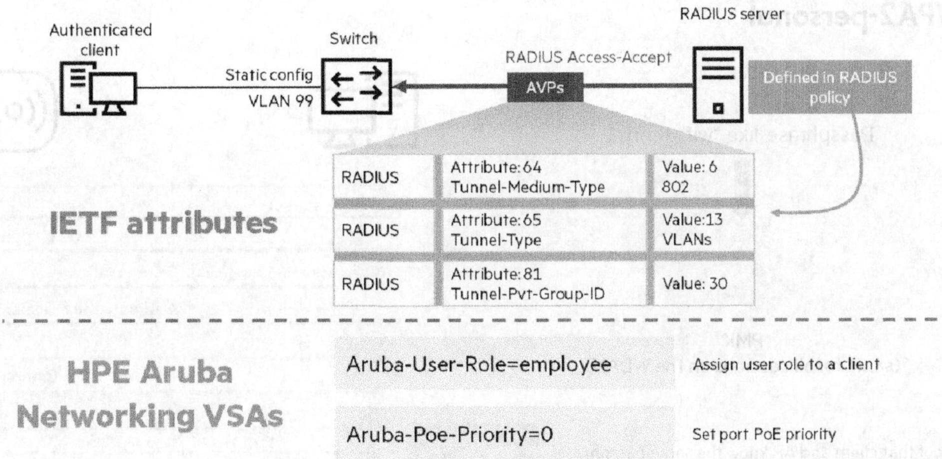

The server communicates these settings to the NAS switch in attribute-value pairs (AVPs) within the RADIUS-Accept message. They are called AVPs because each data item has an attribute type and an associated value. The top of the slide shows three IETF AVPs:

- Attribute Type 64 defines the type of medium over which this communication occurs, with a value of 6, which means an 802-type "tunnel."
- Attribute Type 65 defines the tunnel type, set to 13. This is used to indicate that we intend to assign a VLAN ID.
- Attribute Type 81 defines the actual VLAN ID (in this case, VLAN 30).

Thus, this set of AVPs sent from the RADIUS server to the switch overrides the port's statically configured VLAN 99 with VLAN 30.

The IETF defines standard AVPs in an RFC, and custom attributes are defined specifically by a vendor like HPE Aruba Networking. These are called vendor-specific attributes (VSAs). There are also extended RADIUS attributes.

WPA/WPA2-personal

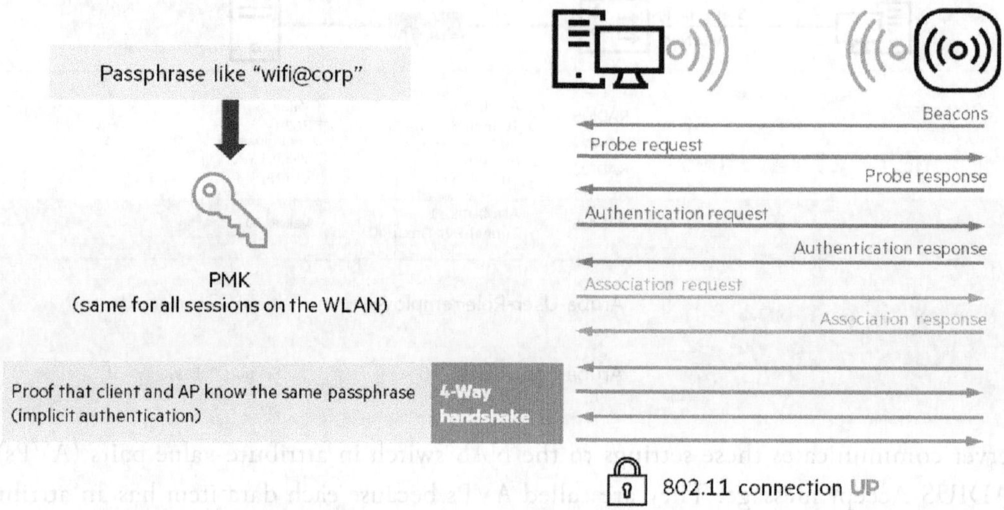

You will next look at WPA/WPA2-Personal WLANs. A client that wants to connect to a WPA/WPA2-Personal WLAN passes through open-system 802.11 authentication and then 802.11 association. It then conducts the four-way handshake. Remember that part of the four-way handshake is proving knowledge of the PMK. For WPA/WPA2-Personal, the PMK is expanded out predictably from the passphrase configured for the WLAN. Thus, the four-way handshake proves that the client and AP use the same passphrase, providing implicit authentication. If the four-way handshake is successful, the connection comes up.

OnConnect for wired non-RADIUS enforcement

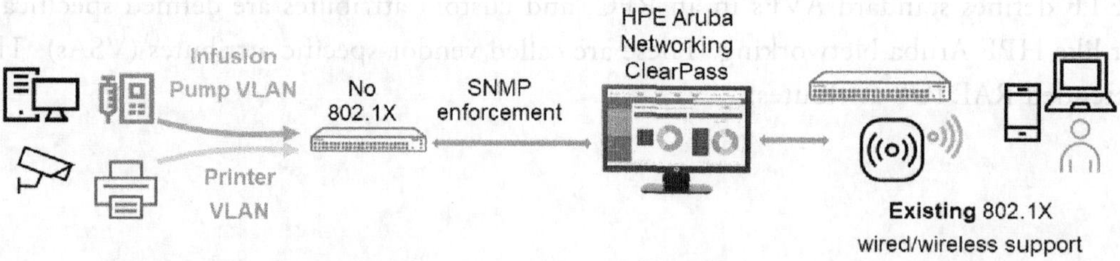

For example, a customer might have an HPE Aruba Networking wireless network but is not yet ready to upgrade to a wired network.

- HPE Aruba Networking ClearPass OnConnect enables access control for non-AAA-ready infrastructures and is easy to configure on legacy switches from many vendors.

- ClearPass uses SNMP to integrate with the legacy infrastructure and learn when a new device connects.
- WMI configuration is used to retrieve the Logged in User information. Whenever a domain-joined Windows client connects to an OnConnect-enabled port with the domain user logged in, Authorization attributes for this user are fetched from authorization sources to determine the role of the user; this information is then used in configuring policy enforcement.
- It then determines whether to let the device connect, using SNMP to enforce that decision on the switch by leaving the port up or taking it down.

The legacy infrastructure uses SNMP to inform ClearPass of the connected device's MAC address.

- ClearPass can then integrate with a Windows directory at the backend to determine which user is logged in on a particular device.
- In this way, it can add user-based information to its decisions.

What is a digital or X.509 certificate?

Digital X.509 certificates bind an identity to a public key

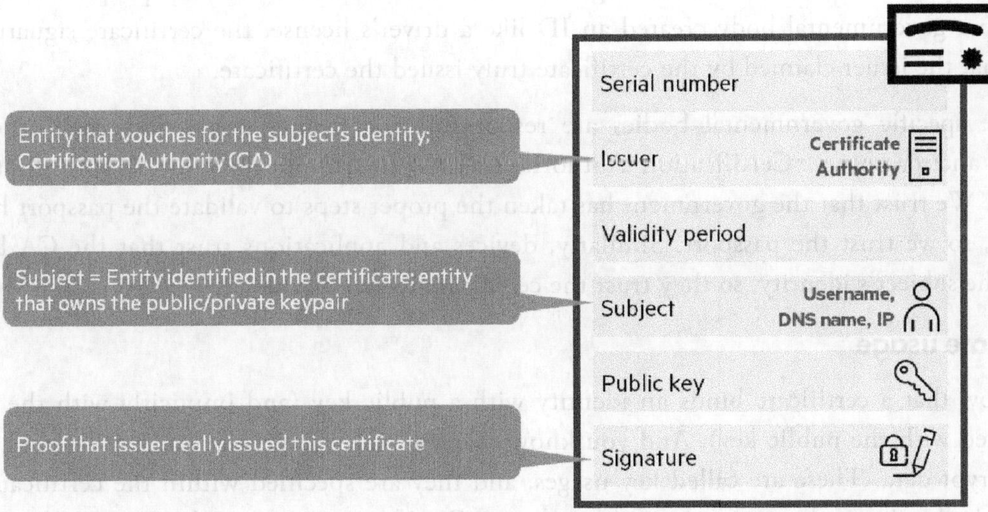

Certificates provide key management: they tie a specific identity to a public key. In other words, they verify that the public key posing as "User1 key" belongs to User1. They do this without User2 having to install User1's public key on his machine in advance. This enables User1 and User2 to set up secure communications on the fly.

Digital certificates follow a standard format based on industry-standard X.509, so you will sometimes hear certificates called X.509 certificates. The figure shows some of the key fields within an X.509 certificate.

The serial number uniquely identifies the certificate among certificates issued by the same CA. The validity period gives the certificate a specified period in which the certificate is valid. The longer a device can retain a certificate, the less secure the certificate becomes. Perhaps the private key is compromised, or the certificate owner is no longer approved for the rights that the certificate provides. Just as you must renew your driver's license or passport periodically, devices must renew their certificates periodically.

The issuer identifies who issued this certificate to the subject. This is a crucial component of certificate trustworthiness. Hackers could create a certificate with their own public key and any identity they want to claim. But unless a trustworthy source has issued the certificate, you would know to reject the certificate.

The subject identifies the entity that owns the certificate and the keypair associated with the certificate.

The public key is the subject's public key.

A certificate also includes the issuer's signature. Much like watermarks help people validate that the proper governmental body created an ID like a driver's license, the certificate signature validates that the issuer claimed by the certificate truly issued the certificate.

Just like specific governmental bodies are responsible for issuing documents such as driver's licenses and passports, Certification Authorities (CAs) are responsible for issuing digital certificates. We trust that the government has taken the proper steps to validate the passport holder's identity, so we trust the passport. Similarly, devices and applications trust that the CA has validated the subject's identity, so they trust the certificate issued by the CA.

Certificate usage

You know that a certificate binds an identity with a public key (and implicitly with the private key paired with the public key). And you know that you can use private and public keys to sign and encrypt data. These are called key usages, and they are specified within the certificate. Certificate signing is another example of a key usage. Certificate signing is data signing but for the specialized purpose of issuing certificates. Only CA certificates should have the certificate signing usage.

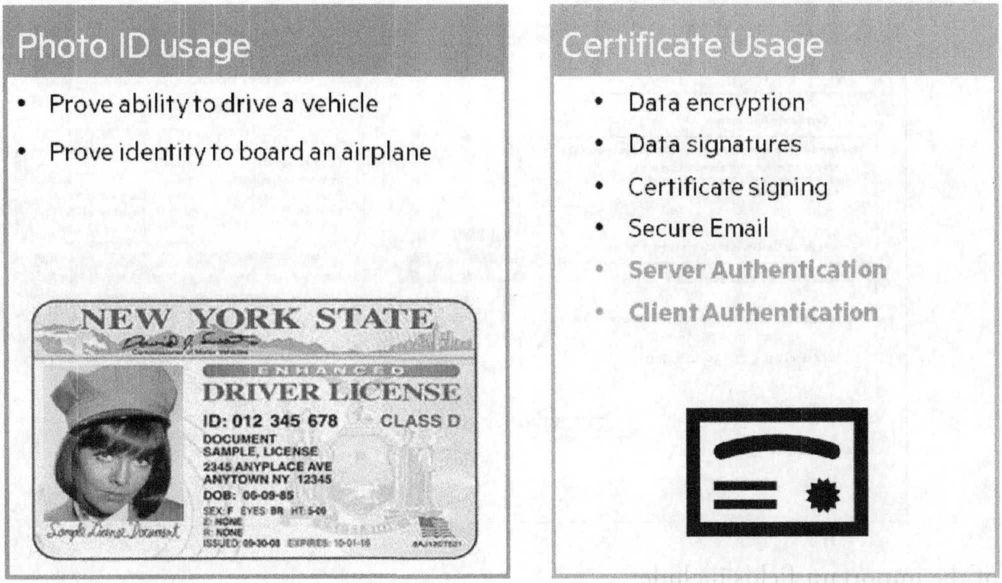

Also, consider a broader viewpoint: why the device must encrypt or sign data and how the device will use the certificate. People might use a driver's license to prove they can drive a car or to prove their identity when boarding an airplane. Likewise, users and devices can use their certificates for multiple purposes, defined in extended key usages (EKUs), which can also be included in the certificate. For example, the public key in a certificate with the Secure Email extended key usage can encrypt an email destined for the certificate owner.

The most relevant EKUs for your purposes are Server Authentication and Client Authentication. To authenticate itself with a certificate, either a server or a client signs data with the private key. By signing the data, the device proves that it owns the private key associated with the certificate and is the source of the signed communications.

Note

It is optional to define key usage – certificates function without a usage defined. But this means the certificate can be used for anything! This widens the possibility of abuse. When defining how certificates shall be created, it is a best practice to lock down how they can be used.

Example certificate

In the figure you see an actual certificate, as shown on a Windows machine.

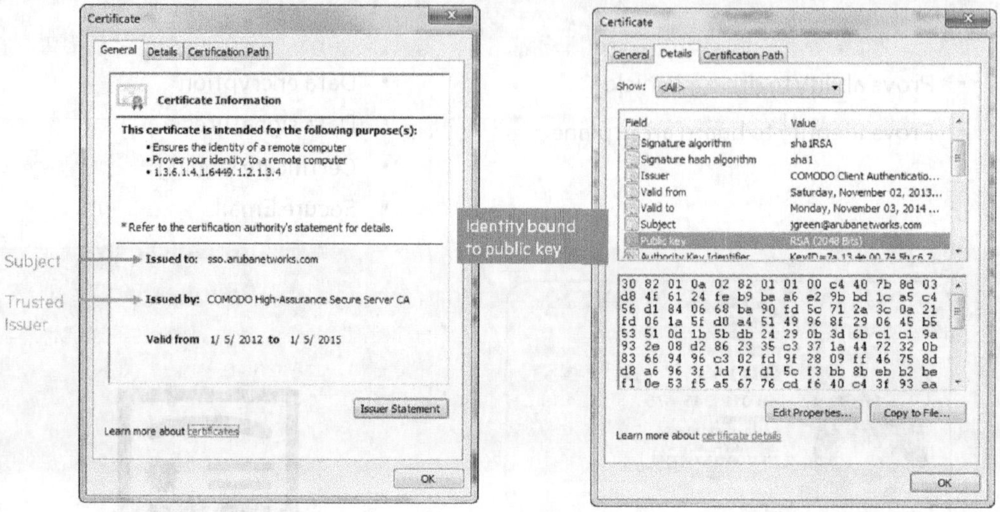

Some of the important fields include:

- Subject name, which identifies the owner of the certificate
- Issuing entity (the name of the CA that signed the certificate)
- Certificate serial number, which is unique to this certificate and issuing entity
- Validity dates (just like passwords should expire, secure certificates should expire too, forcing a renewal in which the entity generates a new keypair and once again proves its identity)
- Public key—Remember that the certificate includes the public key itself.

You can also look through these fields to see the key usages and EKUs.

CAs and the public key infrastructure

Now that you have looked at certificates, you examine the mechanisms for issuing those certificates.

Certificates exist within a public key infrastructure (PKI) hierarchy, enabling devices to determine which certificates they should trust. Every certificate within the PKI is signed by an entity that vouches for the authenticity of the certificate holder's identity.

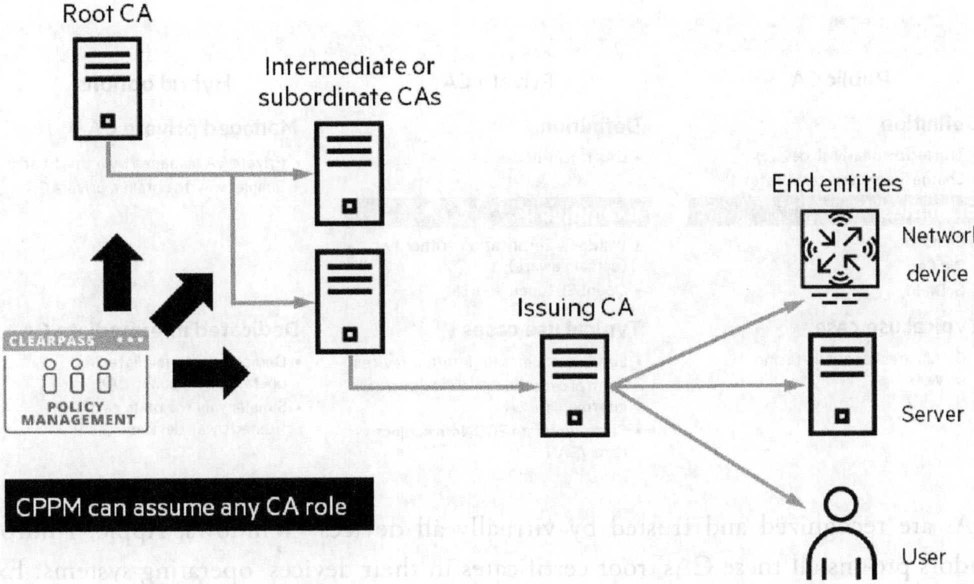

Root CAs are at the top of the hierarchy. They sign their own certificates, called root CA certificates. Root CAs are well-known entities that are widely believed to be trustworthy. For example, DigiCert and Comodo are root CAs. A company can also set up its own internal or private root CA, which is believed to be trustworthy within the company.

As the ultimate source of trust, root CAs must be rigorously secured. The root CA issues just a few certificates, issuing and signing certificates for intermediate CAs. This ensures minimal contact between the root CA and other entities. The root CA is typically kept offline for security purposes. The intermediate CAs, sometimes called subordinate or signing CAs, are responsible for issuing and signing certificates for end entities such as Web servers, network infrastructure devices, and users. The end entity certificates are sometimes called server certificates, user certificates, or client certificates, depending on the type of entity to which they are assigned. They are also sometimes called leaf certificates because they are on the other side of the PKI tree from the root. A PKI could also have multiple tiers of intermediate CAs, as shown in the figure.

Trusted issuers

You now understand how certificates exist and are validated within a PKI. Next, you will consider how devices obtain their certificates. You can have your users and devices obtain certificates from public CAs, your organization's private CA, or both. Hybrid options are now available as well.

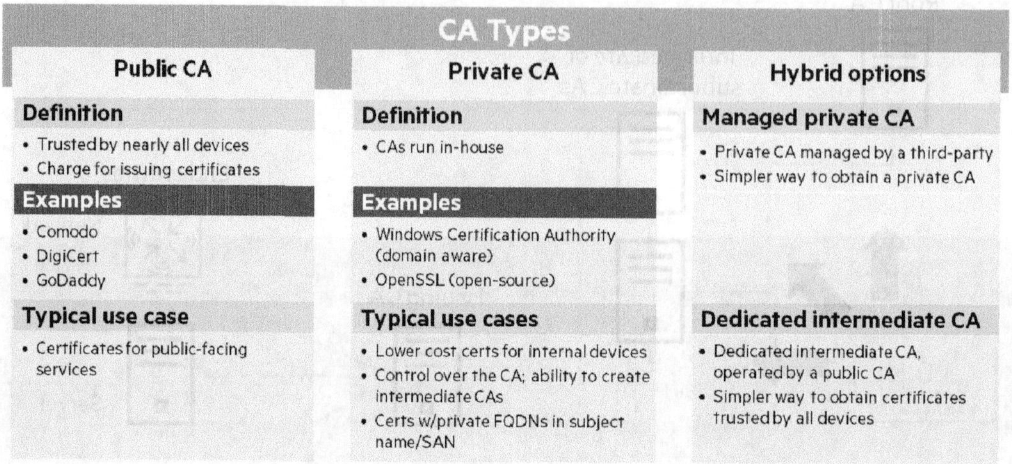

Public CAs are recognized and trusted by virtually all devices. Windows, Apple, Android, and other vendors pre-install these CAs' root certificates in their devices' operating systems. Examples of prominent public CAs include Comodo, DigiCert, and GoDaddy, but many other public CAs exist. You typically install CA-signed certificates on public-facing services so that you know that any devices that interact with those services will trust the certificates automatically. However, you must pay a fee to a public CA to obtain the certificate, which is why organizations are selective in which services use these certificates. In addition, public CAs now only issue certificates with 13-month validity periods.

Companies can set up their private CAs that are specific to their organization. They often use the Windows Certification Authority (CA) because it is domain-aware. However, open-source options such as OpenSSL also exist. A private CA can be a cost-effective choice for issuing certificates to many end-users and endpoints. In addition, the company can maintain more control over which users and devices receive certificates, establishing greater security for internal communications. With a private CA, the company controls the PKI and can create the intermediate CAs required. In addition, network infrastructure devices and non-public-facing services often use private IP addresses and fully qualified domain names (FQDNs). A public CA cannot issue certificates with a private IP address in the subject name or subject alternative name (SAN), based on RFC 1918. Similarly, public CAs do not issue certificates with subject names or SANs that specify private domain names, such as mydevice.example.local. Therefore, a company must use a private CA to obtain certificates for devices and services with private IP addresses or FQDNs. Companies might also use a private CA to issue certificates with longer validity periods.

A company that does not want to manage its own PKI could also obtain a managed private CA; for example, Globalsign offers such an option. In this way, the company can obtain the benefits of a private CA without needing the expertise to establish a CA in-house. Some public CAS, such as DigiCert, also offer a dedicated intermediate CA option. The public CA creates an intermediate CA, which it dedicates to the organization and operates for. The organization can then obtain all the certificates from that intermediate CA. Because the certificates are ultimately backed by the public CA, they will be trusted by nearly all devices inside and outside the organization.

Chain of trust

You can also look at the PKI as a chain of trust, which flows back up through the intermediate CAs to the root CA. Devices and applications can trust the server's certificate because they trust the intermediate CA's certificate. In other words, to validate the server's certificate, applications check the signature in the certificate using the issuing intermediate CA's certificate.

In turn, devices and applications trust the intermediate CA's certificate because they trust the root CA's certificate. Because the root CA certificate is self-signed, applications and devices must accept this certificate as trusted through another mechanism.

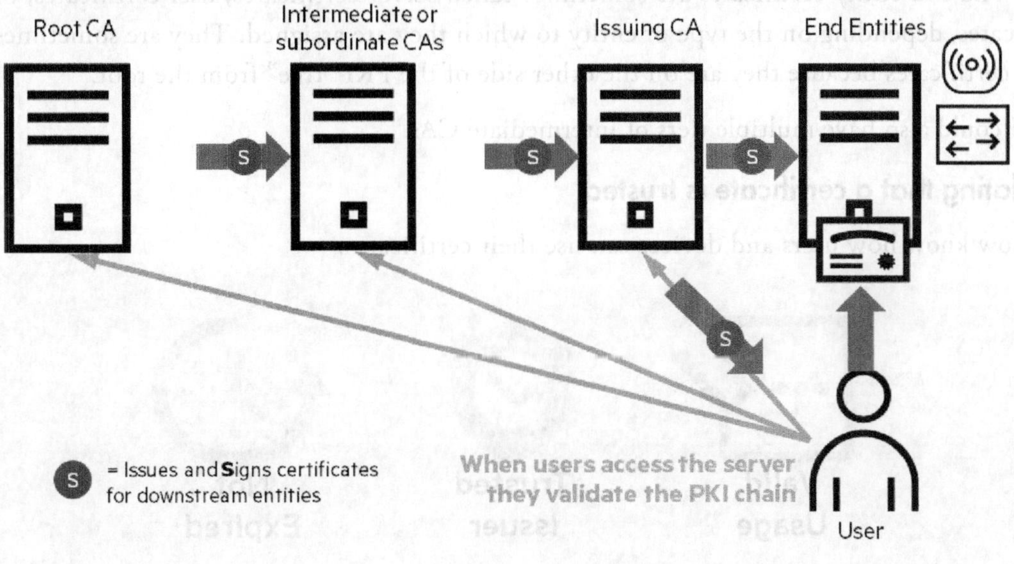

Typically, they are pre-configured to do so. This structure solves the practical problem of using asymmetric encryption. Rather than learning the public key for every entity you might want to communicate over a secure channel in advance, you only need to learn the certificate, including the public key, for a few root CAs.

Every certificate has an "issuer," the entity that vouches for the identity specified in the certificate. It also has a signature, which proves that the issuer truly issued the certificate.

The issuer forms a crucial component of the certificate's trustworthiness. Anyone can create a certificate with any identity they want—however, other devices only "trust" certificates issued by authorities they trust.

Certificates exist within a public key infrastructure (PKI) hierarchy, which enables devices to determine which certificates they should trust.

Root Certification Authorities (CAs) are at the top of the hierarchy. A root CA signs its own certificates, called root CA certificates. Root CAs are well-known entities that are widely believed to be trustworthy. For example, DigiCert and Comodo are well-known CAs. A company can also set up its own internal or private root CA, which is believed to be trustworthy within the company.

As the ultimate source of trust, root CAs must be rigorously secured. The root CA issues just a few certificates, issuing and signing certificates for intermediate CAs. This ensures minimal contact between the root CA and other entities. The root CA is typically kept offline for security purposes.

The intermediate CAs, sometimes called subordinate or signing CAs, are responsible for issuing and signing certificates for end entities such as Web servers, network infrastructure devices, and users. The end entity certificates are sometimes called server certificates, user certificates, or client certificates, depending on the type of entity to which they are assigned. They are sometimes called "leaf" certificates because they are on the other side of the PKI "tree" from the root.

A PKI could also have multiple tiers of intermediate CAs.

Validating that a certificate is trusted

You now know how users and devices can use their certificates.

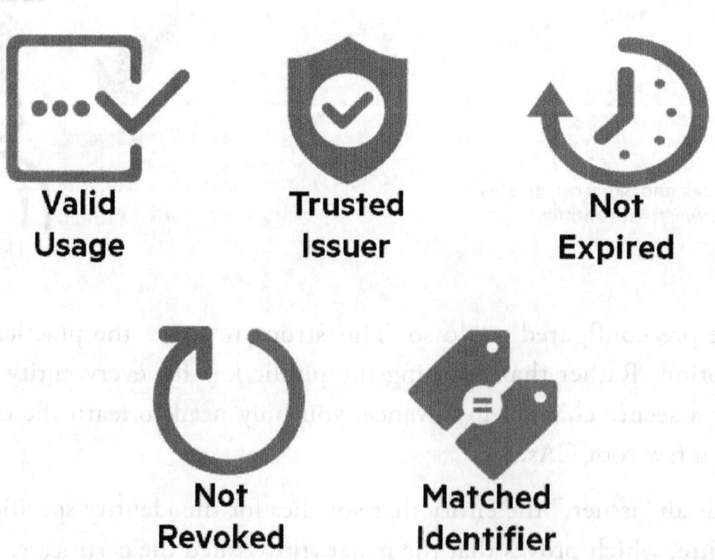

First, the certificate must have a valid usage. In other words, if the certificate owner uses the certificate to authenticate as a client, the clientAuth EKU must be specified within it.

The certificate must also be within its validity period. In addition, the certificate must be issued by a trusted source, must not be revoked by that source, and must have an identifier that checks out with the party the receiver believes it is communicating. These criteria call for a deeper look, which you will take over the next few slides.

Types of ClearPass server certificates

When you install a certificate on ClearPass, you can choose the usage. You can also choose an existing certificate and assign additional usage to it. In this way, you can ensure that ClearPass has a certificate signed by the correct CA for each use case. ClearPass uses server certificates for four purposes. Note that if you are using a Windows CA for any of the certificates, the WebServer template is recommended.

Type	Usage	CN in subject name	SAN	Typical issuing CA
RADIUS/EAP	Authenticating to 802.1X supplicants	FQDN wildcard prohibited	DNS (match CN)	Private
RadSec	Authenticating to RadSec clients	FQDN (or IP address)	Optional (match RADIUS address on clients)	Private or public
HTTPS	Establishing secure Web sessions to the UI, guest portal, Onboard, etc.	FQDN	DNS (match CN plus optional extra)	Public
Database	Establishing secure sessions to the Insight Database	FQDN or IP address	DNS, but must be IP address	Private

RADIUS/EAP

CPPM uses this certificate during the 802.1X process to authenticate itself to 802.1X supplicants (users and clients). This certificate must not use a wildcard for its CN, as some clients will reject RADIUS/EAP certificates with wildcards. The certificate should have a DNS SAN that matches the CN. The CN and SAN should not specify IP addresses to avoid exposing the IP address to users and permit the certificate to remain valid if the IP address changes.The supplicants must trust this certificate. Generally, you should obtain the certificate from a private CA. 802.1X authentication typically applies to corporate devices, which are managed or onboarded, so the organization can ensure that the clients trust the private CA. Using a private CA gives the organization more control over the certificate. Public CAs typically only issue certificates valid for 13 months, but a company can issue its own certificates with longer validity periods, making them easier to maintain.

You can install multiple RADIUS/EAP certificates on CPPM and associate a specific RADIUS/EAP certificate with a specific service. This feature is useful if CPPM authenticates users from multiple organizations and clients in different organizations who trust different CAs. You simply install certificates signed by different CAs on CPPM and then match the correct certificate with the service used by each domain.

RadSec

Because RADIUS itself does not provide rigorous security, CPPM (a RADIUS server) and network devices (RADIUS clients such as APs, gateways, or switches) can use RADIUS Secure (RadSec). RadSec carries RADIUS messages inside a secure TLS session and is useful for cases in which an untrusted network separates CPPM and the network devices. CPPM uses the RadSec certificate to authenticate itself to network devices that use RadSec. The network access devices (NADs) must be manually configured by admins to trust the root CA associated with this certificate.

Often, a private CA works well for issuing the RadSec certificate. The company controls its own devices and can ensure they trust the CA. However, some companies have particular requirements that call for public CA-signed certificates, and those work for RadSec as well.

HTTPS

ClearPass uses this certificate to authenticate itself to clients that access its web interfaces, including its management interface, Guest portal, and Onboard. Any clients that might need to access these interfaces must trust this certificate. If you are providing guest authentication services for users outside of the corporation, those users' devices will not trust the corporate CA. Therefore, the HTTPS certificate should typically be signed by a well-known public CA.

This certificate should also follow other best practices for HTTPS certificates. It is allowed to have a wildcard for the CN, but you may not necessarily have one, as wildcards can be less secure. The certificate must have a SAN with a DNS name that matches the CN; it can also have additional DNS names in the SAN. The CN and SAN should not specify IP addresses to avoid exposing the IP address to users and permit the certificate to remain valid if the IP address changes.

Database

ClearPass uses the database certificate to authenticate clients accessing its Insight Database. When you deploy a cluster, the database certificates of cluster members must have a DNS type SAN but with the server's IP address as the value. Because public CAs might not issue a certificate that meets these criteria, you will often obtain the database certificate from a private CA.

HPE Aruba Networking network infrastructure certificate options

Like ClearPass, HPE Aruba Networking infrastructure devices can use certificates for many usages, depending on the device type and its role in the architecture.

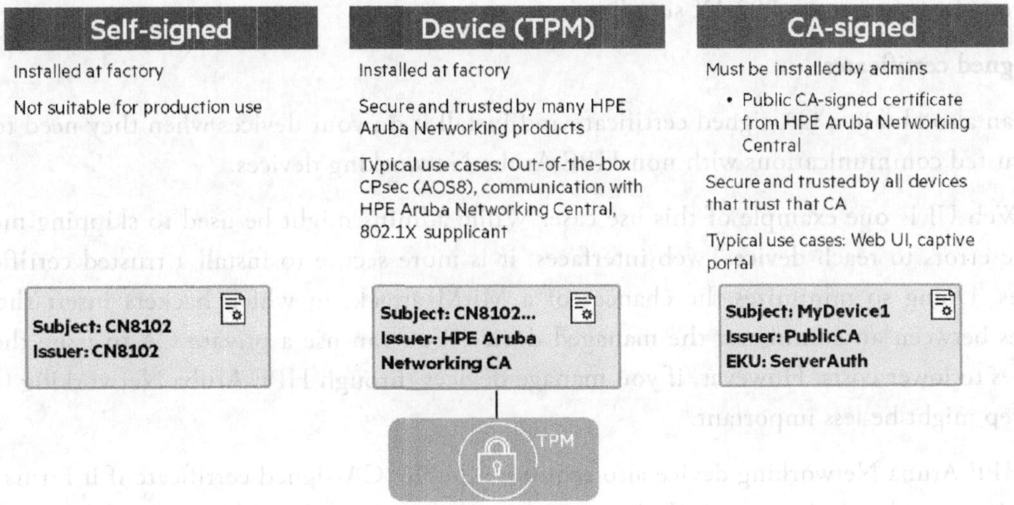

Self-signed certificates

The HPE Aruba Networking devices come with self-signed certificates, which let admins obtain initial access to the web UI using HTTPS. Self-signed certificates are not considered secure and should not be used in production.

Device (TPM) certificates

Current HPE Aruba Networking APs, gateways, and HPE Aruba Networking CX switches also come with factory-installed device certificates. An HPE Aruba Networking or HPE CA issues this certificate, and it is installed at the factory in a tamper-proof hardware Trusted Platform Module (TPM), making its private key highly secure. Admins find the certificates convenient to use as they are pre-installed and have a long validity period, eliminating the need for renewals. Because the certificate is issued by a secure HPE Aruba Networking or HPE CA rather than self-signed, it can be used in production.

Many HPE Aruba Networking products trust device certificates, which can be useful for setting up secure communications between HPE Aruba Networking devices. In an AOS-8 architecture, APs can discover MCs and set up CPsec tunnels (as long as the MC is configured to auto-accept them or has an allow list). In an HPE Aruba Networking security-first, AI-powered networking architecture based on HPE Aruba Networking Central, devices can establish secure connectivity with HPE Aruba Networking Central. HPE Aruba Networking Central typically learns the serial numbers included in the device certificates from Activate.

You do not need to set anything up to let HPE Aruba Networking devices use these certificates in these ways; they do so out of the box.

As you will learn later, you can also set up HPE Aruba Networking APs to use their TPM certificates when they act as 802.1X supplicants.

CA-signed certificates

You can also obtain a CA-signed certificate and install it on your devices when they need to establish trusted communications with non-HPE Aruba Networking devices.

The Web UI is one example of this use case. While admins might be used to skipping most certificate errors to reach devices' web interfaces, it is more secure to install a trusted certificate on devices. Doing so minimizes the chances of a MitM attack, in which hackers insert their own devices between an admin and the managed device. You can use a private CA to issue these certificates to lower costs. However, if you manage devices through HPE Aruba Networking Central, this step might be less important.

The HPE Aruba Networking device also requires a public CA-signed certificate if it forms part of a captive portal solution, even if the captive portal is hosted on an external server such as ClearPass Guest. The HPE Aruba Networking device redirects HTTPS traffic and needs a certificate that clients trust to do so successfully. Note that HPE Aruba Networking Central comes with a public CA-signed certificate, which it can push to devices that it manages, eliminating the need for you to acquire a certificate for this purpose.

HPE Aruba Networking recommends using ClearPass to provide RADIUS server and captive portal capabilities in most cases. However, sometimes, a company will want to host these services on their HPE Aruba Networking mobility devices. The device should use a CA-signed certificate for these functions in this case. A private CA can sign the RADIUS/EAP certificate. However, the devices will need public CA-signed certificates for the captive portal so that guests who access the portal do not see errors.

Note that all of these functions require ServerAuth for the EKU.

CA-signed certificates for RadSec

If you want to implement RadSec between HPE Aruba Networking devices and CPPM, use a CA-signed certificate.

You must obtain certificates from your choice of CA. You should usually use a private CA to issue these certificates to minimize costs. This option lets you choose longer validity periods or wildcards for CNs or SANs.

You then install the certificates on the HPE Aruba Networking devices, such as APs or switches, and select the RadSec usage for the certificates. If you have IAPs, you can install a single certificate for the entire cluster. If you use HPE Aruba Networking Central, you might use the same certificate across a group of APs or other devices.

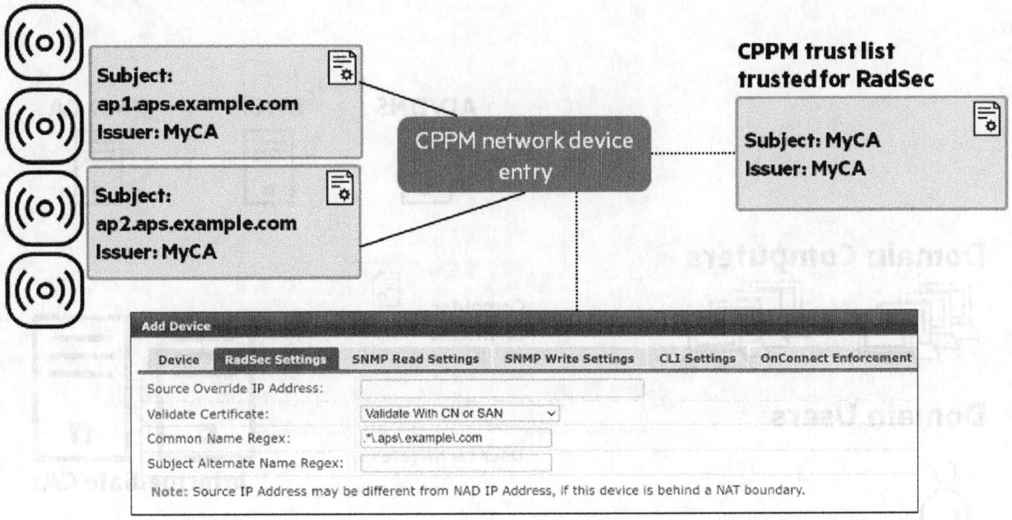

On CPPM, you must complete two steps to ensure that CPPM trusts the certificate. You must add the CA that signed the devices' certificates to CPPM's trust list with the RadSec usage. You must also create one or more network device entries that match your devices' IP addresses. You then use the RadSec Settings tab to specify which certificates CPPM trusts. You have three choices:

No Authorization Checks—CPPM only checks that the certificate is signed by a trusted CA, not expired, and not revoked. HPE Aruba Networking generally recommends against using this option as it is less secure.

Validate With CN or SAN—You must specify the common name or SAN within the certificate or certificates on your devices. You can use regular expressions to match multiple subject names or SANs. In this example, .*\.aps\.example\.com matches any CN that ends with aps.example.com. This option lets you use a single network device entry that matches the CNs or SANs for multiple valid certificates.

RFC Compliant (Serial + Issuer)—You must specify the Issuer's DN from a drop-down menu, which includes the DNs for all RadSec usage trusted CAs on CPPM. You must also specify the serial number for that certificate. You can also put in a regular expression for the certificate's CN or SAN. Some companies' security policies require this option. Using this option, each network device entry can match only one certificate. However, you might be able to install the same certificate on multiple devices.

Certificates within a windows domain

You have been examining how certificates work in general. You will now focus on a particular certificate use case you often encounter in the real world. The company has a Windows domain. Employees are Domain Users. They connect to the network on desktops and laptops and join the domain.

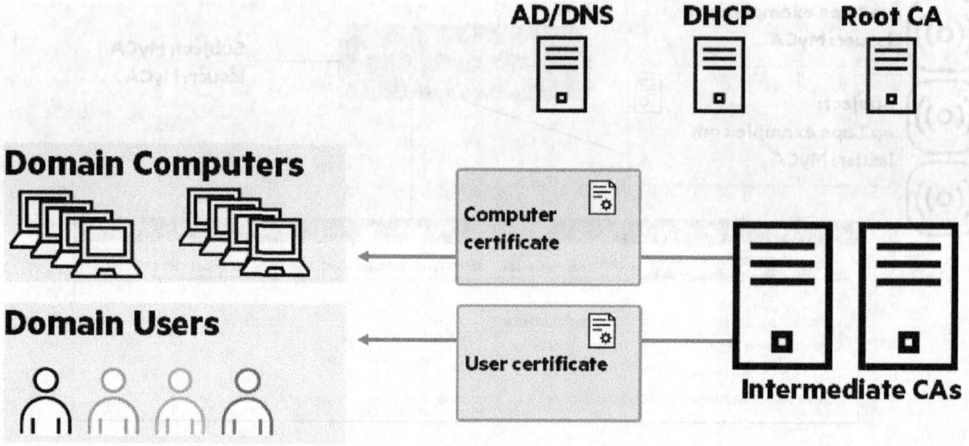

The company has a Windows certificate authority (CA) service consisting of a root CA and several intermediate CAs.

The company uses its CAs to issue certificates to internally-facing servers and network infrastructure devices. The company now wants to deploy certificates to domain computers and users so that they can authenticate. The next few sections take you through how this process could work.

Certificates in a Microsoft Azure AD domain

Some companies now use Azure Active Directory (AD), a cloud-based AD version. Like on-prem AD, Azure AD enables companies to manage their users and devices, organizing them into groups with different privileges. However, unlike a traditional Windows domain, Azure AD can manage mobile devices and Windows PCs and servers. It integrates with Microsoft Intune, a Mobility Device Management (MDM) solution.

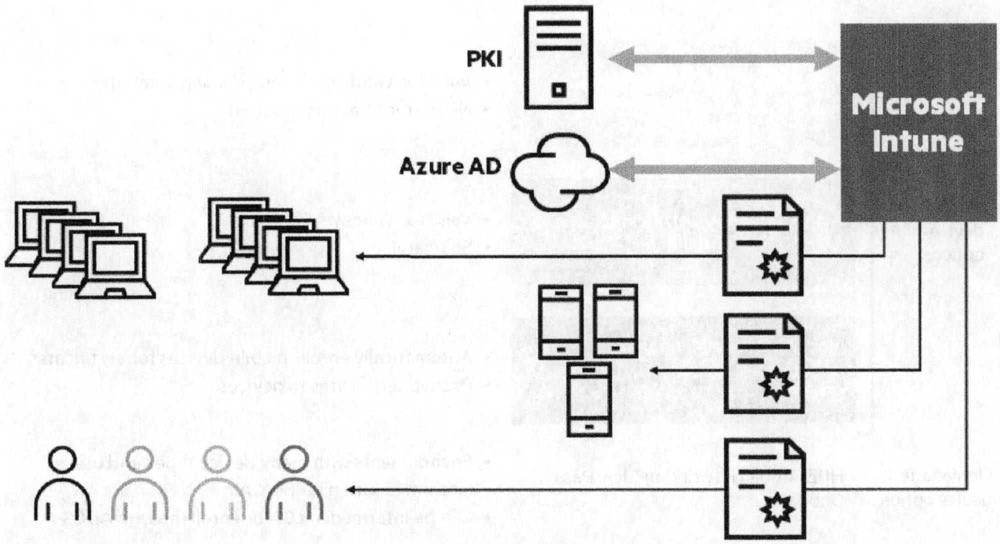

To support 802.1X with EAP-TLS in an Azure AD environment, Intune should deploy certificates to all the devices within the domain. Intune can work with a Windows CA to obtain the certificates, and it supports Simple Certificate Enrollment Protocol (SCEP) and Public Key Cryptography Standards (PKCS) for deploying the certificates.

Options for deploying certificates on user devices

As you see, ensuring that devices have the correct certificate can be challenging. Maintaining a large number of certificates requires specialized software and trained IT staff. In addition, installing certificates on all clients can be difficult in environments with multiple operating systems. There are many options for deploying and managing certificates. In the figure, you see just a few of those options.

Typically, customers set up a private CA to issue the certificates required for EAP-TLS. Microsoft Windows CA is a common option for Windows domains that use Active Directory (AD). AD administrators can set up group policy objects (GPOs) to automatically push down certificates to devices joined to the domain. The Windows CA also supports Active Directory Certificate Services (ADCS) Web Enrollment, which enables users to obtain certificates through a self-service web portal.

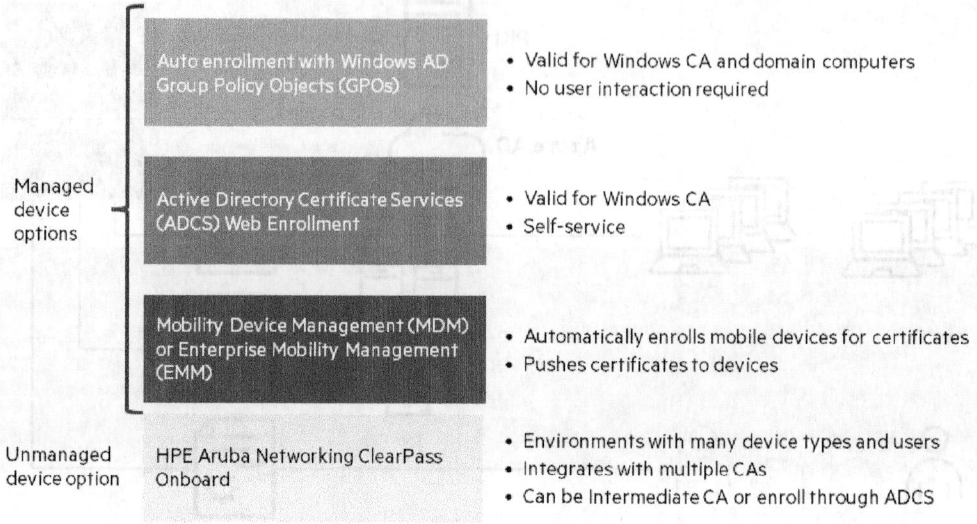

However, the Windows CA options are geared for a world where the only devices are computers joined to the domain. A modern network has far more types of devices, including smartphones and Internet of Things (IoT) devices. Bring Your Own Device (BYOD) environments can make managing certificates even more difficult. Mobility device managers (MDMs) help automatically enroll mobile devices for certificates and push certificates to devices.

If a company needs to deploy certificates to many unmanaged devices, HPE Aruba Networking ClearPass Onboard Software can help. ClearPass Onboard Software helps users easily download the correct certificates to their devices. Onboard can integrate with multiple CAs. It can act as an Intermediate CA, issuing certificates for devices. When Onboard integrates with a Windows CA PKI, Onboard can alternatively act as an intermediary with ADCS, in which case the Windows domain signing CA issues client certificates. Still, the Onboard client makes obtaining the certificate transparent to the user.

Supplemental: Network Device Hardening

This supplemental information can help hone or refresh your knowledge of network device hardening.

Network infrastructure security overview

Network Infrastructure Security is implemented to ensure that only authorized users and devices can access a secure network. Centralized AAA services to maintain user and define profiles are recommended. Policies can be applied, and usage can be tracked from a single pane of glass for more consistent, accurate, and streamlined security control.

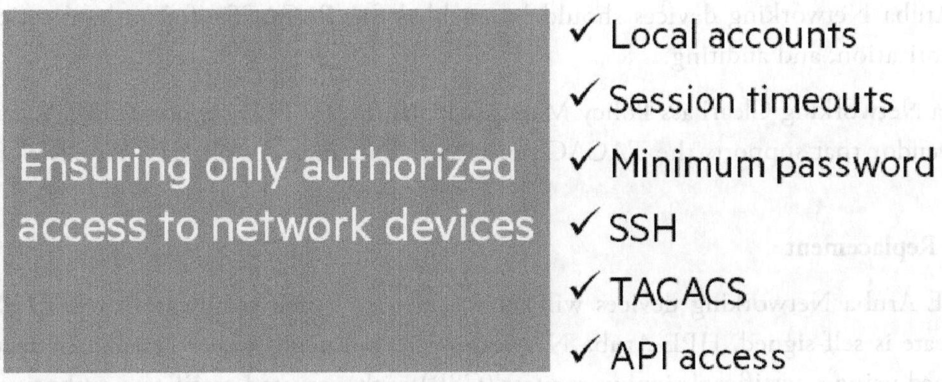

Local Accounts

A default admin account with a secure password is known only to a few managers with strict access and usage requirements. This should be a last resort only, with central authentication no longer reachable.

Use a 64-character complex password stored in a safe or secure password-management application.

All management access should use TACACS+ or RADIUS.

Session Timeouts

Management access to HPE Aruba Networking Central, gateways, or switches should be limited to reduce the number of open sessions and the risk of an unauthorized user gaining access with an idle connection.

The recommended timeout range should be less than five minutes.

Minimum Password

Complex passwords should be required for all user access without exception.

Eight or more characters with a combination of uppercase, lowercase, numerals, and special characters.

SSH

Ensure only the strongest ciphers are used to access network devices. Review the HPE Aruba Networking hardening guide for details on which cipher to select.

TACACS

The security protocol Terminal Access Controller Access-Control System (TACACS+) provides centralized authentication and validation of users managing network devices.

All HPE Aruba Networking devices should be enabled for TACACS+ for authentication, command authorization, and auditing.

HPE Aruba Networking ClearPass Policy Manager Platform (CPPM) supports TACACS+ for any hardware vendor that supports the TACACS protocol. This includes all HPE Aruba Networking devices.

Certificate Replacement

While HPE Aruba Networking devices will have a factory-issued certificate for HTTPS access, this certificate is self-signed. HPE Aruba Networking recommends either getting the factory certificate signed using a certificate signing request (CSR) with a trusted certificate authority (CA) or installing a new signed certificate instead.

Using a wildcard certificate for web management is acceptable on HPE Aruba Networking platforms.

API Access

Any HPE Aruba Networking device's application programming interface (API) can give administrators full access to that device and, therefore, is critical to protect.

APIs are accessed using Representation State Transfer (ReST) protocol over HTTPs. Consider using only signed certificates for the web interface of any device.

Disable APIs or restrict with read-only access if there is no intention to write commands to the API.

An ACL can be utilized so that only the management subnet can access the API.

Best practices checklist

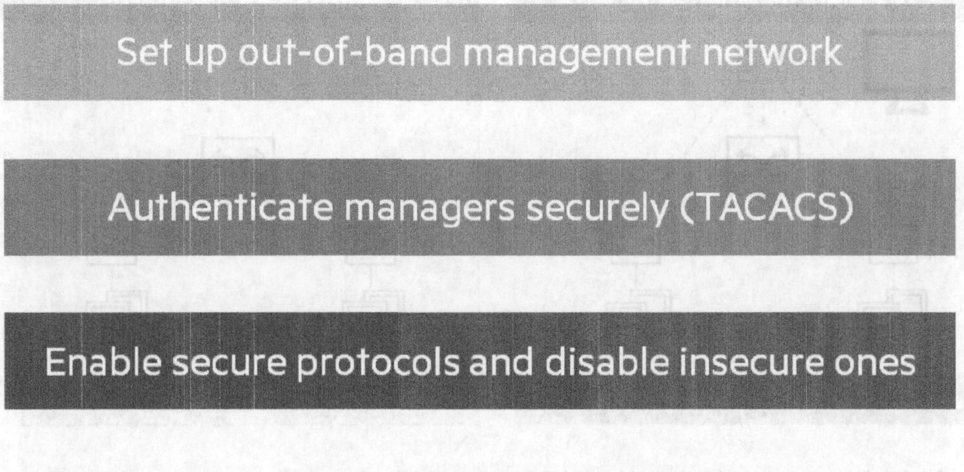

For the rest of this module, you will explore how to complete the best practices outlined here. For more information, visit the HPE Aruba Networking Support site and search for hardening guides.

Review management access on AOS-CX switches

AOS-CX switches offer two main options for accessing the management plane.

The switches support out-of-band management through an Ethernet port called the management (mgmt) port. The management port operates at Layer 3 and has an IP address. Admins can access the CLI or Web UI at the mgmt IP address.

AOS-CX switches also support in-band management. With this method, admins reach the management plane through the same ports to which users and other endpoints connect. They contact the switch at an IP address configured on one of the Layer 3 VLAN interfaces or ports.

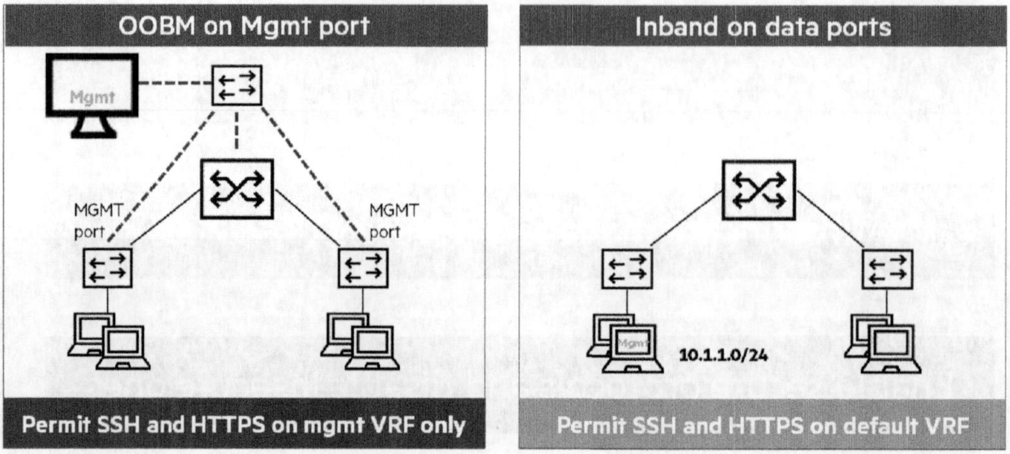

You control which access methods are available out-of-band and in-band by telling the switch on which VRFs particular protocols are enabled. The mgmt port is part of the "mgmt" VRF, while data ports are part of the "default" VRF by default. If you want to confine the switch to out-of-band management, you should enable SSH and HTTPS on the mgmt VRF and disable them on the "default" VRF. This approach makes it easy to specify and determine which protocols the switch accepts from different network segments.

HPE Aruba Networking generally recommends out-of-band management as the more secure option. To achieve the best available security, you should physically separate the OOBM network and the production network. In other words, you should connect the mgmt ports and the management stations to switches dedicated to the management network. Doing so does require extra effort and equipment. However, management switches can be inexpensive since 1GbE will provide all the necessary bandwidth for management.

If the customer requires inband management, you should leave SSH and HTTPS enabled on the default VRF (or whatever VRFs the customer uses).

—

Note that enabling a protocol on a VRF or port applies to traffic using that protocol that is directed to the switch. It does not affect traffic using that protocol but is sent to another destination.

Review management authentication options

You have several options for authenticating managers who reach the switch's management plane. For the best scalability and manageability for authenticating SSH and HTTPS managers, you should generally choose an external authentication option: TACACS+ or RADIUS. Each of these protocols has its advantages; often, a company will have its preference as to which you use.

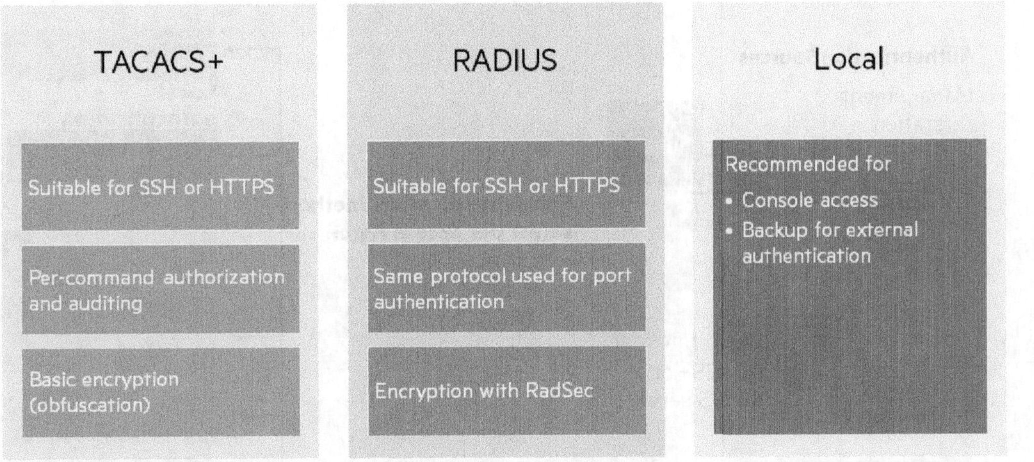

TACACS+ provides the ability to control the commands that managers enter in a granular way. It also provides detailed auditing for each command that managers enter, which is a requirement for some companies. In addition, the TACACS+ packets exchanged between the switch and the authentication server are encrypted by a shared secret. While this encryption does not meet modern standards and can be cracked, it does provide a layer of obfuscation against casual eavesdropping.

RADIUS also has some features that companies might want. RADIUS is the standard authentication protocol used with port authentication like 802.1X. So, if the company is also implementing port authentication, the company might choose to use the same RADIUS server to authenticate managers. RADIUS itself does not encrypt messages between the switch and the authentication server. However, if security is required, you can add RadSec.

Managers can also authenticate against a local list of users. However, HPE Aruba Networking recommends local authentication as a backup method in case the TACACS+ or RADIUS authentication server is unavailable.

Best practices for authentication methods

You can now set up a secure connection to access switch management interfaces. The next step is configuring the switch to enforce secure authentication of users who reach those interfaces.

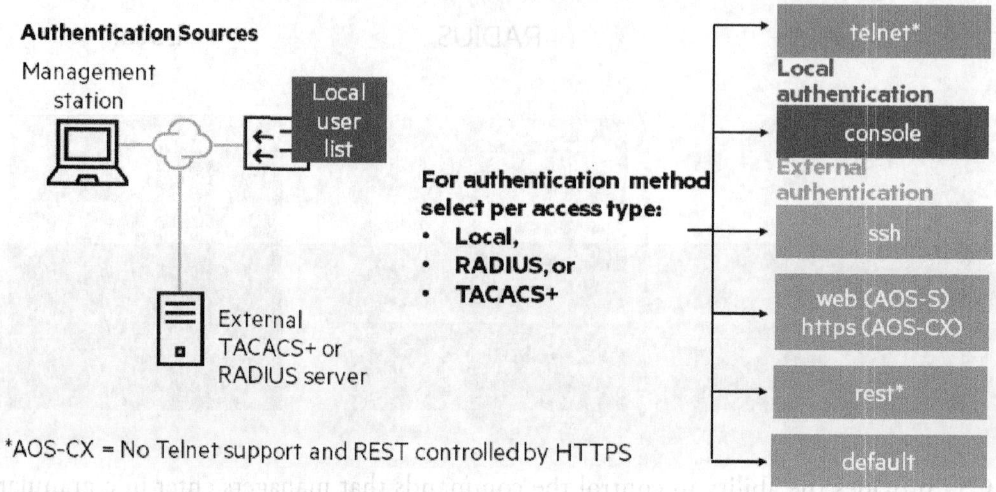

HPE Aruba Networking switches require users to authenticate with a username and password to prevent unauthorized access. HPE Aruba Networking switches allow users to authenticate through a local account, with unique credentials recognized by the switch itself, or through an external AAA server, a RADIUS or TACACS+ server. An external server provides far superior scalability since one set of credentials can work for many different devices. Using an external server also makes it easier to manage multiple admin accounts. Many enterprise servers have built-in mechanisms to force users to create new passwords periodically. At the same time, local credentials will likely stay the same for far too long due to the manual labor involved in renewing them. In addition, if an admin leaves the organization, it is relatively easy to revoke that admin's account on the RADIUS or TACACS+ server. Still, local authentication can be valuable. In the event of a malfunction with a RADIUS or TACACS+ server, local authentication will remain available. Additionally, local authentication requires almost no additional infrastructure.

HPE Aruba Networking switches allow customers to specify the authentication type based on the access type—sometimes called AAA channel. The access types define all of the ways that admins can access the switch management interfaces. On AOS-CX switches, these are:

- Console
- SSH
- HTTPS

HTTPS encompasses both the Web UI and REST.

- On AOS-S switches, these are:
- Console
- Telnet (however, Telnet should be disabled)

- SSH
- Web (for HTTP and HTTPS; however, only HTTPS should be enabled)
- REST

You can choose your preferred authentication type for each access type. You can also choose different RADIUS or TACACS+ server groups for different access types.

The best practice for authentication is local authentication on the console port. Console users will have physical access to the infrastructure, which already provides one layer of security, making local authentication acceptable. Other access types should use RADIUS or TACACS with local as a backup. Because other channels, even if used with OOBM/mgmt ports, are inherently a bit less secure, they should leverage the scalability and security of an external server.

Manager roles

When a manager authenticates, that manager is assigned a role.

AOS-S switch Roles

AOS-S switches maintain two roles. Operators have limited access. They can enter show commands to view statistics and some configuration information. Managers have complete read-write access. They can view all information and make any changes to the configuration.

For web UI authentication, manager access corresponds to full access to the Web UI. Operator access corresponds to a more limited version of the UI, which enables the user to view status and settings but not change them.

You can also configure command-level authorization to assign more nuanced controls, but that configuration is beyond the scope of this guide.

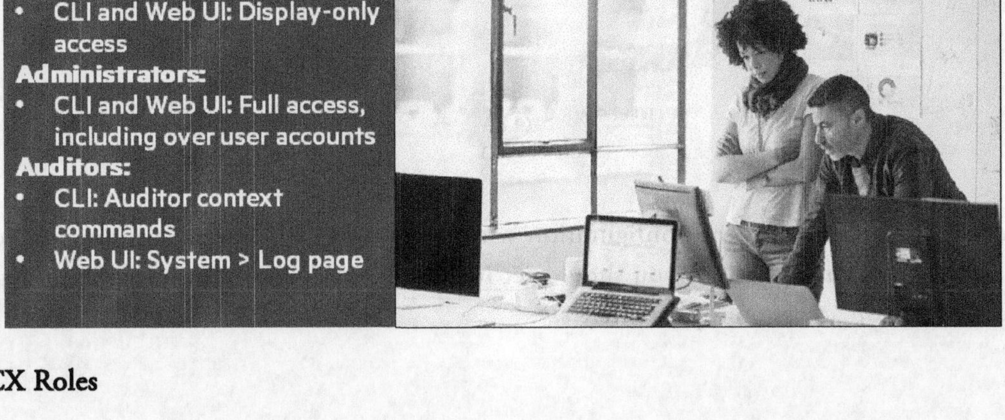

AOS-CX Roles

AOS-CX separates management users into three different default roles: administrators, operators, and auditors. All types of users may access the switch through the CLI or Web UI (or REST API); however, their privileges differ.

Administrators have full privileges, including full read-write access to the CLI and Web UI. They can configure any settings, including user accounts.

Operators have no configuration privileges. They can only view nonsensitive switch configuration information. In the CLI, operators can only issue show commands, and they can only view settings in the Web UI.

Finally, auditors have access to CLI commands only in the auditor context (auditor>). Auditors can view event log contents and local accounting log contents. They can also copy command output to a remote server or a local USB drive.

Admins can create custom manager groups as well, but that configuration is beyond the scope of this guide.

Understanding how credentials are stored in the config

AOS-CX

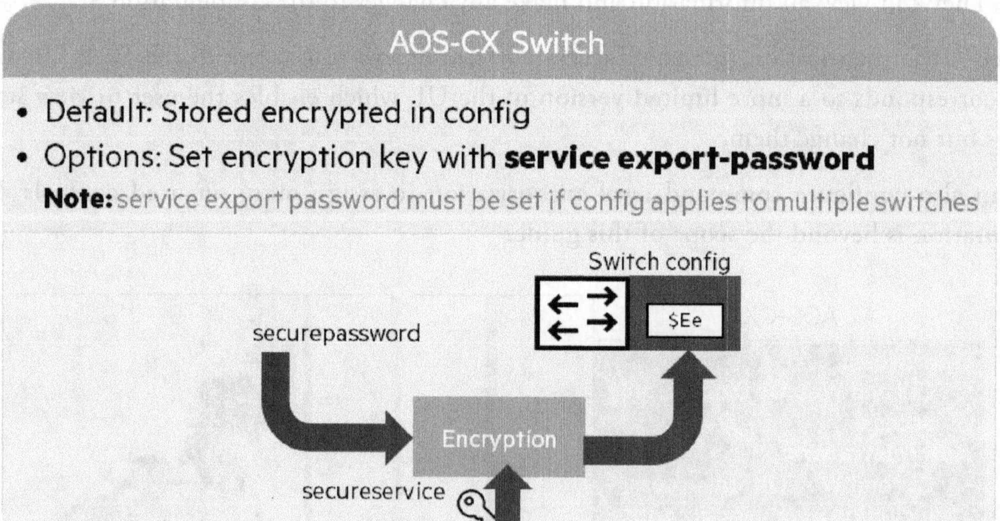

All credentials are stored in the configuration with encryption. The service export password acts as the key for this encryption, so you should set it in the configuration if you want to deploy the same configuration across multiple switches.

Use secure protocols

You have established a secure out-of-band-management network. You have set up authentication. You must also ensure that managers can only use secure protocols.

Access type	Protocol	Security	Enabled by default on AOS-CX
CLI	Telnet	**Insecure**	
	SSH	**Secure**	X (mgmt + default VRF)
File transfer through serial	TFTP	**Insecure**	
	SFTP	**Secure**	X (mgmt + default VRF)
Web	HTTP	**Insecure**	
	HTTPS	**Secure**	X (mgmt + default VRF; self-signed cert)
SNMP	SNMPv1/v2c	**Insecure**	
	SNMPv3	**Secure**	

This table displays insecure protocols that you should disable and secure protocols to use instead. Note that this table only recommends which protocols you should use when the company requires a particular form of access. However, you might choose to shut down a form of access entirely. For example, some companies might not want to use the Web UI on the switches. In this case, you would disable HTTP and HTTPS on the switches.

Telnet vs SSH

Telnet and SSH provide access to the switch CLI over TCP/IP: both support username- and password-based authentication of clients. However, while SSH encrypts communications, Telnet does not, meaning that data delivered over Telnet can be stolen much more easily than data delivered over SSH. In addition, SSH supports mutual authentication, that is, authentication of both the server (in this case, the switch) and the client. Because the client authenticates the switch, a manager such as Alice can know she is truly connecting to the switch and not to hacker Mallory, who might try to insert herself between the management station and the switch. In this way, mutual authentication protects against MITM attacks.

Both Telnet and SSH are enabled on AOS-S switches by default. Given the substantial security advantages of SSH, it is best practice to leave SSH enabled and disable Telnet entirely. Telnet is not supported on AOS-CX switches at all for this very reason. SSH is enabled by default on the "mgmt" and "default" VRF; if you are using the Mgmt port to establish an OOBM network, you can disable SSH on the "default" VRF.

AOS-CX switches further secure SSH by periodically renegotiating new keys for the session after an hour or 1 GB of data is transmitted, whichever comes first. AOS-S switches do the same after every 10 minutes (by default) or when 1048576 KB of data is transmitted. You can adjust these values (see your switches' configuration guide).

TFTP versus SFTP

AOS-S switches can act as TFTP clients to receive files from TFTP servers or send files to those servers. However, TFTP sends files in plaintext and does not provide authentication, so it should be disabled to prevent admins from using it. SFTP is a secure alternative that transfers files over a secure SSH connection, which provides authentication and encryption. SFTP is always enabled when admins are using SSH.

HTTP versus HTTPS

HTTP transmits data in plaintext. HTTPS uses Transport Layer Security (TLS) to establish a secure tunnel for transmitting data. The switch acts as an HTTPS server, which uses a certificate to authenticate to the client and set up a secure tunnel. The manager submits credentials over that tunnel, and the rest of the session is encrypted.

On AOS-S switches, HTTP is enabled by default, but HTTPS is not. You must install a certificate on the switch before you can enable HTTPS. AOS-CX switches do not support HTTP, only HTTPS. HTTPS is enabled by default on the default and mgmt VRFs using the self-signed certificate. However, you should replace the self-signed certificate with a CA-signed one.

SNMPv1/v2c versus SNMPv3

SNMPv1/v2c transmit community strings, originally meant to act as "passwords," in cleartext, so they are insecure. SNMPv3 provides data authentication and privacy with secure algorithms and keys.

AOS-S switches have v1/v2c enabled on them by default with "public" as the community name, providing manager-level access. You will learn how to disable SNMPv1/v2c and enable SNMPv3 instead.

AOS-CX switches also have the public community configured on them by default, but SNMP is not enabled on any VRFs by default. You will learn how to configure SNMPv3 users on the AOS-CX switches and remove the public community for security.

Set up SSH securely

You will now learn to set up secure SSH management for your switches. A brief review of SSH will help.

When the management client requests to open an SSH session with the switch, the switch responds and starts setting up a secure tunnel. The switch and client must agree on algorithms to secure the tunnel. The switch also authenticates itself to the client with its SSH public key. In other words, the switch sends the public key and signs data with it, proving it has the associated private key. The devices establish a secure tunnel if the client accepts the switch's authentication. The management client then authenticates across that tunnel. The client can authenticate with a username and password or with an SSH private/public keypair like the switch. This guide focuses on username and password authentication.

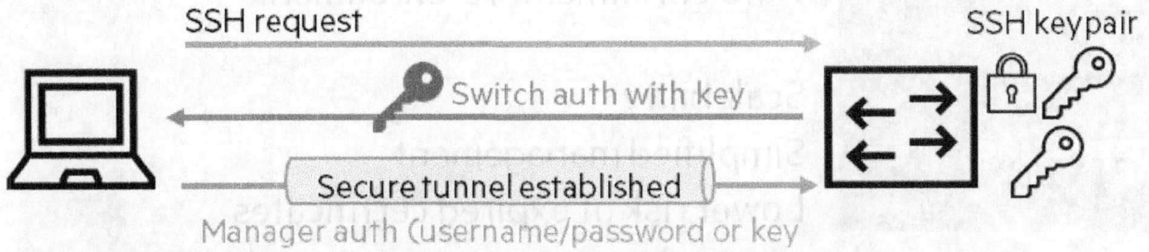

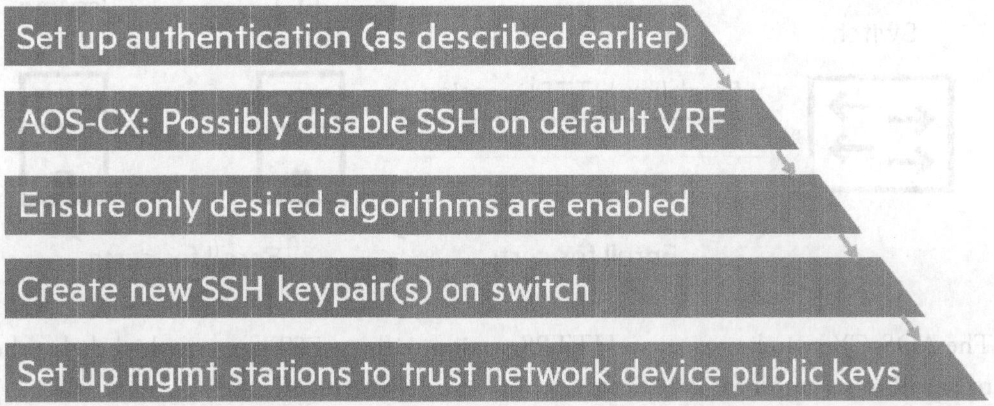

HPE Aruba Networking switches support SSH out of the box, but you need to take steps to ensure that you are using SSH in a secure way.

First, as discussed earlier, you need to configure the desired authentication type, whether you use external authentication, as recommended, or local accounts.

Next, you should disable Telnet and TFTP and make sure SSH is enabled (it is, by default, in the software versions covered in this guide). If you are using the mgmt VRF on an AOS-CX switch, remember to disable SSH on the default VRF.

You should also check the SSH algorithms and disable non-certified ones. You will see which algorithms are uncertified in the lab.

Then, you should generate a new SSH key pair.

Finally, you must set up management clients to trust the switch's public key. The next slide goes into more detail.

Managing certificates with Enrollment over Secure Transport (EST)

You can also use Enrollment over Secure Transport (EST) to automatically manage certificates on AOS-CX switches.

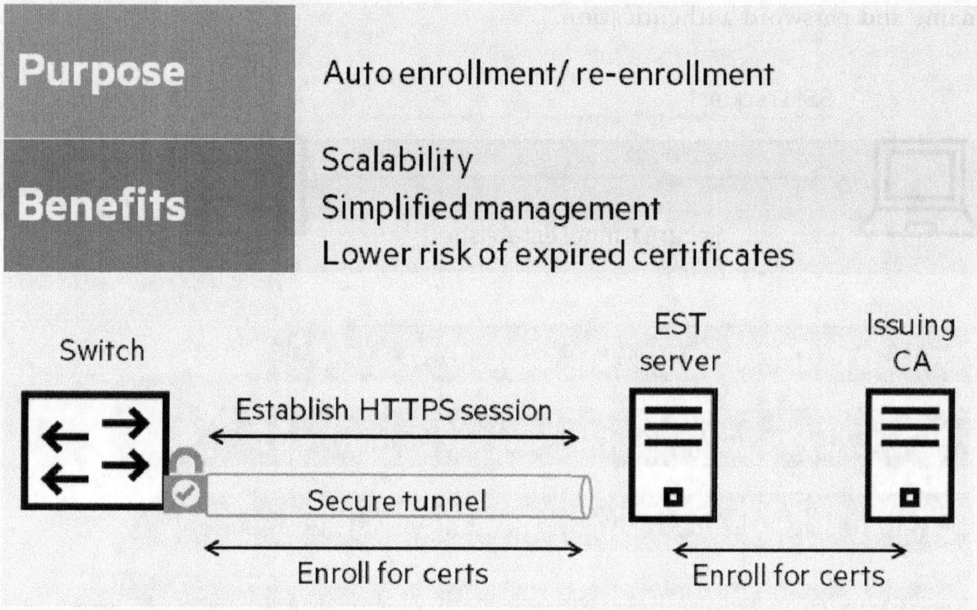

- The AOS-CX switch creates an HTTPS session with an EST server, which helps the switch to enroll or re-enroll for certificates.
- This option is preferred for wider scalability because it simplifies the deployment of certificates.
- Automatic re-enrollment before a certificate expires also decreases the risk that admins will forget to renew a device's certificate, which would cause functions that use that certificate to fail.
- Using EST requires deploying an EST server, which can communicate with the issuing CA.
- Explaining how to set up the EST infrastructure is beyond the scope of this guide. (ClearPass Onboard can act as an EST for endpoints.)

Disable SNMPv1/v2

Simple Network Management Protocol (SNMP) is an application layer protocol that helps network administrators manage and monitor their network. SNMP-managed networks have several key components. Each managed device, such as a switch, has an "agent," a software module that enables management functions. The managing device is called the Network Management Station (NMS) or SNMP server. AirWave is an example of such a server.

While SNMP is a useful tool, its earlier versions suffer from poor security. The first two versions send critical data in clear-text, creating a serious vulnerability for a packet sniffing exploit. Community strings were meant to act as authentication. The agent only accepts write requests with the correct read-write community or read requests with the correct read-only or read-write community. However, because community strings are sent in cleartext, they do not provide true authentication. If hackers can spoof read and write requests, they can collect information about the network and even make configuration changes to it.

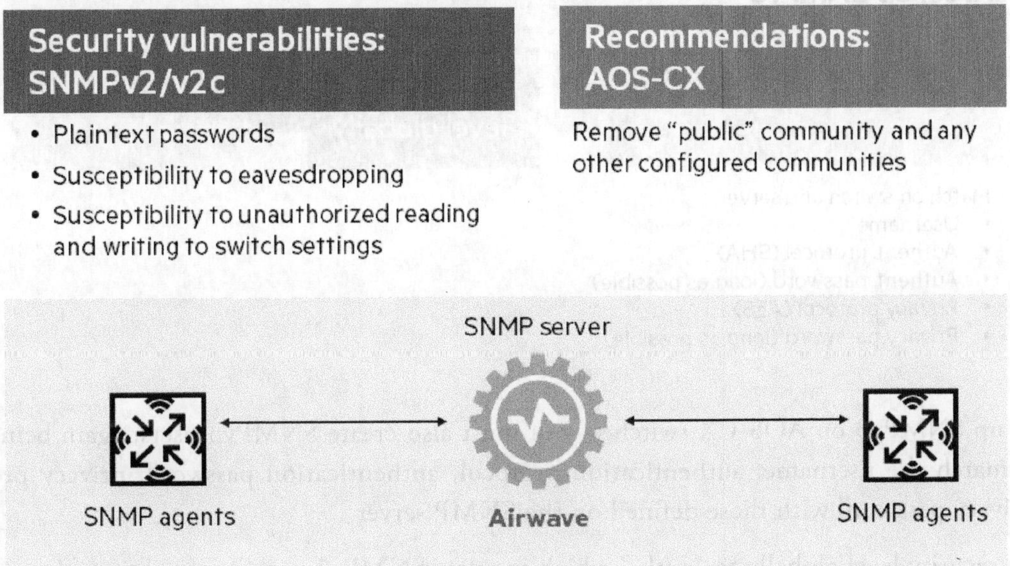

For that reason, best practice is to disable SNMPv1 and v2c.

You can also remove the "public" community. On AOS-CX switches, SNMP is not enabled on any VRFs by default so it is implicitly disabled. If you plan to use SNMPv3, you should remove the "public" community so that it is not available when you enable SNMP on a VRF, which enables all forms of SNMP.Simple Network Management Protocol (SNMP) is an application layer protocol that helps network administrators manage and monitor their network. SNMP-managed networks have several key components. Each managed device, such as a switch, has an "agent," a software module that enables management functions. The managing device is called the Network Management Station (NMS) or SNMP server. AirWave is an example of such a server.

While SNMP is a useful tool, its earlier versions suffer from poor security. The first two versions send critical data in clear-text, creating a serious vulnerability for a packet sniffing exploit. Community strings were meant to act as authentication. The agent only accepts write requests with the correct read-write community or read requests with the correct read-only or read-write community. However, because community strings are sent in cleartext, they do not provide true authentication. If hackers can spoof read and write requests, they can collect information about the network and even make configuration changes to it.

For that reason, best practice is to disable SNMPv1 and v2c.

You can also remove the "public" community. On AOS-CX switches, SNMP is not enabled on any VRFs by default so it is implicitly disabled. If you plan to use SNMPv3, you should remove the "public" community so that it is not available when you enable SNMP on a VRF, which enables all forms of SNMP.

AOS-CX: set up SNMPv3

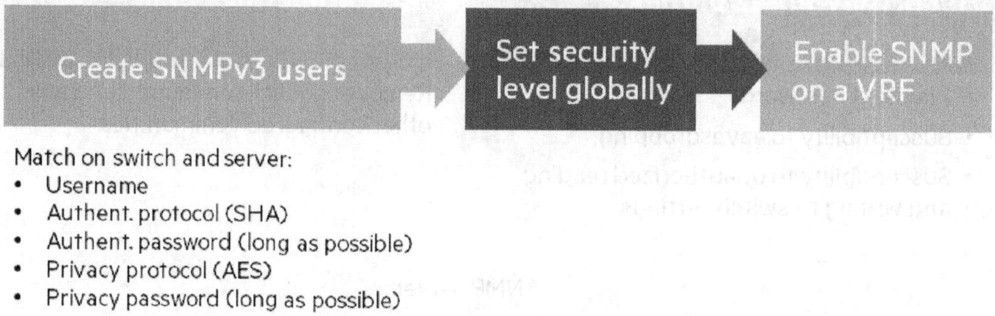

To set up SNMPv3 on AOS-CX switches, you must also create SNMPv3 users, again being careful to match the username, authentication protocol, authentication password, privacy protocol, and privacy password with those defined on the SNMP server.

Set the security level globally to "auth," which requires SNMPv3 users to use data authentication, or "auth-privacy," which requires SNMPv3 users to use data authentication and encryption. The "auth-privacy" option is enabled by default and recommended. All SNMPv3 users are granted access to the full switch MIB except certain excluded access control MIBs.

To finish enabling SNMPv3 on AOS-CX switches, enable SNMP on the desired VRF or VRFs.

Review RADIUS vs RadSec

As you learned, network devices and CPPM communicate with RADIUS while authenticating clients for network access. You can implement RADIUS alone or add a layer of security with RadSec.

RADIUS	RadSec
Key benefits	**Key benefits**
Simple setup	Certificate-based authentication Encryption
Key drawbacks	**Key drawbacks**
Shared secret not cryptographically secure No encryption	Increased complexity
Use when	**Use when**
NAS-to-RADIUS server connections = trusted	NAS-to-RADIUS server connections = untrusted Corporate policies require

RADIUS

RADIUS uses a shared secret to authenticate the RADIUS client (NAD) and RADIUS server to each other. You must configure a matching shared secret on both devices. The NAD and server include a hash of the message and shared secret within all RADIUS messages. If the hash does not check out, indicating that the secret is incorrect or the message was tampered with, the receiving device drops the message. However, the shared secret does not provide rigorous security. Setting a very long, random key can make it more secure. However, RADIUS itself is not considered a secure protocol. It also transmits data in plaintext.

You should only run RADIUS over trusted network connections. For example, if network devices and CPPM are deployed at the same site, you can generally use RADIUS between them. You can also use RADIUS between network devices and CPPM servers at another site if the site-to-site connection is secured with a protocol like IPsec.

RadSec

RadSec establishes a TLS tunnel between the RADIUS client (NAD) and the RADIUS server. During the TLS tunnel establishment, the NAD and RADIUS servers authenticate each other with certificates. After the TLS tunnel comes up, the NAD and RADIUS servers exchange all RADIUS messages over the tunnel, including authentication and accounting. The tunnel provides encryption and data integrity for the messages. Because the tunnel uses TCP rather than RADIUS's typical UDP, the NAD can detect when it loses contact with the RADIUS server more quickly.

When you enable RadSec on the NAD or CPPM, the RADIUS secret is automatically set to "RadSec." Instead, you set up certificates to provide the authentication.

RadSec provides greater security than RADIUS, but the certificate requirements can make it harder to deploy and maintain. Issues can occur such as a NAD's certificate expiring. HPE Aruba Networking recommends RadSec when network devices and CPPMs must communicate over untrusted network connections. Some companies also require RadSec as part of their security policies.

Network devices can also use TACACS+ to authenticate managers to CPPM. You will learn more about TACACS+ later.

Review RADIUS Dynamic Authorization

A RADIUS server like CPPM is capable of making decisions about a user's or device's authorization beyond the initial authentication. For example, ClearPass Onboard Software can detect a client's security posture and tell CPPM to disconnect a client that does not have the proper firewall and antivirus settings. Or CPPM might find out after the initial authentication that the client is an IoT device, which has particular security policies associated with it. Many other examples of such events exist.

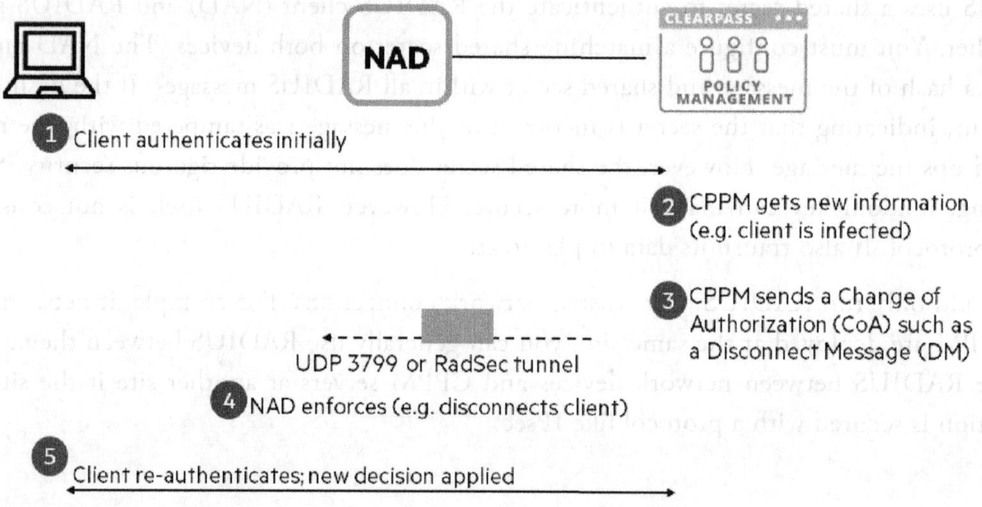

So that CPPM can enforce the new policies, it must force the client to disconnect and re-authenticate. It sends a Change of Authorization (CoA) message to the NAD, such as an AP, switch, gateway, or MC. By default, CPPM sends the message on standard port UDP 3799. But you can change that port, as long as you match the setting on CPPM and the NAD. If CPPM and the NAD are using RadSec, CPPM sends the CoA messages in the RadSec tunnel.

When the NAD receives a CoA from an authorized dynamic authorization server, it enforces the message and disconnects the client. The client will typically try to re-authenticate, and CPPM can then apply its policies to make new decisions about the client's access.

RADIUS Interim Accounting

As an AAA protocol, RADIUS provides authentication, authorization, and accounting. RADIUS accounting logs a broad range of information, such as a client's username, NAD port, and bandwidth consumption in packets and bytes. A NAS can implement RADIUS accounting in these modes:

- Stop—When the client's session terminates, the NAD sends a message that summarizes the session.
- Start/stop—The NAD sends an accounting message at the beginning of the session and then a session summary at the end.
- Stop or start/stop + interim accounting—The NAD sends the start/stop messages and periodic accounting messages.

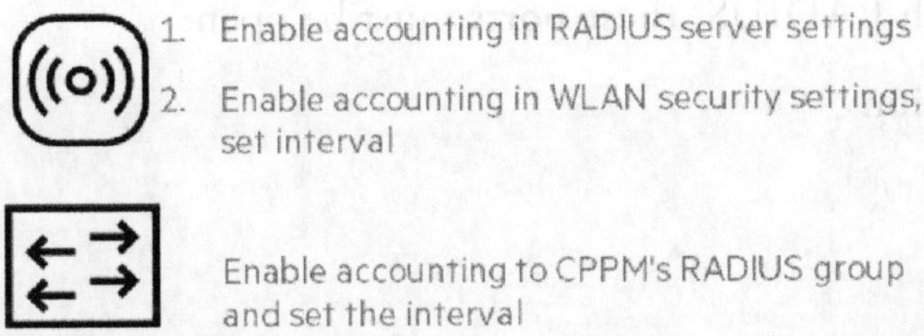

HPE Aruba Networking recommends setting up interim accounting because it provides important context to CPPM about the client's location and network usage. Admins can then view this information in Access Tracker. CPPM can also deliver this information to third-party devices integrated as endpoint context servers.

To enable interim accounting on IAPs, ensure accounting in two locations. Enable it in the RADIUS server settings. Also, enable accounting in the WLAN security settings. Choose to send accounting information to the authentication server defined for the WLAN. Then, set the interval for the interim updates.

On AOS-CX switches, you enable interim accounting for authenticated clients with this command: aaa accounting port-access start-stop interim <minutes> group <group_name>

You must also enable the processing of interim updates in the ClearPass server configuration (go to **Administration > Server Manager > Server Configuration** and Click the server). In the Service Parameters tab, for the RADIUS server settings, set Log-Accounting-Interim-Update-Packets to True.

Standard ports for network device and CPPM communications

If a firewall exists between NADs and CPPM, you must ensure that the firewall permits the required ports.

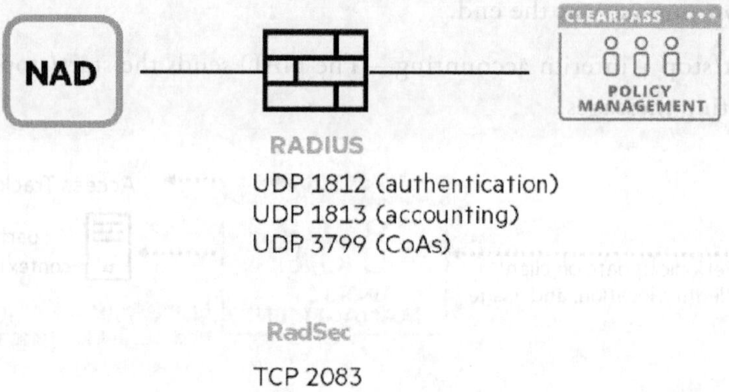

When NADs and CPPM communicate with RADIUS, these ports must be open:

- UDP 1812 for RADIUS authentication
- UDP 1813 for RADIUS accounting
- UDP 3799 for CoAs

For RadSec, only one port is required: TCP 2083. The TCP session carries all RADIUS authentication, accounting, and CoA messages.

Note that these are the standard ports. If you change your configuration to use non-standard ports, the firewall must permit those instead.

Summary of Spanning Tree protections

This table reviews the features that you have examined in this section.

	Recommended feature	Alternative feature
Edge ports	BPDU protection Prevents rogue switches from connecting	Root guard Prevents rogue or misconfigured switches from becoming root
Global	CoPP rate limit for **stp** or **loop-ctrl-mstp**	AOS-Switch: BPDU throttling

You should typically enable BPDU protection on edge ports to prevent rogue switches implementing spanning tree from connecting to the network. Root guard is an alternative technology that prevents rogue or misconfigured switches from becoming root. However, if a switch sends BPDUs that are not superior BPUDs, the switch is allowed to connect.

Globally, the switch should implement CoPP rate limiting on BPDUs and other traffic classes, as you learned in a previous module. BPDU throttling is an alternative, legacy technology on AOS-S switches, which you can apply if you are not using CoPP.

DHCP attacks

Because DHCP was designed to work in the trusted internal network, it does not provide authentication or access controls. A DHCP server cannot verify that the client requesting an address is a legitimate client on the network, and the DHCP client has no way of knowing if the DHCP server that offers it an address is legitimate. As a result, DHCP is vulnerable to attacks from rogue clients and servers.

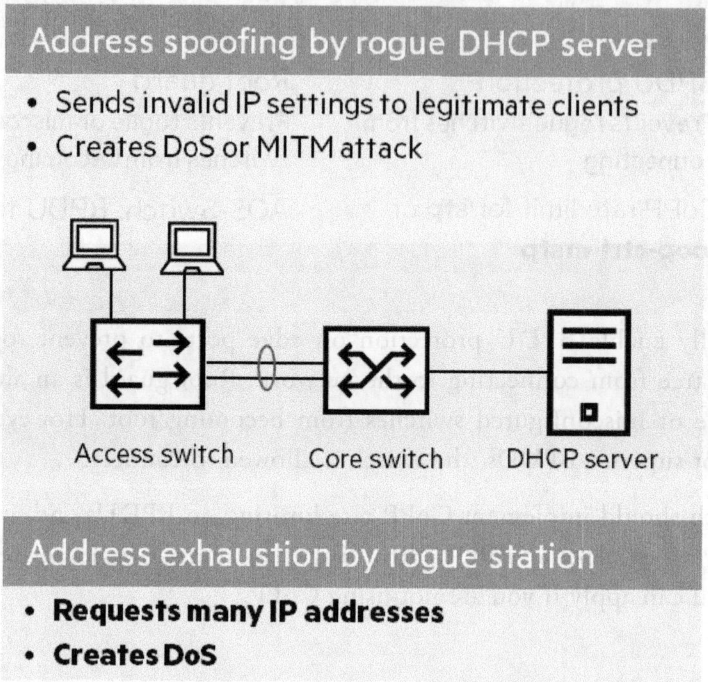

You need to protect your network from two types of common DHCP attacks:

- **Address spoofing**—A rogue DHCP server on the network can assign invalid addresses to network devices. Without valid IP addresses and network gateway addresses, these devices cannot connect to any other IP network device, which prevents users from reaching the resources they need to do their jobs. Or a rogue DHCP server might respond to a client's DHCP discover request with the rogue server's IP address as the default gateway, setting up the client for an MITM attack and creating a potential security breach.
- **Address exhaustion**—An attacker can access the network and request IP addresses until the DHCP server's supply of available addresses is exhausted. This prevents legitimate hosts from receiving IP addresses and accessing the network.

ARP protection

AOS-S switches can use ARP protection to protect your network against ARP poisoning attacks. To enable the feature, you enable the feature on a VLAN. You must also enable the feature globally for it to take effect on the enabled VLANs. When you set this up in the real world, though, you should ensure that you have planned and followed the guidelines in the next slides before enabling ARP protection globally.

IP-to-MAC binding table

IP	MAC
10.1.10.10	18...aa
10.1.10.11	d8...bb
10.1.10.12	d4...cc

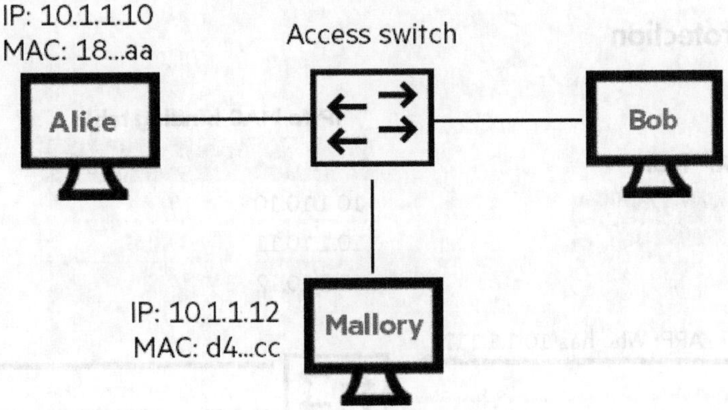

Like the DHCP snooping feature, the ARP protection feature allows you to designate trusted and untrusted ports. By default, all ports in an ARP protection VLAN are untrusted. The switch checks ARP requests and responses on all untrusted ports; it:

- Examines all ARP requests and responses received on that port
- Verifies that each of these packets has a valid IP-to-MAC address binding
- If the binding is valid, the switch updates its local ARP cache or forwards the packet to the appropriate destination. If the binding is invalid, the switch simply drops the packet, preventing other devices from receiving it and being tricked by the faulty information.

Additional checks

In addition to verifying IP-to-MAC address bindings, you can configure the switch to perform three further checks:

- src-mac—The switch checks ARP request and response packets to ensure that the source MAC address in the Ethernet header matches the sender MAC address in the body of the ARP packet. If the two addresses do not match, the switch discards the packet.
- dest-mac—The switch checks each unicast ARP response packet to ensure that the destination MAC address in the Ethernet header matches the target MAC address in the body of the ARP packet. If the two do not match, the switch drops the packet.

- ip—The switch checks the sender and target IP addresses of ARP packets and drops any ARP packets that have an invalid address. Invalid IP addresses include:
- 0.0.0.0 and 255.255.255.255
- All IP multicast addresses and Class E IP addresses

By default, none of these checks are enabled.

Results of ARP protection

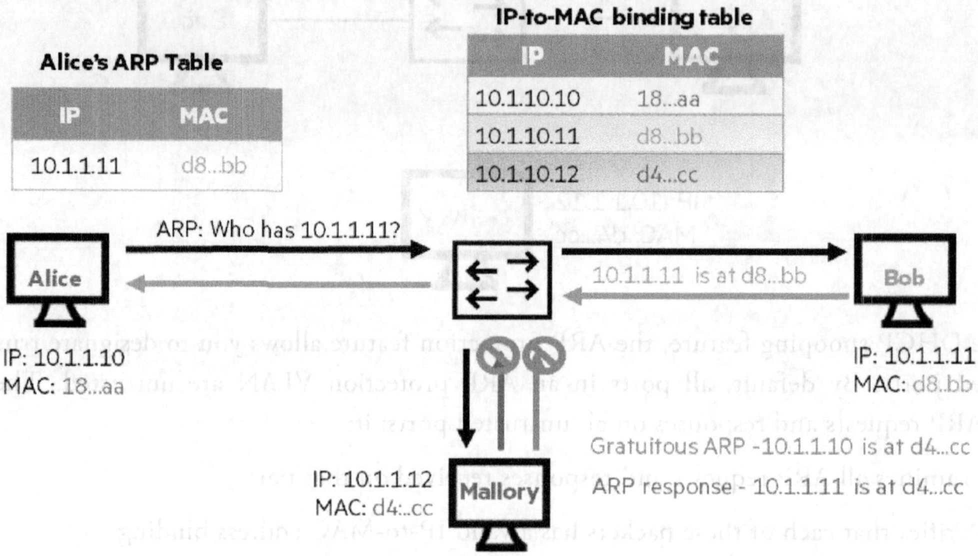

You can see how this protects the network in the figure above:

- Alice broadcasts an ARP request for Bob's MAC address.
- Mallory responds with an ARP response that maps her MAC address to Bob's IP address.
- It also sends a gratuitous ARP posing as Alice. However, the traffic arrives on an untrusted port in a VLAN on which ARP protection is enabled. Therefore, the switch checks both ARP messages and sees that the IP-to-MAC address binding is incorrect. The switch drops the packets.
- Bob sends his legitimate ARP response. This response also arrives on an untrusted port in an ARP protection VLAN. However, its IP-to-MAC address binding is correct, so the switch forwards it as normal.
- Now Alice and Bob can exchange traffic.

DHCP snooping and ARP protection

In most VLANs, devices receive their IP addresses using DHCP, and these addresses can change. The switch must discover the correct MAC to IP bindings automatically and dynamically. DHCP snooping fulfills this function.

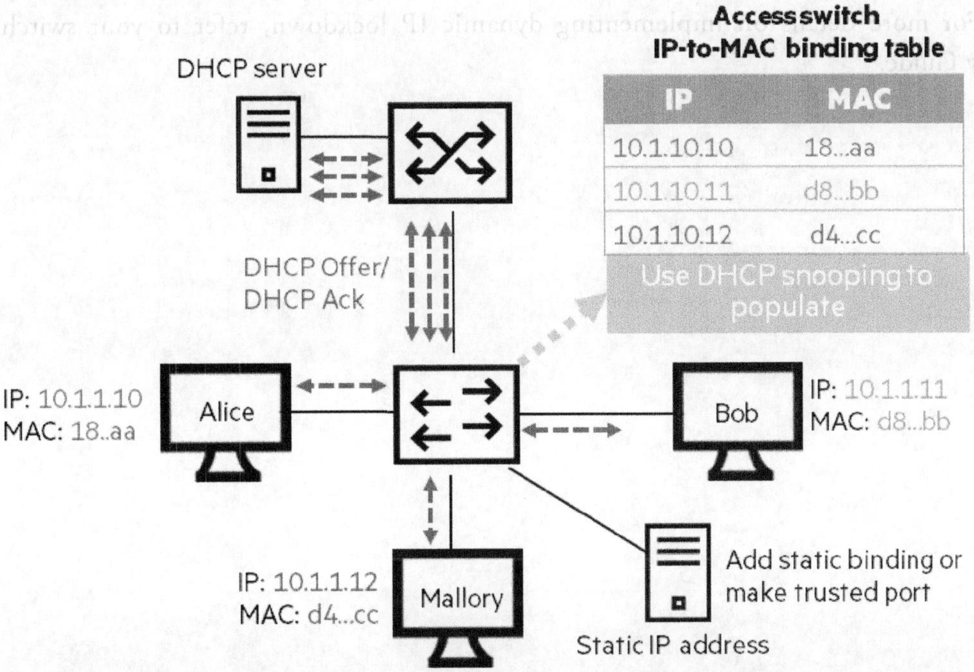

DHCP snooping builds and maintains a DHCP snooping-binding table, which tracks the MAC address, IP address, lease time, VLAN number, and interface information corresponding to each DHCP lease. Using this table, the switch can learn which IP addresses should legitimately send traffic on which ports and which MAC address can legitimately use an IP address. Therefore, you must enable DHCP snooping as part of configuring ARP protection.

If you are not using DHCP to assign IP addresses in the protected VLAN, you can configure static IP-to-MAC address bindings, and the switch will use this information to verify ARP packets. However, most networks do, of course, use DHCP.

Even when using DHCP and DHCP snooping, you may need to add a couple of static IP-to-MAC address bindings to the DHCP snooping table. The static bindings allow the switch to verify IP-to-MAC bindings for devices assigned static IP addresses. You must also specify the interface for the static binding.

In an active network, you should generally wait at least a week after enabling DHCP snooping to enable ARP. This ensures that the table is properly built up.

—

For increased security and protection against any form of IP spoofing, you can implement dynamic IP lockdown (ip source-lockdown). This feature goes further than ARP protection and enforces the correct source IP address and MAC address for all IP traffic on a port. Like ARP protection, dynamic IP lockdown requires DHCP snooping to populate the IP-to-MAC binding table. For more details on implementing dynamic IP lockdown, refer to your switch's Access Security Guide.

8 Overlay Networks

This module reviews options for extending segments in a network over an overlay tunnel.

OBJECTIVES

After completing this chapter, you should be able to:

- ✓ Describe HPE Aruba Networking gateways
- ✓ Explain centralized overlay networks with HPE Aruba Networking gateways
- ✓ Describe distributed overlay networks with HPE Aruba Networking Central NetConductor

HPE Aruba Networking Gateways

This section introduces different gateway (GW) model options and clustering GWs together for redundancy.

When to use gateways

Since the control plane resides in the cloud (HPE Aruba Networking Central), you might question whether you still need gateways (formerly known as controllers). But gateways are still a key element in the AOS-10 architecture. They support tunneled and mixed SSIDs, allowing you to dramatically reduce the number of VLANs spread across the campus, with traffic inspection in a single location.

AOS-10 also uses gateways for dynamic segmentation, allowing for authentication and traffic inspection switches to tunnel users to a gateway. Plus, AOS-10 needs gateways to deploy micro-branch APs, which terminate their IPsec tunnels on gateways over the internet, to provide remote access to small branches.

But what about network simplification? How do gateways help with that?

Note

For scenarios with Docsis or GPON wiring AOS-10 gateway is required

---

HPE Aruba Networking gateways

The main reason you need gateways in a large-scale deployment is to simplify network design and provide scalability. The gateway is responsible for traffic termination, so all wireless traffic is tunneled from the APs to the gateway, which then passes through the firewall function on the gateway, and from there, it is forwarded on its way to the traffic's final destination. This eliminates the excessive span of VLANs. HPE Aruba Networking offers gateways and controllers to support larger enterprises requiring additional scalability, 24x7 operations, and greater security. Gateways are optional: in other words, you can run the APs in standalone mode, but if you need scalability and simplicity, gateways are the solution of choice.

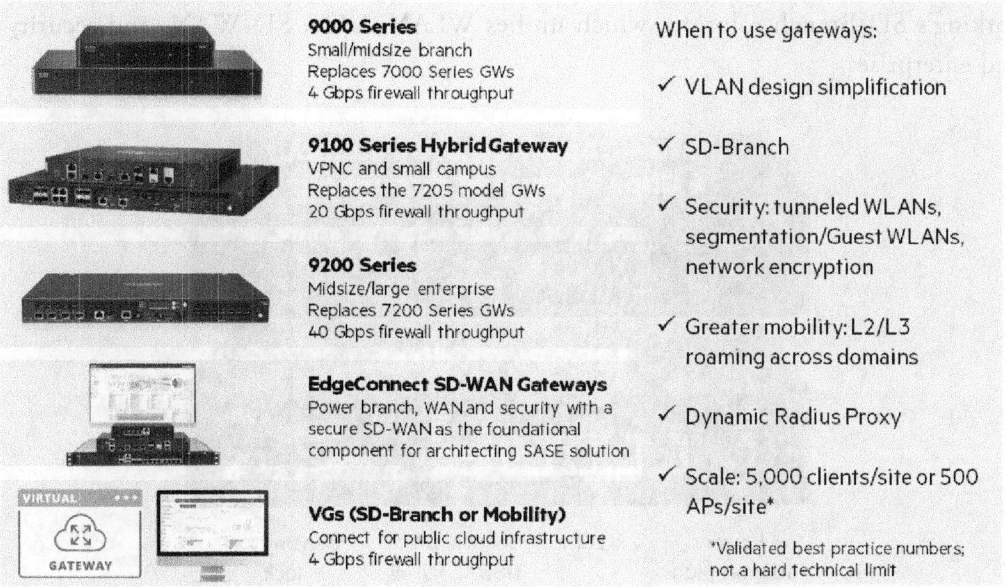

You should use a gateway deployment when any of the following is required:

- Simplified VLAN design
- SD-WAN
- Security: tunneled WLANs, segmentation/guest WLANs, network encryption
- More scalable and robust mobility: layer 2 and layer 3 roaming across domains
- Dynamic Radius Proxy
- Scale: More than 5,000 clients per site or more than 500 APs per site

The SD-WAN virtual gateway (VG) connects the SD-WAN environment to the public cloud environment and simplifies cloud access: the cloud becomes another node in the SD-WAN network. This can replace expensive cloud VPNs

---

Note

Previously, in AOS-8, there were two different terms for the hardware shown in the figure: gateways and mobility controllers (MCs). Gateways provided SD-WAN connectivity, and MCs terminated the APs. Starting in AOS-10, only a single term is used for these devices: gateways.

HPE Aruba Networking 9000 series gateway (9004)

HPE Aruba Networking 9000 Series Gateways provide high performance networking, SD-WAN and security functionality in compact and cost-effective form factors. Ideally suited for branch and small campus networks, the 9000 Series Gateways serve a key role within HPE Aruba

Networking's SD-Branch solution, which unifies WLAN, LAN, SD-WAN, and security for distributed enterprises.

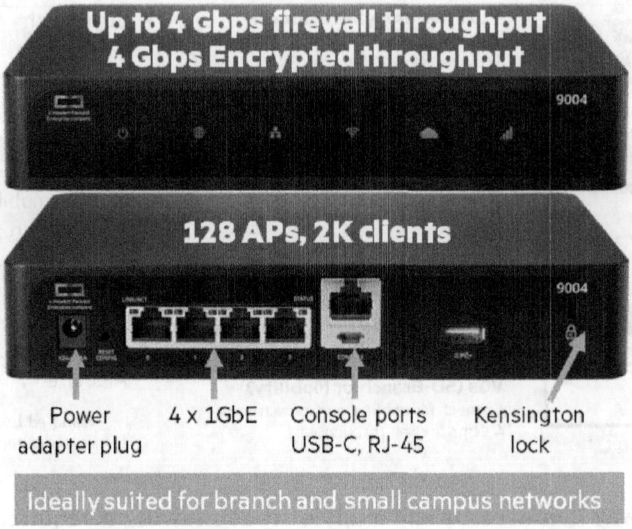

Key features:

- Cloud-managed and purpose-built for branch SD-WAN requirements
- Unified policy enforcement for wired and wireless traffic with dynamic segmentation
- Visibility into over 3,800 applications with no added hardware
- ntegrated LTE option available
- Flexible consumption with advanced subscription licenses

HPE Aruba Networking 9000 series gateway (9012)

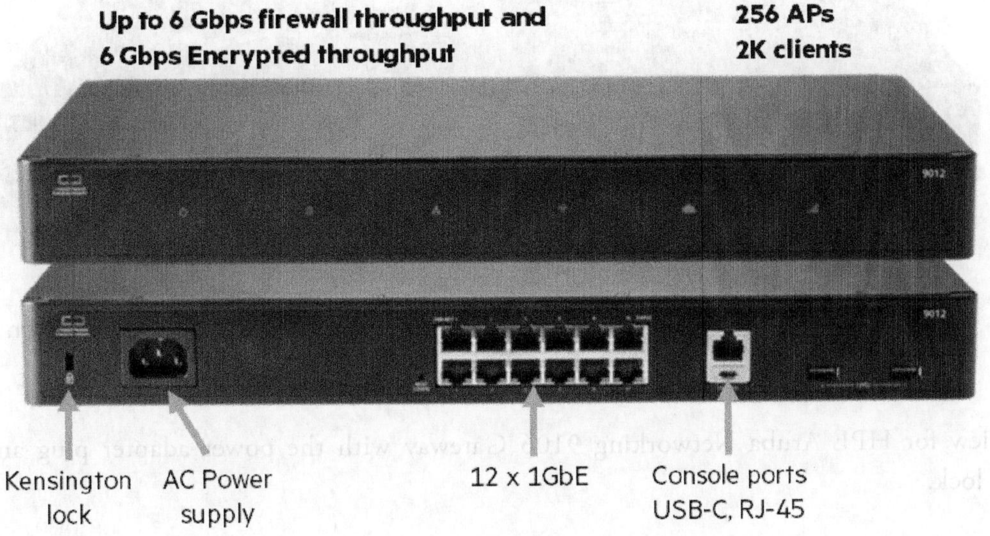

HPE Aruba Networking 9012 Gateway model supports 12x 1GbE ports, up to 6 Gbps firewall throughput and 6 Gbps Encrypted throughput, up to 256 access points and 2K clients.

HPE Aruba Networking 9100 series gateway (9106)

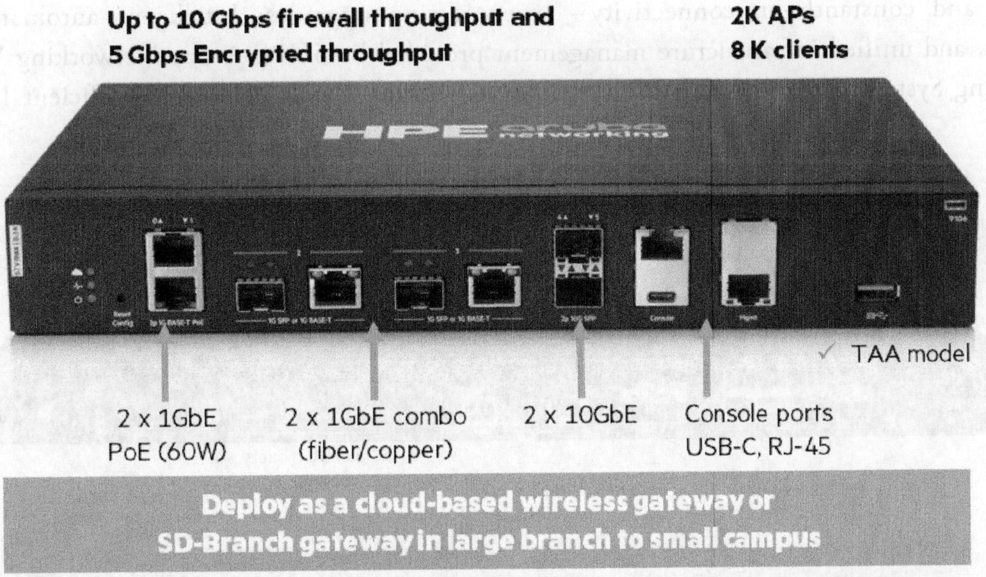

Front-view for HPE Aruba Networking 9106 Gateway with 2x 10GbE SFP+ ports, 2x 1GbE PoE ports (60W), 2x 1GbE combo ports (fiber/copper), USB-C and RJ-45 console ports.

Rear-view for HPE Aruba Networking 9106 Gateway with the power adapter plug and Kensington lock.

HPE Aruba Networking 9100 series gateway (9114)

The HPE Aruba Networking 9100 Series Hybrid Gateway is a cloud-optimized edge appliance designed to extend enterprise services securely and seamlessly to large branch and small campus networks. This purpose-built, high-performance edge appliance delivers Wi-Fi roaming, 24x7 reliability, and constantly-on connectivity— even during upgrades. Intelligent automation, AI insights, and unified infrastructure management provided by 5HPE Aruba Networking Wireless Operating System (AOS-10) and HPE Aruba Networking Central help drive efficient IT operations.

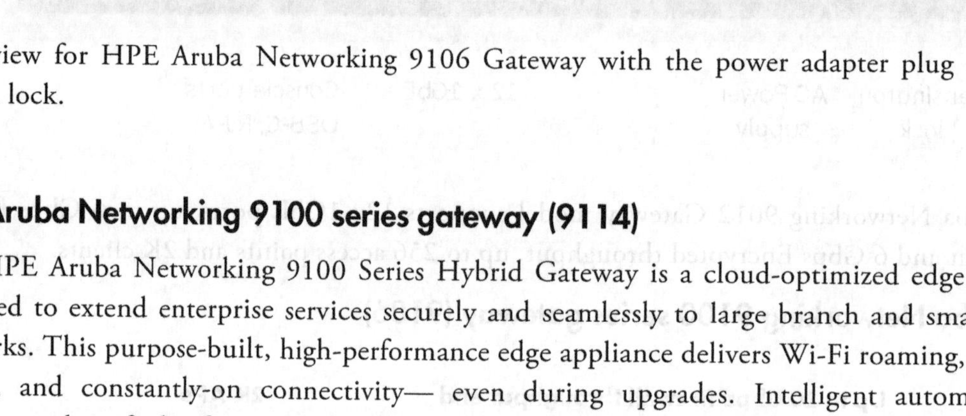

Key features:

- Versatile, scalable platform for cloud-based wireless gateway, SD-Branch device, or VPN Concentrator use cases
- Scalable cloud-native AOS-10 architecture provides connectivity for up to 4K access points and 10K customers
- Streamlined cloud-based management and control with HPE Aruba Networking Central
- Dynamic Segmentation for automated role-based policy enforcement across users, devices, and Internet of Things (IoT)
- High availability with live upgrades, clustering, and N+1 or NxN redundancy
- High-speed wired connectivity with 10GbE fiber ports for up to 20 Gbps throughput
- Zero Touch Provisioning simplifies deployment at branch sites
- Front-view for HPE Aruba Networking 9114 Gateway with 4x 10GbE SFP+ ports, 4x 1GbE compo ports (fiber/copper), USB-C and RJ-45 console ports and expansion slot for future use cases

Rear-view for HPE Aruba Networking 9114 Gateway with AC or DC power supply options, hot-swappable fan modules and optional redundant power supply.

HPE Aruba Networking 9200 series gateway (9240)

The HPE Aruba Networking 9200 Series Campus Gateway delivers enhanced Wi-Fi scalability and security to meet demands at the network edge. Engineered for mission-critical environments, the 9200 Series delivers high-performance, seamless roaming, 24x7 reliability, and always-on connectivity, even during upgrades. Unlike other edge appliances, the 9200 Series provides consistent policy enforcement across wired and wireless and dynamic segmentation for any user or device, even IoT. The 9200 series is designed to replace the 7210, 7220, and 7240 controllers.

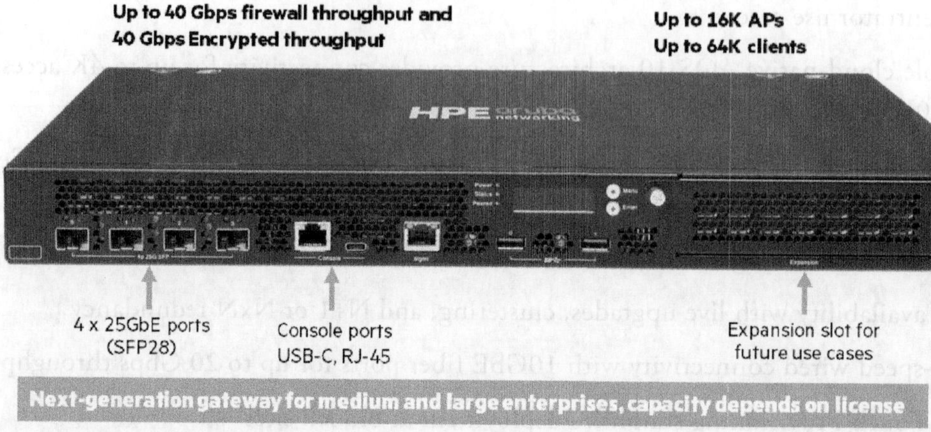

Front-view for HPE Aruba Networking 9240 Gateway with 4x 25GbE SFP28 ports, USB-C and RJ-45 console ports and expansion slot for future use cases.

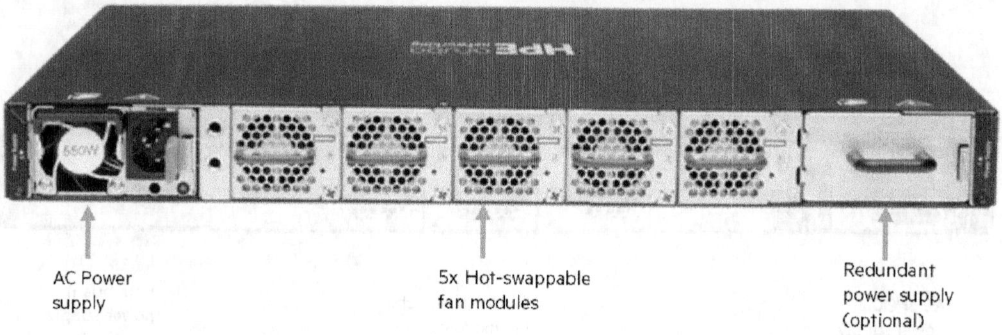

Rear-view for HPE Aruba Networking 9240 Gateway with AC power supply, 5x hot-swappable fan modules and optional redundant power supply.

9200 licensing Base, Silver, and Gold

Performance	Base hardware	w/ Silver license	w/ Gold license
Firewall throughput (Gbps)	20	30	40
Encrypted throughput GRE (Gbps)	20	30	40
Encrypted throughput AES-CBC-128 (Gbps)	20	30	38

Performance	Base hardware	w/ Silver license	w/ Gold license
Encrypted throughput AES-CBC-256 (Gbps)	20	30	39
Encrypted throughput AES-GCM-128 (Gbps)	20	30	40
Encrypted throughput AES-GCM-256 (Gbps)	20	30	40
Encrypted throughput AES-CCM (Gbps)	20	28	20
AOS-10 specifications			
Minimum supported software version	AOS 10.4	AOS 10.6	AOS 10.6
Cluster size	6	6	6
Max clients	32K	48K	64K
Max clients per cluster	128K	192K	256K
Max devices (APs)	4K	8K	16K
Max devices (APs) per cluster	8K	16K	32K
Maximum datapath / firewall sessions	4M	4M	4M

This is a review of the licensing for gateways we covered earlier. The base unit provides 20 Gbps of throughput, 32K maximum clients, 128K maximum clients per cluster, 4K maximum devices (APs) and 8K maximum devices (APs) per cluster. Applying a hardware capacity license will increase the performance and scalability of the 9200. With the Silver license, the maximum throughput is 30 Gbps and 48K maximum clients, 192K maximum clients per cluster, 8K maximum devices (APs) and 16K maximum devices (APs) per cluster. And finally, with the Gold license, you can achieve 40 Gbps throughput and 64K maximum clients, 256K maximum clients per cluster, 16K maximum devices (APs) and 32K maximum devices (APs) per cluster.

Additional information can be found in the datasheet.

Note

9240 base unit lines up with a 7220 Controller

HPE Aruba Networking virtual gateways

The HPE Aruba Networking gateway's virtualized instance is called a virtual gateway. Virtual gateways allow customers to bring their public cloud infrastructure to the SD-WAN fabric and facilitate connectivity between branches and the public cloud.

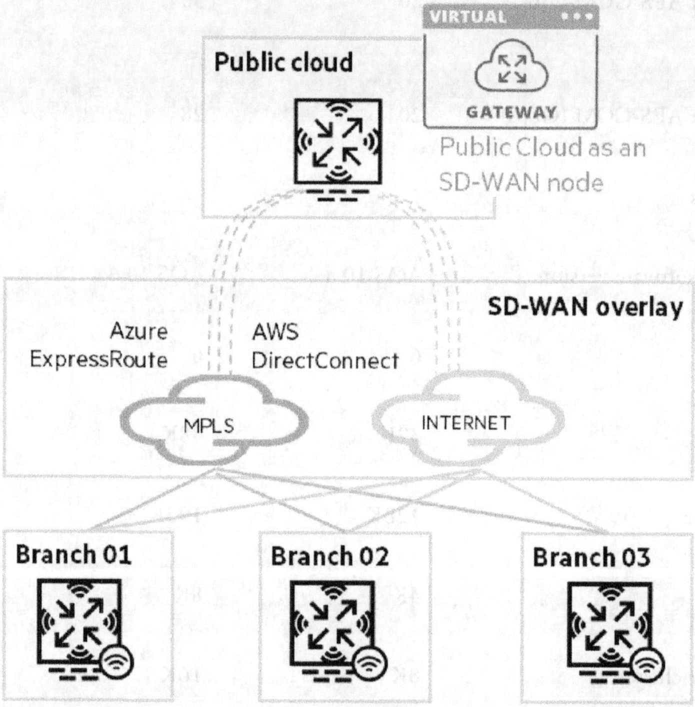

The inclusion of virtual gateways in the HPE Aruba Networking SD-WAN architecture offers the following benefits:

Ability to directly connect a branch to cloud instances, thereby improving access to the instances hosted on a public cloud infrastructure.

Resilient connectivity from the cloud by utilizing multiple transport links; for example, the link between an Internet Gateway and a virtual gateway.

Centralized control and policy management across branches, data centers, and cloud end-points.

Implementing 's SD-WAN solution allows your public cloud environment to appear as an SD-WAN node. This greatly enhances the benefits of integrating with public cloud infrastructure. The HPE Aruba Networking VG can terminate thousands of IPsec VPN tunnels, allowing many branch locations to directly access workloads in public clouds.

The HPE Aruba Networking SD-WAN simplifies routing and resiliency to virtual private cloud (VPC) environments offered by AWS and Microsoft Azure. For example, in Amazon Web Services (AWS), the HPE Aruba Networking VG connects individual customer VPCs with the customer's campus network via Direct Connect or the Internet Gateway by extending the SD-WAN fabric.

Gateway review

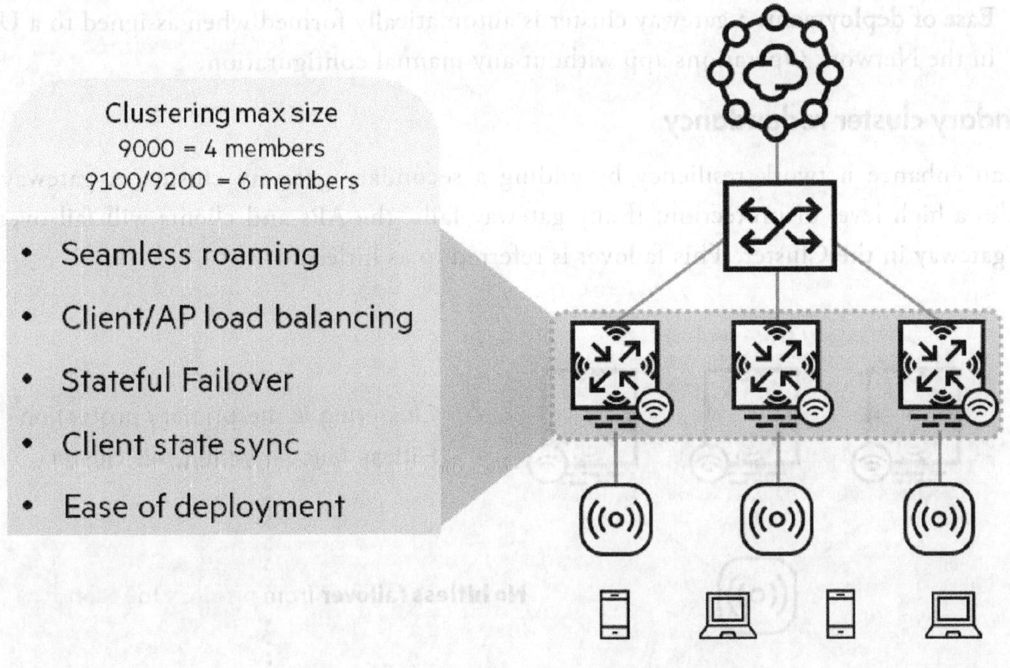

A gateway cluster combines multiple HPE Aruba Networking gateways operating as a single entity to provide high availability and service continuity to the WLAN clients in a network. A gateway cluster combines multiple HPE Aruba Networking gateways operating as a single entity to provide high availability and service continuity to the WLAN clients in a network.

Gateway clusters provide full redundancy to access points (APs) and WLAN clients in case of a failover. AOS-10 supports gateway clusters and provides the following features and benefits:

- **Seamless roaming**: when a client roams between APs within the WLAN boundary, the clients remain anchored to the same gateway in a cluster for a seamless user experience.

- **Client/AP Load Balancing**: if there is excessive workload among the devices, the gateways in a cluster balance AP and client load to ensure faster connectivity to clients.
- **Stateful Failover**: when a gateway device fails, APs and endpoints failover to another gateway in the cluster without any service disruption.
- **Client State Sync**: synchronizes high-value client sessions such as voice, video, FTP, and IGMP (used for IPv4 multicast groups) and MLD (used for IPv6 multicast groups) between active and standby gateways in a cluster. This allows the connected devices to failover to the standby gateway seamlessly. However, AOS 10.x only supports this synchronization for the first failover. For subsequent failovers, it will not synchronize the session to the recent standby. Instead, it synchronizes a new session.
- **Ease of deployment**: a gateway cluster is automatically formed when assigned to a UI group in the Network Operations app without any manual configuration.

Secondary cluster redundancy

You can enhance network resiliency by adding a secondary gateway cluster. A gateway cluster provides a high level of protection: if any gateway fails, the APs and clients will fail over to any other gateway in the Cluster. This failover is referred to as hitless.

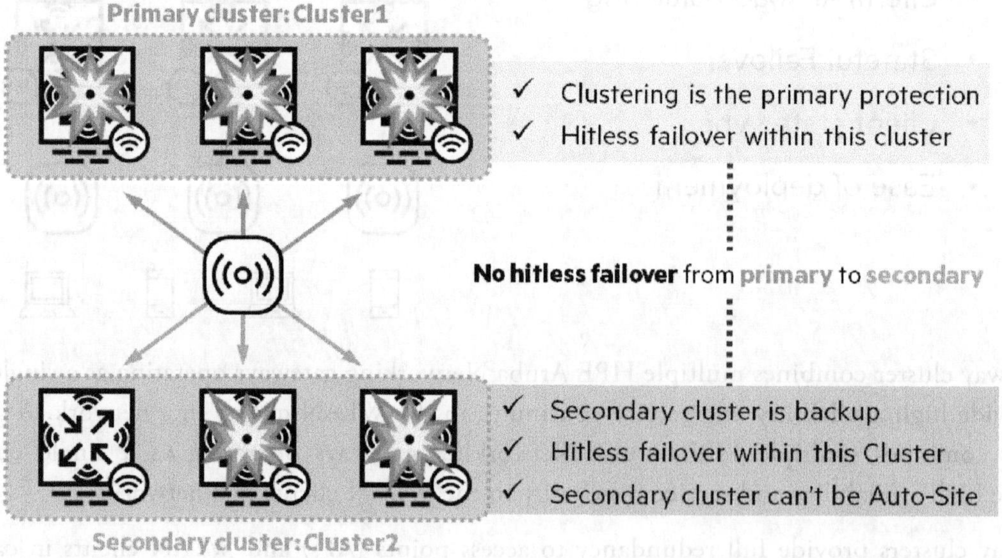

What if all Gateways in a Cluster fail? If you set up a Secondary Gateway Cluster, APs and clients could failover to that Cluster; this failover is not hitless, and some brief downtime is expected (i.e., expect a brief period of disruption).

Note

A secondary gateway cluster cannot be auto-site enabled.

Centralized Overlay Networks with HPE Aruba Networking Gateways

This section explores using gateways as the hub for centralized network overlays in dynamic segmentation and user-based tunnels (UBTs).

Traditional network design

- Localized VLANs
- Extended trunking
- Results in flattened topologies
- Security concerns

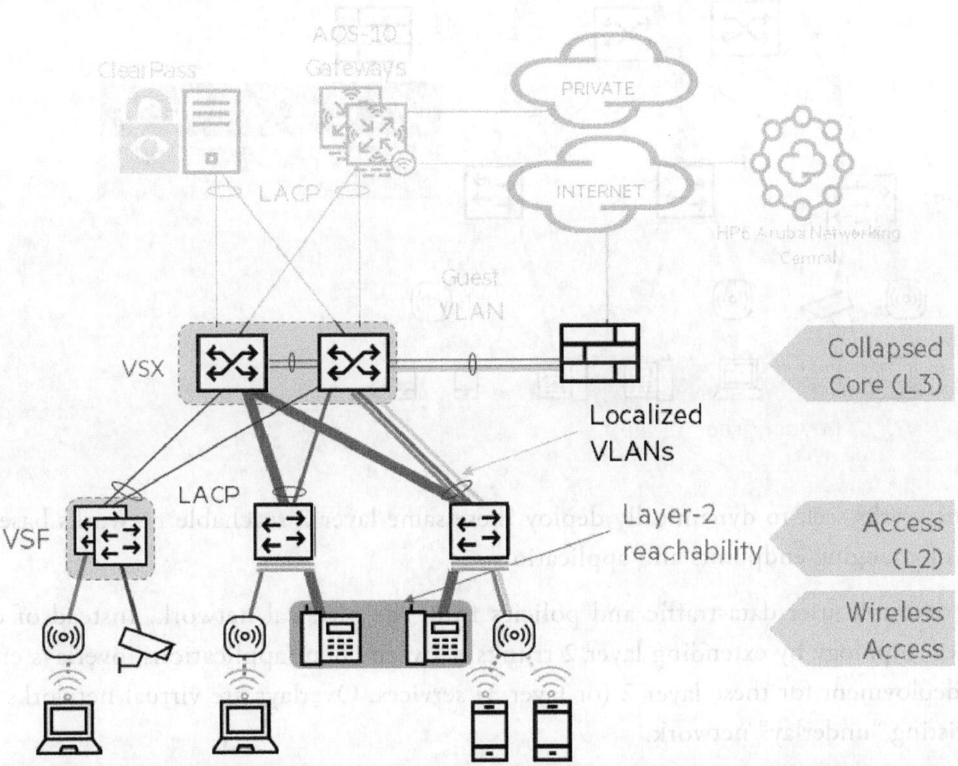

Traditional network designs called for designing applications localized within a VLAN off one gateway in one campus location. Applications and endpoints tend to expect layer 2 reachability. This meant having to try to extend trunk links to carry VLANs that could result in flattened topologies, loops, and severe security concerns.

Network design with centralized overlays

- Centralized using GWs GRE tunnels
- Enable same layer-2 reachability dynamically
- Decouple user traffic from the underlay

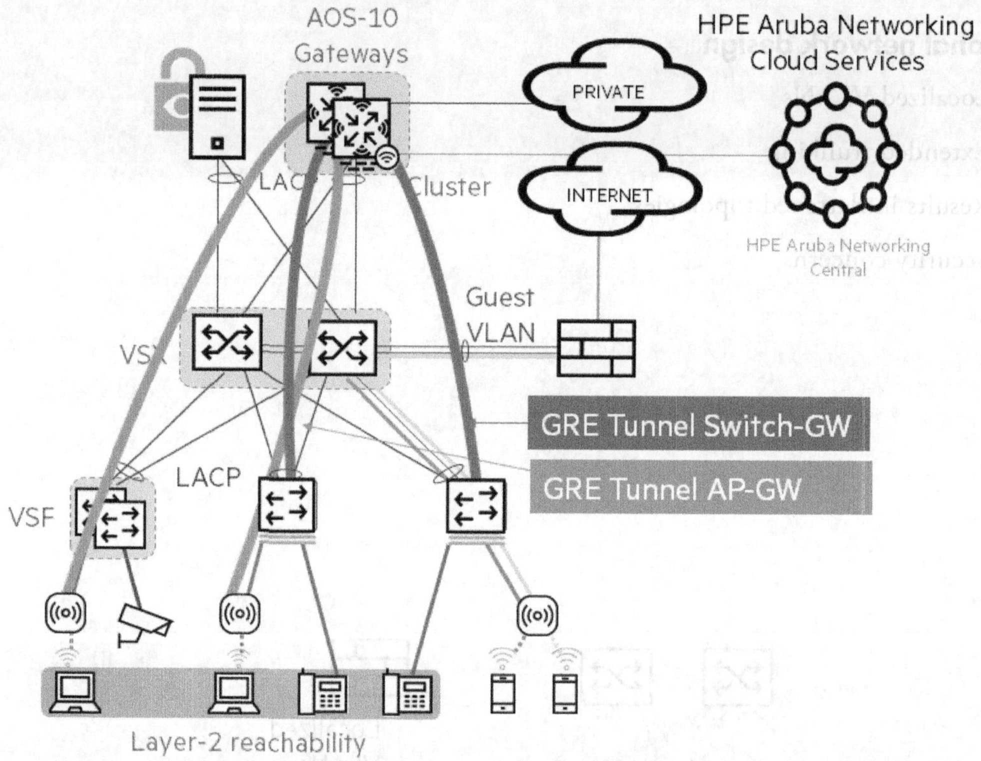

Overlay networks seek to dynamically deploy these same layer 2 reachable networks based on the demands of changing endpoints and applications.

Overlays decouple user data traffic and policies from the physical network. Instead of changing the physical topology by extending layer 2 trunks between client applications, overlays enable on-demand deployment for these layer 2 (or layer 3) services. Overlays are virtual networks building on the existing "underlay" network.

The underlay network should be built with stability, scalability, and predictability. This ensures the overlay can establish the necessary tunneling to create a virtual layer-2 or layer-3 overlay between any two endpoints on the campus network.

Multi-cluster deployment - MultiZone

Different WLANs terminate on different gateway clusters in a multi-cluster deployment called MultiZone.

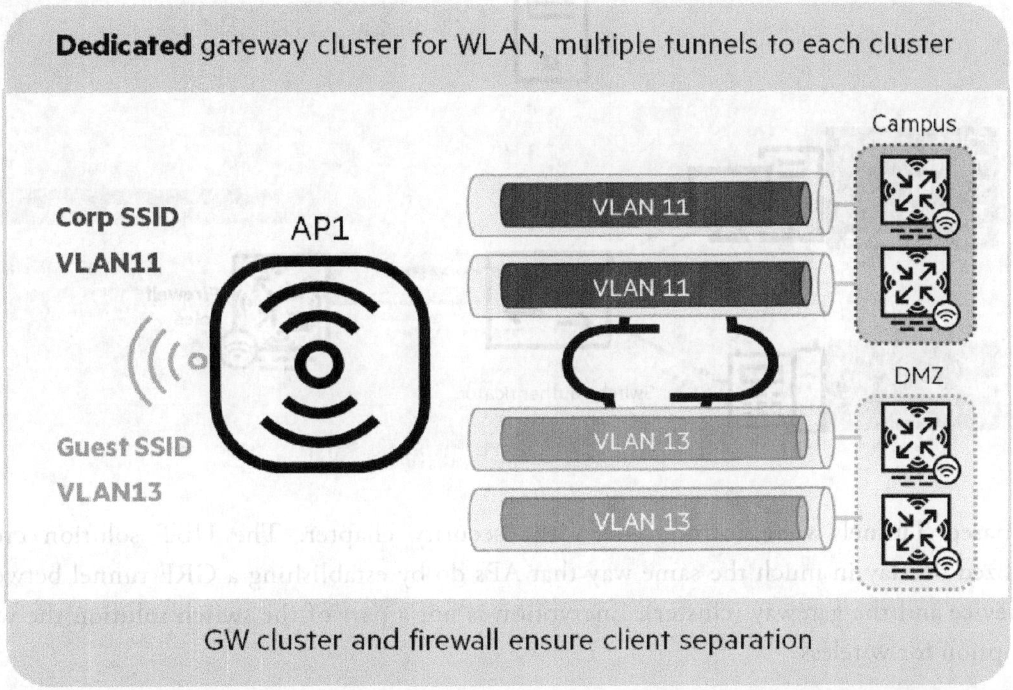

This deployment is recommended if you want to isolate gateways that terminate different types of users, such as employees and guests. Client separation is provided by the firewall and assigning user traffic to a dedicated gateway cluster.

Centralized overlays – switch UBTs

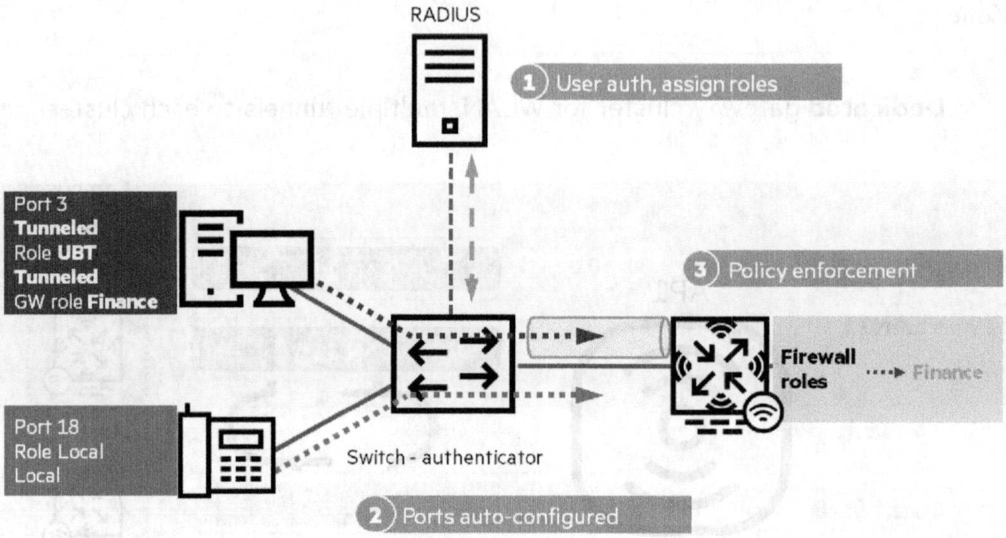

User Based Tunnels were introduced in the security chapter. The UBT solution creates a centralized overlay in much the same way that APs do by establishing a GRE tunnel between the edge device and the gateway (cluster). Encryption is not a part of the switch solution the way it is as an option for wireless.

PAPI communication

The switch uses two protocols to connect to a gateway. PAPI is the protocol that runs the control plane and GRE is used to encapsulate the user's traffic (data plane).

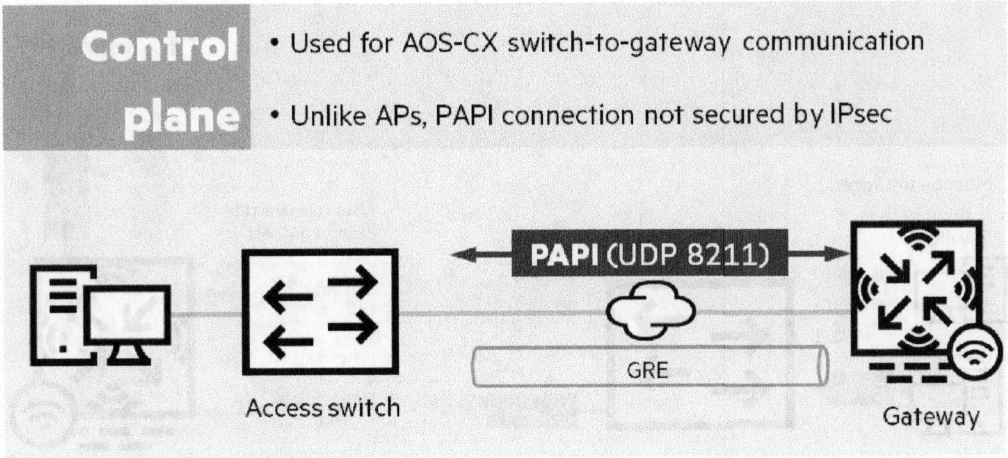

PAPI runs on UDP 8211 and this protocol is the same used by AP-to-Gateway communications. However, where APs use IPsec to protect the PAPI connection between the AP and HPE Aruba Networking Gateway, AOS-CX switches do not currently support IPsec.

UBT reserved VLAN mode: default

The gateway divides the arriving tunneled traffic into one tunnel per tunneled-node port. It decapsulates the traffic and applies various controls. The switch communicates the gateway role to the gateway which applies that role to the user traffic. Based on that role, the gateway controls traffic with firewall policies and other policy actions. Finally, the gateway forwards the packet towards its destination. The gateway can apply different firewall rules to different devices in the same VLAN, again, based on dynamic settings applied during authentication.

Keep in mind that the switch is simply using the untagged VLAN on the tunneled-node port. It is not switching the tunneled-nodes' traffic in this VLAN, and therefore you don't need to tag it on the switches' uplinks as you normally would. Instead, the effective VLAN exists on the gateway and at the distribution or core switch. In keeping with this design, the switch is restricted from implementing many normal features—such as an IP address—on any untagged VLAN on a tunneled-node port. HPE Aruba Networking recommends reserving the tunneled-node VLANs for tunneled-node ports only. Client traffic in reserved VLAN mode must arrive untagged!

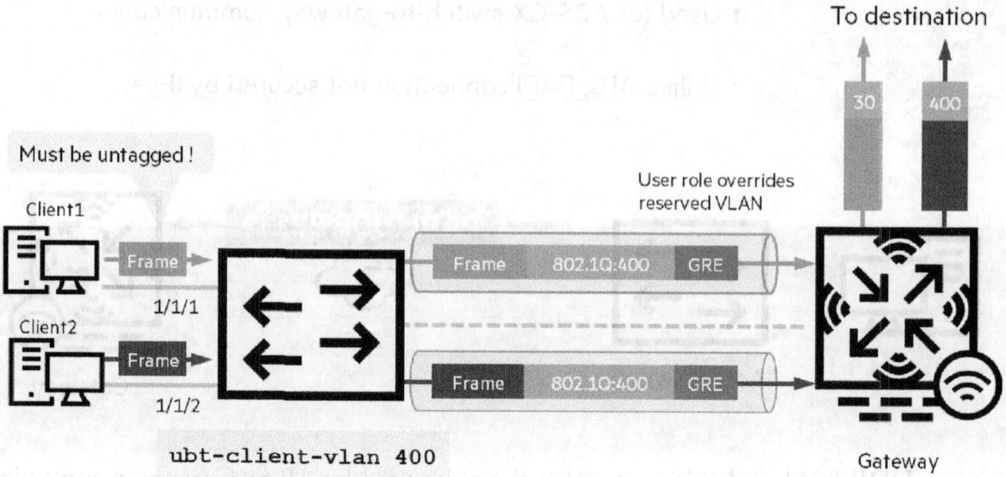

You should also understand that the gateway can override the default VLAN assignment after it authenticates the device. For example, it might apply 802.1X authentication to tunneled-nodes and receive a dynamic VLAN assignment from ClearPass. The gateway would then forward the tunneled-node device traffic in this VLAN, rather than the ubt-client-vlan VLAN configured on the switch. If you don't override the user VLAN on the gateway, then the traffic will be forwarded upstream with ubt-client-vlan configured on the switch. The gateway replicates the broadcast or multicast packet for each user in the user VLAN.

Note that the switch will establish one GRE tunnel per port, even if multiple clients connect on the same port. The maximum number of clients per port in reserved mode is 256. The reserved VLAN is also known as UBT 2.0 and is supported from AOS-CX 10.04.0020 and later versions.

UBT extend VLAN mode

Extend VLAN mode requires user VLANs to exist on the switch and on the gateway. In this mode traffic from the client may arrive untagged or tagged. Notice that in this mode tunneled client traffic will be 802.1Q VLAN tagged. In this mode, the switch creates a multicast tunnel between switch, and the device designated gateway (DDG) that is used to maintain multicast and broadcast traffic sent from gateway to switch.

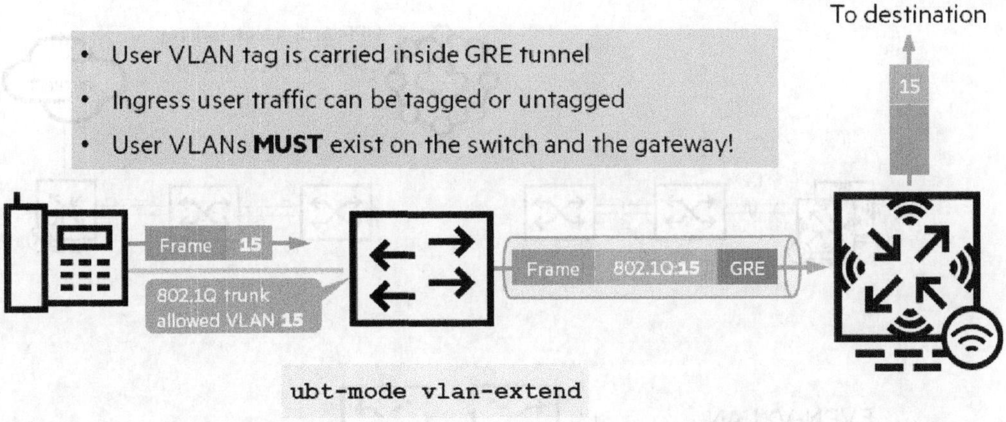

The switch forwards the multicast and broadcast traffic to the UBT clients that are part of the same VLAN. The unicast, broadcast, or multicast traffic from switch to controller is always sent through UDG tunnel. This mode is also known as UBT 1.0 and is supported from AOS-CX 10.06 and later versions.

Distributed Overlay networks with HPE Aruba Networking Central NetConductor

Distributed overlays can create tunnels between any supported virtual tunnel endpoints (VTEPs), such as HPE Aruba Networking campus switches. Let's take a look at the technology.

Network design with distributed overlays

- Campus Access fabric orchestrated by HPE Aruba Networking Central NetConductor
- Distributed overlay networks across the campus access fabric
- Dynamic VXLAN tunnels between virtual tunnel endpoints (VTEPs)
- Enable layer-2 reachability dynamically over layer-3 boundaries
- Decouple user traffic from the underlay
- End-to-end segmentation using EVPN-VXLANs (VNBT) or static VXLANs
- VXLAN-GBP segments user traffic by roles

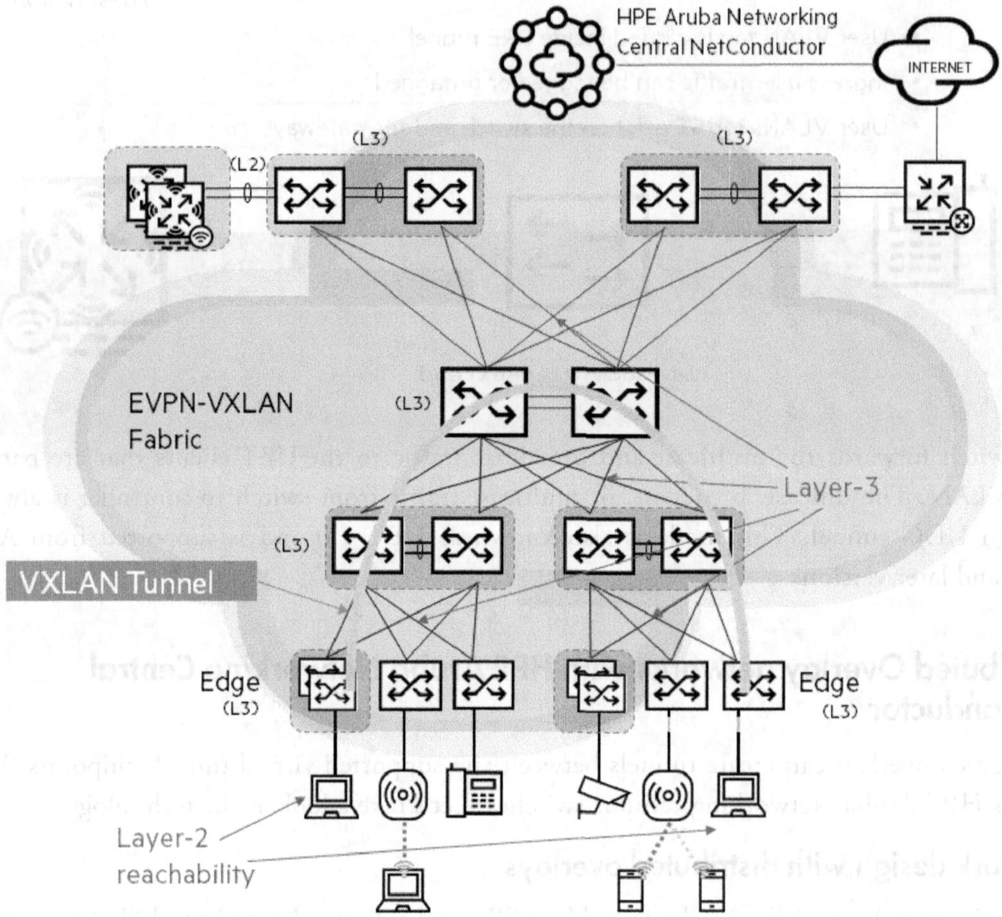

Traditional tunneling protocols like GRE and IPsec have been used in campus and WAN networks for years. A better option to support multitenancy and scaled distributed overlays is the Virtual Extensible LAN (VXLAN) protocol.

HPE Aruba Networking allows the choice to design for centralized or distributed overlays. User-based tunneling (UBT) is a centralized overlay architecture where traffic is tunneled to a gateway or gateway cluster using GREs. Distributed overlays, in comparison, use EVPN-VXLANs to enable dynamic tunnel creation between VXLAN Tunnel Endpoint (VTEP) switches anywhere on the campus. When deploying either architecture, it is recommended to implement "Colorless Ports" to automate client onboarding and access control.

VXLAN tunnels can be statically configured or dynamically created using EVPN-VXLANs, sometimes referring to dynamic segmentation such as Virtual Network Based Tunneling (VNBT).

EVPN supports full policy-based micro-segmentation based on global rules using VXLAN-GBP (Group Based Policies). Group Based Policies (GBP) segment user traffic in a network by grouping the users into roles based on user authentication at the source or ingress VXLAN tunnel endpoint (VTEP). Source-based roles will remain effective even if a device authenticates at a different location or is assigned a different IP address. Users can be locally authenticated, MAC authenticated, or 802.1X authenticated clients. After authentication, a Group Policy ID is mapped based on the role applied.

Static VXLAN vs EVPN

You can use two methods to deploy a VXLAN infrastructure:

- Static VXLANs
- Ethernet VPN (EVPN) over a VXLAN

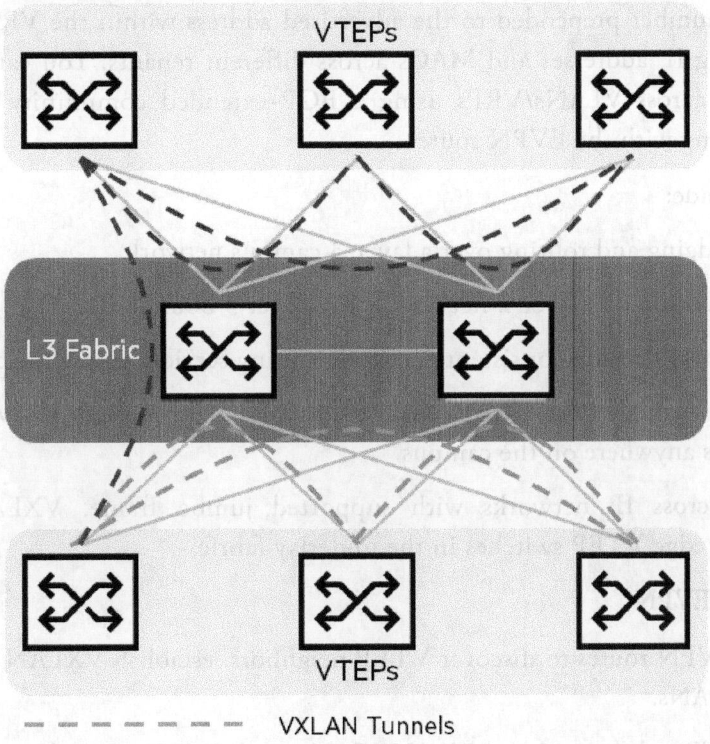

Static VXLANs

Static VXLAN uses flood and learn (or ingress replication) to learn the MAC addresses of the remote hosts, which involves manual configurations of remote VTEPs in the flood list. Tunnels must be manually created, which is challenging in large networks. It is also difficult to maintain. Static VXLANs, therefore, are typically used in small deployments.

Ethernet VPN (EVPN) over VXLAN overview

Ethernet VPN (EVPN) is a standards-based BGP control plane that advertises MAC addresses, MAC and IP bindings, IP prefixes, etc. The initial EVPN standard (RFC 7432) defined the BGP EVPN control plane and specified an MPLS data plane. The control plane, containing an MPLS data plane, extends to consider additional data plane encapsulation models, including VXLAN, NVGRE, and MPLS over GRE. These are covered in detail in IETF RFC 8365. This section focuses on EVPN and its operation with a VXLAN data plane for building overlay networks.

You can use MP-BGP EVPN to dynamically discover remote VTEPs and advertise MAC address and MAC/IP bindings in the VXLAN overlay, thus eliminating the flood and learn and the need for creating static tunnels. MP-BGP supports a new EVPN Network Layer Reachability Information (NLRI) carried in BGP, using multiprotocol BGP extensions with a newly defined address family identifier (AFI) and subsequent address family identifier (SAFI). A route distinguisher (RD) is a unique number prepended to the advertised address within the VRF that ensures support for overlapping IP addresses and MACs across different tenants. You can selectively import and export routes across VLANs/VRFs using a BGP-extended community called route target (RT) advertised along with the EVPN routes.

Other benefits include:

- Uniform bridging and routing over a layer-3 campus network.
- Efficient extension of a layer-2 network across layer 3 boundaries
- Anycast gateways ensure consistent first-hop routing services across the campus.
- End-to-end segmentation using VXLAN-Group Based Policies (VXLAN-GBP) to apply these policies anywhere on the campus.
- Transport across IP networks with supported jumbo frame. VXLAN must only be deployed on edge VTEP switches in the underlay fabric.

Auto-discovery via EVPN

VTEPs use BGP EVPN routes to discover VTEP neighbors, establish VXLAN tunnels, and assign the tunnels to VXLANs.

IMET route – VTEPs advertise their VXLAN IDs through IMET routes. If two VTEPs have the same VXLAN ID, they automatically establish a VXLAN tunnel and assign it to the VXLAN. This method can create the replication list for forwarding broadcast, unknown, unicast, and multicast (BUM) traffic.

MAC/IP advertisement route and IP prefix advertisement route – In the EVPN deployment, VTEPs advertise MAC/IP advertisement routes or IP prefix advertisement routes, which carry export targets. When a VTEP receives a route, it compares the export targets of the route with the local import targets.

The full coverage and configuration of EVPNs with VXLANs is beyond the scope of this guide.

VXLAN packet structure

VXLAN encapsulates layer 2 Ethernet frames in UDP packets. These VXLAN tunnels provide layer 2 virtualized network services to connected endpoints. A VTEP is the function within a switch that handles the origination or termination of VXLAN tunnels. Similar to a traditional VLAN ID, a VXLAN network identifier (VNI) identifies an isolated layer 2 segment in a VXLAN overlay topology. Symmetric Integrated Routing and Bridging (IRB) capability enables the overlay networks to support contiguous layer-2 forwarding.

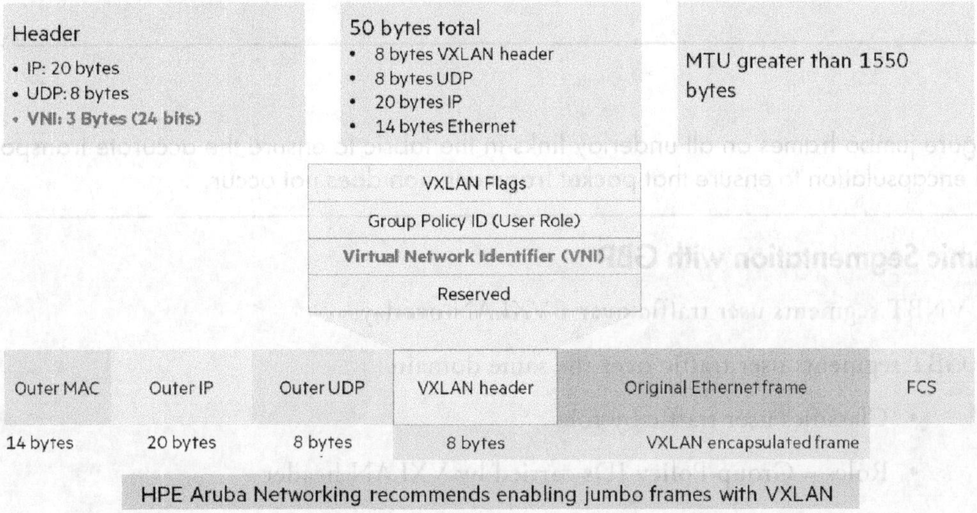

An overlay network is implemented using VXLAN tunnels that provide both layer 2 and layer 3 virtualized network services to endpoints connected to the campus. The VNI associates tunneled traffic with the correct corresponding layer 2 VLAN or layer 3 route table so the receiving VTEP can forward the encapsulated frame appropriately. The symmetric IRB capability allows the overlay networks to support contiguous layer 2 forwarding and layer 3 routing across leaf nodes.

A VTEP encapsulates a frame in the following headers:

- IP header: IP addresses in the header can be VTEPs or VXLAN multicast groups in the transport network. Intermediate devices between the source and destination forward VXLAN packets based on this outer IP header.
- UDP header for VXLAN: The default VXLAN destination UDP port number is 4789.
- VXLAN header: VXLAN information for the encapsulated frame.
- 8-bit VXLAN flags: The first bit signals if a GBP ID has been set on the packet and the fifth bit signals if the VNI is valid. All other bits are reserved and set to "0".
- 16-bit VXLAN group policy ID: The group ID identifies the policy enforced on tunneled traffic. (This will be discussed later in the module.)
- 24-bit VXLAN Network Identifier: Specifies the virtual network identifier (VNI) of the encapsulated frame.
- 24-bit reserved field

Note

Configure jumbo frames on all underlay links in the fabric to ensure the accurate transport of additional encapsulation to ensure that packet fragmentation does not occur.

Dynamic Segmentation with GBP

- VNBT segments user traffic over a VXLAN overlay
- GBT segments user traffic over the same domain
    - Classifies user traffic by role
    - Roles = Group Policy IDs carried by VXLAN header
    - Allows micro-segmentation of user traffic on same VLAN
    - GWs' traffic can exit one VXLAN tunnel and enter another

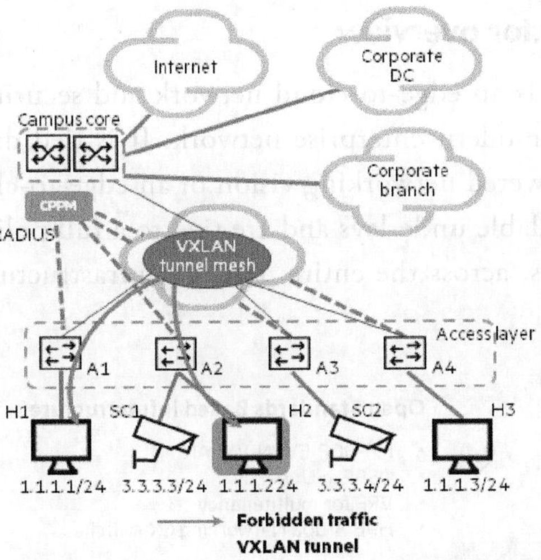

Client	MAC	VLAN/Subnet	L2-VNI	Role
H1	M1	100/1.1.1.0/24	1000	Guest
H2	M2	100/1.1.1.0/24	1000	Admin
H3	M3	100/1.1.1.0/24	1000	Guest
SC1	M4	300/3.3.3.0/24	3000	Security Camera
SC2	M5	300/3.3.3.0/24	3000	Security Camera

Group Based Policies (GBP), as mentioned, allows policy-driven segmentation by grouping users into roles based on user authentication at the source (ingress) VTEP. Allows micro-segmentation of user traffic on the same VLAN by classifying the user traffic based on user role. Roles are translated into Group Policy IDs carried by the VXLAN header to match Group Policy IDs at the egress VTEP where enforcement occurs.

Macro segmentation refers to a policy that affects group-to-group traffic. Micro-segmentation refers to policies that affect group traffic to or from a user, even within the same subnet. Group Policy ID-based segmentation can extend between wired and wireless networks. The figure shows an example of GBPs used within the same subnet.

Dynamic Segmentation or Virtual Network Based Tunneling (VNBT) helps segment user traffic over a VXLAN overlay-based network, and GBP helps by segmenting user traffic in the same domain based on user role (determined during authentication). Segmentation using the GBP feature performs the following actions:

Allows micro-segmentation of the user traffic on the same VLAN by classifying the user traffic based on user role. The role is then converted into Group Policy IDs and carried in the VXLAN header.

At egress, the switch determines if the traffic from the source role (as carried in the Group Policy ID) is permitted to the destination role (as determined from the destination MAC) and either forwards or drops the traffic accordingly.

In a gateway scenario, the traffic on a VXLAN tunnel may get terminated and enter another VXLAN tunnel. In such scenarios, the Group Policy ID must be transported to the new tunnel for the final destination to enforce role-based policies.

HPE Aruba Networking Central NetConductor overview

HPE Aruba Networking Central NetConductor is an edge-to-cloud network and security framework designed to tackle these problems for the modern enterprise network. It is tied directly to the HPE Aruba Networking security-first, AI-powered networking vision of an edge-to-cloud network. Intelligent overlays are built on highly available underlays and are tied to a full policy-based micro-segmentation model, based on global roles, across the entire network infrastructure of the customer.

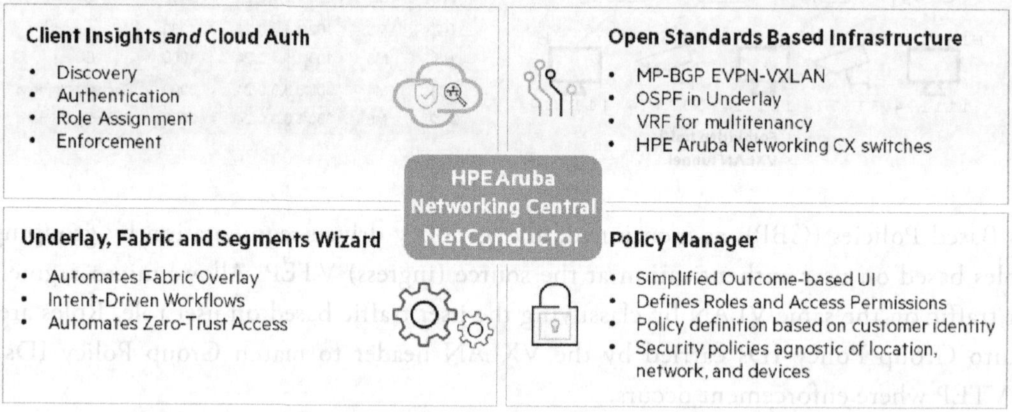

Role-based policies abstract policy from the underlying network and enable flexible and simplified policy definition and enforcement. This is enhanced by the full automation of the underlay, orchestration of the overlay, a single pane of glass for management and monitoring, and a rich array of complementary services. The NetConductor framework has evolved to enhance the policy and orchestration components to deliver true intent-based network evolution and optimization.

The following are the main pillars of NetConductor:

Role-based Segmentation

NetConductor provides the ability to deploy a zero-trust enforcement model using role-based segmentation. Traditional policies use location specific entities like IP addresses or subnets to define security policies. Role-based policies abstract policy from the underlying network by assigning roles to endpoints or users and using roles to enforce policies. Role-based policies can be enforced in a distributed manner at different parts of the network. NetConductor also provides the ability to automate and simplify policy definition for IoT devices with behavior-based profiling using AI or ML based classification. This greatly simplifies policy definition and ensures consistent policy enforcement across wired and wireless campus networks, the datacenter, and across the WAN.

Intelligent Overlays

Overlay networks provide the ability to deploy flexible services based on ever-changing demands of the endpoints and applications. Decoupling of overlay network from the physical topology enables on-demand deployment of layer 2 and layer 3 services irrespective of underlay physical topology. Overlay networks also enable the ability to carry endpoint or user role information across the network without requiring all devices in the path to understand or manage the roles. NetConductor provides customers the flexibility to choose between centralized overlays or distributed overlays to address their unique requirements. The centralized overlay provides simplified operations and advanced security features for distributed enterprise and smaller campus deployments. For large enterprise campus deployments, NetConductor provides the ability to use distributed overlays for wired and wireless endpoints. This enables large enterprises to deploy a standards-based and scalable overlay network. Both overlay models support the Colorless Ports feature, which enables automated client on-boarding and access control for ease of operations.

Automation and AI Ops

One of the primary requirements for enterprise campus network is simplicity of deployment, maintenance, and troubleshooting. With cloud-based management provided by HPE Aruba Networking Central, enterprise devices can be on-boarded and managed in a matter of minutes. Intent-based workflows enable architects to deploy and provision the network without the need for technical expertise in the networking protocols and the command line interface. HPE Aruba Networking Central also provides unified policy orchestration for the global network across wired, wireless, and SD-WAN. NetConductor also enhances end user experience and reduces help desk calls with real time problem identification and AI or ML driven actionable insights.

HPE Aruba Networking Central NetConductor orchestration

HPE Aruba Networking Central NetConductor can simplify network design and operations. Traditionally, location-/network-specific constructs such as IP addresses or subnets were used to define security policies. This led to complexity and inflexibility in the network due to the lack of client mobility brought about by these segmentation requirements. IT teams also lose the opportunity for automation as they have to pre-provision the network based on these VLANs and subnet constructs.

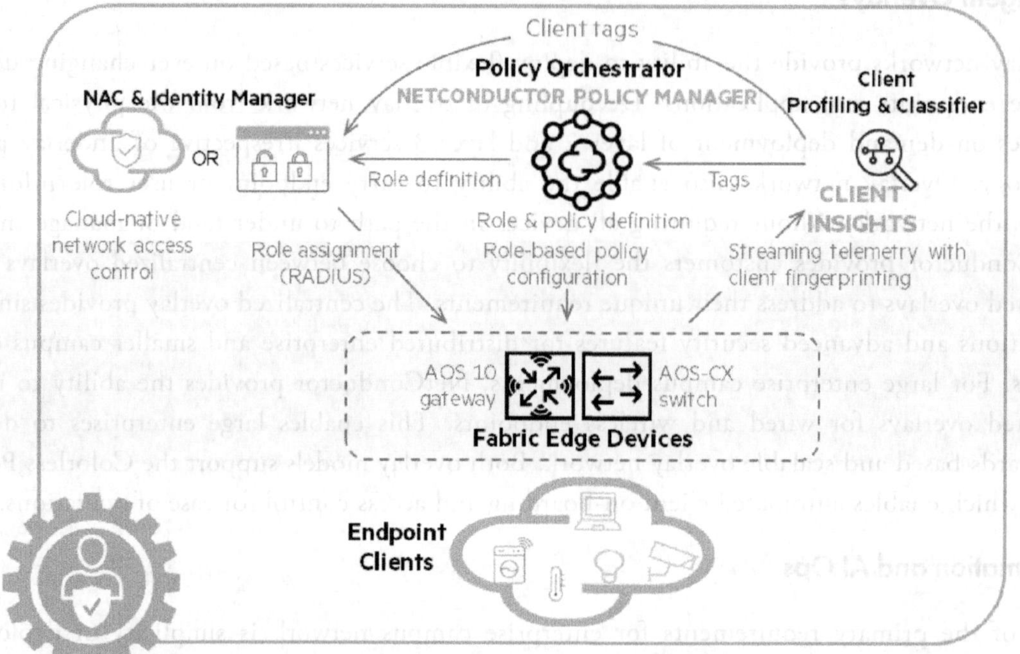

Role-based policies allow the policy to be abstracted from the underlying network by assigning identity-based Roles to endpoints and users. These identities are derived by authentication via Identity stores such as Active Directory or by profiling how these endpoints behave with Client Insights. This naturally leads to the simplified design, deployment, and operations of these less complex networks, as the network no longer needs to be segmented using traditional network constructs such as VLANs, and the administrator needs to pre-provision the network to accommodate user onboarding.

HPE Aruba Networking Central NetConductor allows the network administrator, working with the security team, to implement role-based security policies across the global network using intent-based workflows. The NetConductor policy manager simplifies the task of administrators when defining policy for clients using HPE Aruba Networking User Roles, which allows clients to be grouped based on their identity. This condenses what typically would be thousands of complex IP or VLAN-based security rules in a firewall to a few dozen role-based security rules. This enables administrators to focus on defining policies based on their desired business intent and eliminates constant updating of policies because of client and application VLAN/IP changes.

HPE Aruba Networking Central NetConductor orchestrates policies with these easy steps:

1. **Global Policy Orchestration** — The NetConductor policy manager is a new service that acts as the single interface for the management of network and security policies in the network. Network administrators can define roles in HPE Aruba Networking Central and assign role-

based permissions for these roles in a common policy language. For example, an organization that wants to implement a business-driven policy can do so from HPE Aruba Networking Central in a few simple clicks: "Allow all the video cameras in the network to access DVR recorders, but not the printers. Oh, and by the way, ensure interns and contractors cannot access these cameras."

2. **NAC (Network Access Control) and Identity Management** — NetConductor provides customers the flexibility to pick their NAC solution of choice, whether that is ClearPass, our market-leading on-premises Network Access Control (NAC) solution, or Cloud Auth, the first integrated and cloud-native NAC and identity management solution, which builds on ClearPass NAC market leadership and streamlines the protection of distributed enterprise networks by working seamlessly across wired, wireless, and WAN connections. The roles defined in HPE Aruba Networking Central are automatically synchronized to either Cloud Auth or ClearPass, where customers can then map their users and devices to these roles. HPE Aruba Networking also supports third-party NAC solutions using standards-based mechanisms (VSAs) to assign roles to clients, giving customers the flexibility of choice and the opportunity to migrate in a phased manner.
3. **Policy Enforcement on Network Edge Devices** — NetConductor enables 's AOS-10 gateways and AOS-CX switches to enforce role-based policies in line. Role-based policies defined in HPE Aruba Networking Central are pushed to the edge infrastructure devices. As clients and devices authenticate onto the network, these clients are assigned a role based on authentication from the NAC. The infrastructure edge devices enforce role-based policies either in a distributed manner at the edge via AOS-CX switches or at the centralized AOS-10 gateway, depending on the organization's choice of overlays. This enhances security from internal threats within the network that would have otherwise gone undetected by traditional perimeter firewalls.
4. **Client Profiler and Threat Analysis** — Client Insights is a service that discovers, profiles, and continuously monitors devices. It provides AI-powered user and device discovery to ensure all users and devices on the network—even rogue IoT devices—are seen and controlled. In the era of IoT and BYOD devices, the need to eliminate blind spots related to these devices on the network has become very relevant. HPE Aruba Networking Central NetConductor provides the ability to automate and simplify policy definition for IoT devices with behavior-based profiling using AI- or ML-based classification. This simplifies policy definition and ensures consistent policy enforcement across the entire network.

HPE Aruba Networking Central NetConductor provisioning

Automated Fabric Provisioning on AOS-CX and AOS-10 with HPE Aruba Networking Central

HPE Aruba Networking Central NetConductor introduces the ability to deploy a distributed overlay fabric on a campus network with AOS-10 gateways and AOS-CX switches. This capability is a standards-based implementation leveraging EVPN on the control plane and VXLAN-GBP on the data plane, enabling multi-vendor interoperability and phased migration strategies. The CX portfolio is further differentiated by optimized TCAMs for role-based policies, dynamic tunnel creation based on client authentication, and optimized host-route learning mechanisms that provide greater scale and enhanced usability compared to competitive products.

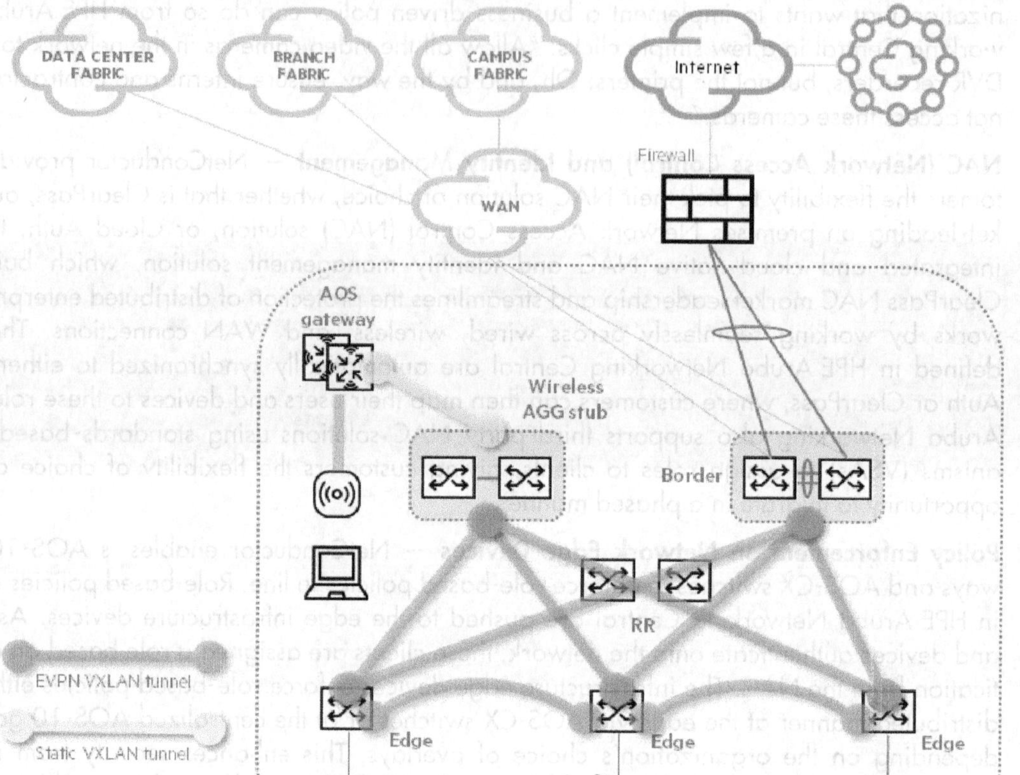

Automated Deployment of Distributed Overlay with HPE Aruba Networking Central NetConductor Fabric Wizard

The HPE Aruba Networking Central NetConductor fabric wizard simplifies the deployment of fabric over an existing traditional network. The wizard uses simple workflows to translate customers' business intent into fabric-wide configurations, eliminating users needing detailed knowledge of concepts such as EVPN, VNIs, VTEPs, and VRFs. For example, manually provisioning a BGP-EVPN fabric across 200 switches within a campus—which entails creating switch-specific configurations, provisioning these switches individually, and then checking and validating the configurations—can take a week; the fabric wizard accomplishes provisioning in less than 10 minutes, without introducing human error.

Distributed Overlay with eVPN requires an HPE Aruba Networking Central Advanced license on CX switches.

More details are available here: https://arubanetworking.hpe.com/techdocs/VSG/docs/020-campus-deploy/esp-campus-deploy-153-distributed-overlay/

EVPN-VXLAN fabric considerations

EVPN-VXLAN fabric:

- Use OSPF for L3 connectivity (loopbacks/VTEPs) in Underlay
- MP-BGP EVPN-VXLAN for Overlay Control Plane
- MP-BGP EVPN routes for MAC/IP bindings
- ARP suppression
- Leverage route reflectors (RRs)
- Virtual VTEPs in VSX clusters
- Allow jumbo frames end-to-end

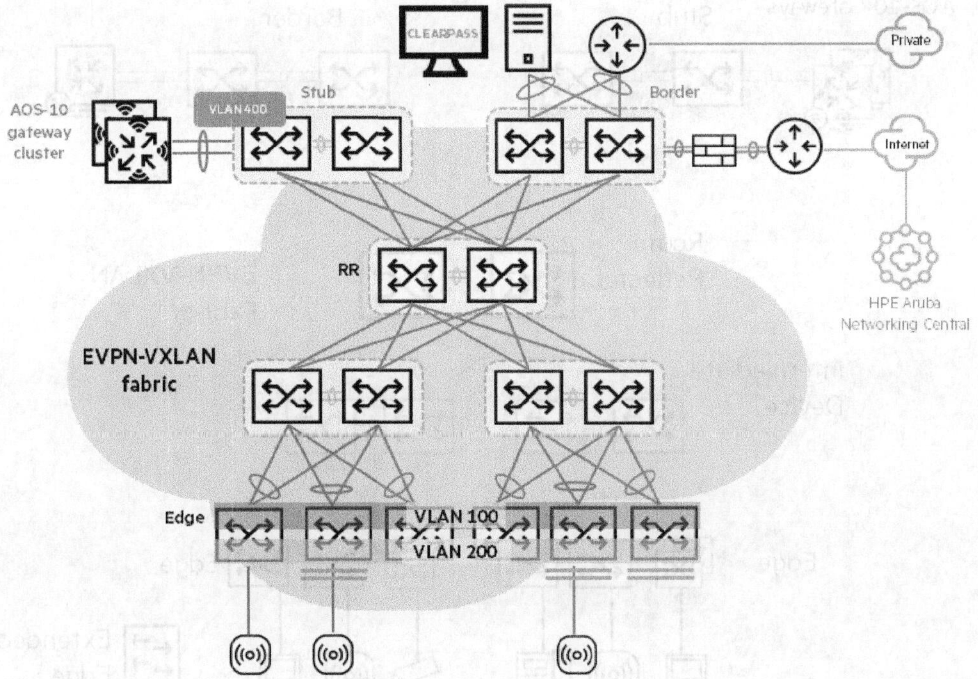

OSPF provides network reachability across the campus by routing subnets between Layer 3 hops. This often meant that when an application requires Layer 2 reachability, devices would have to be physically located in the same switching domain. A distributed overlay can overcome this limitation using an EVPN-VXLAN. The goal is to treat the existing routed network as an underlay over which a virtual private network (overlay) distributes traffic as if in the same Layer 2 subnet.

EVPN-VXLAN components

- Ethernet VPN (EVPN): BGP control plane for overlays allowing virtual connectivity between Layer 2 and Layer 3 domains over IP.
- Virtual extensible LANs (VXLAN): Common network virtualization tunneling protocol that expands the Layer 2 network address space from 4,000 to 16 million networks.

Distributed fabric personas

A Distributed Fabric is formed by assigning personas to various devices in the network. The list below describes the purpose of each persona.

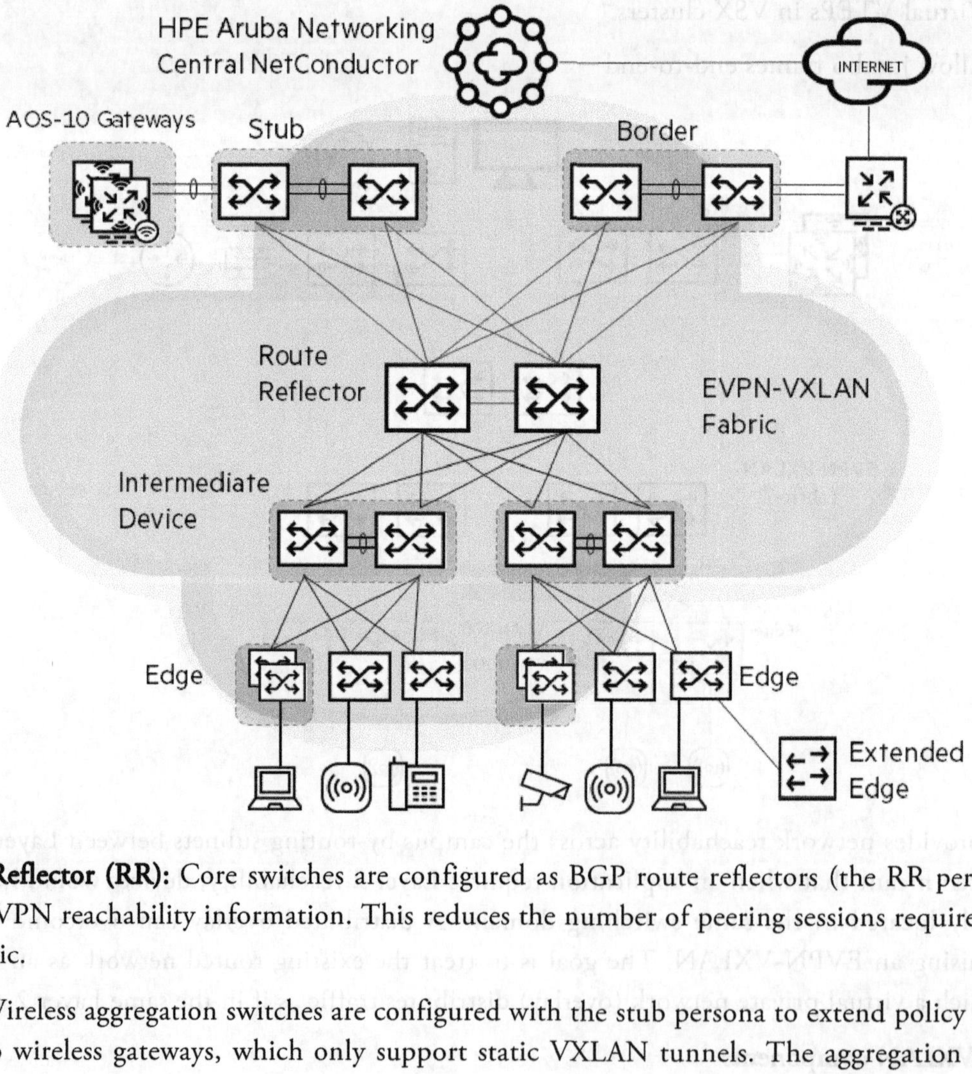

Route Reflector (RR): Core switches are configured as BGP route reflectors (the RR persona) to share EVPN reachability information. This reduces the number of peering sessions required across the fabric.

Stub: Wireless aggregation switches are configured with the stub persona to extend policy enforcement to wireless gateways, which only support static VXLAN tunnels. The aggregation switches carry GPID values from the campus fabric VXLAN tunnels forward into static VXLAN tunnels configured between the aggregation switches and the gateways. Edge switches can also host the stub persona (stub+edge) when extending the fabric to an extended edge switch which only support static VXLAN tunnels to extend the fabric beyond the edge.

Border: Internet edge switches use the border persona to provide connectivity between the fabric and services outside the fabric.

Edge: The edge persona is applied to access switches that provide primary VXLAN tunnel ingress/egress and policy enforcement for endpoint traffic into or out of the fabric. Clients connect directly to the edge switch.

Extended Edge: The extended edge persona is used to extend the fabric beyond the access layer. Static VXLAN tunnels are orchestrated between edge and extended edge. The extended edge switches performs authentication, authorization, VLAN and role assignment, and group-based policy enforcement. Clients connect directly to the extended edge switch.

Intermediate Devices: Wired aggregation switches are underlay devices with no fabric persona assigned. They do not run a VTEP and must support jumbo frames.

All devices in a NetConductor distributed campus fabric require Advanced Central subscriptions.

Supported devices for distributed fabric personas

Personas for AOS-CX switches and AOS-10 gateways in HPE Aruba Networking Central NetConductor

- Extended Edge: 6200, 6300, 6400
- Edge: 6300, 6400, 8100
- Stub: 6300, 6400, 8360
- Border: 6400, 8325, 8360, 8400, 9300, 9300S, 10000
- Route Reflector: 6300, 6400, 8100, 8325, 8360, 8400, 9000, 9000S, 10000
- WLAN gateway: 7xxx, **9xxx**

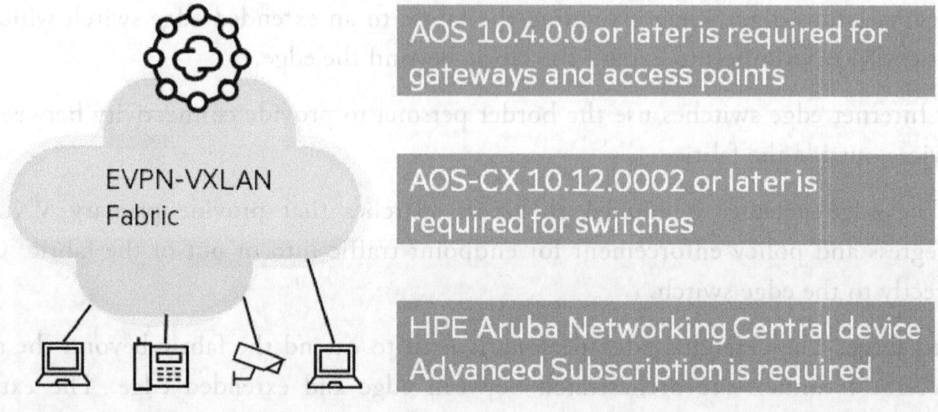

The information above shows the AOS-CX switches and AOS-10 gateways, supported software versions, and the supported personas in HPE Aruba Networking Central. Check more details in the documentation.

Reference Architectures in Validated Solution Guide:

https://arubanetworking.hpe.com/techdocs/VSG/docs/010- campus- design/esp- campus- design-060-reference-architectures/#netconductor

Product documentation:

https://arubanetworking.hpe.com/techdocs/central/latest/content/nms/apps/acn/supported-devices/supp-dev.htm

Learning check

Check your understanding of overlay networks by answering the following questions.

Question #1

Which of the following accurately describes HPE Aruba Networking gateways?

- O Gateways are not recommended for environments with more than 5,000 APs.
- O AOS-10 gateways are required for dynamic segmentation with centralized overlay.
- O The 72xx series gateways are a replacement for the 92xx series.
- O AOS-10 virtual gateways (VGWs) require HPE Aruba Networking Cloud Gateway Platinum license for AWS (Amazon Web Services) cloud-hosted version.

Question #2

Which one of the following is not related to traditional network design?

- O Localized VLANs

- ○ Extensive VLAN trunking
- ○ Risk of flattened topologies, loops, and security concerns
- ○ Decoupled user traffic from the underlay

Question #3

Which two options best describe the difference between centralized versus distributed overlays? (Select two)

- ☐ Centralized overlays tunnel traffic from APs and switches to a gateway (or gateway cluster) location.
- ☐ Centralized overlays leverage datagram transport layer security (DTLS) to secure AP and CX switch as it is tunneled over the underlay to the gateway cluster.
- ☐ Distributed overlays tunnel traffic from VXLAN tunnel endpoints (VTEPs) to other VTEPs.
- ☐ Distributed overlays must use ethernet VPN or EVPN-VXLANs to establish tunnels between VXLAN tunnel endpoint (VTEP) switches on the campus.
- ☐ With AOS-CX switches, you can use the routing protocol Interswitch to Interswitch (IS-IS) for EVPNs to dynamically discover remote VTEPs and advertise MAC address and MAC/IP bindings in the VXLAN overlay.

Supplemental: VXLANs

Check out this additional review information on VXLANs.

VXLAN review

VXLAN is an IP-based layer 2 overlay technology. It encapsulates layer 2 traffic inside an IP packet for improved function and scalability. Basically, a logical layer 2 topology is built on top of a physical layer 3 topology. Therefore, a VXLAN in the logical layer 2 can easily span across a campus network.

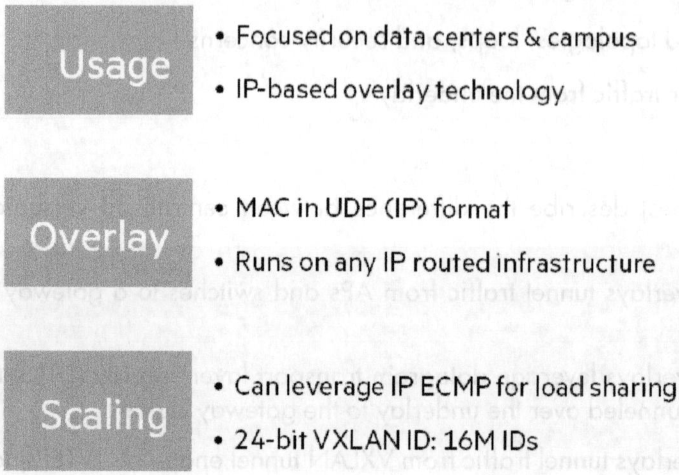

A format referred to as MAC-in-UDP is used for this encapsulation. Once the network encapsulates this layer 2 traffic, that traffic can take advantage of any existing IP routed infrastructure inside the data center or between data centers. Therefore, all of the redundancy, resiliency, and load-sharing capabilities typical of routed infrastructures are available to VXLAN services.

The VXLAN ID is a 24-bit value – double the number of bits available with 802.1Q. This means over 16 million VXLAN IDs are available, compared to 4,000 VLAN IDs for traditional VLANs. VXLAN provides new functionalities that would be impossible with classic campus networking, such as GBP, which will be discussed later.

VXLAN terminology

- **VXLAN**: like a classic VLAN without an L3 IP interface; totally isolated
- **VNI**: 24-bit VXLAN Network Identifier

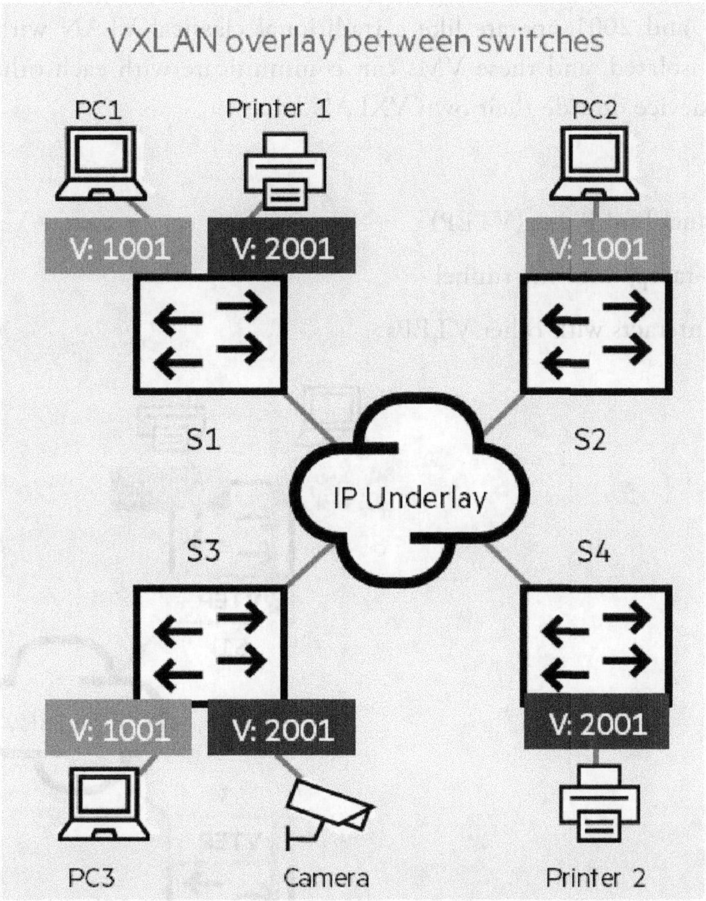

This example topology shows a transport IP network composed of several switches interconnected in a routed environment. You have several client hosts connected to this infrastructure.

In this example, you have not fully deployed the VXLAN functionality, with no VXLAN configuration on the switches. However, the network configuration includes VLAN IDs, and you have the switches configured to map VLAN IDs to VNIs (virtual network identifiers) acting as VTEPs (entry points into the VXLANs). (You'll learn more about this in the upcoming slides.)

The S1 switch was configured with VXLAN 1001, which was bound to the VLAN 1001 (where PC1 connects). The printer was assigned a VLAN 2001 that maps to VXLAN 2001. Similar configurations exist on the other switches as well for their edge devices.

Switches will encapsulate the edge devices' layer 2 frames into an IP packet, including the appropriate VNI. The result is that PCs 1 through 3 are functionally on the same layer 2 multipoint subnet. Similarly, virtual network 2001 contains hosts Printer 1, Printer 2, and the security camera. VXLANs 1001 and 2001 operate like a traditional classical VLAN without any layer 3 IP interface. They are isolated, and these VMs can communicate with each other but cannot communicate with any device outside their own VXLAN.

VXLAN VTEP

- VXLAN Tunnel End Point (VTEP)
- Entry or "on-ramp" into the tunnel
- Device that interacts with other VTEPs

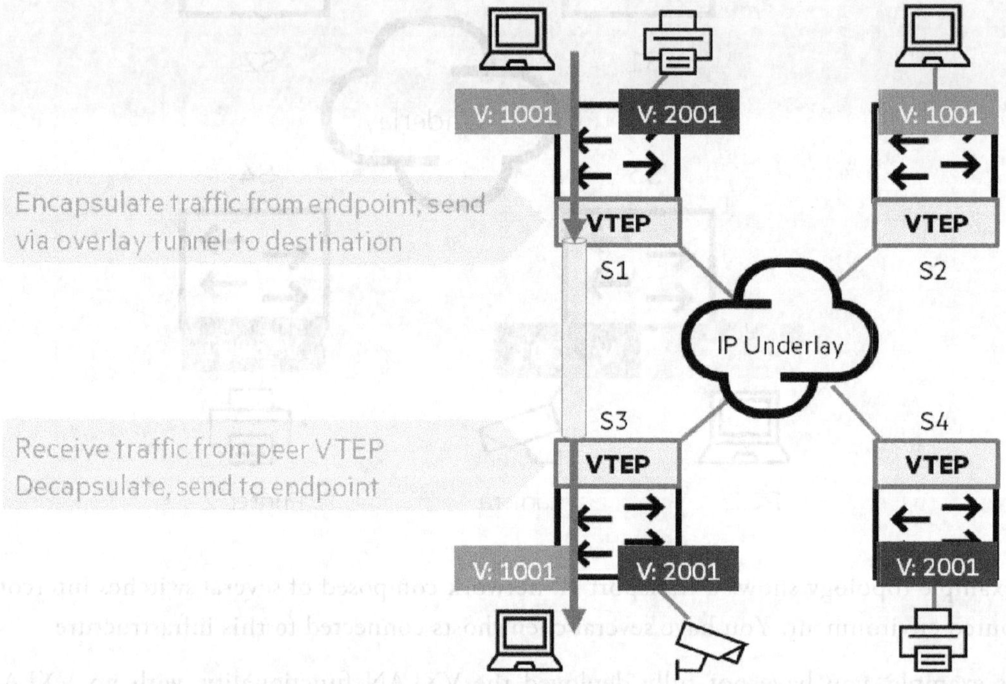

The VXLAN tunnel endpoint (VTEP) provides an entry point into the VXLAN while interacting with the other VTEPs to provide connectivity. In topology, each physical switch receives a role assignment. They are responsible for encapsulating traffic from the VXLANs into the tunnel and forwarding the resulting packets to their destination.

The receiving destination VTEP must decapsulate the packet and deliver it toward the intended target. The VTEP provides an on-ramp to the VXLAN by performing this frame encapsulation and decapsulation service. VXLANs only need to be deployed at the edge switches of the transport network. Devices within the transport network perform typical layer 3 forwarding.

VXLAN tunnel

- VXLAN tunnel: tunnel between two VTEPs
- Jumbo frames: MTU greater than 1550 bytes

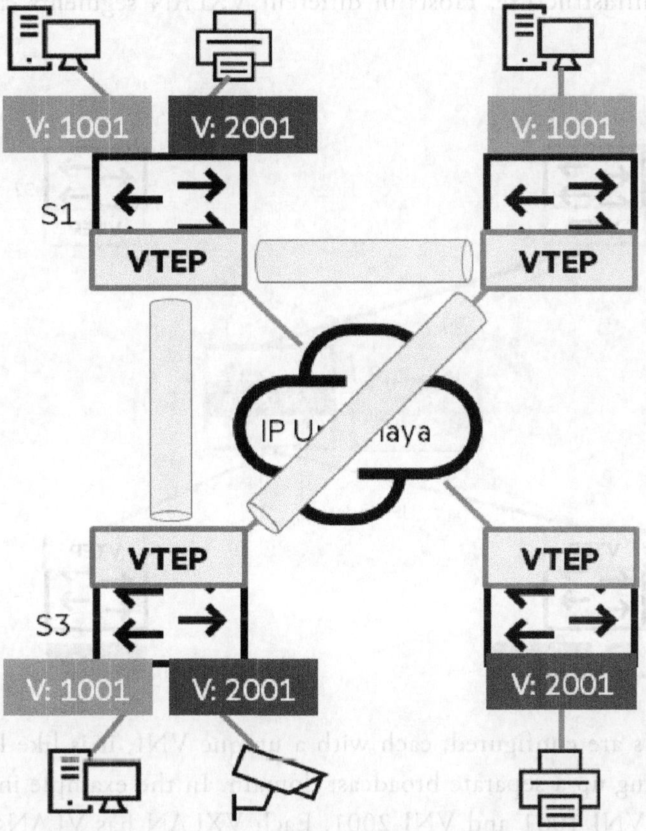

One VXLAN VTEP or tunnel endpoint communicating with another establishes a VXLAN tunnel, a requirement for each destination VTEP in the same VXLAN. In this example, switches S1, S2, and S3 are the physical VTEPs assigned to VNI 1001, so each switch maintains a tunnel with the other two. Clients assigned to VNI 2001 connect to switches S1, S3, and S4, with similar interconnected tunnels.

The tunnel is merely a mechanism of transport through the IP network. Multiple VNIs can use one tunnel because each VTEP encapsulated packet contains the associated VNI. The network then uses this VNI to send packets to the appropriate tunnel endpoints. This is why traffic from multiple VXLANs can use the same tunnel. The underlying IP transport network uses IP unicasts to deliver broadcast, multicast, and unknown unicast frames for each VXLAN.

VNI "Virtual Switch"

The VNI creates a virtual switched domain on an IP-based underlay network. A VXLAN segment differentiates individual logical networks so that several isolated layer 2 VXLAN networks coexist on a common layer 3 infrastructure. Hosts in different VXLAN segments cannot communicate with each other directly.

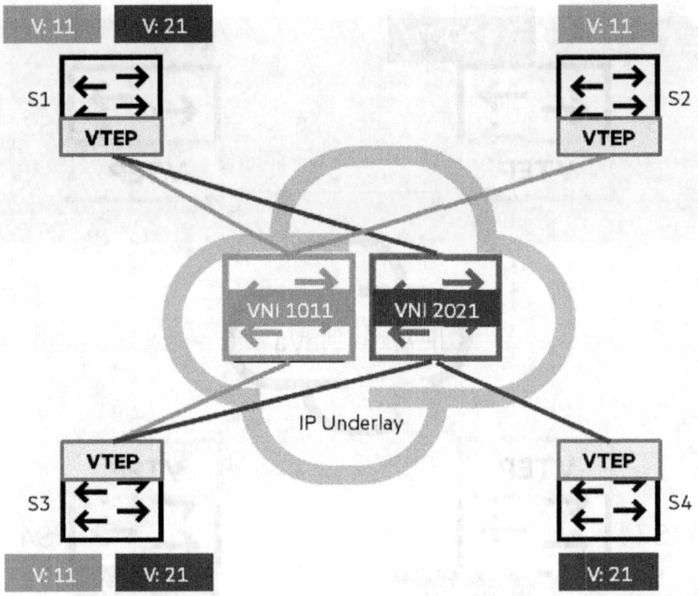

When multiple VXLANs are configured, each with a unique VNI, it is like having multiple virtual switches, each making up a separate broadcast domain. In the example in the slide, there are 2 VXLANs configured: VNI 1001 and VNI 2001. Each VXLAN has VLANs mapped on VTEP devices to connect your endpoints to these virtual switches.

9 Network Optimization

With the campus design almost complete, this module covers optimizing the campus network.

OBJECTIVES

After completing this chapter, you should be able to:

- ✓ Design for QoS best practices
- ✓ Compose a QoS strategy for Wi-Fi
- ✓ Implement UXI for the campus

QoS Best Practices

This section begins the module by examining QoS best practices for a campus network.

Application classification

When collecting information about a customer's requirements, it is very important to use this information to assess their applications. Network control traffic should always have high priority, but generally, devices automatically prioritize this traffic, so you do not need to plan to prioritize it.

- You need to identify real-time applications (e.g., voice and video) that require special handling to ensure a good quality of service (QoS).
- You might also need to classify traffic based on how critical it is to the company's mission.
- This list shows several common classes listed in the order of priority, with the highest priority at the top. You can also see the types of applications associated with these classes.
- Many customers do not require every traffic class. Often, you simply need to make a plan to provide the proper QoS-to-voice and interactive video traffic, as well as perhaps streaming

video. Everything else receives best-effort forwarding.

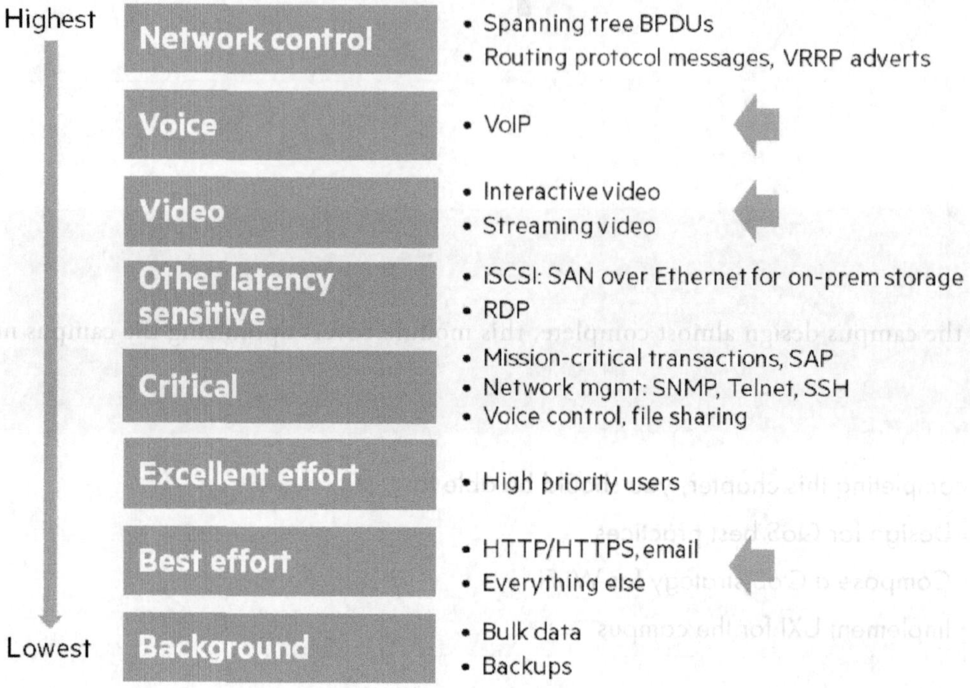

Eventually, you will associate each traffic class with a forwarding queue. In the wireless world, devices use one or four queues depending on their capabilities. HPE Aruba Networking CX switches typically have eight queues. The advantage of using fewer classes, and therefore queues, is simplicity and – for wired switches – greater buffering space per queue. In addition, the simpler the design, the easier the solution is to deploy and maintain.

For customers who do require more classes, you know a bit more about classifying and prioritizing different types of traffic.

As you see, network and internetwork control traffic should have the highest priority. This traffic includes control protocols transmitted by network infrastructure devices themselves. For example, internetwork control traffic includes open shortest path first (OSPF) and other routing protocol messages. Virtual router redundancy protocol (VRRP) advertisements provide another example of internetwork control traffic. The BDPUs used by spanning tree protocol (STP) are network control traffic. This traffic requires the highest priority because the network cannot ensure delivery for any traffic if these protocols stop functioning correctly.

Multimedia applications like voice and video require the next highest priority. Voice typically requires higher priority than video because it occurs in real time. You can further divide video into interactive video and streaming video. Interactive video might require higher priority because real-time communications require timely delivery.

The iSCSI protocol carries SCSI communications between an initiator – typically a server – and a storage target. This technology enables a server to mount volumes from disks hosted on a remote storage array. While you often find iSCSI deployed within data centers, some customers have on-prem storage and run iSCSI traffic over HPE Aruba Networking 5400R and 3810 switches. The iSCSI communications require high priority to prevent excessive latency or packet loss.

Remote (RDP) can also be latency-sensitive.

Mission-critical traffic might include applications associated with transactional databases and SAP applications. Companies may define network management traffic like SNMP, Telnet, and SSH as critical. Non-voice and non-video Unified Communications (UC) traffic often fits in this class. This traffic includes voice control (the traffic for setting up a call) and file sharing.

Classifying traffic into the next two classes depends less on the application type and more on the scenario. For example, a hotel might offer a premium and a best-effort service. All traffic from users who purchase the premium service should receive critical or excellent effort.

Finally, you can identify traffic that takes much bandwidth but does not require timely delivery as background traffic. For example, FTP uses TCP, which can tolerate loss and speed up and slow down based on congestion. The company might have periodic backups that need to be executed but can be completed anytime during the day without negative effects. You could also define applications the customer does not care about as background traffic.

Quality of Service overview

QoS enables networks to provide better or special service to a set of users and/or applications, perhaps to the detriment of other users and/or applications. With QoS, you are trying to make the best possible use of available bandwidth.

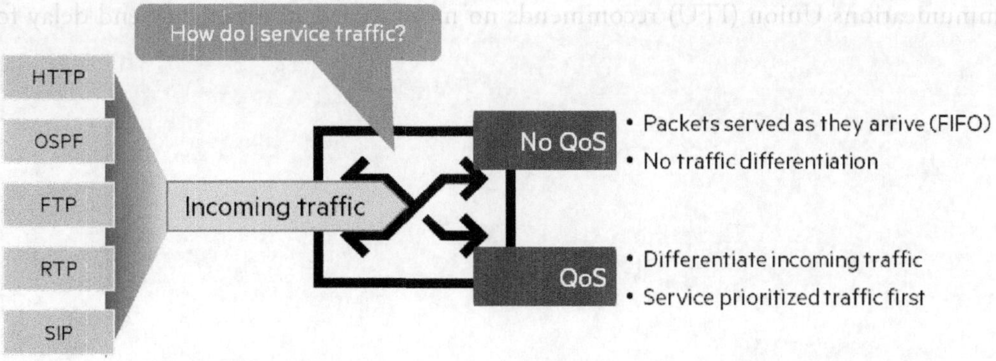

Different traffic types compete for the same network resources such as bandwidth and fast processing. Without QoS, packets are served as they arrive, typically first-in first out (FIFO). Your network management traffic may be more important to you than some users watching videos during their lunch break, but, by default, there is no differentiation between this traffic.

On the highway, an ambulance gets higher priority than other vehicles. Likewise, QoS is not being fair when dividing the network resources, but rather selecting certain traffic types for improved service, perhaps at the expense of other traffic. QoS does not make the road wider, it just decides who goes first and consequently who must wait.

Note

Understand that QoS is not a standalone service or product, but rather a concept that supports a framework of protocols, techniques, and network attributes that span across the infrastructure.

Voice requirements for low loss, latency, and jitter

Now that you have classified the customers' applications, you can begin to consider the needs. You need to take steps to protect the higher-priority classes of traffic from the ill effects of congestion: loss, latency, and jitter. Since voice applications tend to be the most sensitive to all three of these issues, as an example, you will learn about how these issues affect voice.

Packet loss occurs when congestion prevents the network from delivering all packets. While voice applications can tolerate a few lost packets, too many lost packets will degrade quality and eventually drop the call.

Latency is the time that passes between the sending of a transmission and its arrival at the receiving station. Within a campus network, significant latency typically occurs, not due to the transmission time for signals but because congestion causes a packet buffer before transmission. Latency can also occur if roaming or re-authentication disrupts an endpoint's connection. Latency can cause issues for voice because it introduces a slight delay after a person begins to speak. This delay can cause people to talk over each other and sometimes create an echo. The International Telecommunications Union (ITU) recommends no more than 150 ms end-to-end delay for voice traffic.

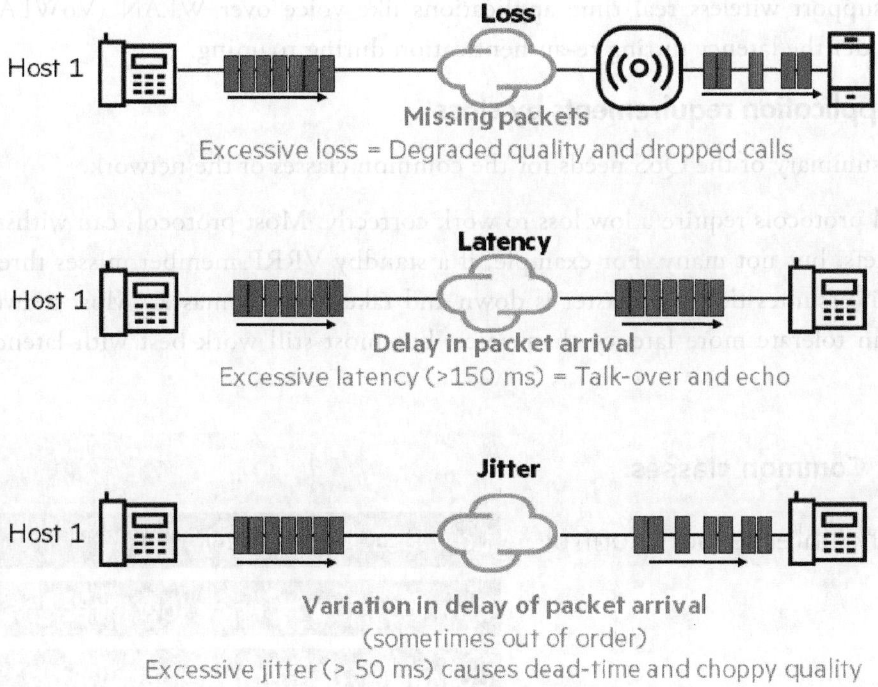

Because congestion typically occurs in unpredictable bursts, latency can vary from packet to packet within a stream. Jitter is the name of this effect, and it can have a much worse effect on call quality than simple latency. This is because real-time applications' design requires a consistent packet rate, and jitter causes smaller and greater gaps between packets. It can sometimes even cause packets to arrive out of order. For instance, a VoIP telephone sends a very small packet every 20-30 milliseconds (ms). If the interval between packets grows to 50 ms due to jitter, the transmission will be unsatisfactory for many users.

VoIP is particularly sensitive to many of these issues because it is a real-time UDP application. TCP-based applications can recover from packet loss because they use acknowledgments. However, resending the packets does slow down the application. UDP applications do not use acknowledgments and cannot recover lost packets. In a real-time conversation, by the time it recovers the packet, it is too late for it to be useful.

You should also be aware that some latency is inevitable. While the speed of light in a vacuum is 186,000 miles per second (267,000 km per second), signals move through fiber or copper cable at approximately 125,000 miles per second (200,000 km per second). Consequently, a fiber network stretching halfway around the Earth introduces a delay of approximately 70 milliseconds under the best circumstances, with an additional delay introduced by the physical transport medium. However, within a campus, the latency introduced by the physical media is negligible. Instead, you need to worry about the latency introduced by congestion.

If you plan to support wireless real-time applications like voice over WLAN (VoWLAN), you should also consider the latency during re-authentication during roaming.

Summary of application requirements by class

Here, you see a summary of the QoS needs for the common classes of the network.

Network control protocols require a low loss to work correctly. Most protocols can withstand one or two lost packets, but not many. For example, if a standby VRRP member misses three VRRP advertisements, it assumes that the master is down and takes over as master. Most network control protocols can tolerate more latency than voice, but most still work best with latency under one second.

Common classes:	Needs
Network/ internetwork control	Low loss, low latency
Voice	Low loss, low latency, low jitter
Video	Low jitter, guaranteed bandwidth
Critical	Low loss, low latency
Excellent effort	Varying needs, high priority
Best effort	Varying needs, normal priority
Background	Varying needs, low priority

Like voice traffic, video traffic is susceptible to jitter, which can cause a choppy quality. Streaming video can tolerate some latency due to buffering mechanisms. However, excessive latency can negatively affect interactive video for the same reasons as voice. Video traffic also requires guaranteed levels of relative bandwidth, which you should consider when planning capacity.

Critical applications also require low loss and low latency (of course, the applications we mentioned earlier can also be critical. Here, you are examining applications of other critical types due to the purpose for which the business uses them).

Depending on the customer scenario, the last three classes typically include various applications. You should remember that the network should deliver excellent-effort traffic before best-effort traffic, which should arrive before background traffic.

If the network uses iSCSI, remember that iSCSI is a TCP application that can recover lost packets and respond to congestion. Nonetheless, the underlying SCSI communication was designed for direct links dedicated to data transfer, and iSCSI operates best when the application experiences low latency, loss, and jitter. In addition, iSCSI can consume a high amount of bandwidth, so you should provision links accordingly.

Meet application requirements

As you have seen, in a campus network, congestion causes excessive packet loss, latency, and jitter. Therefore, the first step toward ensuring good QoS is minimizing congestion by planning adequate capacity. Nonetheless, even a well-designed network features oversubscription and the possibility of momentary bursts of congestion, which can cause excessive damage to time-sensitive solutions. Therefore, you must also implement QoS mechanisms to protect this sensitive traffic and ensure users have a positive experience running all necessary applications.

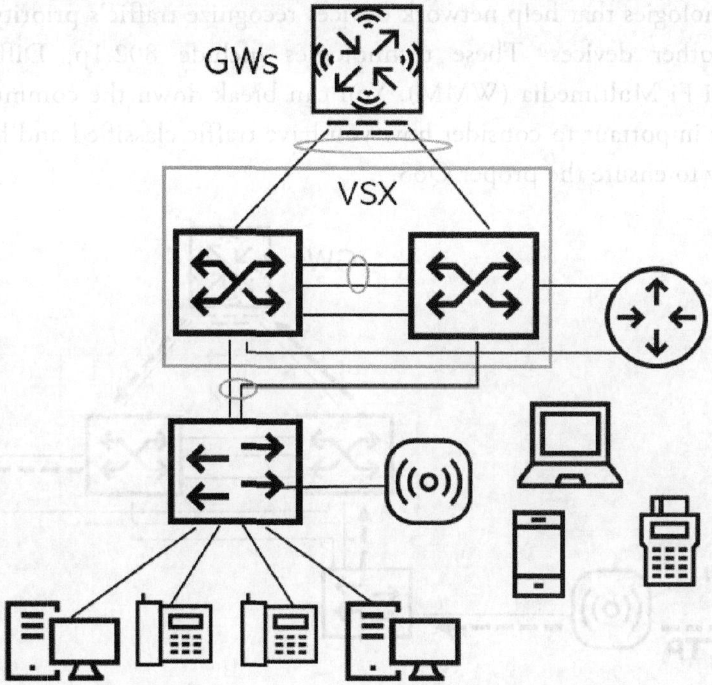

You will often find switch-to-switch links oversubscribed, and thus, switch-to-switch links provide less bandwidth for outgoing traffic than the total amount of traffic that could be incoming from other devices – often many times less. Usually, endpoints must send and receive only a small fraction of their bandwidth, so oversubscription makes sense. However, fluctuating traffic patterns and bursts of traffic can cause congestion moment-to-moment. The lower-bandwidth link acts as a bottleneck, and if traffic continues to arrive more quickly than the port can forward, the port's buffers fill, dropping traffic.

In other words, even when infrastructure links typically provide ample bandwidth, implementing QoS technologies is important because even brief periods of congestion can degrade the quality of time-sensitive applications. QoS helps to protect high-priority traffic during such periods of network congestion. For example, the wired infrastructure might provide enough bandwidth all the time except at 9 a.m. when all the employees arrive. Implementing the proper technologies can also help to improve user experience over wireless connections. These can be subject to greater unpredictability than wired connections due to the shared medium and nature of wireless signals.

Traffic prioritization

Traffic prioritization enables the network infrastructure to place different types of traffic in different queues and then forward traffic in higher priority queues preferentially. In this example, one wireless device sends voice traffic, and another sends data traffic. Each infrastructure device in the traffic's path prioritizes the voice traffic to minimize loss, latency, and jitter. You will learn about several technologies that help network devices recognize traffic's priority and communicate that priority to other devices. These technologies include 802.1p, Differentiated Services (DiffServ), and Wi-Fi Multimedia (WMM). You can break down the communications into several segments. It is important to consider how you have traffic classified and handled at each segment of its journey to ensure the proper QoS.

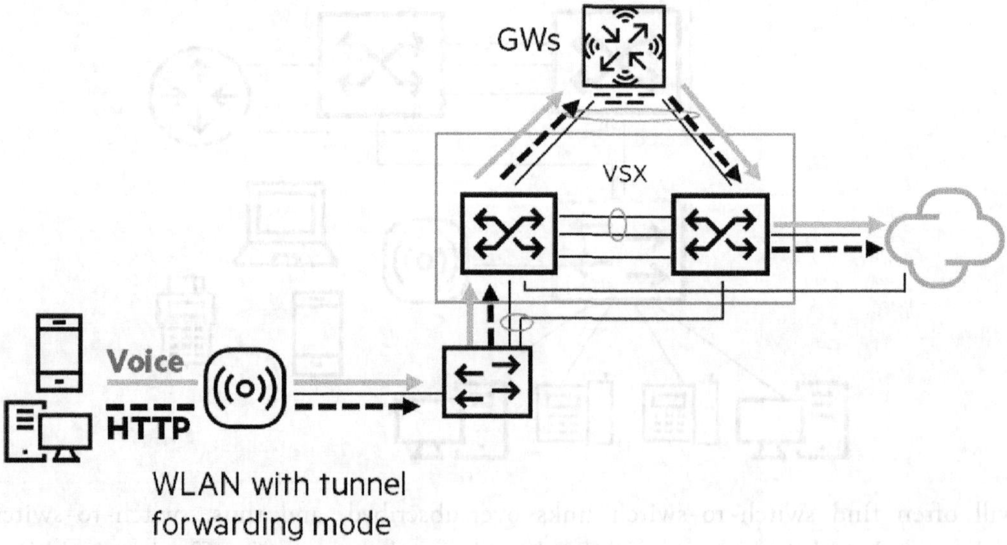

First, consider QoS between the wireless device, the AP, a wired endpoint, and its switch. If the wireless devices lack support for WMM or another QoS technology, the traffic cannot receive priority handling on this journey segment. In this example, this journey segment is less important for wired devices, such as the computers connected to the access layer switch. Each of these devices has its full-duplex link.

Next, consider the AP's link to the switch and the switch-to-switch links across the network. Congestion can occur on these links because they are typically oversubscribed, with multiple downstream devices sharing the links' bandwidth. You should remember that every device in the path can affect the QoS by correctly or incorrectly prioritizing the traffic.

Finally, consider traffic in both directions of a conversation and ensure that you have return traffic properly classified and prioritized.

Capacity planning

Including capacity planning and strategies to meet your customer's needs is important.

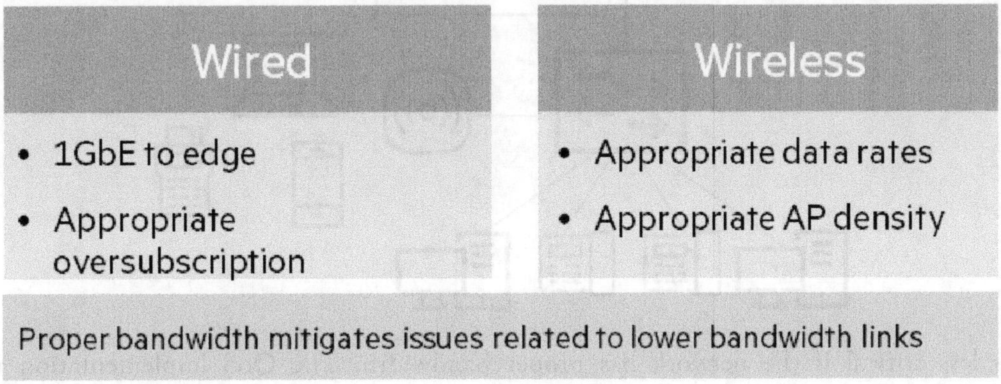

Most modern campuses should feature Gigabit to the wired edge. You should plan appropriate oversubscription on switch-to-switch links. Oversubscribing the links too much can lead to poor performance. Wireless capacity planning depends a great deal on a proper RF design. You must ensure clients can operate at high data rates by providing a good signal across the campus. In high-density environments, you also need to deploy enough APs such that the right number of clients share the bandwidth. HPE Aruba Networking recommends 40-60 users per radio in most typical enterprise indoor deployments. However, user density per AP radio also depends on the AP model.

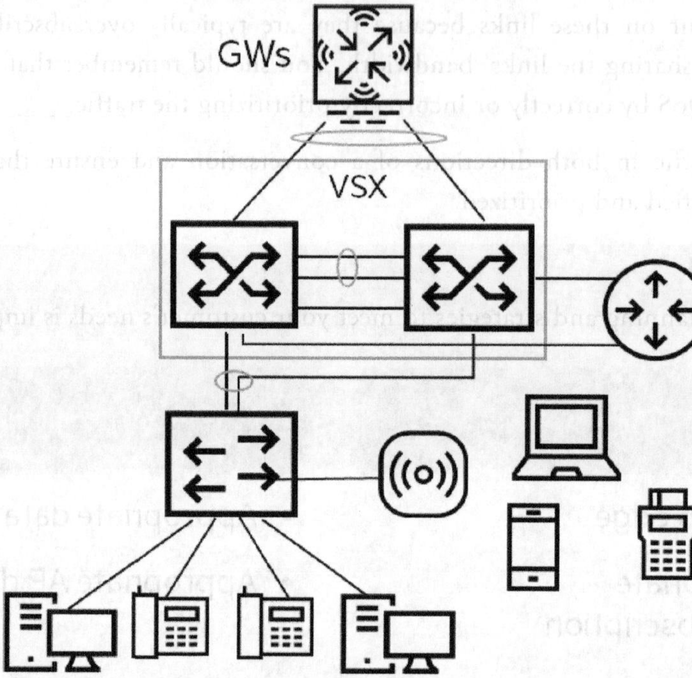

QoS is less critical if the network has proper bandwidth. The QoS implementation requires coordination of all equipment from one end to the other end of the connection. When you make a capacity-based AP plan, removing support for the lower data rates is important, helping to ensure that clients connect to an AP that offers a stronger signal and higher data rate.

The definition of appropriate oversubscription for the wired network varies based on customer circumstances and application usage. However, appropriate oversubscription at the wired edge ranges from about 10:1 for higher bandwidth applications to about 20:1 or 24:1 for light users. Up to 4:1 oversubscription is appropriate in a 3-tier topology between the aggregation layer and the core. The core should have no oversubscription. In a 2-tier topology, the core might introduce up to 4:1 in connections to other network segments, such as the data center.

Over-provisioning

Class of Service (CoS) is a method for classifying network traffic within eight service classes. This is when you need to prioritize traffic.

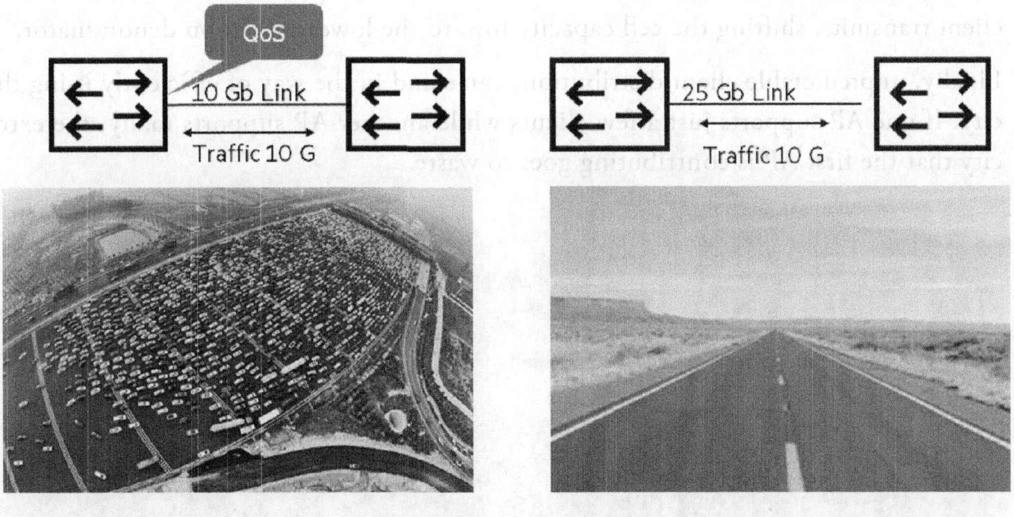

It is best to make an effort for all traffic to receive equal treatment in a first-come, first-served manner. Suppose the traffic load is low in relation to the capacity of the network links. In that case, there is no need for the administrative complexity and costs of maintaining a more complex end-to-end policy. This is sometimes called over-provisioning, as all link speeds are much higher than peak loads on the network. This is the norm with backbone links in the 10G, 25G, 40G, 50G, and 100G range. Very few networks go through the process of implementing QoS. QoS comes into play when there is competition for bandwidth.

Wireless QoS

This section narrows the focus to wireless QoS.

Issues that prevent efficient use of wireless capacity

Even when you have planned AP locations and density well, various issues can prevent full use of the available wireless capacity.

- Each RF channel is a shared medium. If two nearby APs operate on the same or overlapping channel, the system offers the same capacity as it would with a single AP. The situation can worsen if the APs are just far enough apart that some of their clients cannot hear each other. The network will experience a large number of collisions and retransmissions that slow down communications. Other sources of non-RF interference can also take up airtime or cause errors in transmissions.
- It can also be difficult to determine the maximum capacity of a wireless cell because all clients within the cell rarely use the same data rate. Slower clients reduce the capacity of the entire cell. Since the wireless medium is shared, every device must wait while the slow

client transmits, shifting the cell capacity toward the lowest common denominator.

- Finally, unpredictable client distributions can stand in the way of efficiently using the capacity. If one AP supports just a few clients while another AP supports many, the extra capacity that the first AP is contributing goes to waste.

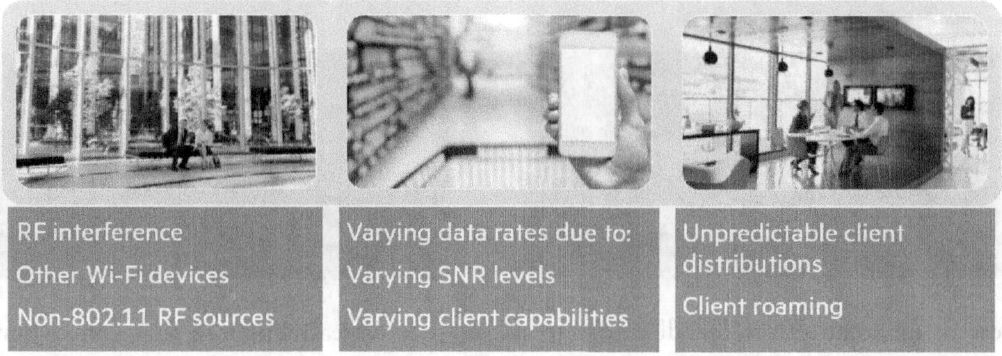

Different clients support different data rates due to varying signal levels, and their data rates may vary as they move. A sticky client remains connected to an AP even after roaming relatively far away and dropping to a low data rate. In addition, clients have widely varying capabilities. 802.11n clients do not support the highest 802.11ac rates, while legacy 802.11a/g clients operate at much lower speeds. Furthermore, many 802.11n and even 802.11ac clients have limited antennas, so they do not support the higher data rates available through more MIMO spatial streams.

QoS provides predictable traffic flow behavior when congestion occurs. Normally, most networks will rarely experience congestion if designed with enough capacity. Even wireless often operate well below capacity, considering that 802.11 is inherently a shared medium; wireless clients share a channel space. A challenge in today's wireless networks is to enable more simultaneous clients on wireless. New features like Wi-Fi 6E bring MU-MIMO and OFDMA and take advantage of the new 6 GHz band to perform layering to accomplish this goal.

Even with these features, overload conditions can and do occur. The QoS mechanisms grant import traffic a higher priority while making fewer resources available to lower-priority traffic. Suppose a hospital wants to prioritize life-saving medical imaging for surgery. In that case, doing so may be at the risk of delaying or dropping YouTube video traffic on a guest network.

CSMA/CA

Carrier-sense, multiple-access with collision avoidance is a slightly evolved version of the wired CSMA/CD. CSMA/CA was designed to deal with forwarding Wi-Fi transmissions before sending their frame. Wi-Fi 6 and BSS coloring greatly reduce contention on wireless.

WMM adjusts two CSMA/CA parameters according to the QoS priority of the transmitted frame.

- Random back-off timer in a shared medium. CSMA/CA requires that each device monitor the shared wireless channel for other transmissions
- Arbitration inter-frame space

Higher-priority frames are assigned shorter back-off times and arbitration inter-frame spaces than low-priority frames.

How WMM works

DSCP is most typically responsible for classifying packets in the wired network, while WMM is responsible for WiFi QoS classification and implementation. WMM maps traffic to four access categories:

- AC_BK: background traffic
- AC_BE: best effort traffic
- AC_VI: video traffic
- AC_VO: voice traffic

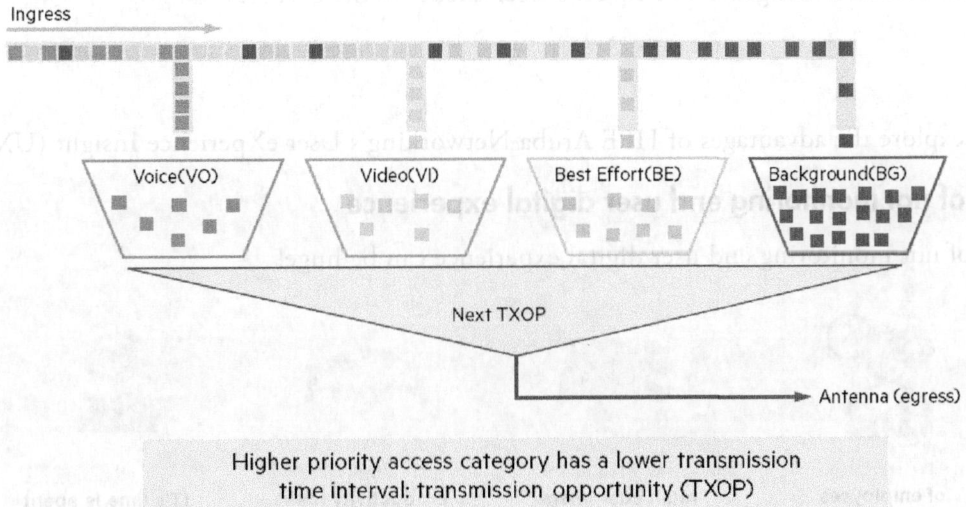

These access categories map to the four queues a WMM device requires.

A TXOP is a bounded time interval during which a station can send as many frames as possible.

Features to optimize wireless capacity

The HPE Aruba Networking wireless solutions offer several features to help overcome these issues and optimize the capacity the complete wireless solution offers.

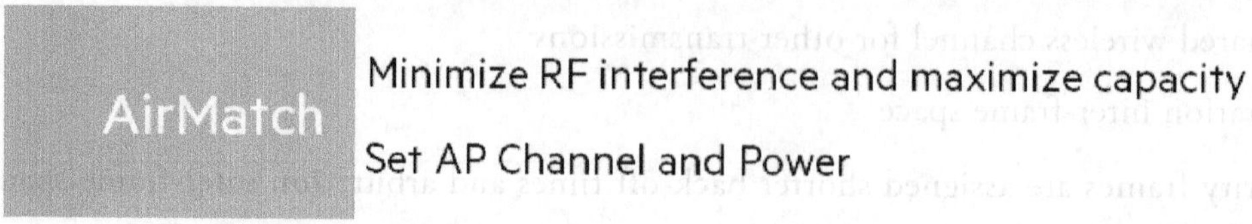

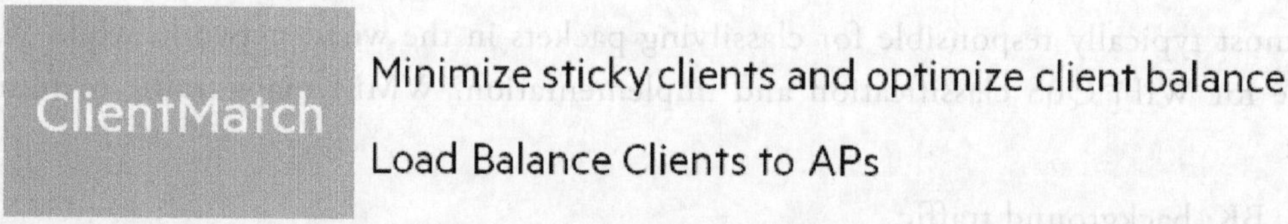

AirMatch helps to minimize co-channel interference between APs and maximize the capacity. ClientMatch provides intelligent load balancing and steering to help even out the client distribution on APs as much as possible while maintaining good signals. This feature can also help to eliminate sticky clients, increasing the number of clients that operate at a high data rate. ClientMatch also uses centralized intelligence to help APs steer clients to the best AP.

UXI

Now you explore the advantages of HPE Aruba Networking's User eXperience Insight (UXI)

The cost of not monitoring end user digital experience

The cost of not monitoring end user digital experience can be huge!

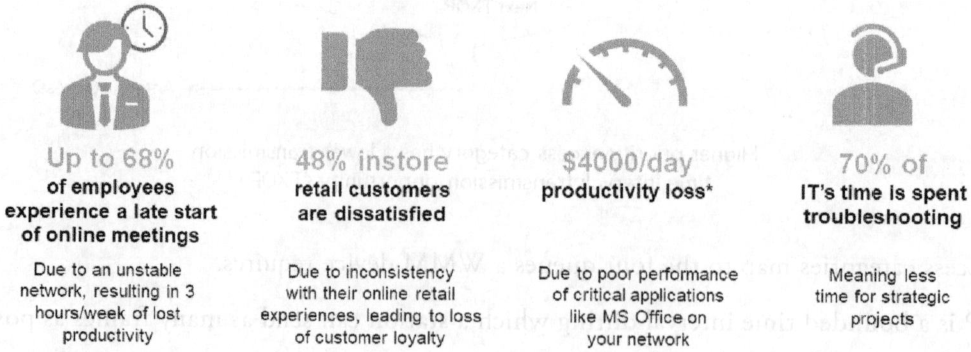

It affects employee productivity, hampers end users experience and loyalty, impacts revenue generating opportunities and consumes a huge chunk of IT teams' time – trying to find the root cause of the issue.

HPE Aruba Networking User Experience Insight (UXI)

AI-powered insights alert IT staff to network performance issues as soon as they begin. Single pane of glass integrates with HPE Aruba Networking Central in the Cloud using an API. Monitors critical services such as corporate ERM or office applications like Microsoft 365. Provides the last 30 days of performance data as well as PCAP files.

- Cloud-based 24/7 service assurance solution for users, networks, and applications.
- UXI sensors enable continuous monitoring and testing across wired, wireless, and WAN.
- Deployed as a network user to continuously monitor the network's performance, connectivity, and responsiveness.

HPE Aruba Networking User Experience Insight (UXI) Agent

When HPE Aruba Networking User Experience Insight (UXI) Sensor and Agent are used together, the IT teams get a holistic view of end-user experience and can: baseline network performance, test network health 24X7, trend network performance data, do efficient network planning, and export network and application performance data on a big data platform via UXI Data Push Destination feature. IT team gets additional insight into device related stats and end-users get a self-service portal that displays important network and device related information that are frequently asked by IT teams for troubleshooting issues.

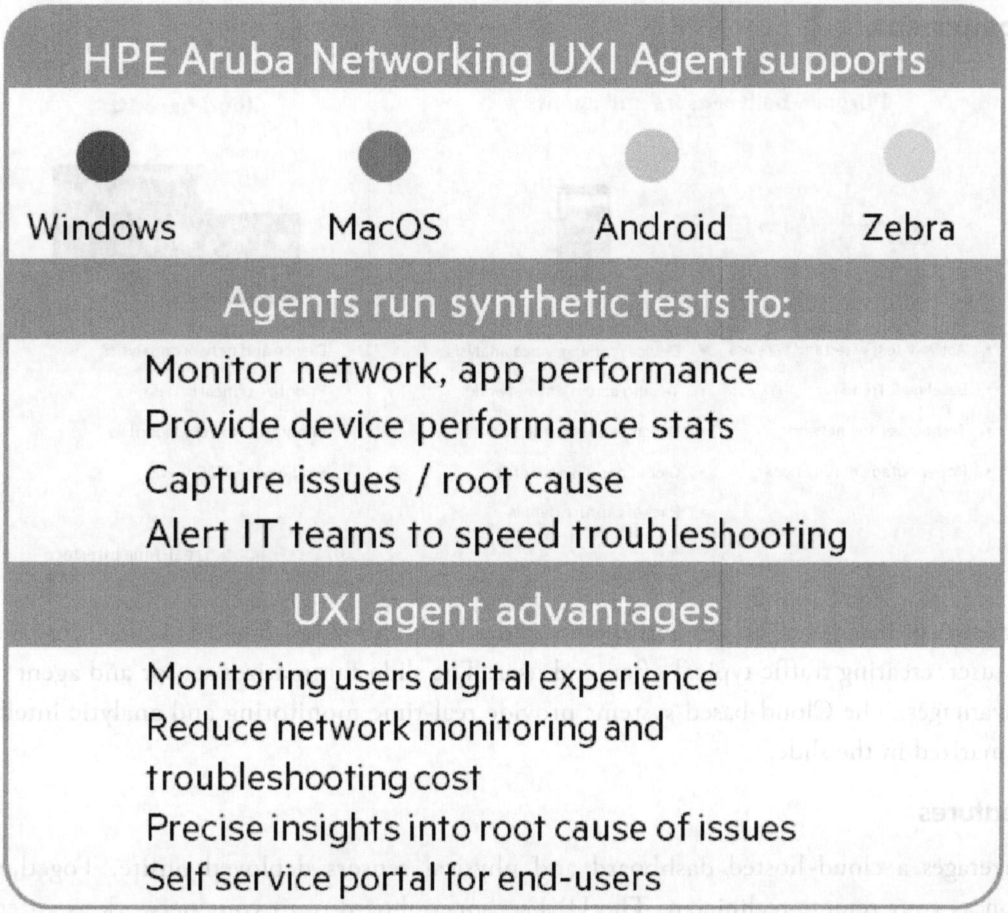
HPE Aruba Networking UXI Agent supports
Windows
MacOS
Android
Zebra
Agents run synthetic tests to:
Monitor network, app performance
Provide device performance stats
Capture issues / root cause
Alert IT teams to speed troubleshooting
UXI agent advantages
Monitoring users digital experience
Reduce network monitoring and troubleshooting cost
Precise insights into root cause of issues
Self service portal for end-users

UXI components

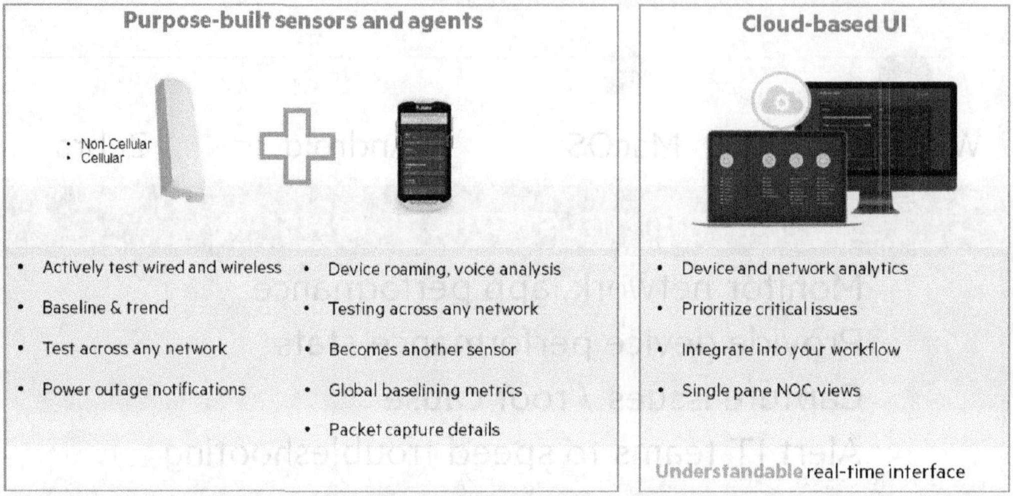

UXI consists of front-end sensors and agents, and a back-end cloud-based UI. The sensors act like an end-user, creating traffic typical of an end user. The slide summarizes sensor and agent features and advantages. The Cloud-based systems provide real-time monitoring and analytic intelligence, as summarized in the slide.

UXI features

UXI leverages a cloud-hosted dashboard and physical sensors deployed onsite. Together, they function as your remote technician. The UXI sensors onboard onto your network as an end-user and actively monitor wireless and wired connections, providing a detailed performance analysis from an edge perspective via a highly intuitive cloud-hosted UXI dashboard. Similar to the sensor, HPE Aruba Networking now has sensor software which can be installed on a Zebra Android-based device. HPE Aruba Networking has integrated UXI with Zebra's Wireless Insights engine to provide real roaming statistics—did it roam, why did it roam or why did it fail to roam?—and true SIP call measurement to be able to provide call metrics like if a call dropped who dropped the call. UXI can measure end-to-end VoIP.

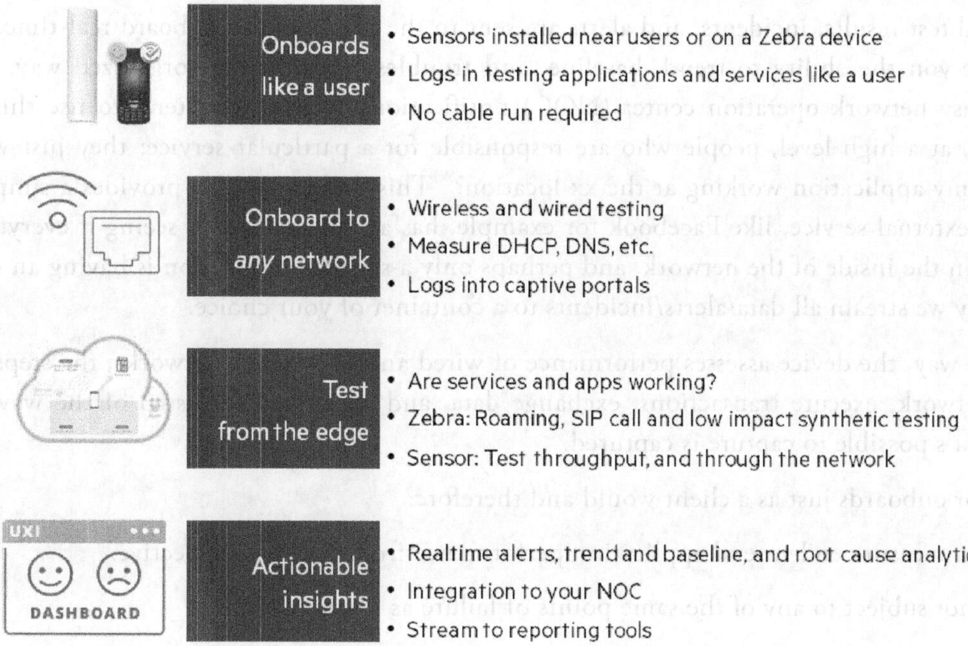

Note

An overview of UXI can be found here: https://www.hpe.com/psnow/doc/a00048302enw.

Here is the process of setting up UXI:

Step 1: Sensors onboard via cellular or standard PoE. You just plug in the sensor and it looks for the network it's supposed to test. As soon as it finds that network, the cellular function (if being used) turns off. (Note: the agent just onboards to a wireless network, not a wired network.)

Step 2: Just like a normal user the sensor associates to a network, goes through all of the same DHCP, DNS, and gateway (DORA) process. If there is a captive portal involved, the sensor will log in and log out of that captive portal. UXI measures load, and redirect timings (and success) of that captive portal.

Step 3: Configure the sensor/agent to test services and applications which are critical to that location. First, the sensor will test internal applications: things on the inside of the firewall. Maybe printers, authentication servers, the default gateway, and etcetera. In the case of Zebra, the sensor may perform a test using a basic ping/webserver of a work app or voice analysis of calls. (Note: the agent does not do web application tests, or throughput tests because it would impact the performance of the device itself)

Step 4: All test results, incidents, and alerts are sent to the UXI cloud dashboard real-time. These views give you the ability to trend, baseline, and troubleshoot in a very organized way. AIOps enables easy network operation center (NOC) notifications. People who tend to use this dashboard are, at a high level, people who are responsible for a particular service: they just want to know "is my application working at the xx location?" This dashboard also provides a simple way to see an external service, like Facebook for example, has an outage and/or seeing if everything is working on the inside of the network, and perhaps only a specific application is having an outage. Alternately we stream all data/alerts/incidents to a container of your choice.

Along the way, the device assesses performance of wired and/or wireless networks, the steps to get onto a network, execute transactions, exchange data, and etcetera. Every step of the way, every metric that's possible to capture is captured.

The sensor onboards just as a client would and therefore:

- Can measure end-to-end application performance from client's perspective.
- Is not subject to any of the same points of failure as the network .
- Is vendor neutral.

UXI provides a single pane of glass into the metrics that are gathered. UXI and HPE Aruba Networking Central/AI Insight provide assurance of connectivity to and through the network with optimal performance and Wi-Fi coverage. As you would expect from the names and descriptions, individually they provide insights into how the network is performing and should perform. Collectively, they provide holistic intelligence from the perspective of the customers/end-users and the network infrastructure. Note that AI Insight requires an HPE Aruba Networking infrastructure; there are other requirements too, like ideally having over 2,000 APs.

UXI leverages Selenium IDE for Web Application Testing (WAT) to gain a deeper understanding of web application performance from the end-user perspective. User Experience Insight offers test templates for checking web server port availability, SSL certificates, and HTTP status codes. This is effective for measuring and correlating network performance with application performance.

UXI sensor portfolio

Sensor platform lineage:

	Model	UX-G6	UX-G6C	UX-G6E	UX-G6EC
Chipsets	Arm CPU	2.0GHz	2.0GHz	Quad-Core 1.5Ghz	Quad-Core 1.5Ghz
	Wi-Fi 2G/5G Radio	.11ax	.11ax	Dual radio 802.11ax 2.4GHz, 5GHz & 6Ghz	
	BLE	Yes	Yes	Yes + Zigbee	Yes + Zigbee
	Cellular as back-haul	-	Yes	-	Yes
Storage	eMMC	8GB	8GB	8GB	8GB
	RAM	2GB	2GB	2GB	2GB
Power	TPM2.0	1	1	1	1
	1GBE PHY	1	1	1	1
	PoE 802.3af	1	1	1	1
	Power Backup	-	≥30s	-	>=60s
	Power Consumption (limit)	12W	12W	12W	12W
	Ambient T & V sensing	1	1		
SKU		R7H75A	R7H76 A	S0U51A	S0U52A

UXI sensor portfolio consists of 4 sensors (Considering the F series sensors are EOL now)

The G5 sensors supports Wi-Fi 5 and earlier environments

And G6 sensors supports Wi-Fi 6 and earlier environments

- Ship to any country and plug into power and it will self-configure. No technical person needed on site.
- Supports out of band connectivity

UXI use cases

The figure shows actual, proven use cases, where UXI helped to quickly identify and resolve issues.

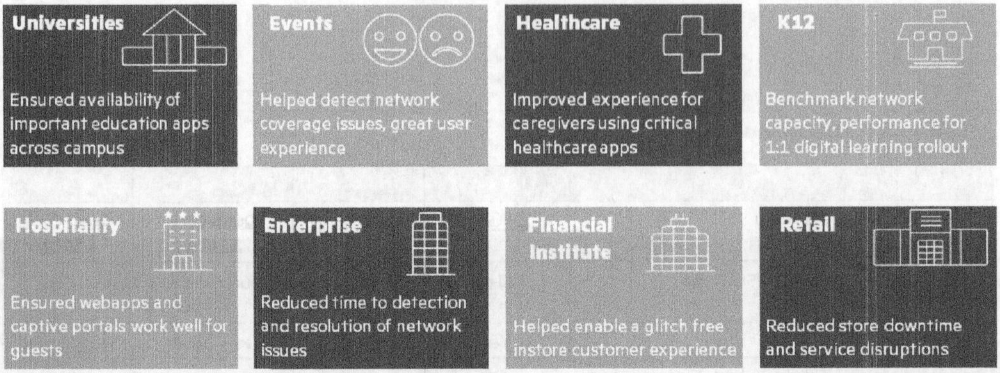

Learning check

Check your understanding of network optimization by answering the following questions.

Question #1

What is considered the first step towards ensuring good QoS in a campus network?

- ○ Mark packets for prioritization at the access layer of the network.
- ○ Minimizing congestion by planning adequate capacity.
- ○ Enable deep packet inspection (DPI) on gateways for application matching.
- ○ Implement QoS policies to reduce packet loss, jitter, and latency for VoIP traffic.

Question #2

You have completed an AP design with proper density and location considerations. Which of the following would NOT impact the full capacity of your current access points?

- ○ Untested clients that fail to support the expected 802.11ax standard
- ○ Non-802.11 RF sources
- ○ Clients gathering in unpredicted locations
- ○ Ratification of a new Wi-Fi 8 standard

Question #3

HPE Aruba Networking User Experience Insight (UXI) provides what benefit to HPE Aruba Networking customers?

- ○ AI-powered insights alert IT staff to network performance issues as soon as they begin.
- ○ 24/7 service assurance solution for users, networks, and applications through an on-premises-based server solution.

- UXI sensors enable continuous monitoring and testing across 802.11 wireless, wired, ZigBee, and Bluetooth networks.
- UXI provides assurance for on-site services such as Wi-Fi, DHCP, DNS, and gateway reachability, and requires UXI-Cloud license only for cloud services accessibility testing.

Supplemental: Wired QoS

Use this supplemental material to study more details about wired QoS.

Using L2 and L3 classification

These tables show default markings or classifications for the CoS (left side of the slide) and DSCP values (right side of the slide).

VLAN Protocol ID	802.1p priority	1	VLAN-ID

802.1p	Traffic Type (802.1Q)
7	Network control
6	Internetwork control
5	Voice (<10ms latency)
4	Video (<100ms latency
3	Critical applications
2	Excellent effort
0	Best effort
1	Background

Version	ToS Field	Rest of IPv4 header

DSCP	Name	Examples
56	CS6	Network control
46	EF	Telephony (VoIP)
40	CS5	Signaling
34,36,38	AF41 / AF42/ AF43	Multimedia conferencing
32	CS4	Real-time interactive
26,28,30	AF31 / AF32 / AF33	Multimedia streaming
24	CS3	Broadcast video
18,20,22	AF21 / AF22 / AF23	Low-latency data
16	CS2	OAM
00	CS0 / BE / DF	Best effort
10,12,14	AF11 / AF12 / AF13	Bulk data
08	CS1	Low-priority data

The right table in the slide recommends the type of applications you might assign to each AF class. For example, you could assign multimedia conferencing applications to AF class 4. These applications are interactive, so they need very low latency. You could assign multimedia streaming, which can use buffering and has less strict latency requirements, to AF class 3. You can assign AF class 2 to applications that require low latency but less than multimedia. For example, interactive network management applications such as SSH could use this class. Finally, AF class 1 is for bulk data that can tolerate delays: it has a priority of less than 0.

You can use the CS values for legacy IP precedence values if needed. Also, RFC 3246 defines Expedited Forwarding (EF) class 46 to be used for applications more sensitive to jitter and delay. HPE Aruba Networking often recommends this DSCP for voice-over IP (VoIP) traffic.

DiffServ Code Point (DSCP)

DiffServ Code Point (DSCP) is a layer 3 marking using the 8-bit Type of Service (ToS) field in IPv4 or the IPv6 traffic class byte. DSCP uses the upper 6 bits of the byte where there are 64 possible codepoints to assign priority. Originally, for QoS, IP Precedence utilized the three most significant bit values. This gave each packet a priority of 0 through 7 and could coincide with the 3-bit Class of Service (CoS) found in the layer two priority field of a vlan header.

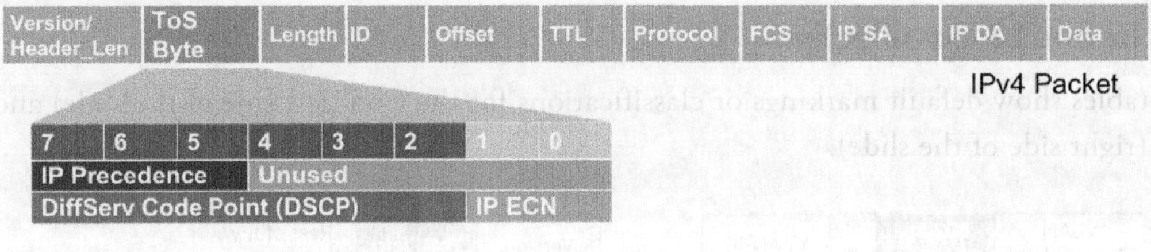

Using markings at layer 3 is typically a preferred strategy, given that the ToS field often remains unchanged. In contrast, layer two ethernet headers carrying CoS markings are dropped at the first hop gateway.

DSCP Markings diagram

RFC 2472 defined the 6 lead bits as DSCP assigned and the last 2 bits as currently unused (CU). The RFC has been expanded upon several times to define these DSCP values as they may be used for per-hop behaviors (PHBs) of traffic as it passes through network devices towards its destination.

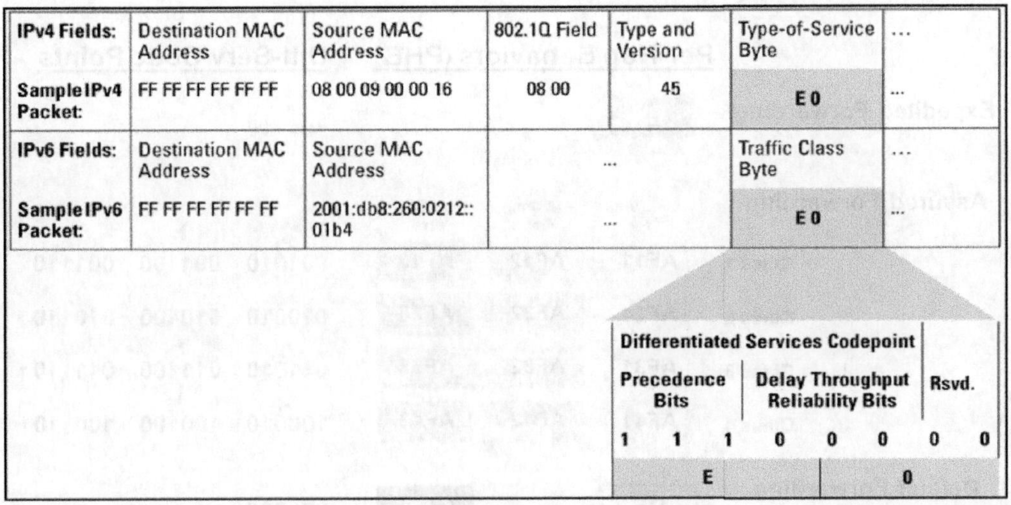

DSCP relationship with per-hop behaviors

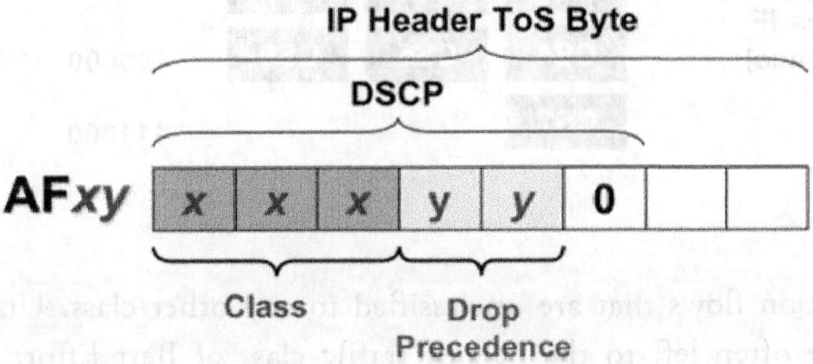

DSCP relationship with per-hop behaviors

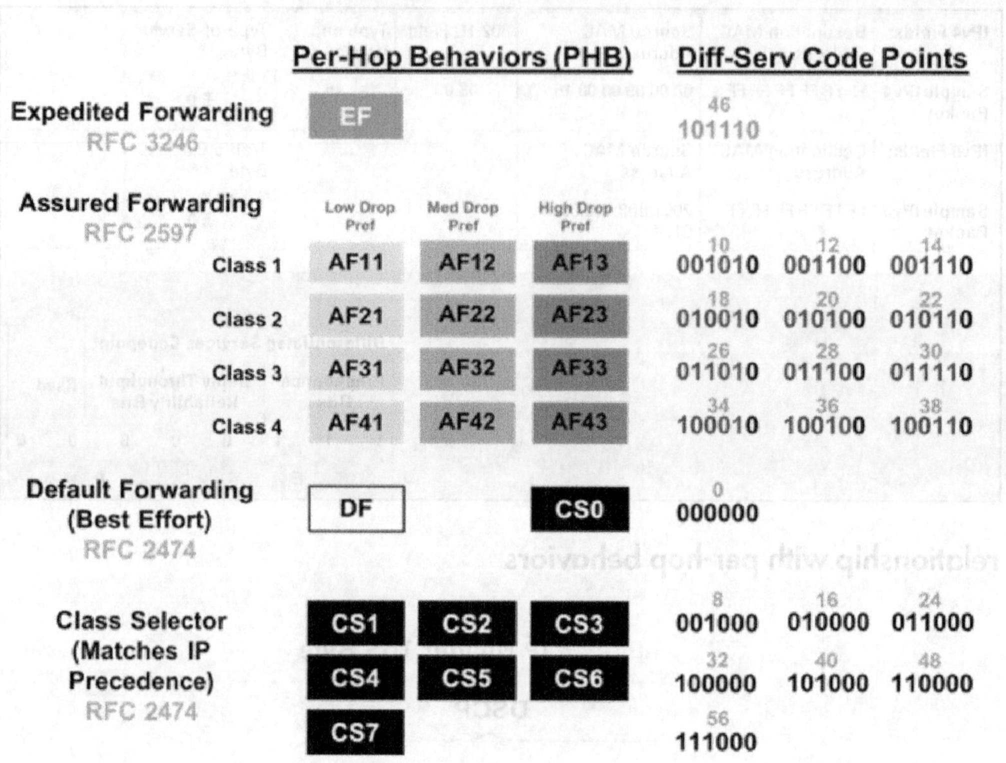

Default traffic class:

Used for all application flows that are unclassified to any other class. Untrusted markings from user applications are often left to the default traffic class of Best Effort (BE) or Default Fault (DF).

DF or CS0 or DSCP [000 000 = 0]

Voice should receive the highest user-assigned class of Expedited Forwarding

EF or DSCP [101 110 = 46]

Collaboration applications such as real-time video conferences are placed into the assured forwarding (AF31) class.

AF31 or CS3 or DSCP [011 010 = 26]

Low-latency data and critical business applications may be given an AF21 class.

AF21 or DSCP [010 010 = 18]

Example: six apps divided across a four class LAN queueing model

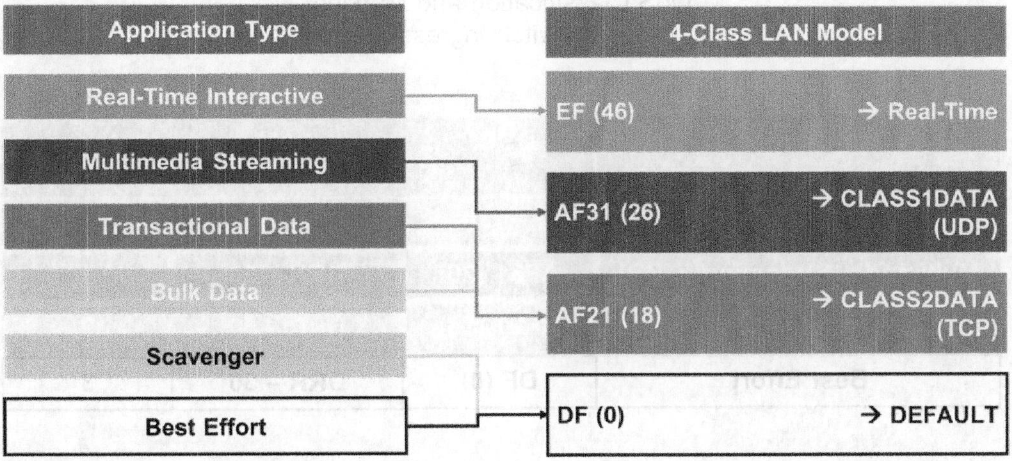

Example of half a dozen applications divided across a four class LAN queueing model.

Traffic prioritization and queuing

Traffic Prioritization and Queuing

HPE Aruba Networking switches support up to eight queues. Once traffic has been classified and marked, the switch can group per queue on egress ports. Normally, traffic will be scheduled and sent upon arrival, regardless of the queue. However, when congestion occurs, the egress port will schedule the traffic based on its queue and scheduling policy.

A strict priority queue (SP) is a scheduling policy that empties this queue before any other queues are serviced. For this reason, SP is ideal for quickly forwarding sensitive real-time VoIP traffic (EF) that otherwise will break calling if it suffers drops of more than 1%, jitter of greater than 30ms, or one-way latency delays exceeding 150ms. There is a danger of putting too much traffic in the SP queue and starving the remaining queues from being serviced. Sizing the queue and enforcing policing or shaping is best practice.

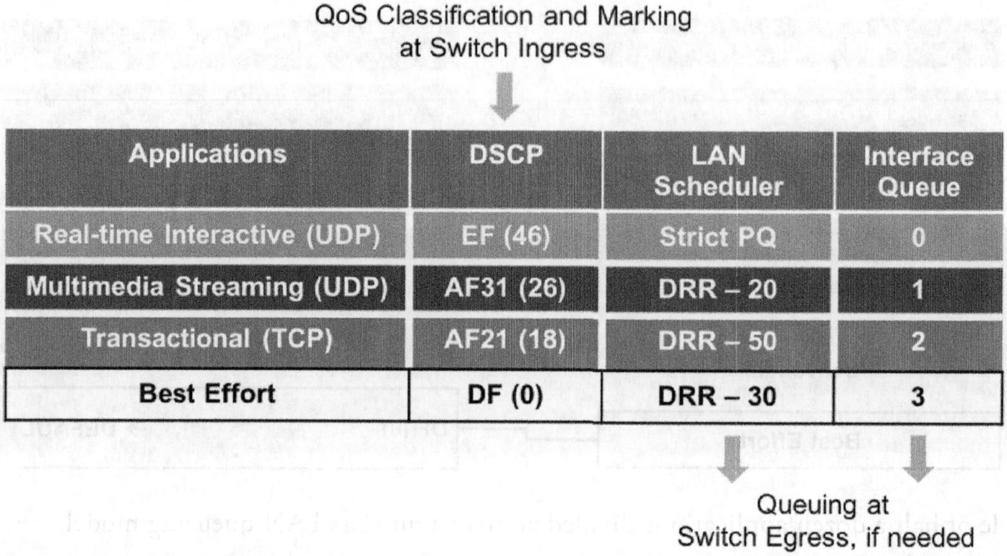

The assured forwarding classes are serviced using a deficit weighed round robin (DWRR). This is the default scheduling profile used for all CX switches apart from the 8400. The 8400 uses a Weighed Fair Queue (WFQ) scheduling strategy.

DWRR applies weighted values assigned to traffic queues to guarantee a weighted percentage of bandwidth. Each schedule round starts with the highest queue being serviced down to the lowest. Each queue can send at least its guaranteed bandwidth. If a queue has more traffic than can be serviced, it sends its guaranteed amount but must hold the remaining traffic in the buffer until bandwidth refreshes for the next round.

Example of Marking at the switch ingress

If the higher queue has less traffic than the queue amount, that amount is made available to lower queues based on their weighting. This means that should there only be Best Effort traffic without any higher queues needed, 100% of the link can be used by that default traffic class.

Traffic classification and marking: WMM

Classification will group traffic applications with similar characteristics into classes or groups to be treated similarly on the network. These groups allow scheduling policies to apply precedence should queuing occur by marking frames and packets. Scheduling algorithms rely on these markings to quickly identify the priority of a given frame queued in the egress port without repeating the classification process. Markings can be done at:

- layer 2 in the 802.1p field or Class of Service of a frame
- layer 3 in the Type of Service or TOS field (Preferred deployment option)
- layer 2 wireless frame as part of Wireless Multimedia (WMM) 802.11e

The WMM is a subset of the 802.11e standard that classifies wireless applications into four defined access categories:

- Voice or AC_VO
- Video or AC_VI
- Best Effort or AC_BE
- Background AC_BK

Wi-Fi Multimedia (WMM) details

A QoS policy for Wi-Fi should define which traffic classification fits into each of these four WMM queues. As traffic is bridged from wireless to the wired network, so should the four WMM queues map to existing DSCP and IEEE 802.1p values to enable an end-to-end consistence QoS policy.

WMM to DSCP Mapping table

Note that WMM is supported in all of HPE Aruba Networking's Wi-Fi products.

Packets are put in order of their priority queues as they are sent to the MAC layer of the wireless interface. They are then serviced using EDCA mechanism which gives higher Access Classes such as AC_VO and AC_VI a distinct advantage by having a short AIFS and smaller CW.

Wi-Fi Multimedia (WMM)

A certification program created by the Wi-Fi for QoS over wireless networks.

WMM is supported on all HPE Aruba Networking products.

Four available priority queues.

- Voice
- Video
- Best effort
- Background

Implementation uses shortened wait times between packets.

Tagging of packets using DSCP and IEEE 802.1p markings should map to wired LAN QoS markings.

WMM to DSCP mapping

WMM Access Category	Description	DSCP
Voice Priority	Real-time interactive	46
Video Priority	Multimedia streaming	26
Best Effort Priority	Transactional	18
Background Priority	Best effort	0

WMM to DSCP Mappings

QoS provides predictable traffic flow behavior when congested occurs. Normally most networks will rarely experience congestion if designed with enough capacity on. Even wireless often operates well below capacity when you consider that 802.11 is inherently a shared medium; wireless clients share a channel space. A challenge in today's wireless networks is to enable more simultaneous clients on wireless. New features like Wi-Fi 6E bring MU-MIMO, OFDMA, and take advantage of the new 6 GHz band to perform layering to accomplish this goal.

Even with these features, overload conditions can and do occur. The QoS mechanisms in place grant import traffic a higher priority while making fewer resources available to lower-priority traffic. If a hospital wants to prioritize life saving medical imaging for surgery, then doing so may be at the risk of delaying or dropping YouTube video traffic on a guest network.

Wi-Fi Multimedia operation

Wi-Fi Multimedia is a standardized certification program created by the Wi-Fi Alliance for Quality of Service over wireless networks.

It uses four WMM queues, which are implemented through the Enhanced Distributed Channel Access (EDCA) mechanism.

- This mechanism adjusts the wait times between packets.

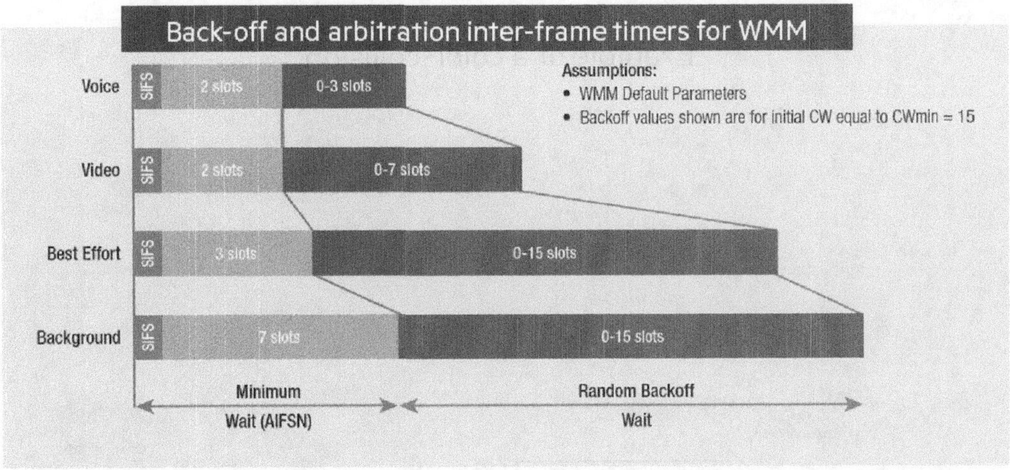

Wireless frames have a **Distributed Interframe Space (DIFS)** as the minimum medium idle time used for contention-based services.

- The more a wireless station must wait, the less likely a collision will occur.
- However, WMM replaces this with a differentiated Arbitration Interframe Space (AIFS), a shortened wait time, to provide priority queuing.

Should two wireless stations be in contention to send their frames, the station with the higher priority frame will have a shorter AIFS and gain earlier access to the medium than lower-priority frames or normal traffic.

A station with a low-priority frame must set the back-off timer to a large random number range when waiting to access the medium.

- However, another station with a high-priority frame can use a much smaller random number range.
- Statistically, this will ensure that high-priority frames are transmitted with a shorter delay, more often, and with a lower chance of being dropped compared to lower-priority frames.

BSS coloring

BSS coloring, by default, enables an HPE Aruba Networking AP to set an available color tag. This color indicator notifies the AP that other radios transmit the same BSS on the same channel. When the detected frame has a different BSS color from its own, then the STA considers that frame as an inter-BSS frame from an overlapping BSS.

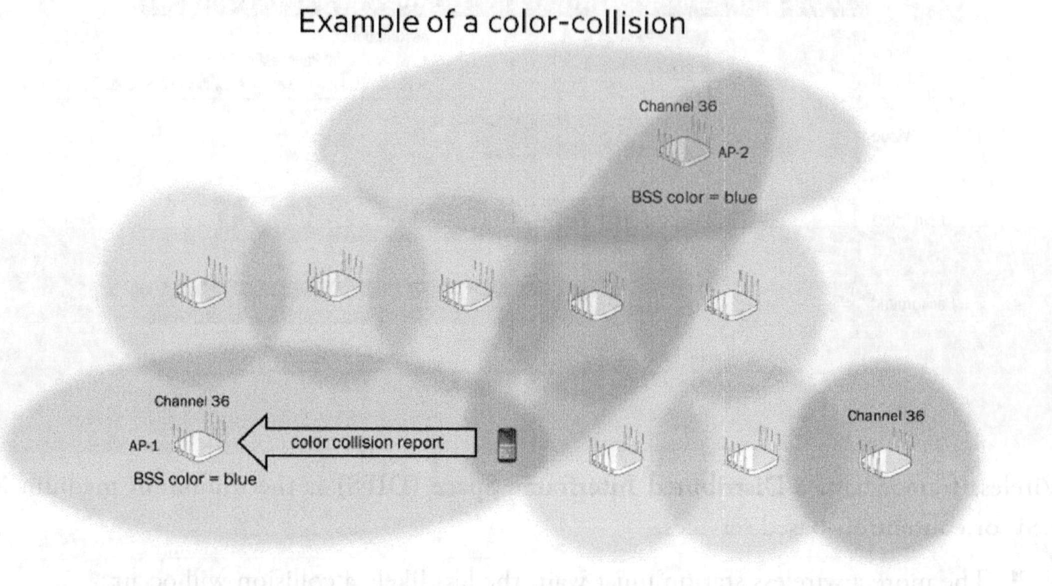

A color collision occurs when an AP hears a transmission from an access point or client in a different BSS with the same color as the AP. An 802.11AX client can also send a color collision report to its associated access point should it detect a color collision or an Overlapping Basic Service Set (OBSS). The AP can then notify clients of a new BSS color change.

Example of a color collision.

The goal of BSS coloring is to improve Spatial Reuse, avoiding transmissions from an OBSS and therefore being able to transmit wireless frames simultaneously. Without BSS coloring, OBSS requires APs and clients to avoid collisions by waiting to transmit when the channel is clear.

WLAN multicast and broadcast – DMO

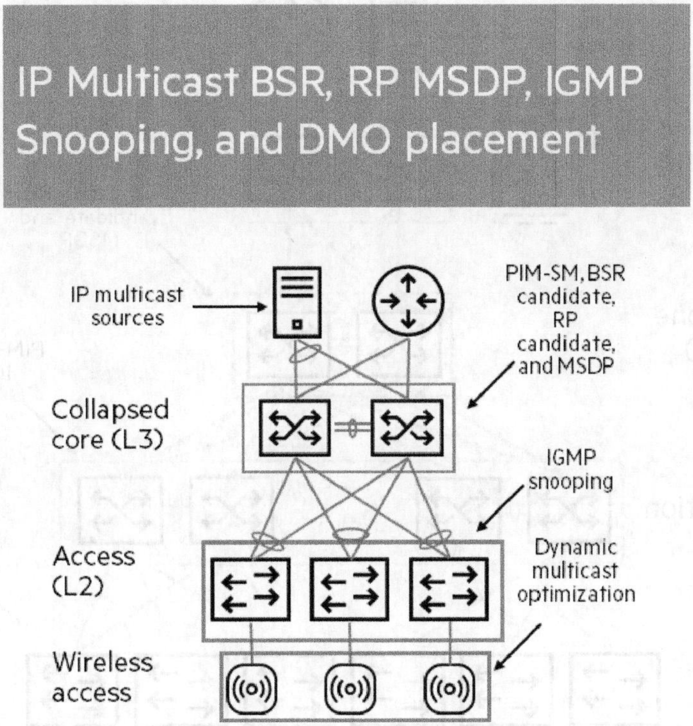

- Multicast traffic over WLAN must be transmitted at the lowest basic rate
- Dynamic Multicast Optimization (DMO)
- IP Multicast BSR, RP MSDP, IGMP Snooping, and DMO placement
- Unicast frames acknowledged by the client
- It can be retransmitted if lost over the air
- Use the highest possible data rate supported by individual clients

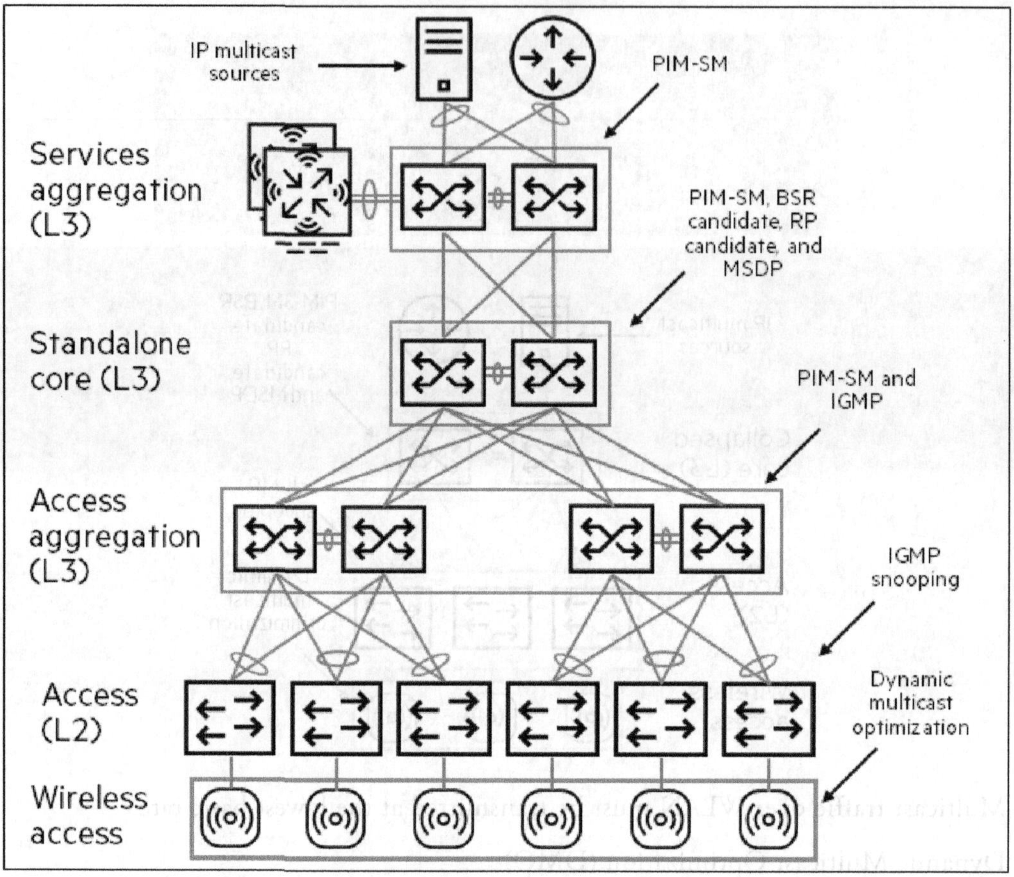
IP multicast sources
PIM-SM
Services aggregation (L3)
PIM-SM, BSR candidate, RP candidate, and MSDP
Standalone core (L3)
PIM-SM and IGMP
Access aggregation (L3)
IGMP snooping
Access (L2)
Dynamic multicast optimization
Wireless access

Broadcast to unicast conversion

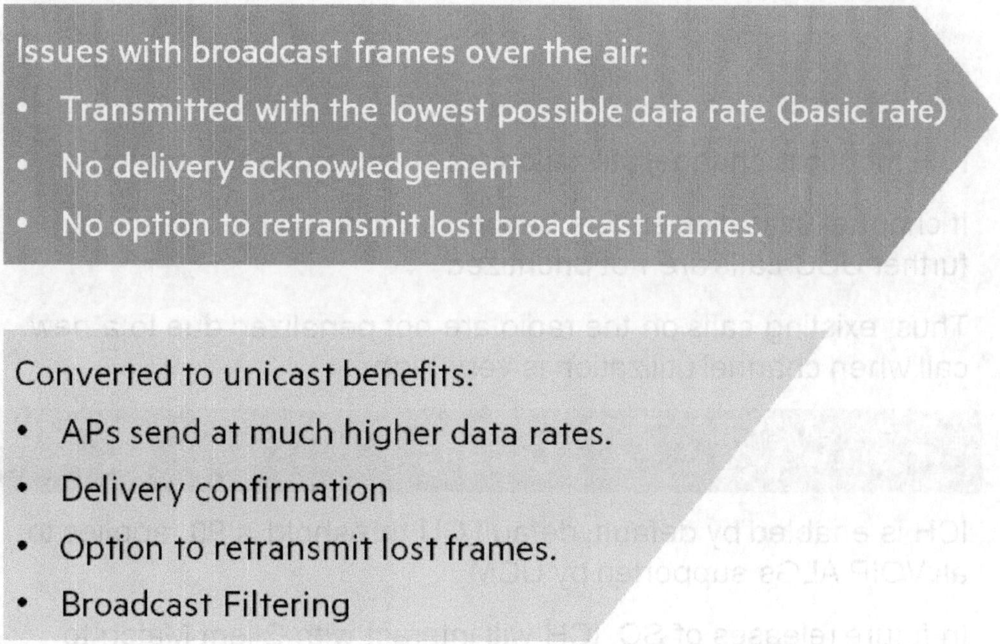

Broadcast to unicast conversion does, as suggested; converts broadcast packets into unicast frames. This provides the following features:

- Optimizes airtime usage.
- Issues with broadcast frames over the air
- Transmitted with the lowest possible data rate (basic rate)
- No delivery acknowledgment
- No option to retransmit lost broadcast frames.
- Converted to unicast benefits:
- APs send at much higher data rates.
- Delivery confirmation
- Option to retransmit lost frames.
- Broadcast Filtering
- Configure ARP filtering to reduce broadcast transmissions.

What is Intelligent Call Handling (ICH)

Objectives

ICH monitors channel utilization for all radios on the controller

If channel utilization shoots beyond a configurable threshold, further UCC calls are not prioritized

Thus, existing calls on the radio are not penalized due to a new call when channel utilization is very high

Details

ICH is enabled by default, default CU threshold is 90, applies to all VOIP ALGs supported by UCM

In future releases of SC. ICH will interact with Client Match to move clients between APs to reduce AP utilization and improve call quality

Intelligent Call Handling (ICH) replaced Call Admission Control (CAC) for AOS-8 and AOS-10. ICH requires appropriate cell overlap and matching cell size to the voice client transmit power:

- Ensuring there is no overutilization of AP radio and ethernet bandwidth, prioritizing voice end-to-end with QoS
- Using mobile voice clients that support all the standards to support modern "Client Match" and WMM
- Making use of Aruba Central's UCC Service to both ensure the voice datapath is QoS marked and call quality is monitored

How ICH works

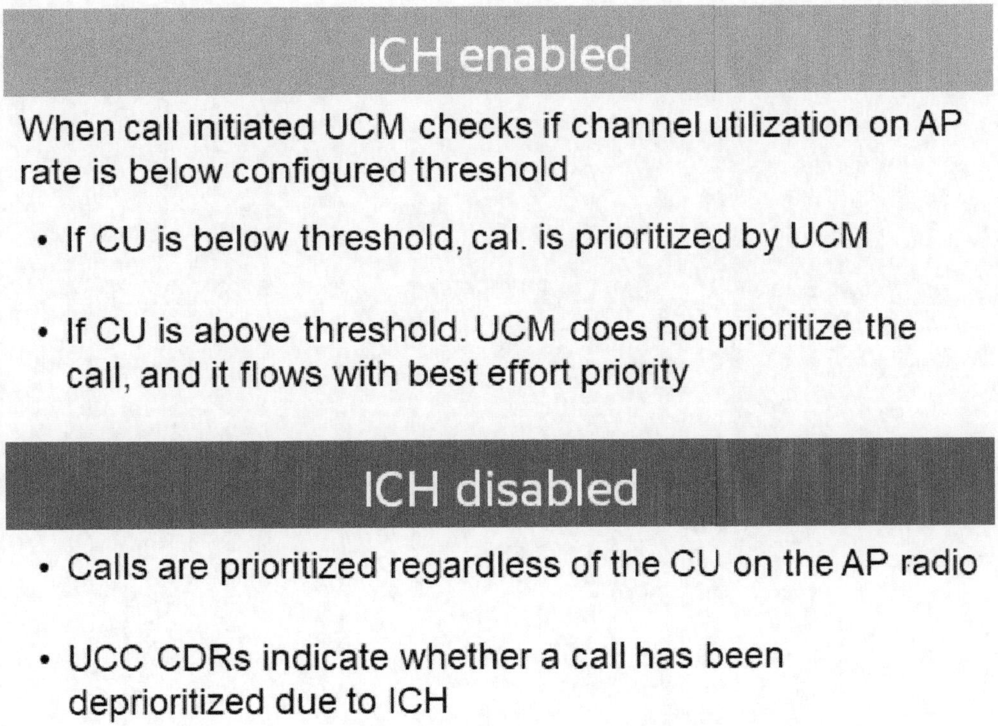

The information above offers a brief overview of ICH operation.

10 Conclusion

Next steps

Now that you have completed this course, what should you do next?

Get certified

Consider taking the certification/accreditation exam that goes with this course. Why?

- Increase your skills and technical confidence with HPE Networking technologies and become the go-to leader for strategic networking and digital transformation projects.
- Showcase your talent by adding HPE Certifications to your resume and become recognized as the leading visionary at the Edge.

This course prepares you to take the HPE Network Campus Access Architect certification/accreditation exam.

To learn more about certification/accreditation, go to https://certification-learning.hpe.com/tr/technical-certifications-networking.

Advance your knowledge

This course is part of a series of courses. Having completed this course, you might choose to take the Campus Access Expert (Switching) and/or Campus Access Expert (Mobility) courses.

HPE values your feedback to help improve our courseware. If you found any errors (spelling, grammatical, technical) or want to provide other feedback regarding the content of this guide, please send an email to aruba.training.feedback@hpe.com with details. Thanks for helping us improve our courses!

11 HPE7-A11 practice exam

This chapter provides sample test questions based on the style and difficulty you would find on the real HPE7-A11 exam for the HPE Network Campus Access Architect certification/accreditation.

OBJECTIVES

The purpose of this practice exam is to:

- ✓ Reinforce the material you have learned.
- ✓ Become familiar with the type and style of questions found on the HPE Network Campus Access Architect exam.

Exam overview

The HPE Network Campus Access Architect validates your ability to architect a cost-optimized solution that meets customer requirements. It assesses your skills in translating technical requirements into HPE Aruba Networking solutions and creating a bill of materials for technical decision-makers. The solutions may involve migration or deployment strategies within an existing architecture. While you may not be responsible for the entire solution, you are expected to contribute to its development.

Ideal candidate for this exam

Typical candidates for this exam are IT professionals with experience designing and/or contributing to the design of large-scale networks

Exam contents

This exam has 70 questions. Here are the types of questions to expect:

- Multiple choice (multiple responses)
- Multiple choice (single response)
- Drag-and-drop

Advice to help you take this exam

- Complete the training and review all materials and documents before you take the exam.
- Exam items are based on expected knowledge acquired from job experience, an expected level of industry-standard knowledge, or other prerequisites (events, supplemental materials, and so forth).
- Successful completion of the course or study materials alone does not ensure you will pass the exam.

Read the entire question and consider all options before you answer. If the question includes an exhibit, study the exhibit and read the question again. Select the answer that fully responds to the question. If the question asks for more than one answer, select all correct answers. There is no partial credit.

Exam objectives and exam information

This section covers the objectives for the HPE7-A11 exam, as well as information about the exam.

HPE7-A11 exam objectives

Table 11-1 breaks down the objectives covered on the exam.

Table 11-1: HPE7-A11 exam objectives

Percentage of exam	Sections/Objectives
23%	1. Discover customers' requirements. a. Identify the technical stakeholders and map their roles (4%) b. Identify, collect, analyze, and document the current environment (8%) c. Interpret the technical requirements (7%) d. Document assumptions (4%)
30%	2. Analyze requirements. a. Identify technologies to satisfy requirements (9%) b. Create a functional diagram or workflow (10%) c. Create wired and wireless topology and preliminary BoM (11%)

Percentage of exam	Sections/Objectives
30%	3. Architect the solution. a. Write technical proposal document (7%) b. Finalize BOM (7%) c. Write solution overview and executive brief (5%) d. Document Implementation and migration guidelines (6%) e. Conduct technical and business review (5%)
17%	4. Propose the solution. a. Contribute to the proposal of the solution (10%) b. Defend the solution (7%)
0%	New technologies

Exam information

Table 11-2 summarizes the information to register for the exam. You need an HPE Learner ID and a Pearson Vue login and password. No reference material is allowed at the testing site. This exam may also contain beta test items for experimental purposes. During the exam, you can make comments about the exam items. HPE welcomes these comments as part of our continuous improvement process. To register for the exam, visit https://www.pearsonvue.com/hpe.

Table 11-2: HPE7-A11 exam information

Exam ID	HPE7-A11
Exam type	Proctored
Exam duration	2 hours
Exam length	70 questions
Passing score	68%
Delivery languages	English
Supporting resources	*HPE Aruba Networking Certified Professional - Campus Access Architect*, Rev. 25.42 course, available in physical, virtual, and eLearning formats.

Note

More and current information concerning the HPE7-A11 exam can be found here: https://certification-learning.hpe.com/tr/datacard/exam/HPE7-A11.

Practice exam

The following questions will help you to measure your understanding of topics that may be covered in the exam. Read all the choices carefully since there may be more than one correct answer. Choose all correct answers for each question.

Practice exam questions and answers

The following questions will help you to measure your understanding of topics that may be covered in the exam. Read all the choices carefully since there may be more than one correct answer. Choose all correct answers for each question.

Questions

1. As part of an HPE Aruba Networking campus access layer design project, you are meeting with various stakeholders from the customer's IT organization. Your objective is to gather **technical input** that directly supports design decisions such as **switch port capacity**, **uplink requirements**, and **user connectivity needs**.

   Which two of the following stakeholder-role pairings are most relevant to these design activities?

   (Select 2)

   A. **IT Infrastructure Manager** – Supplies physical site details, rack space availability, power and cabling constraints, and switch port projections.

   B. **Help Desk Supervisor** – Shares common user access issues and recurring support tickets that may indicate access layer shortcomings.

   C. **Chief Financial Officer (CFO)** – Provides budget oversight and prioritizes spending efficiency across IT initiatives, including infrastructure projects.

   D. **Director of Security Compliance** – Advises on data protection and access control requirements that may affect network segmentation or authentication strategies.

   E. **Chief Operating Officer (COO)** – Offers insight into organizational structure and operational workflows that could influence network priorities

2. You are designing a wireless intrusion detection and prevention (WIDS/WIPS) strategy for your client's enterprise network using HPE Aruba Networking and AOS-10. Your design includes the deployment of dedicated Air Monitors (AMs).

   Which of the following statements reflect HPE Aruba Networking best practices or recommendations regarding the design, deployment and implementation of WIDS/WIPS?

   (Select 2)

   A. There is no HPE Aruba Networking advised legal risk to recommending enabling "40MHz Intolerance" in your radio profile settings.

B. Automatic "Wireless containment" is recommended only when AIOps is enabled with an Advanced Security subscription.

C. "Wired containment" is disabled by default, enabling it is recommended only after consulting with security and legal stakeholders.

D. As a starting point, it is recommended to deploy four or five Air Monitors per 10,000 sq ft (930 sq m).

E. AOS-10 APs operating in "radio mode" are hybrid AMs capable of scanning rare channels, but at a reduced rate.

3. Review the exhibit.

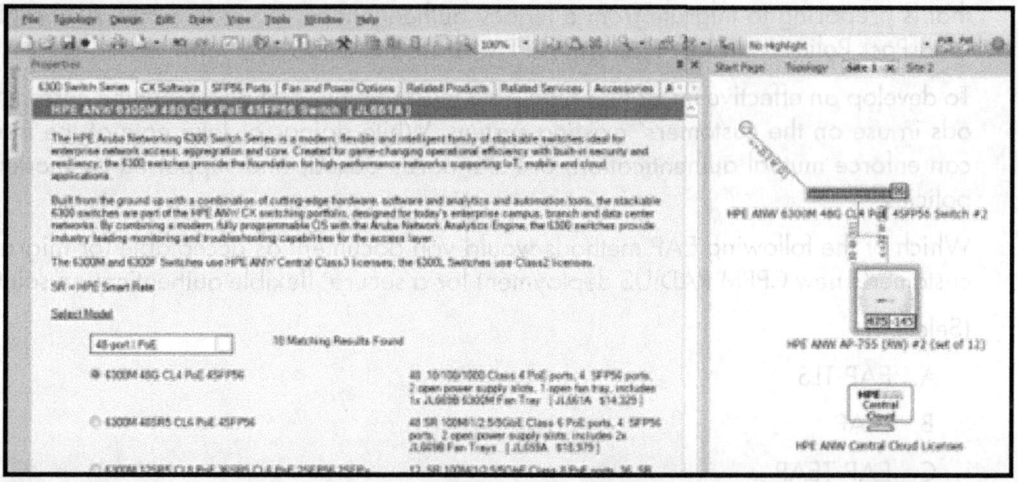

You are modeling a customer's existing network using Intangi IRIS express mode. You are proposing the replacement of legacy Access Points (APs) with HPE Aruba Networking 750 Series Campus Access Points.

You added the customers' existing access switch; when you added a set of 12 APs and connected them to the access switch, you observed red highlighting around the "HPE ANW AP-755" icon.

Which of the following statements are accurate regarding the exhibits Site-1 block view?

(Select 3)

A. This AP model requires being connected to a switch supporting the CL8 feature set to enable PoE support.

B. Enabling N+1 redundancy on the switch model shown is feasible and would support the set of 12 APs.

C. To support the requested PoE budget, you must add an additional power supply to the switch to which these APs are connected.

D. To generate an appropriate BOM from "Site-1", you should set the switch attribute Status to "existing".

E. This AP model requires being connected to a switch supporting the CL6 feature set to enable PoE support.

F. This AP model supports being connected to a switch supporting the CL4 feature set to enable PoE support.

G. This AP model requires being connected to a switch supporting Smart Rate to enable PoE support.

4. You are the network architect responsible for assessing an existing enterprise environment that is preparing to migrate from a legacy authentication server to a HPE Aruba Networking ClearPass Policy Manager Platform (CPPM)-based RADIUS solution.

   To develop an effective migration plan, you must identify and document the current EAP methods in use on the customers' existing system. While doing so, take note of the methods that can enforce mutual authentication, are standards-based, and supported by modern security policies.

   Which of the following EAP methods would you document as acceptable for migration to the customer's new CPPM RADIUS deployment for a secure, flexible authentication solution?

   (Select 4)

   A. EAP-TLS

   B. LEAP

   C. EAP-TEAP

   D. EAP-FAST

   E. EAP-PEAP

   F. EAP-GTC

   G. EAP-TTLS

5. Review the exhibit.

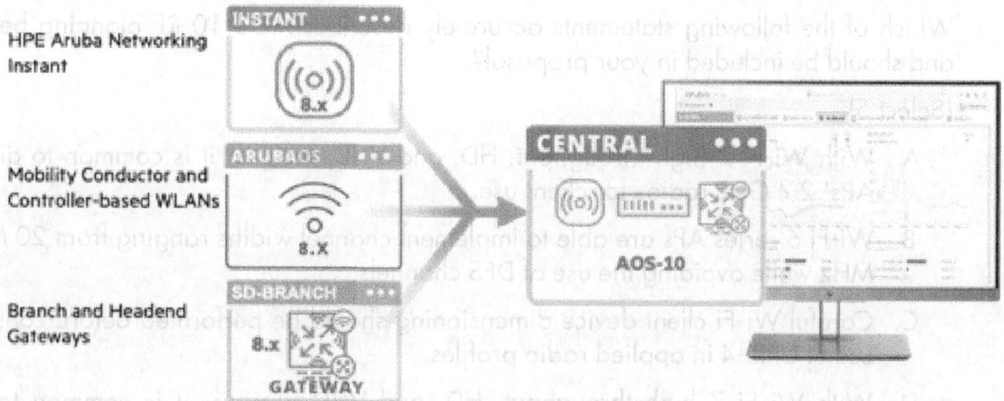

You are developing an implementation plan to migrate a customer from HPE Aruba Networking AOS-8 to AOS-10. Which of the following are valid technical requirements for the proposed solution?

Note

Refer to the latest release notes, as requirements may change.

(Select 3)

A. In some situations, AOS-10 APs must be permitted TCP port 2083 access through the internet perimeter firewall.

B. It is common practice to use the latest version of HPE Aruba NetworkingAirWave to upgrade HPE Aruba Networking APs and GWs from AOS-8 to AOS-10 firmware.

C. Organizations with networks based on single-stack IPv6 may need to implement dual-stack (IPv4 + IPv6) for the AOS-10 devices.

D. All HPE Aruba Networking 3xx, 5xx, 6xx, and 7xx Series Access Points can be migrated from an Instant AOS-8 cluster to AOS-10.

E. HPE Aruba Networking 7005 mobility controllers do not support AOS-10 versions greater than 10.3.1.1, therefore it can't be clustered.

F. HPE Aruba Networking APs being migrated to AOS-10 must be permitted HTTPS (TCP port 443) access through the internet perimeter firewall.

G. If an AOS-8 MCR is added to a new HPE Aruba Networking Central On-Premises Group, its existing configuration can be migrated to the group.

6. You are preparing a technical proposal for a customer designing a Wi-Fi 7 high-density (HD/VHD) wireless environment using HPE Aruba Networking AOS-10. This section of the proposal will document radio channel planning and optimization considerations.

Which of the following statements accurately describes AOS-10 RF planning best practices and should be included in your proposal?

(Select 5)

A. With Wi-Fi 7 high-throughput, HD, and VHD designs, it is common to disable some APs' 2.4 GHz radios for client use.

B. Wi-Fi 6 series APs are able to implement channel widths ranging from 20 MHz to 160 MHz while avoiding the use of DFS channels.

C. Careful Wi-Fi client device dimensioning should be performed before considering the use of UNII-4 in applied radio profiles.

D. With Wi-Fi 7 high-throughput, HD, and VHD designs, it is common to adjust the default radio profile channel bandwidth settings.

E. AOS-10 radio profiles have IEEE 802.11h Dynamic Frequency Selection (DFS) channels disabled by default.

F. The Wi-Fi 802.11be standard is capable of operating over all of the following frequency bands: 2.4, 5 and, 6 GHz.

G. When customizing radio profiles, customers should disable channels 169, 173 and, 177 to ensure maximum client compatibility.

7. You are acting as the network architect for a distributed enterprise preparing to implement HPE Aruba Networking Cloud Authentication. The customer has requested a cloud-native authentication solution, but has not yet provided full details about their current identity infrastructure or endpoint capabilities.

   To move forward with initial design planning, you must document reasonable assumptions about the customer's existing equipment and their willingness to align with cloud-native authentication requirements.

   Which of the following assumptions would be valid to document at this early stage?

   (Select 3)

   A. Your customer has or is willing to make use of cloud identity providers such as Microsoft Entra ID, Google Workspace or Okta.

   B. All Wi-Fi clients are able to support the required client-based HPE Aruba Networking Cloud Central "Onboard App".

   C. Your customer endpoint client infrastructure supports supplicants based on client certificates.

   D. The customer is willing to phase out non-EAP-TLS clients, allowing MAC-auth only for limited legacy devices.

   E. Your customer has or is willing to make use of Active Directory Domain Services authentication through CPPM domain integration.

F. Your customer endpoint infrastructure supports EAP-TLS or EAP-PEAP based authentication.

G. Your client is willing to support Active Directory Domain Services authentication with CPPM domain integration.

8. Review the exhibit.

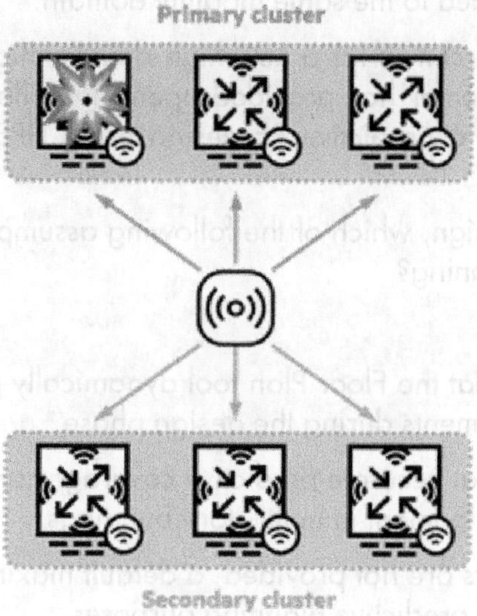

You are the network architect reviewing the technical requirements for a customer's HPE Aruba Networking AOS-10 Access Point (AP) redundancy strategy. The customer requires that client connectivity be minimally disrupted during failover between primary and secondary gateway clusters.

To proceed with design validation, you must document key assumptions that support a seamless failover experience.

(Select 3)

A. Your secondary gateway cluster uplinks to their aggregation switches must be tagged to the same VLAN IDs as the primary cluster.

B. Ensure all APs are able to establish IPsec connectivity between themselves and each member of their associated clusters.

C. Your secondary gateway cluster must be created in a backup group; this "parallel" cluster must have the same name as the primary cluster.

D. Configure the APs for high-availability by assigning them both a primary and a secondary cluster, enabling cluster failover redundancy.

E. The AOS-10 Tunnel Orchestrator sets up an IPsec map (state synchronization) between the primary and secondary cluster leaders.

F. Configure your primary and secondary clusters in a pair of linked sites, and add both clusters to the same group with cluster mode "auto-site" enabled.

G. Configure the same anycast IP address on the loopback of both the primary and secondary cluster members.

H. Create an organizational mobility domain; ensure both the primary and secondary clusters are added to the same mobility domain.

9. You are the network architect for a new high-density, high-throughput Wi-Fi 7 campus wireless design. Your customer has provided open office floor plans for multiple buildings and asked you to provide a predictive RF design using HPE Aruba Networking Central's Floor Plan tool.

   Before starting the design, which of the following assumptions would be valid to document at this early stage of planning?

   (Select 5)

   A. It is assumed that the Floor Plan tool dynamically calibrates RF propagation using live signal measurements during the design phase.

   B. It is assumed that accurate predictive coverage can be achieved without accounting for floor-to-floor alignment in multi-story buildings.

   C. If ceiling heights are not provided, a default maximum height of 8 meters (25 feet) will be assumed for predictive planning purposes.

   D. It is assumed that there are no hidden or undocumented internal RF interference sources, such as intercom systems or electrical panels.

   E. It is assumed that significant RF interference sources will only be identified during a site validation survey and are not included in initial planning.

   F. It is assumed that floor plan images provided by the customer are to scale and have been verified by a qualified professional or architect.

   G. It is assumed that the environment region slider in the HPE Aruba Networking Central Floor Plan tool will be set to 'Cubicles (3.0)' for initial modeling.

   H. It is assumed that AP placement recommendations will automatically adjust based on user density estimates extracted from the floor plan image.

   I. It is assumed that all floor plan images provided are CAD files with layers of metadata that classify all internal walls with RF attenuation values.

10. Your customer has employees using Microsoft Windows 11 laptops — a mix of BYOD and corporate-owned devices. They are concerned that older Windows versions have reduced security and want tighter control over network access. Their requirement is to block or restrict network access for any device that:

- is running a Windows version older than Windows 11
- does not have an active antivirus agent running

Which of the following statements accurately describes HPE Aruba Networking features and capabilities relevant to this requirement?

(Select 2)

A. ClearPass Device Insight supports these profiling requirements via its Microsoft NAP agent Augmentation method.

B. HPE Aruba Networking Cloud Central Cloud Auth combined with its "Onboard App" supports these profiling requirements.

C. ClearPass OnGuard Software persistent agent supports agent-based profiling for these checks.

D. HPE Aruba Networking Central Client Insights, combined with unified infrastructure NADs, supports these profiling requirements.

E. HPE Aruba Networking Secure Service Edge (SSE) ZTNA supports agent-based profiling to enforce these checks.

11. You are the lead architect of a successful and growing HPE Aruba Networking (ANW) Partner and are responsible for growing and expanding the knowledge of your team that responds to customer RPFs for campus network engagements.

Your junior staff members need to understand the nature of the customers' existing equipment as they prepare responses to those requests for proposals.

Toward that effort, help your team match the following Wi-Fi features to their appropriate standard. **Your options for Wi-Fi standards are Wi-Fi 4, Wi-Fi 5, Wi-Fi 6, Wi-Fi 6E, and Wi-Fi 7.**

Note

Not all descriptions will match a Wi-Fi standard

Description	Fill in your Wi-Fi standard answers below
802.11be	
First introduced 6 GHz	
First introduced 802.11ax	

Description	Fill in your Wi-Fi standard answers below
First introduced 1024-QAM	
First introduced BSS coloring	
Only supports 5 GHz	
Supports Wave 1 & 2	
802.11n	
First introduced 60 GHz	

12. You are the lead architect of a successful and growing HPE Aruba Networking Partner and are responsible for growing and expanding the knowledge of your team that responds to customer RPFs for campus network engagements.

    Your junior staff members need to understand the nature of the customers' existing equipment as they prepare responses to those requests for proposals.

    Toward that effort, help your team match the following IEEE designations to their official technical Physical Layer (PHY) protocol acronym. **Your choices for PHY protocol acronym are HT, VHT, HE, EHT, UHT, and XHT.**

    Note

    Not all acronyms will be used — some are distractors.

IEEE designation	Your answers for a matching Protocol acronym
802.11n	
802.11ac	
802.11ax	
802.11be	

13. You are the lead architect of a successful and growing HPE Aruba Networking Partner and are responsible for growing and expanding the knowledge of your team that responds to customer RPFs for campus network engagements.

    Your junior staff members need to understand the nature of the customers' existing equipment as they prepare responses to those requests for proposals.

    Toward that effort, help your team understand which of the following statements accurately describe the HPE Aruba Networking Wi-Fi 7 Access Points' features?

    (Select 3)

    A. All models support 6 GHz, offering up to 320 MHz channel width.

    B. The AP-730 supports 2x5GE, while the AP-740 only supports 1x5GE.

    C. Multiple models support MACsec from the Access Point to the edge switches.

    D. The 7xx Series Access Points introduced the 802.11ax Wave 2 standard.

    E. All models have 3 radios supporting either 5 + 5 + 6 GHz or 5 + 6 + 6 GHz.

    F. The 7xx Series Access Points support the 802.11be standard.

    G. At full power draw, some AP models require switches that support Class 8 PoE.

14. You are the lead architect of a successful and growing HPE Aruba Networking Partner and are responsible for growing and expanding the knowledge of your team that responds to customer RPFs for campus network engagements.

    Your junior staff members need to understand the nature of the HPE Aruba Networking Secure Service Edge (SSE) cloud services so they comprehend how this product and all of its components can meet customer requirements.

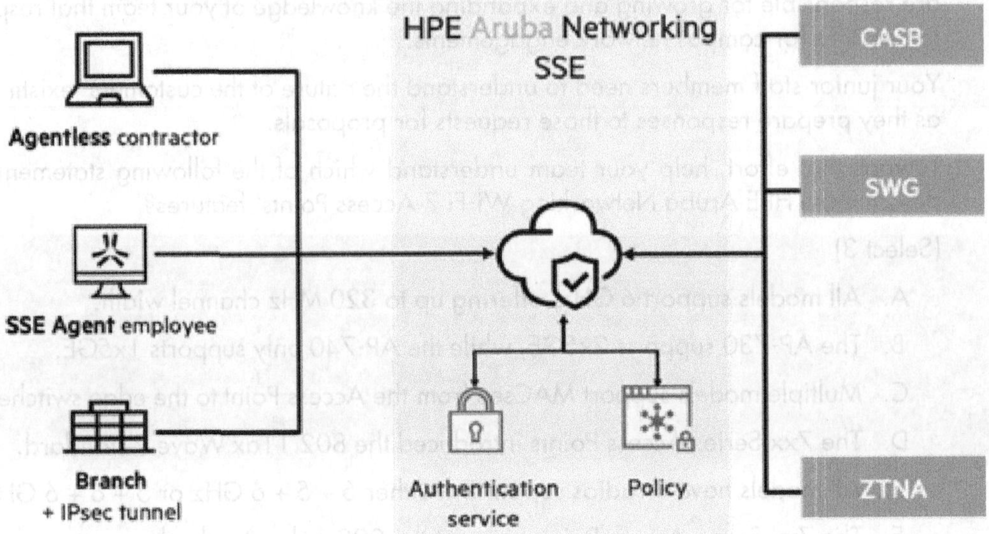

Toward that effort, help your team match the appropriate HPE Aruba Networking SSE cloud services to their correct descriptions. The cloud services are SWG, ZTNA, DEM, and CASB.

A. Makes use of the SSE Connector to broker least-privilege access to private resources and applications.________

B. Protects users' access to Software-as-a-Service (SaaS) applications using SSE public points of presence.________

C. Establishes rules that permit or block users' access to website categories based on organizational internal policies.____

D. Provides a comprehensive view of the entire network path, from a user's endpoint device to the application server._____

15. You are the lead architect of a successful and growing HPE Aruba Networking Partner and are responsible for growing and expanding the knowledge of your team that responds to customer RPFs for campus network engagements.

    Your junior staff members need to understand the nature of the HPE Aruba Networking AOS-10 Gateway High Availability features so they comprehend how this product and all of its components can meet customer requirements.

    Toward that effort, help your team match the following mobility gateway cluster features to their partial descriptions. Not all descriptions will be used.

Feature	Description
Seamless roaming	Makes use of a designated standby gateway.

Feature	Description
Client/AP load balancing	Enables state synchronization across different clusters.
Stateful failover	The clients remain anchored to the same gateway.
Client state sync	Enables APs to perform client authentication.
Auto clustering	Supports a group of one for ease of configuration.
	Supports synchronization for the first failover only.
	Helps improve client connectivity, particularly during periods of high demand.

16. You are the lead architect of a successful and growing HPE Aruba Networking Partner and are responsible for growing and expanding the knowledge of your team that responds to customer RPFs for campus network engagements.

    Your junior staff members need to understand the nature of the HPE Aruba Networking ClearPass Policy Manager Platform(CPPM) Onboard so they comprehend how this product and all of its components can meet customer requirements.

    Toward that effort, help your team list valid design considerations when recommending Policy Manager in a design proposal?

    (Select 2)

    A. You plan to use the Onboard CA's EST server to issue device certificates to HPE Aruba Networking Unified Infrastructure devices.

    B. Onboard supports being configured as a Self-signed CA or as an Intermediate CA with the issuing role.

    C. Onboard does not support acting as a Registration Authority that proxies signing requests to an external CA.

    D. You wish to use the Onboard EST OpsRamp integration to issue and manage RadSec certificates on third-party NADs.

    E. When hosting multiple PKI Certificate Authorities, Policy Manager Onboard requires a Foundation + Advanced subscription.

17. You are the lead architect of a successful and growing HPE Aruba Networking Partner and are responsible for growing and expanding the knowledge of your team that responds to customer RPFs for campus network engagements.

    Your junior staff members need to understand the nature of the HPE Aruba Networking User Experience Insight (UXI) so they comprehend how this product and all of its components can meet customer requirements.

    Toward that effort, explain which of the following statements accurately describe UXI capabilities or features.

(Select 5)

A. Cloud-based 24/7 service assurance solution for users, networks, and applications.

B. Client agent subscriptions are included as part of Unified Infrastructure Advanced subscription agreements.

C. Integration with HPE Aruba Networking SSE allows UXI to monitor Data Protection Policy violations.

D. Deployed as a network endpoint device to continuously monitor the network's performance, connectivity, and responsiveness.

E. Provides the last 30 days of performance data and PCAP files for analysis and troubleshooting.

F. Each physical sensor requires a one-, three-, or five-year HPE Aruba Networking Central Cloud subscription.

G. UXI sensors rely on passive traffic monitoring and do not generate active synthetic test traffic.

H. UXI sensors enable continuous monitoring and testing across wired, wireless, and WAN.

I. The UXI agent for Linux performs the same functions as a sensor but with reduced performance capabilities.

18. You are the lead architect of a successful and growing HPE Aruba Networking Partner and are responsible for growing and expanding the knowledge of your team that responds to customer RPFs for campus network engagements.

    Your junior staff members need to understand the nature of the distributed overlay support in HPE Aruba Networking Central so they comprehend how this product and all of its components can meet customer requirements. Which of the following statements accurately this feature set.

    (Select 3)

    A. Group-Based Policy (GBP) IDs are directly assigned to an endpoint's VTEP using a vendor-specific attribute.

    B. Least privilege access policies are enforced by Aruba-User-Roles at the ingress VTEP.

    C. NetConductor can be used to orchestrate the extension of Layer 2 networks across Layer 3 boundaries.

    D. NetConductor eliminates the need for a control-plane protocol such as EVPN in VXLAN overlay networks.

    E. A VSX cluster can host Anycast gateways to ensure consistent first-hop routing services across EVPN-VXLAN networks.

F. Least privilege access policies are enforced by Group-Based Policy (GBP) at the egress VTEP.

19. You are the lead architect of a successful and growing HPE Aruba Networking Partner and are responsible for growing and expanding the knowledge of your team that responds to customer RPFs for campus network engagements.

    Your junior staff members need to understand the nature of AOS-CX queuing and scheduling mechanisms so, they comprehend how these features can meet customer requirements.

    Which of the following statements accurately describes the queuing and scheduling mechanisms — Strict Priority (SP), Weighted Fair Queuing (WFQ), and Deficit Weighted Round Robin (DWRR) — on AOS-CX switches commonly deployed in enterprise campus three-tier network architecture?

    (Select 4)

    A. On all AOS-CX switches, SP scheduling can only be assigned to the highest-priority egress queue in the queue mapping.
    B. The AOS-CX 8400 switch supports SP and WFQ scheduling, whereas the 9300 supports SP along with either WFQ or DWRR scheduling.
    C. To protect SP from starving lower-priority queues, it is recommended that you also set a max-bandwidth limitation on the SP queue.
    D. WFQ is based on per-class bandwidth fairness, while DWRR scheduling operates based on queue weights and packet sizes.
    E. SP is typically applied to a mid-level queue mapping designated for traffic marked by DSCP 46 (EF), commonly used for voice.
    F. On all AOS-CX switches that support DWRR, DWRR must be manually enabled because FIFO is the default queuing mechanism.
    G. On the AOS-CX 6300, 6400, 8325, and 8360 switches, DWRR is the default weighted scheduler, whereas WFQ is not supported.

20. You are designing a secure system that uses digital certificates to protect sensitive communication between a client and a server.

    Place the following steps in the correct order (1 to 5) to reflect the workflow for using certificates to enable secure, authenticated data exchange.

Workflow step description	Steps in progressive order
Certificates are issued to the client and server, binding their identities to their public keys and specifying key usages (e.g., Client/Server Authentication).	

Workflow step description	Steps in progressive order
During a communication session, the server presents its certificate to the client. The client validates the certificate chain back to the trusted Root CA.	
The Root Certification Authority (CA) securely signs a certificate for an Intermediate (Issuing) CA, which then signs certificates for end entities such as clients and servers.	
The client and server use their private keys to sign and decrypt messages, ensuring authentication, data integrity, and confidentiality of the data flow.	
The server and client generate key pairs and submit certificate signing requests (CSRs) to the Intermediate (Issuing) CA, providing proof of identity.	

21. You are analyzing the high availability requirements of a customer's proposed HPE Aruba Networking CX Switch Operating System 3-tier campus network design. Your colleague, who is responsible for the technical drawing to be included in the proposal, asks you to explain the high availability potential of various different models, so they can better understand how to create these diagrams using IRIS.

    Which of the following statements are accurate?
    (Select 5)

    A. 6400 switches support VSX clustering.
    B. 6200 switches support VSF stacking between any switches of the same series.
    C. 6300 switches support VSX clustering.
    D. both 6200 and 6100 switches are capable of supporting 6 node VSF stack.
    E. 6400 switches support 2-10 node VSF stacking.
    F. 4100i switches support VSF between all same series switches.
    G. 6300 switches support 2-10 node VSF stacking.
    H. 5420 switches support either VSF stacking or VSX clustering.

22. Review the exhibit.

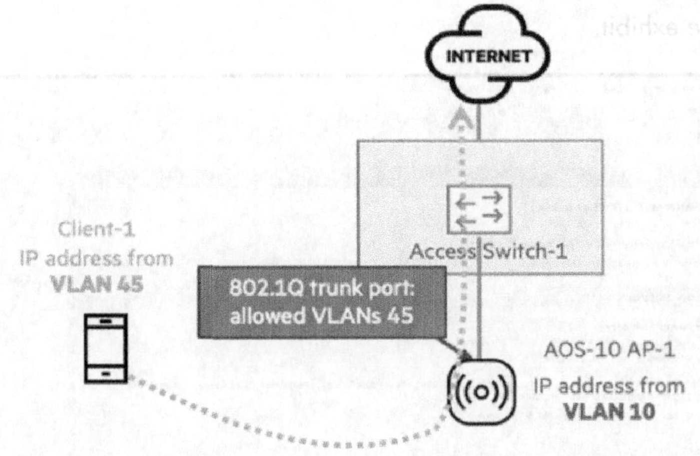

You are designing an HPE Aruba Networking Cloud Central deployment with an AOS-10 AP operating in bridged mode, connected to an AOS-CX campus access layer switch (there is a 3-tier network architecture, however only the access layer is shown in the exhibit).

Which of the following are statements are valid recommendations that help ensure users are blocked from accessing any destination where its web category is rated "Bot Nets"? (Select 4)

A. Within your ACL allowlist, allow access to FQDNs as per the BrightCloud FQDN list documentation.

B. Install the appropriate Foundation + Security subscriptions, then enable IDS/IPS in IDS mode.

C. Create an access control rule using the appropriate Web Category, and set the action to Deny.

D. Ensure that the AOS-CX switch (Switch 1) enforces client-mode with a max-client count of 0 for AP 1.

E. Create an access control rule with an appropriate "App Category" with the action set to Deny to all destinations.

F. Ensure that the VLAN assigned to Client-1, VLAN 45, is extended from Access Switch-1 to AOS-10 AP-1.

G. Install the appropriate Foundation subscription(s), then enable AppRF DPI = all.

H. Install the appropriate Foundation + Security subscriptions, then enable IDS/IPS in IPS mode.

I. It is recommended to enable jumbo frame support between AP 1 and the campus access switch (Switch 1).

J. Ideally, the AOS-CX switch should trust QoS markings from the AP; however, it must enforce device mode for the AP. (T)

23. Review the exhibit.

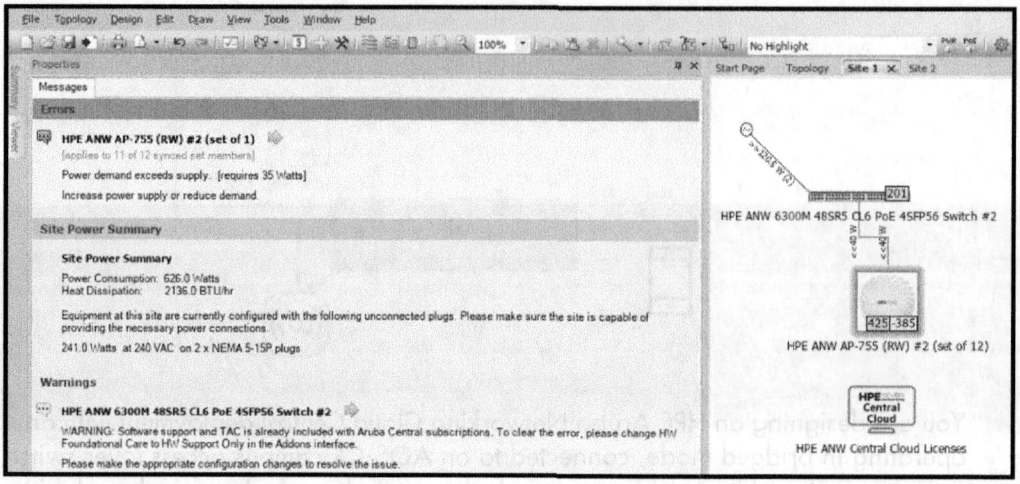

You've been asked to create a preliminary Bill of Materials (BOM) using Intangi IRIS for a new customer design. As you build Site-1, you encounter the Warnings message shown in the Messages tab (refer to the exhibit). In IRIS, where should you click next to resolve the indicated warning?

(Select 1)

A. Click the "$" in the Intangi IRIS toolbar.
B. Click the "hammer and wrench" in the Intangi IRIS toolbar.
C. Click the "HPE Aruba NETWORKING Central Cloud" icon.
D. Click the "+" in the Intangi IRIS toolbar.
E. Click the "HPE ANW AP-755" icon.
F. Click the "HPE ANW 6300M" icon.

24. Review the exhibit.

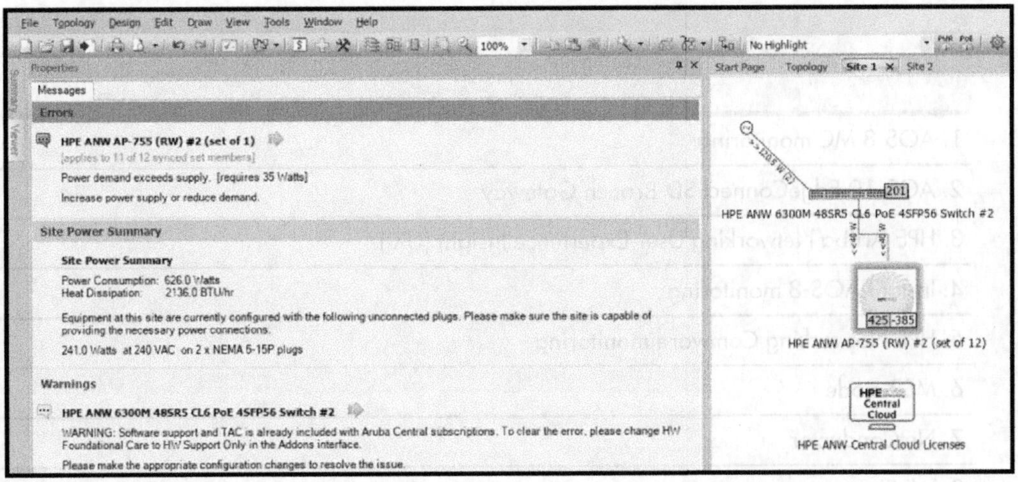

You've been asked to create a preliminary Bill of Materials (BOM) using Intangi IRIS for a new customer design, along with a network topology of the proposed solution. You were about to add another switch so you could incorporate VSF high availability into the design, but before you got to that step, you encountered red highlighting around the "HPE ANW AP-755" AP in the "Site-1" tab or IRIS. Review the related Error in the "Messages" tab in Iris.

Which of the following actions could help you resolve the indicated issue?

(Select 2)

A. Select the switch connected to the AP and adjust the model selection to support the CL8 feature.

B. Select the connecting line between the AP and the switch, right-click, and choose the "Clone Connection" option.

C. Select the connecting line between the AP and the switch, and adjust the AP's "quantity multiplier" attribute.

D. Select the connecting line between the AP and the switch, press Ctrl and "+" keys simultaneously.

E. Select the switch connected to the AP and modify its Fan and Power Options settings.

25. You have been asked to write a technical proposal for a key customer that is planning a major network migration project, seeking a unified Single Pane of Glass (SPOG), and robust Security Orchestration, Automation, and Response (SOAR) integration. Your proposed solution is based on HPE Aruba Networking Central. The crucial architectural decision remains whether to implement the SaaS (Cloud) or the On-premises deployment model.

Your current task is to create a technical proposal to aid in making this decision; you must compare the feature sets. For each HPE Aruba Networking Central feature listed below, specify the deployment model where it is supported. **Your options include: On-Premises, SaaS, Both — some features have no match**.

HPE Aruba Networking Central feature to matching platform	Your platform option answer
1. AOS-8 MC monitoring	
2. AOS-10 EdgeConnect SD-Branch Gateway	
3. HPE Aruba Networking User Experience Insight (UXI)	
4. Instant AOS-8 monitoring	
5. HPE Networking Comware monitoring	
6. MSP mode	
7. NetConductor	
8. Wi-Fi Spectrum Boost	
9. Aruba Analytics and Location Engine (ALE)	

26. **You are preparing a technical proposal document for a customer. Which of the following EIRP radio profile settings should you recommend as a starting point for HPE Aruba Networking Cloud Central AirMatch AOS-10 APs managed in a typical campus office deployment?**

    (Select 3)

    A. **2.4 GHz band, set the minimum to 6 dBm and maximum to 12 dBm**

    B. 5 GHz band, set the minimum to 15 dBm and maximum to 21 dBm

    C. 6 GHz band, set the minimum to 15 dBm and maximum to 21 dBm

    D. 2.4 GHz band, set the minimum to 3 dBm and maximum to 9 dBm

    E. 5 GHz band, set the minimum to 18 dBm and maximum to 21 dBm

27. Review the exhibit.

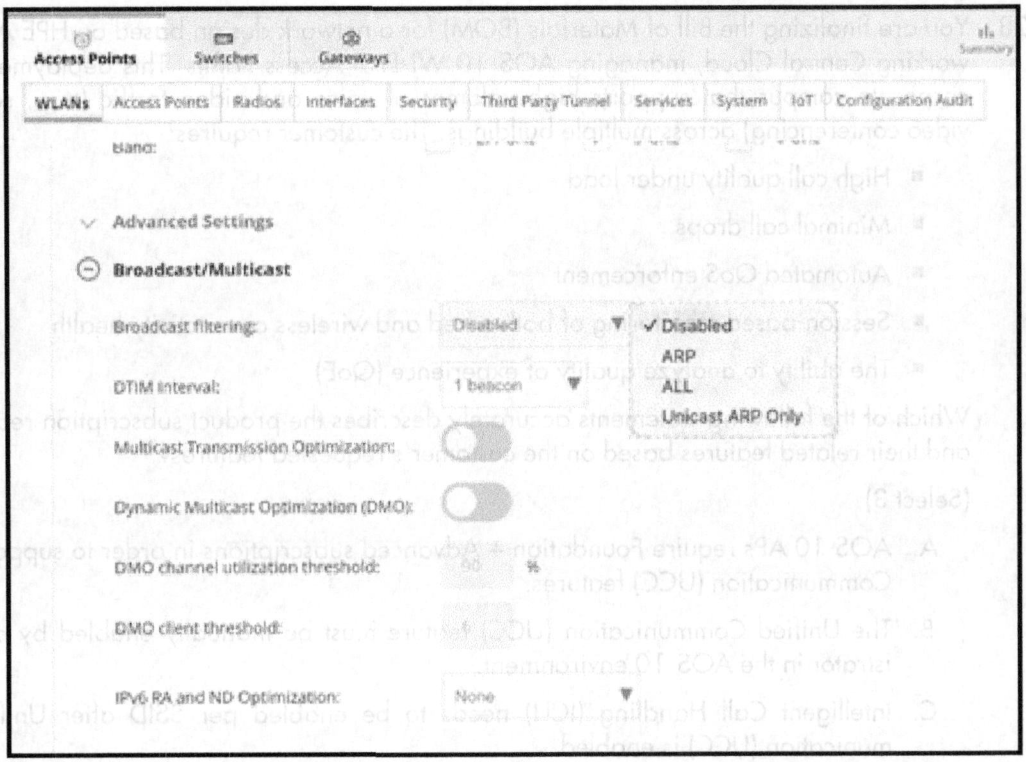

You are designing a high-density wireless deployment usingHPE Aruba Networking AOS-10 APs for a customer environment that includes voice, video, and general enterprise traffic. During the design phase, the customer expresses concern about potential performance degradation caused by excessive broadcast and multicast traffic, particularly in crowded conference areas and open workspaces. You plan on including technical advice on how to optimize related configuration settings.

Which of the following statements accurately describe AOS-10 configuration features? (Select 1)

A. Broadcast filtering must be enabled before Dynamic Multicast Optimization (DMO) can be optimized.

B. Raising the DMO client threshold from 6 to 24 is recommended for high-throughput environments.

C. Disabling 802.11k and 802.11v is recommended to prevent sticky client behavior in high-density deployments.

D. Setting Broadcast Filtering = ARP enables the AP to proxy ARP requests, converting them from broadcast to unicast.

28. You are finalizing the Bill of Materials (BOM) for a network design based on HPE Aruba Networking Central Cloud, managing AOS-10 Wi-Fi 7 Access Points. This deployment is for a corporate campus that supports high volumes of voice and video traffic (e.g., softphones, video conferencing) across multiple buildings. The customer requires:

    - High call quality under load
    - Minimal call drops
    - Automated QoS enforcement
    - Session-based monitoring of both wired and wireless connectivity health
    - The ability to analyze quality of experience (QoE)

    Which of the following statements accurately describes the product subscription requirements and their related features based on the customer's requested features?

    (Select 3)

    A. AOS-10 APs require Foundation + Advanced subscriptions in order to support Unified Communication (UCC) features.
    B. The Unified Communication (UCC) feature must be manually enabled by an administrator in the AOS-10 environment.
    C. Intelligent Call Handling (ICH) needs to be enabled per SSID after Unified Communication (UCC) is enabled.
    D. After enabling UCC, customers can customize DSCP priority tags for both SIP protocol traffic as well as custom applications.
    E. The Unified Communication (UCC) feature is enabled automatically when appropriate subscriptions are tied to AOS-10 APs.
    F. AOS-10 APs only require Foundation subscriptions in order to support Unified Communication (UCC) features.
    G. Intelligent Call Handling (ICH) feature is bundled with Unified Communication (UCC) and requires no extra subscription.
    H. Once UCC has been enabled globally, administrators can import custom ALG definitions via an XML file.

29. As the network architect, you are finalizing the Bill of Materials for an HPE Aruba Networking ClearPass Policy Manager Platform deployment.

    The customer is refreshing their network access control solution and requires a full ClearPass design and licensing proposal.

    Their current environment includes:

    - 5,000 wired IoT devices authenticated via MAC authentication.
    - 5,000 wireless IoT devices using MPSK (one unique key per device).

- 9,000 corporate Windows 11 laptops, joined to Active Directory, authenticating via EAP-TLS (machine and user certs), and requiring posture checks using a client-based WMI agent.
- 500 BYOD users, each onboarding up to 3 personal devices, all receiving certificates via ClearPass Onboard Software PKI.
- Up to 1,000 guest users per day, using ClearPass Guest with self-registration.

Which of the following license combinations most accurately reflects what should be included in the IRIS-generated Bill of Materials (BoM) to support this ClearPass deployment?

(Select 1)

A. Access = 21,500, OnGuard = 9000, Onboard = 500

B. Access = 21,000, OnGuard = 9000, Onboard = 1500

C. Access = 20,500, Guest = 1000, Device Insight = 9000, Onboard = 500

D. The requested solution cannot be completely fulfilled by CPPM

30. Your customer wants a proposal for HPE Aruba Networking ClearPass Policy Manager Platform. Specifically, they want to know how implementing Policy Manager Insight will impact their Policy Manager cluster design and the number of required nodes and licensing. Which of the following statements accurately represent Policy Manager Insight licensing requirements, features, and best practices related to your customer query?

(Select 3)

A. For large-scale installations, it is typical to enable ClearPass Device Insight on a dedicated Policy Manager subscriber node.

B. The Insight primary role should always be enabled on a Policy Manager cluster subscriber node, if one exists.

C. To enable Insight, you must enable Insight on a specific cluster node, then also enable it as a primary on the same node.

D. Insight is provided as part of the Policy Manager Access Upgrade subscription license.

E. Insight can be enabled on three or more Policy Manager cluster nodes for redundancy purposes.

F. The Insight primary and standby roles are enabled by default on the Policy Manager primary and standby publishers.

G. Insight reports and alerts are always created and managed on the Policy Manager cluster publisher.

H. Insight is included with the Policy Manager Access subscription; no add-on subscription is required.

Answers

1. As part of an HPE Aruba Networking campus access layer design project, you are meeting with various stakeholders from the customer's IT organization. Your objective is to gather **technical input** that directly supports design decisions such as **switch port capacity**, **uplink requirements**, and **user connectivity needs**.

   Which two of the following stakeholder-role pairings are most relevant to these design activities?

   (Select 2)

   A. **IT Infrastructure Manager** – Supplies physical site details, rack space availability, power and cabling constraints, and switch port projections.

   B. **Help Desk Supervisor** – Shares common user access issues and recurring support tickets that may indicate access layer shortcomings.

   C. **Chief Financial Officer (CFO)** – Provides budget oversight and prioritizes spending efficiency across IT initiatives, including infrastructure projects.

   D. **Director of Security Compliance** – Advises on data protection and access control requirements that may affect network segmentation or authentication strategies.

   E. **Chief Operating Officer (COO)** – Offers insight into organizational structure and operational workflows that could influence network priorities

   Answers A and B are correct. **Managers** can provide you with critical business requirements and projected growth of end-users and employees. They also may have the flexibility to make purchasing decisions within their department and **may have some technical resources** such as access to floor plans. (This would be especially true for an IT Manager.)**Technical Employees** such as network admins are an excellent resource for technical requirements and information about the existing network design and pain points to be resolved.

2. You are designing a wireless intrusion detection and prevention (WIDS/WIPS) strategy for your client's enterprise network using HPE Aruba Networking and AOS-10. Your design includes the deployment of dedicated Air Monitors (AMs).

   Which of the following statements reflect HPE Aruba Networking best practices or recommendations regarding the design, deployment and implementation of WIDS/WIPS?

   (Select 2)

   A. There is no HPE Aruba Networking advised legal risk to recommending enabling "40MHz Intolerance" in your radio profile settings.

   B. Automatic "Wireless containment" is recommended only when AIOps is enabled with an Advanced Security subscription.

   C. "Wired containment" is disabled by default, enabling it is recommended only after consulting with security and legal stakeholders.

D. As a starting point, it is recommended to deploy four or five Air Monitors per 10,000 sq ft (930 sq m).

E. AOS-10 APs operating in "radio mode" are hybrid AMs capable of scanning rare channels, but at a reduced rate.

Answers A and C are correct.

References:

- https://arubanetworking.hpe.com/techdocs/central/latest/content/nms/monitoring-network/rapids/cont-methods.htm
- https://arubanetworking.hpe.com/techdocs/aos/wifi- design- deploy/security/wids-wips/protection/

3. Review the exhibit.

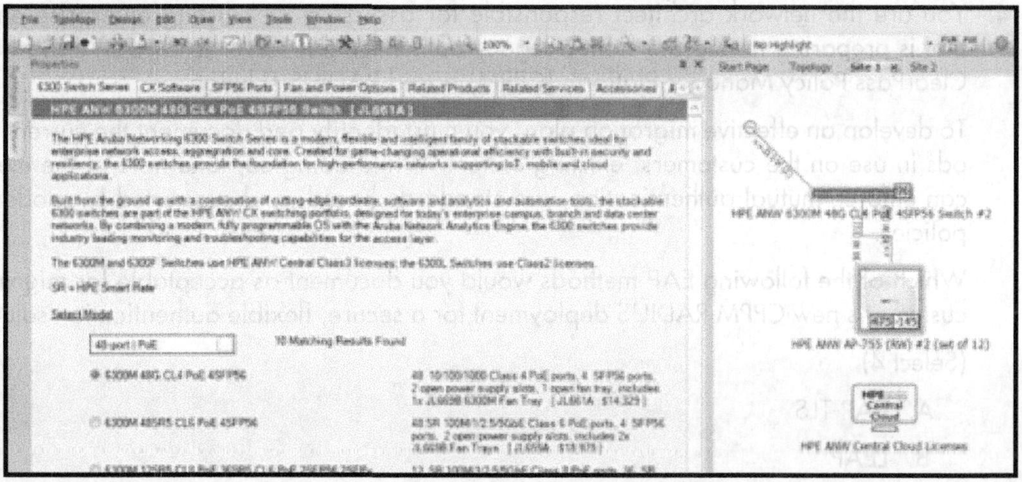

You are modeling a customer's existing network using Intangi IRIS express mode. You are proposing the replacement of legacy Access Points (APs) with HPE Aruba Networking 750 Series Campus Access Points.

You added the customers' existing access switch; when you added a set of 12 APs and connected them to the access switch, you observed red highlighting around the "HPE ANW AP-755" icon.

Which of the following statements are accurate regarding the exhibits Site-1 block view? (Select 3)

A. This AP model requires being connected to a switch supporting the CL8 feature set to enable PoE support.

B. Enabling N+1 redundancy on the switch model shown is feasible and would support the set of 12 APs.

C. To support the requested PoE budget, you must add an additional power supply to the switch to which these APs are connected.

D. To generate an appropriate BOM from "Site-1", you should set the switch attribute Status to "existing".

E. This AP model requires being connected to a switch supporting the CL6 feature set to enable PoE support.

F. This AP model supports being connected to a switch supporting the CL4 feature set to enable PoE support.

G. This AP model requires being connected to a switch supporting Smart Rate to enable PoE support.

Answers B, D, and F are correct.

4. You are the network architect responsible for assessing an existing enterprise environment that is preparing to migrate from a legacy authentication server to a HPE Aruba Networking ClearPass Policy Manager Platform (CPPM)-based RADIUS solution.

To develop an effective migration plan, you must identify and document the current EAP methods in use on the customers' existing system. While doing so, take note of the methods that can enforce mutual authentication, are standards-based, and supported by modern security policies.

Which of the following EAP methods would you document as acceptable for migration to the customer's new CPPM RADIUS deployment for a secure, flexible authentication solution?

(Select 4)

A. EAP-TLS

B. LEAP

C. EAP-TEAP

D. EAP-FAST

E. EAP-PEAP

F. EAP-GTC

G. EAP-TTLS

Answers A, C, E, and G aare correct. Each of these methods provide mutual authentication—both the server and supplicant authenticate. The supplicant must authenticate the server to avoid man-in-the-middle (MITM) attacks in which the supplicant connects to a hacker device rather than the legitimate network. EAP-TLS is the preferred method and the one that you should implement for enterprise networks.

5. Review the exhibit.

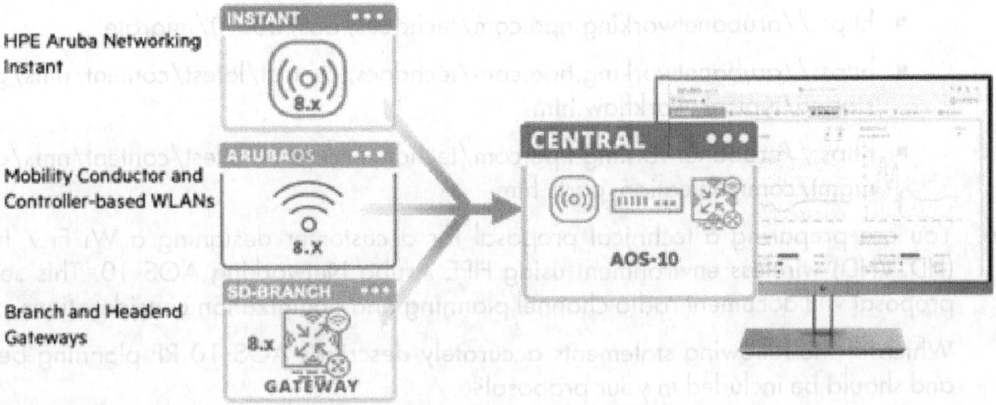

You are developing an implementation plan to migrate a customer from HPE Aruba Networking AOS-8 to AOS-10. Which of the following are valid technical requirements for the proposed solution?

Note

Refer to the latest release notes, as requirements may change.

(Select 3)

A. In some situations, AOS-10 APs must be permitted TCP port 2083 access through the internet perimeter firewall.

B. It is common practice to use the latest version of HPE Aruba NetworkingAirWave to upgrade HPE Aruba Networking APs and GWs from AOS-8 to AOS-10 firmware.

C. Organizations with networks based on single-stack IPv6 may need to implement dual-stack (IPv4 + IPv6) for the AOS-10 devices.

D. All HPE Aruba Networking 3xx, 5xx, 6xx, and 7xx Series Access Points can be migrated from an Instant AOS-8 cluster to AOS-10.

E. HPE Aruba Networking 7005 mobility controllers do not support AOS-10 versions greater than 10.3.1.1, therefore it can't be clustered.

F. HPE Aruba Networking APs being migrated to AOS-10 must be permitted HTTPS (TCP port 443) access through the internet perimeter firewall.

G. If an AOS-8 MCR is added to a new HPE Aruba Networking Central On-Premises Group, its existing configuration can be migrated to the group.

Answers A, C, and F are correct.

References:

- https://arubanetworking.hpe.com/techdocs/aos/aos10/migrate
- https://arubanetworking.hpe.com/techdocs/central/latest/content/nms/get-started/typical_workflow.htm
- https://arubanetworking.hpe.com/techdocs/central/latest/content/nms/device-mgmt/communication_ports.htm

6. You are preparing a technical proposal for a customer designing a Wi-Fi 7 high-density (HD/VHD) wireless environment using HPE Aruba Networking AOS-10. This section of the proposal will document radio channel planning and optimization considerations.

   Which of the following statements accurately describes AOS-10 RF planning best practices and should be included in your proposal?

   (Select 5)

   A. With Wi-Fi 7 high-throughput, HD, and VHD designs, it is common to disable some APs' 2.4 GHz radios for client use.

   B. Wi-Fi 6 series APs are able to implement channel widths ranging from 20 MHz to 160 MHz while avoiding the use of DFS channels.

   C. Careful Wi-Fi client device dimensioning should be performed before considering the use of UNII-4 in applied radio profiles.

   D. With Wi-Fi 7 high-throughput, HD, and VHD designs, it is common to adjust the default radio profile channel bandwidth settings.

   E. AOS-10 radio profiles have IEEE 802.11h Dynamic Frequency Selection (DFS) channels disabled by default.

   F. The Wi-Fi 802.11be standard is capable of operating over all of the following frequency bands: 2.4, 5 and, 6 GHz.

   G. When customizing radio profiles, customers should disable channels 169, 173 and, 177 to ensure maximum client compatibility.

   Answers A, C, D, F, and G are correct. Review Chapter 3 - RF Planning, especially Supplemental Section: Review RF Fundamentals and related material.

7. You are acting as the network architect for a distributed enterprise preparing to implement HPE Aruba Networking Cloud Authentication. The customer has requested a cloud-native authentication solution, but has not yet provided full details about their current identity infrastructure or endpoint capabilities.

   To move forward with initial design planning, you must document reasonable assumptions about the customer's existing equipment and their willingness to align with cloud-native authentication requirements.

   Which of the following assumptions would be valid to document at this early stage?

   (Select 3)

A. Your customer has or is willing to make use of cloud identity providers such as Microsoft Entra ID, Google Workspace or Okta.

B. All Wi-Fi clients are able to support the required client-based HPE Aruba Networking Cloud Central "Onboard App".

C. Your customer endpoint client infrastructure supports supplicants based on client certificates.

D. The customer is willing to phase out non-EAP-TLS clients, allowing MAC-auth only for limited legacy devices.

E. Your customer has or is willing to make use of Active Directory Domain Services authentication through CPPM domain integration.

F. Your customer endpoint infrastructure supports EAP-TLS or EAP-PEAP based authentication.

G. Your client is willing to support Active Directory Domain Services authentication with CPPM domain integration.

Answers A, C, D, E, and F are correct. Please reference Chapter 7 - Campus Security, focusing on the section titled ClearPass Policy Manager and Cloud Auth.

8. Review the exhibit.

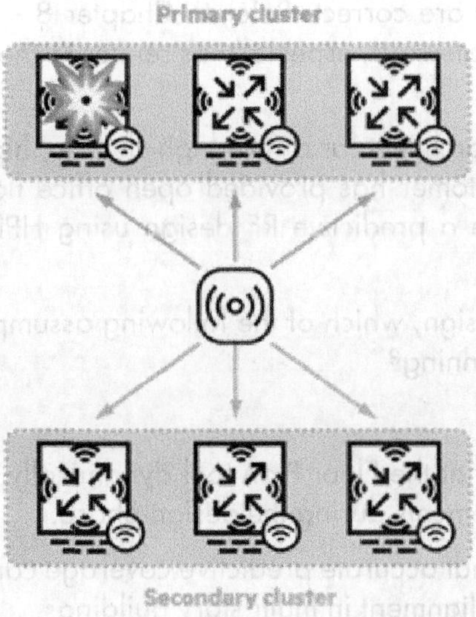

You are the network architect reviewing the technical requirements for a customer's HPE Aruba Networking AOS-10 Access Point (AP) redundancy strategy. The customer requires that client connectivity be minimally disrupted during failover between primary and secondary gateway clusters.

To proceed with design validation, you must document key assumptions that support a seamless failover experience.

(Select 3)

A. Your secondary gateway cluster uplinks to their aggregation switches must be tagged to the same VLAN IDs as the primary cluster.

B. Ensure all APs are able to establish IPsec connectivity between themselves and each member of their associated clusters.

C. Your secondary gateway cluster must be created in a backup group; this "parallel" cluster must have the same name as the primary cluster.

D. Configure the APs for high-availability by assigning them both a primary and a secondary cluster, enabling cluster failover redundancy.

E. The AOS-10 Tunnel Orchestrator sets up an IPsec map (state synchronization) between the primary and secondary cluster leaders.

F. Configure your primary and secondary clusters in a pair of linked sites, and add both clusters to the same group with cluster mode "auto-site" enabled.

G. Configure the same anycast IP address on the loopback of both the primary and secondary cluster members.

H. Create an organizational mobility domain; ensure both the primary and secondary clusters are added to the same mobility domain.

Answers A, B, and D are correct. Refer to Chapter 8 - Overlay Networks section titled HPE Aruba Networking Gateways, especially as relates to the topic titled Secondary cluster redundancy.

9. You are the network architect for a new high-density, high-throughput Wi-Fi 7 campus wireless design. Your customer has provided open office floor plans for multiple buildings and asked you to provide a predictive RF design using HPE Aruba Networking Central's Floor Plan tool.

   Before starting the design, which of the following assumptions would be valid to document at this early stage of planning?

   (Select 5)

   A. It is assumed that the Floor Plan tool dynamically calibrates RF propagation using live signal measurements during the design phase.

   B. It is assumed that accurate predictive coverage can be achieved without accounting for floor-to-floor alignment in multi-story buildings.

   C. If ceiling heights are not provided, a default maximum height of 8 meters (25 feet) will be assumed for predictive planning purposes.

   D. It is assumed that there are no hidden or undocumented internal RF interference sources, such as intercom systems or electrical panels.

E. It is assumed that significant RF interference sources will only be identified during a site validation survey and are not included in initial planning.

F. It is assumed that floor plan images provided by the customer are to scale and have been verified by a qualified professional or architect.

G. It is assumed that the environment region slider in the HPE Aruba Networking Central Floor Plan tool will be set to 'Cubicles (3.0)' for initial modeling.

H. It is assumed that AP placement recommendations will automatically adjust based on user density estimates extracted from the floor plan image.

I. It is assumed that all floor plan images provided are CAD files with layers of metadata that classify all internal walls with RF attenuation values.

Answers C, D, E, F, and G are correct. Please see Chapter 1 - Information Gathering, with a focus on the section titled Physical Environment.

10. Your customer has employees using Microsoft Windows 11 laptops — a mix of BYOD and corporate-owned devices. They are concerned that older Windows versions have reduced security and want tighter control over network access. Their requirement is to block or restrict network access for any device that:

- is running a Windows version older than Windows 11
- does not have an active antivirus agent running

Which of the following statements accurately describes HPE Aruba Networking features and capabilities relevant to this requirement?

(Select 2)

A. ClearPass Device Insight supports these profiling requirements via its Microsoft NAP agent Augmentation method.

B. HPE Aruba Networking Cloud Central Cloud Auth combined with its "Onboard App" supports these profiling requirements.

C. ClearPass OnGuard Software persistent agent supports agent-based profiling for these checks.

D. HPE Aruba Networking Central Client Insights, combined with unified infrastructure NADs, supports these profiling requirements.

E. HPE Aruba Networking Secure Service Edge (SSE) ZTNA supports agent-based profiling to enforce these checks.

Answers C and E are correct. Please see Chapter 7 - Campus Security, including the sections titled HPE Aruba Networking SASE and ClearPass Policy Manager & Cloud Auth, and Profiling and NAC.

11. You are the lead architect of a successful and growing HPE Aruba Networking (ANW) Partner and are responsible for growing and expanding the knowledge of your team that responds to customer RPFs for campus network engagements.

Your junior staff members need to understand the nature of the customers' existing equipment as they prepare responses to those requests for proposals.

Toward that effort, help your team match the following Wi-Fi features to their appropriate standard. **Your options for Wi-Fi standards are Wi-Fi 4, Wi-Fi 5, Wi-Fi 6, Wi-Fi 6E, and Wi-Fi 7.**

Note

Not all descriptions will match a Wi-Fi standard.

Description	Fill in your Wi-Fi standard answers below
802.11be	Wi-Fi 7
First introduced 6 GHz	Wi-Fi 6E
First introduced 802.11ax	Wi-Fi 6
First introduced 1024-QAM	Wi-Fi 6
First introduced BSS coloring	Wi-Fi 6
Only supports 5 GHz	Wi-Fi 5
Supports Wave 1 & 2	Wi-Fi 5
802.11n	Wi-Fi 4
First introduced 60 GHz	no match

Answers are as shown. Please see Chapter 3 - RF Planning in the section titled Supplemental Section: RF Fundamentals.

12. You are the lead architect of a successful and growing HPE Aruba Networking Partner and are responsible for growing and expanding the knowledge of your team that responds to customer RPFs for campus network engagements.

    Your junior staff members need to understand the nature of the customers' existing equipment as they prepare responses to those requests for proposals.

    Toward that effort, help your team match the following IEEE designations to their official technical Physical Layer (PHY) protocol acronym. **Your choices for PHY protocol acronym are HT, VHT, HE, EHT, UHT, and XHT.**

Note

Not all acronyms will be used — some are distractors.

IEEE designation	Protocol acronym
802.11n	HT
802.11ac	VHT
802.11ax	HE
802.11be	EHT
distractor	UHT
distractor	XHT

Answers as shown above. Please see Chapter 4, at the beginning of the section titled High-Density Challenges.

13. You are the lead architect of a successful and growing HPE Aruba Networking Partner and are responsible for growing and expanding the knowledge of your team that responds to customer RPFs for campus network engagements.

    Your junior staff members need to understand the nature of the customers' existing equipment as they prepare responses to those requests for proposals.

    Toward that effort, help your team understand which of the following statements accurately describe the HPE Aruba Networking Wi-Fi 7 Access Points' features? (Select 3)

    A. All models support 6 GHz, offering up to 320 MHz channel width.
    B. The AP-730 supports 2x5GE, while the AP-740 only supports 1x5GE.
    C. Multiple models support MACsec from the Access Point to the edge switches.
    D. The 7xx Series Access Points introduced the 802.11ax Wave 2 standard.
    E. All models have 3 radios supporting either 5 + 5 + 6 GHz or 5 + 6 + 6 GHz.
    F. The 7xx Series Access Points support the 802.11be standard.
    G. At full power draw, some AP models require switches that support Class 8 PoE.

    Answers B, C, and F are correct. Please see Chapter 3 - RF Planning, in the section titled Selecting APs and Antennas.

14.

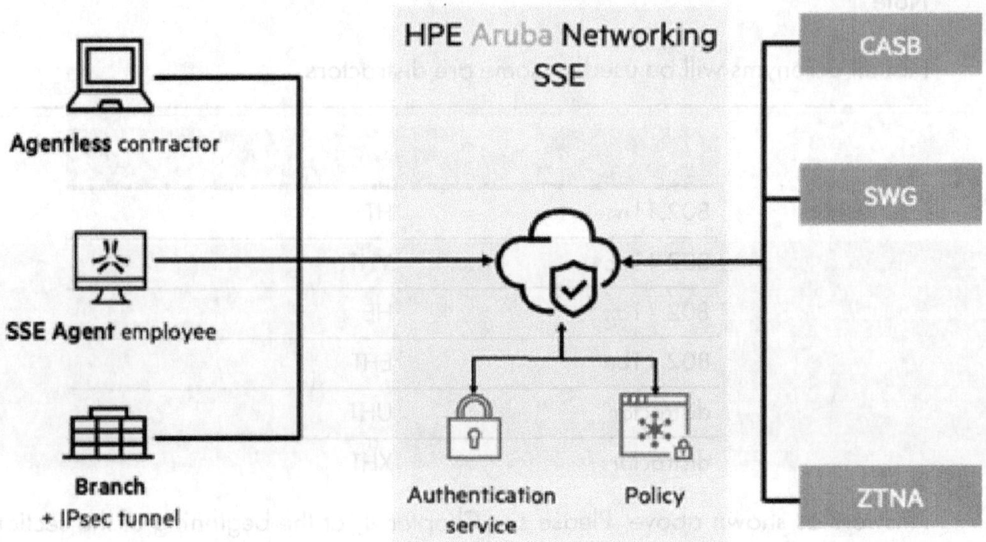

You are the lead architect of a successful and growing HPE Aruba Networking Partner and are responsible for growing and expanding the knowledge of your team that responds to customer RPFs for campus network engagements.

Your junior staff members need to understand the nature of the HPE Aruba Networking Secure Service Edge (SSE) cloud services so they comprehend how this product and all of its components can meet customer requirements.

Toward that effort, help your team match the appropriate HPE Aruba Networking SSE cloud services to their correct descriptions. The cloud services are SWG, ZTNA, DEM, and CASB.

A. Makes use of the SSE Connector to broker least-privilege access to private resources and applications. **ZTNA**

B. Protects users' access to Software-as-a-Service (SaaS) applications using SSE public points of presence. **CASB**

C. Establishes rules that permit or block users' access to website categories based on organizational internal policies. **SWG**

D. Provides a comprehensive view of the entire network path, from a user's endpoint device to the application server. **DEM**

Answers as shown above. Please see Chapter 7 - Campus Security, in the section titled HPE Aruba Networking SASE.

15. You are the lead architect of a successful and growing HPE Aruba Networking Partner and are responsible for growing and expanding the knowledge of your team that responds to customer RPFs for campus network engagements.

Your junior staff members need to understand the nature of the HPE Aruba Networking AOS-10 Gateway High Availability features so they comprehend how this product and all of its components can meet customer requirements.

Toward that effort, help your team match the following mobility gateway cluster features to their partial descriptions. Not all descriptions will be used.

Feature	Description
Seamless roaming	The clients remain anchored to the same gateway.
Client/AP load balancing	Helps improve client connectivity, particularly during periods of high demand.
Stateful failover	Makes use of a designated standby gateway.
Client state sync	Supports synchronization for the first failover only.
Auto clustering	Supports a group of one for ease of configuration.
distractor	Enables APs to perform client authentication.
distractor	Enables state synchronization across different clusters.

Answers are as shown above.

Please see Chapter 8 - Overlay Networks in the section titled HPE Aruba Networking Gateways. Focus especially on the topic titled Gateway review. You can also review the following reference:
https://arubanetworking.hpe.com/techdocs/central/2.5.8/content/aos10x/cfg/gw- clust-tun-orch/gw-cl-architct.htm

16. You are the lead architect of a successful and growing HPE Aruba Networking Partner and are responsible for growing and expanding the knowledge of your team that responds to customer RPFs for campus network engagements.

Your junior staff members need to understand the nature of the HPE Aruba Networking ClearPass Policy Manager Platform(CPPM) Onboard so they comprehend how this product and all of its components can meet customer requirements.

Toward that effort, help your team list valid design considerations when recommending Policy Manager in a design proposal?

(Select 2)

A. You plan to use the Onboard CA's EST server to issue device certificates to HPE Aruba Networking Unified Infrastructure devices.

B. Onboard supports being configured as a Self-signed CA or as an Intermediate CA with the issuing role.

C. Onboard does not support acting as a Registration Authority that proxies signing requests to an external CA.

D. You wish to use the Onboard EST OpsRamp integration to issue and manage RadSec certificates on third-party NADs.

E. When hosting multiple PKI Certificate Authorities, Policy Manager Onboard requires a Foundation + Advanced subscription.

Answers A and B are correct. Please see Chapter 7 - Campus Security, in the section titled Supplemental: AAA Overview, focusing on those topics related to Certificate Authorities and digital certificates.

17. You are the lead architect of a successful and growing HPE Aruba Networking Partner and are responsible for growing and expanding the knowledge of your team that responds to customer RPFs for campus network engagements.

Your junior staff members need to understand the nature of the HPE Aruba Networking User Experience Insight (UXI) so they comprehend how this product and all of its components can meet customer requirements.

Toward that effort, explain which of the following statements accurately describe UXI capabilities or features.

(Select 5)

A. Cloud-based 24/7 service assurance solution for users, networks, and applications.

B. Client agent subscriptions are included as part of Unified Infrastructure Advanced subscription agreements.

C. Integration with HPE Aruba Networking SSE allows UXI to monitor Data Protection Policy violations.

D. Deployed as a network endpoint device to continuously monitor the network's performance, connectivity, and responsiveness.

E. Provides the last 30 days of performance data and PCAP files for analysis and troubleshooting.

F. Each physical sensor requires a one-, three-, or five-year HPE Aruba Networking Central Cloud subscription.

G. UXI sensors rely on passive traffic monitoring and do not generate active synthetic test traffic.

H. UXI sensors enable continuous monitoring and testing across wired, wireless, and WAN.

I. The UXI agent for Linux performs the same functions as a sensor but with reduced performance capabilities.

Answers A, D, E, F, and H are correct. Please see Chapter 9 - Network Optimization, in the section titled UXI.

18. You are the lead architect of a successful and growing HPE Aruba Networking Partner and are responsible for growing and expanding the knowledge of your team that responds to customer RPFs for campus network engagements.

    Your junior staff members need to understand the nature of the distributed overlay support in HPE Aruba Networking Central so they comprehend how this product and all of its components can meet customer requirements. Which of the following statements accurately this feature set.

    (Select 3)

    A. Group-Based Policy (GBP) IDs are directly assigned to an endpoint's VTEP using a vendor-specific attribute.

    B. Least privilege access policies are enforced by Aruba-User-Roles at the ingress VTEP.

    C. NetConductor can be used to orchestrate the extension of Layer 2 networks across Layer 3 boundaries.

    D. NetConductor eliminates the need for a control-plane protocol such as EVPN in VXLAN overlay networks.

    E. A VSX cluster can host Anycast gateways to ensure consistent first-hop routing services across EVPN-VXLAN networks.

    F. Least privilege access policies are enforced by Group-Based Policy (GBP) at the egress VTEP.

    Answers C, E, and F are correct. Please see Chapter 8 - Overlay Networks, in the section titled Distributed Overlay networks with HPE Aruba Networking Central NetConductor.

19. You are the lead architect of a successful and growing HPE Aruba Networking Partner and are responsible for growing and expanding the knowledge of your team that responds to customer RPFs for campus network engagements.

    Your junior staff members need to understand the nature of AOS-CX queuing and scheduling mechanisms so, they comprehend how these features can meet customer requirements.

    Which of the following statements accurately describes the queuing and scheduling mechanisms — Strict Priority (SP), Weighted Fair Queuing (WFQ), and Deficit Weighted Round Robin (DWRR) — on AOS-CX switches commonly deployed in enterprise campus three-tier network architecture?

    (Select 4)

    A. On all AOS-CX switches, SP scheduling can only be assigned to the highest-priority egress queue in the queue mapping.

    B. The AOS-CX 8400 switch supports SP and WFQ scheduling, whereas the 9300 supports SP along with either WFQ or DWRR scheduling.

    C. To protect SP from starving lower-priority queues, it is recommended that you also set a max-bandwidth limitation on the SP queue.

D. WFQ is based on per-class bandwidth fairness, while DWRR scheduling operates based on queue weights and packet sizes.

E. SP is typically applied to a mid-level queue mapping designated for traffic marked by DSCP 46 (EF), commonly used for voice.

F. On all AOS-CX switches that support DWRR, DWRR must be manually enabled because FIFO is the default queuing mechanism.

G. On the AOS-CX 6300, 6400, 8325, and 8360 switches, DWRR is the default weighted scheduler, whereas WFQ is not supported.

Answers A, C, D, and G are correct. Please see Chapter 9 - Network Optimization, in the section titled Supplemental: Wired QoS.

20. You are designing a secure system that uses digital certificates to protect sensitive communication between a client and a server.

Place the following steps in the correct order (1 to 5) to reflect the workflow for using certificates to enable secure, authenticated data exchange.

Workflow step description	Steps in progressive order
The Root Certification Authority (CA) securely signs a certificate for an Intermediate (Issuing) CA, which then signs certificates for end entities such as clients and servers.	Step 1
The server and client generate key pairs and submit certificate signing requests (CSRs) to the Intermediate (Issuing) CA, providing proof of identity.	Step 2
Certificates are issued to the client and server, binding their identities to their public keys and specifying key usages (e.g., Client/Server Authentication).	Step 3
During a communication session, the server presents its certificate to the client. The client validates the certificate chain back to the trusted Root CA.	Step 4
The client and server use their private keys to sign and decrypt messages, ensuring authentication, data integrity, and confidentiality of the data flow.	Step 5

Answers are as shown above. Please see Chapter 7 - Campus Security, in the section titled Supplemental: AAA Overview. Focus on topics related to digital certificates, CAs, and PKIs.

21. You are analyzing the high availability requirements of a customer's proposed HPE Aruba Networking CX Switch Operating System 3-tier campus network design. Your colleague, who is responsible for the technical drawing to be included in the proposal, asks you to explain the high availability potential of various different models, so they can better understand how to create these diagrams using IRIS.

Which of the following statements are accurate?

(Select 5)

A. 6400 switches support VSX clustering.

B. 6200 switches support VSF stacking between any switches of the same series.

C. 6300 switches support VSX clustering.

D. both 6200 and 6100 switches are capable of supporting 6 node VSF stack.

E. 6400 switches support 2-10 node VSF stacking.

F. 4100i switches support VSF between all same series switches.

G. 6300 switches support 2-10 node VSF stacking.

H. 5420 switches support either VSF stacking or VSX clustering.

Answers A, C, D, F, and G are correct. See Chapter 5 - Wired Network Design, in the section titled 2-Tier versus 3-Tier Design. You should also review the following reference: https://www.youtube.com/watch?v=6EsrZzPs0kA

22. Review the exhibit.

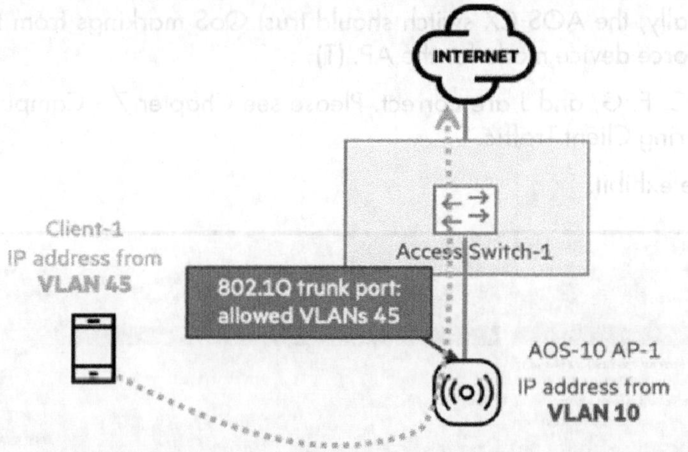

You are designing an HPE Aruba Networking Cloud Central deployment with an AOS-10 AP operating in bridged mode, connected to an AOS-CX campus access layer switch (there is a 3-tier network architecture, however only the access layer is shown in the exhibit).

Which of the following are statements are valid recommendations that help ensure users are blocked from accessing any destination where its web category is rated "Bot Nets"?

(Select 4)

A. Within your ACL allowlist, allow access to FQDNs as per the BrightCloud FQDN list documentation.

B. Install the appropriate Foundation + Security subscriptions, then enable IDS/IPS in IDS mode.

C. Create an access control rule using the appropriate Web Category, and set the action to Deny.

D. Ensure that the AOS-CX switch (Switch 1) enforces client-mode with a max-client count of 0 for AP 1.

E. Create an access control rule with an appropriate "App Category" with the action set to Deny to all destinations.

F. Ensure that the VLAN assigned to Client-1, VLAN 45, is extended from Access Switch-1 to AOS-10 AP-1.

G. Install the appropriate Foundation subscription(s), then enable AppRF DPI = all.

H. Install the appropriate Foundation + Security subscriptions, then enable IDS/IPS in IPS mode.

I. It is recommended to enable jumbo frame support between AP 1 and the campus access switch (Switch 1).

J. Ideally, the AOS-CX switch should trust QoS markings from the AP; however, it must enforce device mode for the AP. (T)

Answers C, F, G, and J are correct. Please see Chapter 7 - Campus Security, in the section titled Securing Client Traffic.

23. Review the exhibit.

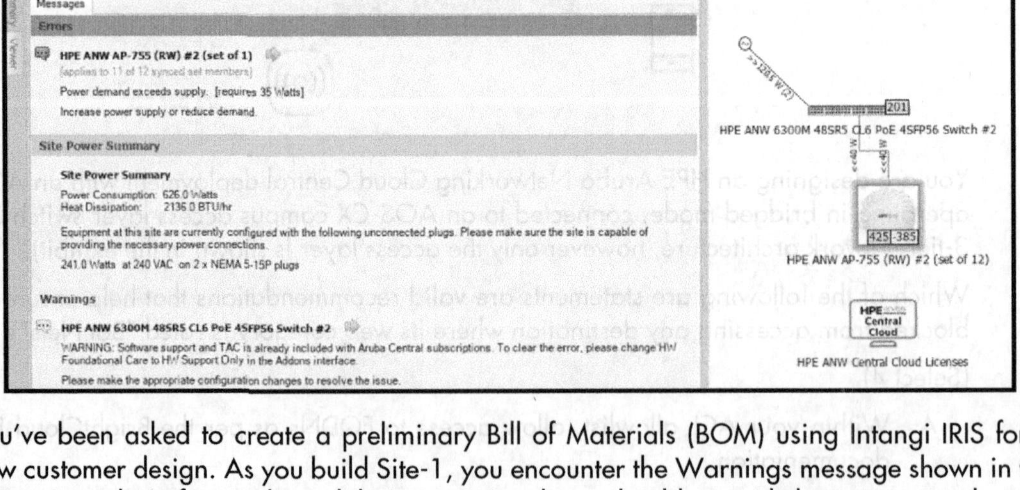

You've been asked to create a preliminary Bill of Materials (BOM) using Intangi IRIS for a new customer design. As you build Site-1, you encounter the Warnings message shown in the Messages tab (refer to the exhibit). In IRIS, where should you click next to resolve the indicated warning?

(Select 1)

A. Click the "$" in the Intangi IRIS toolbar.

B. Click the "hammer and wrench" in the Intangi IRIS toolbar.

C. Click the "HPE Aruba NETWORKING Central Cloud" icon.

D. Click the "+" in the Intangi IRIS toolbar.

E. Click the "HPE ANW AP-755" icon.

F. Click the "HPE ANW 6300M" icon.

Answer D is correct. Spend some time using the IRIS tool to ensure that you are familiar with its basic functions and usage.

24. Review the exhibit.

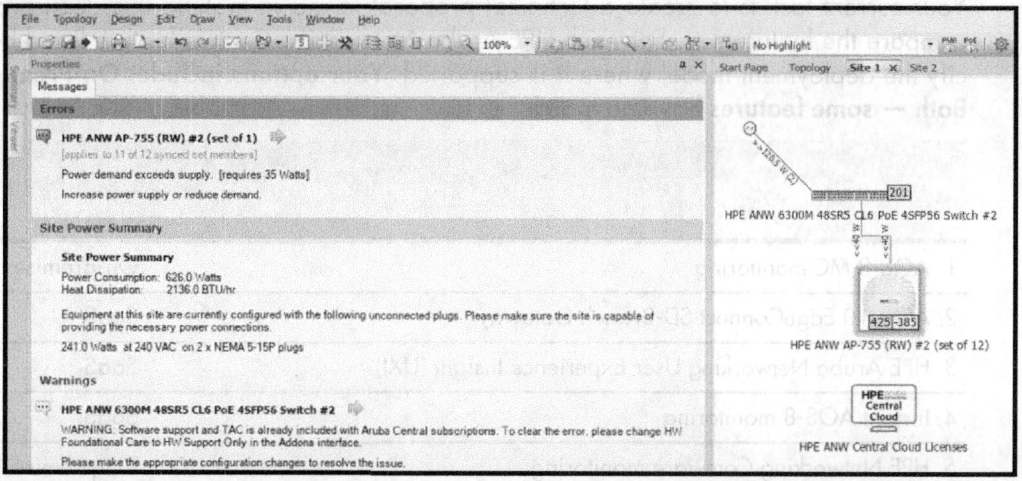

You've been asked to create a preliminary Bill of Materials (BOM) using Intangi IRIS for a new customer design, along with a network topology of the proposed solution. You were about to add another switch so you could incorporate VSF high availability into the design, but before you got to that step, you encountered red highlighting around the "HPE ANW AP-755" AP in the "Site-1" tab or IRIS. Review the related Error in the "Messages" tab in Iris.

Which of the following actions could help you resolve the indicated issue? (Select 2)

A. Select the switch connected to the AP and adjust the model selection to support the CL8 feature.

B. Select the connecting line between the AP and the switch, right-click, and choose the "Clone Connection" option.

C. Select the connecting line between the AP and the switch, and adjust the AP's "quantity multiplier" attribute.

D. Select the connecting line between the AP and the switch, press Ctrl and "+" keys simultaneously.

E. Select the switch connected to the AP and modify its Fan and Power Options settings.

Answers B and D are correct. Spend some time using the IRIS tool to ensure that you are familiar with its basic functions and usage.

25. You have been asked to write a technical proposal for a key customer that is planning a major network migration project, seeking a unified Single Pane of Glass (SPOG), and robust Security Orchestration, Automation, and Response (SOAR) integration. Your proposed solution is based on HPE Aruba Networking Central. The crucial architectural decision remains whether to implement the SaaS (Cloud) or the On-premises deployment model.

Your current task is to create a technical proposal to aid in making this decision; you must compare the feature sets. For each HPE Aruba Networking Central feature listed below, specify the deployment model where it is supported. **Your options include: On-Premises, SaaS, Both — some features have no match.**

HPE Aruba Networking Central feature to matching platform	Your platform option answer
1. AOS-8 MC monitoring	On-premises
2. AOS-10 EdgeConnect SD-Branch Gateway	SaaS
3. HPE Aruba Networking User Experience Insight (UXI)	SaaS
4. Instant AOS-8 monitoring	SaaS
5. HPE Networking Comware monitoring	On-premises
6. MSP mode	SaaS
7. NetConductor	Both
8. Wi-Fi Spectrum Boost	Neither
9. Aruba Analytics and Location Engine (ALE)	Neither

Answers are as shown above. Please see Chapter 2 - HPE Aruba Networking Architecture, in the section titled HPE Aruba Networking Central On-premises.

26. You are preparing a technical proposal document for a customer. Which of the following EIRP radio profile settings should you recommend as a starting point for HPE Aruba Networking Cloud Central AirMatch AOS-10 APs managed in a typical campus office deployment?

(Select 3)

A. 2.4 GHz band, set the minimum to 6 dBm and maximum to 12 dBm

B. 5 GHz band, set the minimum to 15 dBm and maximum to 21 dBm

C. 6 GHz band, set the minimum to 15 dBm and maximum to 21 dBm

D. 2.4 GHz band, set the minimum to 3 dBm and maximum to 9 dBm

E. 5 GHz band, set the minimum to 18 dBm and maximum to 21 dBm

Answers A, B, C are correct. Please see Chapter 3 - RF Planning, in the section titled Deployment Models. Focus on Transmit power settings.

27. Review the exhibit.

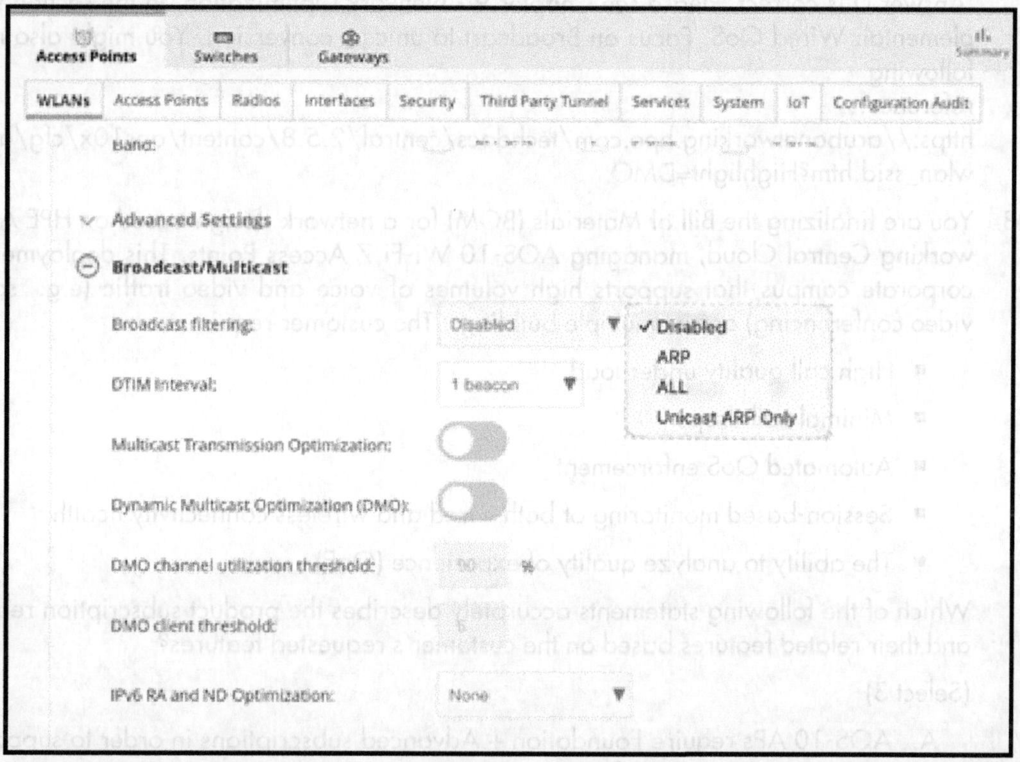

You are designing a high-density wireless deployment usingHPE Aruba Networking AOS-10 APs for a customer environment that includes voice, video, and general enterprise traffic. During the design phase, the customer expresses concern about potential performance degradation caused by excessive broadcast and multicast traffic, particularly in crowded conference areas and open workspaces. You plan on including technical advice on how to optimize related configuration settings.

Which of the following statements accurately describe AOS-10 configuration features?

(Select 1)

A. Broadcast filtering must be enabled before Dynamic Multicast Optimization (DMO) can be optimized.

B. Raising the DMO client threshold from 6 to 24 is recommended for high-throughput environments.

C. Disabling 802.11k and 802.11v is recommended to prevent sticky client behavior in high-density deployments.

D. Setting Broadcast Filtering = ARP enables the AP to proxy ARP requests, converting them from broadcast to unicast.

Answer D is correct. Please see Chapter 9 - Network Optimization, in the section titled Supplemental: Wired QoS. Focus on Broadcast to unicast conversion. You might also review the following
reference:
https://arubanetworking.hpe.com/techdocs/central/2.5.8/content/aos10x/cfg/aps/conf_wlan_ssid.htm?Highlight=DMO

28. You are finalizing the Bill of Materials (BOM) for a network design based on HPE Aruba Networking Central Cloud, managing AOS-10 Wi-Fi 7 Access Points. This deployment is for a corporate campus that supports high volumes of voice and video traffic (e.g., softphones, video conferencing) across multiple buildings. The customer requires:

- High call quality under load
- Minimal call drops
- Automated QoS enforcement
- Session-based monitoring of both wired and wireless connectivity health
- The ability to analyze quality of experience (QoE)

Which of the following statements accurately describes the product subscription requirements and their related features based on the customer's requested features?

(Select 3)

A. AOS-10 APs require Foundation + Advanced subscriptions in order to support Unified Communication (UCC) features.

B. The Unified Communication (UCC) feature must be manually enabled by an administrator in the AOS-10 environment.

C. Intelligent Call Handling (ICH) needs to be enabled per SSID after Unified Communication (UCC) is enabled.

D. After enabling UCC, customers can customize DSCP priority tags for both SIP protocol traffic as well as custom applications.

E. The Unified Communication (UCC) feature is enabled automatically when appropriate subscriptions are tied to AOS-10 APs.

F. AOS-10 APs only require Foundation subscriptions in order to support Unified Communication (UCC) features.

G. Intelligent Call Handling (ICH) feature is bundled with Unified Communication (UCC) and requires no extra subscription.

H. Once UCC has been enabled globally, administrators can import custom ALG definitions via an XML file.

Answers A, B, and G are correct. Review the following resource:
https://arubanetworking.hpe.com/techdocs/central/latest/content/nms/apps/ucc/ucc.htm

29. As the network architect, you are finalizing the Bill of Materials for an HPE Aruba Networking ClearPass Policy Manager Platform deployment.

The customer is refreshing their network access control solution and requires a full ClearPass design and licensing proposal.

Their current environment includes:

- 5,000 wired IoT devices authenticated via MAC authentication.
- 5,000 wireless IoT devices using MPSK (one unique key per device).
- 9,000 corporate Windows 11 laptops, joined to Active Directory, authenticating via EAP-TLS (machine and user certs), and requiring posture checks using a client-based WMI agent.
- 500 BYOD users, each onboarding up to 3 personal devices, all receiving certificates via ClearPass Onboard Software PKI.
- Up to 1,000 guest users per day, using ClearPass Guest with self-registration.

Which of the following license combinations most accurately reflects what should be included in the IRIS-generated Bill of Materials (BoM) to support this ClearPass deployment?

(Select 1)

A. Access = 21,500, OnGuard = 9000, Onboard = 500

B. Access = 21,000, OnGuard = 9000, Onboard = 1500

C. Access = 20,500, Guest = 1000, Device Insight = 9000, Onboard = 500

D. The requested solution cannot be completely fulfilled by CPPM

Answer A is correct. Please see Chapter 7 - Campus Security, in the section titled ClearPass Policy Manager & Cloud Auth.

30. Your customer wants a proposal for HPE Aruba Networking ClearPass Policy Manager Platform. Specifically, they want to know how implementing Policy Manager Insight will impact their Policy Manager cluster design and the number of required nodes and licensing. Which of the following statements accurately represent Policy Manager Insight licensing requirements, features, and best practices related to your customer query?

(Select 3)

A. For large-scale installations, it is typical to enable ClearPass Device Insight on a dedicated Policy Manager subscriber node.

B. The Insight primary role should always be enabled on a Policy Manager cluster subscriber node, if one exists.

C. To enable Insight, you must enable Insight on a specific cluster node, then also enable it as a primary on the same node.

D. Insight is provided as part of the Policy Manager Access Upgrade subscription license.

E. Insight can be enabled on three or more Policy Manager cluster nodes for redundancy purposes.

F. The Insight primary and standby roles are enabled by default on the Policy Manager primary and standby publishers.

G. Insight reports and alerts are always created and managed on the Policy Manager cluster publisher.

H. Insight is included with the Policy Manager Access subscription; no add-on subscription is required.

Answers C, E, and H are correct. Please see Chapter 7 - Campus Security, in the section titled ClearPass Policy Manager & Cloud Auth.

A Learning check answers

Overview

This appendix contains the answers to all the Learning Check sections found in the previous chapters.

Chapter 1: Information Gathering

This section contains the Learning Check answers for Chapter 1.

Learning Check answers

Question 1: The best question here is "What network issues are impacting the business the most?"

Question 2: For wireless network design, important information includes civic address and coordinates, included floors, building materials, and details about emergency exits and escape routes.

Question 3: Migrating to AOS-10 is possible with 600, 500, and some 300 series APs.

Chapter 2: HPE Aruba Networking security-first, AI-powered networking

This section contains the Learning Check answers for Chapter 2.

Learning Check answers

Question 1: ESP's four core attributes are HPE Aruba Networking Central, AIOps, Unified Infrastructure, and Zero Trust Networking.

Question 2: HPE Aruba Networking CX switches could be managed in HPE Aruba Networking Central which is a cloud management platform. You can also manage HPE Aruba Networking APs and Gateways in HPE Aruba Networking Central.

Question 3: Gold, silver, and bronze licensing are available for all hardware gateways to expand capacity.

Chapter 3: RF Planning

This section contains the Learning Check answers for Chapter 3.

Learning Check answers

Question 1: For Wi-Fi 6/6E/7, target a cell size of -55 dBm to -65 dBm.

Question 2: The 7xx series APs are all HPE Aruba Networking Wireless AOS-10.

Question 3: Spectrum clearing including active survey yields the most accurate results.

Chapter 4: Wireless High-Density Design

This section contains the Learning Check answers for Chapter 4.

Learning Check answers

Question 1: A very high-density (VHD) WLAN is an area with so many users and APs that RF channel reuse cannot be expected.

Question 2: The three recommended best practices relate to minimum RSSI, 20MHz channels at -55dBm, and Wi-Fi 7 clients with a poor signal.

Question 3: You should compute the associated device capacity (ADC) by frequency band.

Question 4:

Step 1: 4 000 * 0.85 * 2.5 = 8,500 (total devices)

Step 2: 8 500 * 0.3 = 2,550 (30% 6 GHz)

Step 3: 8 500 * 0.5 = 4,250 (50% 5GHz)

Step 4: 8 500 * 0.2 = 1,700 (20% 2.4GHz)

Chapter 5: Wired Network Design

This section contains the Learning Check answers for Chapter 5.

Learning Check answers

Question 1: One reason to design a 2-tier campus LAN instead of using a 3-tier is if the campus has small buildings with only a few wiring closets.

Question 2: You should implement VSX at the core when you connect your GW cluster to the core.

Question 3: MDF is the point where traffic egresses the campus network, IDFs distribute that connectivity throughout the building.

Question 4: 6:1 is best suited for this scenario.

Chapter 6: Network Segmentation

This section contains the Learning Check answers for Chapter 6.

Learning Check answers

Question 1: VoIP traffic requires special handling through class of service (COS) quality of service assignment tagged in a VLAN header.

Question 2: You should follow the current design unless you have a good reason to change.

Question 3: Key considerations when leveraging a single VLAN relate to the DHCP server scopes, routers' ability to handle large ARP tables, and switches' ability to handle large bridge and MAC tables.

Chapter 7: Campus Security

This section contains the Learning Check answers for Chapter 7.

Learning Check answers

Question 1: Unified infrastructure with AirWave on-premises management is not a part of HPE Aruba Networking security-first, AI-powered networking security.

Question 2: In this scenario, 8,000 Access licenses and 3,750 Client Onboarding Software licenses are appropriate.

Question 3: Enable ClearPass Insight on the Publisher and Standby Publisher for high availability. Just be sure to move the authentication process to other nodes.

Chapter 8: Overlay Networks

This section contains the Learning Check answers for Chapter 8.

Learning Check answers

Question 1: AOS-10 gateways are required for dynamic segmentation with centralized overlay.

Question 2: Decoupled user traffic from the underlay is not related to traditional network design.

Question 3: Centralized overlays tunnel traffic from APs and switches to a central gateway (or gateway cluster) location, while Distributed overlays tunnel traffic from VXLAN tunnel endpoints (VTEPs) to other VTEPs.

Chapter 9: Network Optimization

This section contains the Learning Check answers for Chapter 9.

Learning Check answers

Question 1: The first step towards ensuring good QoS in a campus network is minimizing congestion by planning adequate capacity.

Question 2: In this scenario, ratification of a new Wi-Fi 8 standard would not impact AP capacity.

Question 3: One nice benefit of UXI is that AI-powered insights alert IT staff to network performance issues as soon as they begin.